Plant Microbiology

Plant Microbiology

Michael Gillings and Andrew Holmes
Department of Biological Sciences, Macquarie University, Sydney, NSW 2109, Australia

First published 2004

A CIP catalogue record for this book is available from the British Library.

ISBN 1 85996 224 6

Garland Science/BIOS Scientific Publishers
4 Park Square, Milton Park, Abingdon, Oxon, OX14 4RN, UK and
29 West 35th Street, New York, NY 10001–2299, USA
World Wide Web home page: www.bios.co.uk

Garland Science/BIOS Scientific Publishers is a member of the Taylor & Francis Group.

Distributed in the USA by
Fulfilment Center
Taylor & Francis
10650 Toebben Drive
Independence, KY 41051, USA
Toll Free Tel.: +1 800 634 7064; E-mail: taylorandfrancis@thomsonlearning.com

Distributed in Canada by
Taylor & Francis
74 Rolark Drive
Scarborough, Ontario M1R 4G2, Canada
Toll Free Tel.: +1 877 226 2237; E-mail: tal_fran@istar.ca

Distributed in the rest of the world by
Thomson Publishing Services
Cheriton House
North Way
Andover, Hampshire SP10 5BE, UK
Tel.: +44 (0)1264 332424; E-mail: salesorder.tandf@thomsonpublishingservices.co.uk

Library of Congress Cataloging-in-Publication Data

Plant microbiology / edited by Michael Gillings and Andrew Holmes.
p. cm.
ISBN 1-85996-224-6
1. Plants--Microbiology. 2. Plant-microbe relationsips. 3. Rhizobium. I. Gillings, Michael. II. Holmes, Andrew J. III. Title.

QR351.P58325 2004

577.8′52--dc22 2004003943

Production Editor: Catherine Jones
Typeset by Saxon Graphics Ltd, Derby, UK
Printed and bound by MPG Books, Bodmin, UK

Contents

Abbreviations

ABC	ATP-Binding Cassette
ACL	acyl carrier protein
AFLP	amplified fragment length polymorphism
AHL	acyl homoserine lactone
AM	arbuscular mycorrhizal
AOGCM	Atmospheric–Ocean General Circulation Model
APG	Angiosperm Phylogeny Group
BLAST	Basic Local Alignment Search Tool
bp	basepair
CGA	community genome array
CLPP	community level physiological profiling
DF	diffusible factor
DGGE	denaturing gradient gel electrophoresis
DSF	diffusible extracellular factor
EC	enzyme commission
ENSO	El Nino – Southern Oscillation
EPS	exopolysaccharide
FAME	fatty acid methyl ester
FGA	functional gene array
GMP	genetically modified plant
HGT	horizontal gene transfer
HR	hypersensitive response
INVAM	International Culture Collection of Arbuscular and Vesicular Mycorrhizal Fungi
IS	insertion sequence
ITS	internal transcribed spacer
LCO	lipo-chito-oligosaccharide
LPS	lipopolysaccharide
LRR	leucine-rich repeat
MGS	metabolic group-specific
MYA	million years ago
NAO	North Atlantic Oscillation
NCBI	National Center for Biotechnological Information
NU	nodulation unit
ORF	open reading frame
OTU	operational taxonomic unit
PCR	polymerase chain reaction
PGS	phylogenetic group-specific
POA	phylogenetic oligonucleotide array
R	resistance
RFLP	restriction fragment length polymorphism

RISA	ribosomal intergenic spacer analysis
RNB	root nodule bacteria
ROS	reactive oxygen species
SAM	S-adenosylmethionine
SOI	Southern Oscillation Index
SSCP	single-strand conformation polymorphism
SSU	small subunit
TC	transport classification
Ti	tumour-inducing
T-RFLP	terminal RFLP
TTSS	type III secretion systems
VCG	vegetative compatibility group

Contributors

Bartsev, A. LBMPS, l'Université de Genève, 1 ch de l'Impératrice, 1292 Chambésy/Genéve, Switzerland

Benson, D.R. Dept. Molecular and Cell Biology, University of Connecticut, Storrs, CT 06268–3044, USA

Broughton, W.J. LBMPS, l'Université de Genève, 1 ch de l'Impératrice, 1292 Chambésy/Genéve, Switzerland

Chakraborty, S. CSIRO Tropical Agriculture, CRC for Tropical Plant Pathology, University of Queensland, Qld 4072, Australia

Cother, E. NSW Agriculture, Forest Road, Orange NSW 2800, Australia

Faeth, S.H. Department of Biology, Arizona State University, Tempe, AZ 85287–1501, USA

Fegan, M. CRC for Tropical Plant Protection, The University of Queensland, St. Lucia, Queensland, 4072, Australia

Gerzabek, M.H. Institute of Soil Research, University of Agricultural Sciences Vienna, A-1180 Vienna, Austria

Gillings, M. Molecular Prospecting Group and Key Centre for Biodiversity, Department of Biological Sciences, Macquarie University, Sydney, NSW, 2109, Australia

Hayward, C. CRC for Tropical Plant Protection, The University of Queensland, St. Lucia, Queensland, 4072, Australia

Helander, M. Section of Ecology, Department of Biology, FIN-20014 University of Turku, Finland

Holmes, A.J. School of Molecular and Microbial Biosciences, The University of Sydney, Sydney, New South Wales, 2006, Australia

Husband, R. Department of Biology, University of York, PO Box 373, York, YO10 5YW, UK

Kandeler, E. Institute of Soil Science, University of Hohenheim, D-70599 Stuttgart, Germany

Kobayashi, H. LBMPS, l'Université de Genève, 1 ch de l'Impératrice, 1292 Chambésy/Genéve, Switzerland

Leslie, J.F. Department of Plant Pathology, 4002 Throckmorton Plant Sciences Center, Kansas State University, Manhattan, Kansas 66506–5502, USA

Nester, E.W. Department of Microbiology, Box 357242, University of Washington, Seattle, WA 98195–7242, USA

Pangga, I.B. CSIRO Tropical Agriculture, CRC for Tropical Plant Pathology, University of Queensland, Qld 4072, Australia

Pemberton, C.L. Department of Biochemistry, University of Cambridge, Tennis Court Road, Building O, Downing Site, Cambridge, CB2 1QW, UK

Potter, D. Dept. Pomology, Univ. of California, Davis, One Shields Avenue, Davis, California 95616, USA

Saikkonen, K. MTT Agrifood Research Finland, Plant Production Research, Plant Protection, FIN-31600 Jokioinen, Finland

Salmond, G.P.C. Department of Biochemistry, University of Cambridge, Tennis Court Road, Building O, Downing Site, Cambridge, CB2 1QW, UK

Sessitsch, A. ARC Seibersdorf research GmbH, Div. of Environmental and Life Sciences, A-2444 Seibersdorf, Austria

Setubal, J.C. University of Campinas, Bioinformatics Laboratory Institute of Computing, CP 6176, Campinas, SP 13083–970, Brazil

Slater, H. Department of Biochemistry, University of Cambridge, Tennis Court Road, Building O, Downing Site, Cambridge, CB2 1QW, UK (now at: New Phytologist Central Office, Bailrigg House, Lancaster University, Lancaster, LA1 4YE, UK)

Smalla, K. Federal Biological Research Centre for Agriculture and Forestry, Institute for Plant Virology, Microbiology and Biosafety, D-38104 Braunschweig, Germany

Summerell, B.A. Royal Botanic Gardens and Domain Trust, Mrs. Macquaries Road, Sydney, New South Wales, 2000, Australia

Vanden Heuvel, B.D. Dept. Pomology, Univ. of California, Davis, One Shields Avenue, Davis, California 95616, USA

Watkin, E. L. J. School of Biomedical Sciences, Curtin University of Technology, GPO Box U1987, Perth, WA 6102, Australia

Wood, D.W. University of Washington, Department of Microbiology, 1959 NE Pacific Street, Box 357242, Seattle, WA 98195, USA

Zhang, L.-H. The Institute for Molecular Agrobiology, 1 Research Link, The National University of Singapore, 117604, Singapore

Preface

The last decade has seen major changes in the way that we investigate the varied interactions between micro-organisms and plants. The widespread adoption of molecular methods in plant microbiology has given us the opportunity to investigate these biological systems with unparalleled precision and sensitivity.

We have known for a long time about the many mutually beneficial relationships between plants and micro-organisms. However, the detailed analysis of these relationships has often eluded us because of our inability to bring some of the microbial symbionts into pure culture, and to tease apart the complex biochemical interplay between mutualists. The first chapters in this book demonstrate how far we have come in the characterisation of plant mutualisms. We are now in a position to answer questions about the diversity, ecology and community structure of the most abundant of terrestrial mutualisms, that of mycorrhizal fungi and their plant hosts. There has been enormous progress in the characterisation of the molecular signals and other factors controlling host specificity in rhizobial associations. In a similar vein, the biology and ecology of fungal endophytes and of *Frankia* are yielding up their secrets. We now also have the tools to ask questions about how agricultural practices, and in particular the use of transgenic organisms, might affect rhizosphere communities.

The identification and characterisation of plant pathogens has been a major focus of plant microbiology for both economic reasons and international quarantine. A diverse array of molecular methods is now available for diagnosis and detection of pathogens. These methods have often revealed unsuspected diversity and led to rearrangements of taxonomic schemes. The second part of the book deals with some examples of diversity of pathogens in the fungal and bacterial worlds. It also summarises one of the most important revolutions in our understanding of bacterial communities, the discovery of quorum sensing. We must now view bacteria as communities of interacting cells, able to coordinate their biochemical activities in a manner dependent on the number of cells in the local environment. This discovery offers deep insights into the mechanisms by which bacteria cause disease in plants, and also offers opportunities for new methods of disease control. We must also bear in mind that regardless of our current understanding, global climate change will have major impacts on the distribution and severity of plant diseases.

The rapid improvements in high-throughput DNA sequencing and analysis are also poised to rapidly expand our understanding of plant microbiology. Already the entire genome sequence of several plant pathogens has been obtained, and more are at an advanced stage. We also now have the ability to investigate the last real biological frontier, the vast diversity of micro-organisms that have yet to be discovered, let alone characterised. There are now standard methods to recover microbial genes from environmental samples, such as soils and sediments, without recourse to standard laboratory culture. However, characterisation of the biochemistry and physiology of microbial species may continue to rely on culturable organisms, and for this reason, the continued existence and support for international microbial culture collections must be a high priority.

The science of plant microbiology is a diverse and complex one, and we realise that there are many areas that have not been examined in this book. Nevertheless, we hope that these contents convey some of the rapid progress and excitement inherent in current investigations of micro-organisms and their interactions with plants.

Michael Gillings and Andrew Holmes

1

The diversity, ecology and molecular detection of arbuscular mycorrhizal fungi

Rebecca Husband

1.1 Introduction

The vast majority of land plants rely on interactions with root symbionts to ensure adequate nutrient uptake. Of these root symbionts the most widespread and common are the arbuscular mycorrhizal (AM) fungi, which form associations with *ca* 60% of plant species (Smith and Read, 1997). The AM association is also arguably the most successful root symbiosis in evolutionary terms. Fossil evidence and molecular clock estimates indicate that the AM symbiosis originated at least 400 MYA (million years ago) and has not changed appreciably since (Simon *et al.*, 1993a; Taylor *et al.*, 1995). The association is almost universally distributed in early plant taxa with loss of the symbiosis only occurring more recently in *ca* 10% of plant families (Tester *et al.*, 1987; Trappe, 1987). It is therefore hypothesised that the AM symbiosis was instrumental in ensuring the successful colonisation of land by plants (Simon *et al.*, 1993a).

The AM symbiosis is widely accepted to be mutualistic. The most obvious benefit to the fungus is a ready supply of carbon, whilst the plant gains access to nutrients (most notably phosphorus) that might not be available from root uptake alone (Smith and Read, 1997). Other benefits to the plant include improved water relations and protection against pathogens (Newsham *et al.*, 1995). Plants associated with AM fungi often exhibit increased growth and survival (Smith and Read, 1997) although critically the level of benefit depends on a variety of factors including the particular host–fungal combination (Helgason *et al.*, 2002; Streitwolf-Engel *et al.*, 1997; van der Heijden *et al.*, 1998a). The differential effects individual AM fungi have on plant performance are therefore proposed to have a major influence on ecosystem functioning, from

Plant Microbiology, Michael Gillings and Andrew Holmes

affecting productivity (Klironomos *et al.*, 2000), to altering competitive interactions (Gange *et al.*, 1993; Grime *et al.*, 1987; Hartnett and Wilson, 1999; O'Connor *et al.*, 2002), and influencing the overall diversity of plant communities (van der Heijden *et al.*, 1998b).

Yet despite the important role AM fungi play in ecosystem functioning, little is known of the community structure and ecology of the fungi themselves. To date, fewer than 200 AM fungal species have been identified (Morton and Benny, 1990). This apparent low global diversity of AM fungi compared to their associated host plant communities has led to the widespread belief, only now being challenged, that AM fungi are a functionally homogeneous group (Smith and Read, 1997). Indeed, with few exceptions, the majority of ecological studies have ignored the functional diversity of individual AM fungi and grouped all the fungi into a single class. To be fair, this is because AM fungi are notoriously difficult to study. Firstly, AM fungi are obligate biotrophs and we have yet to find a way of culturing them independently of their plant hosts. Secondly, AM fungi exhibit a very low morphological diversity, making the reliable identification of different species difficult. Fungal structures formed internally within the host root possess few diagnostic characters; therefore the taxonomy of AM fungi is based on the subcellular structures of asexual spores. The spores of AM fungi are relatively large, easy to extract from the soil and have enough characteristics to enable identification to species level by experienced personnel (see the International Culture Collection of Arbuscular and Vesicular Mycorrhizal Fungi (INVAM) website http://invam.caf.wvu.edu for more information).

Estimating the diversity of AM fungi in the field has therefore traditionally relied on detecting the spores present in the soil. Since the spores tend to be ephemeral a complementary approach involves growing 'trap' plants in field soils in the greenhouse and analysing the spores produced. Unfortunately, these methods are problematic because fungal sporulation rates are influenced both by the environment and the host plant species (Bever *et al.*, 1996; Eom *et al.*, 2000; Morton *et al.*, 1995), therefore spore counts are not a direct measure of diversity. Furthermore, the population of spores in the soil may bear little relation to the AM fungal populations colonising roots (Clapp *et al.*, 1995). One of the most important methodological advances in the study of AM communities has been the application of the polymerase chain reaction (PCR) to directly identify the AM fungi *in planta*. Recovering sequence information from the field gives us direct access, without relying on culturing, to the AM fungi present in roots, the ecologically significant niche. Sequence information also provides us with the means to consistently distinguish between morphologically similar taxa. Thus molecular tools now present us with new opportunities for understanding the role of AM fungi in the environment. In this chapter some of the methods available to study AM fungal diversity in the field are outlined and the key findings from such studies are reviewed.

1.1.1 The taxonomy of AM fungi

Based on morphological characters, fewer than 200 AM fungal species have been described (Morton and Benny, 1990). They can be divided into five families, the Acaulosporaceae (*Acaulospora*, *Entrophospora*), Gigasporaceae (*Gigaspora*,

Scutellospora) and Glomaceae (*Glomus*) (Morton and Benny, 1990) plus the newly described lineages (*Archaeosporaceae*) and Paraglomaceae (*Paraglomus*) (Morton and Redecker, 2001). Traditionally these families have been placed in the order Glomales, phylum Zygomycota, but in recent years both morphological and molecular evidence have indicated that the phylum Zygomycota as currently defined cannot be sustained. Firstly, many of the organisms assigned to it, including AM fungi, are not known to have a sexual stage, i.e. they do not form zygosporangia (Benny, 1995). Secondly, phylogenetic analyses based on sequence data have demonstrated that the lineages ascribed to the Zygomycota do not share a common ancestor, i.e. the phylum Zygomycota is polyphyletic (O'Donnell *et al.*, 2001; Tehler *et al.*, 2000). Consequently, based on a small subunit (SSU) ribosomal gene (rDNA) phylogeny, Schüßler *et al.* (2001b) have proposed a new classification for the AM fungi, removing them from the Zygomycota and placing them in a new phylum the Glomeromycota. The analyses of Schüßler *et al.* (2001b) indicate that the AM fungi can be separated into a monophyletic clade that is not closely related to any of the Zygomycota lineages, but is instead probably diverged from the same common ancestor as the Ascomycota and Basidiomycota.

At the lower taxonomic level, the SSU rDNA phylogeny indicates a large genetic diversity within the genus *Glomus* that is not reflected by any of the morphological characters available (Schwarzott *et al.*, 2001). Sequences for the AM fungi within this genus can have genetic distances as large as that between the families Acaulosporaceae and Gigasporaceae and as a result the classification of Schüßler *et al.* (2001b) also proposes a new family and order ranking (see *Figure 1.1*). The proposed order Diversisporales contains the two traditional family groupings of the Acaulosporaceae and Gigasporaceae, plus a new family Diversisporaceae *fam. ined.* consisting of some of the AM fungi originally classified within the genus *Glomus*. The remaining 'classical' *Glomus* spp. have been placed in the order Glomerales, which clearly separates into two distinct family-ranked clades, *Glomus* Group A and B. Two further orders are proposed, the Paraglomerales (single family Paraglomeraceae) and the Archaeosporales (two families, Archaeosporaceae and Geosiphonaceae). As currently defined, the family Archaeosporaceae is paraphyletic. The type species for the Geosiphonaceae is a non-mycorrhizal fungus, *Geosiphon pyriforme*, which forms an association with cyanobacteria. Previously it had been proposed that *Geosiphon* represented the ancestral precursor to AM fungi (Gehrig *et al.*, 1996), but the recent rDNA data reveal it is in fact closely related to the Archaeosporaceae and consequently the phylum Glomeromycota, as defined, includes both mycorrhizal and non-mycorrhizal fungi.

The classification of Schüßler *et al.* (2001b) will necessarily evolve as more information becomes available. The lack of convincing morphological characters for many of the groupings and the presence of multiple sequence variants within single spores of AM fungi (discussed in Section 1.2), means that more detailed work is needed before the taxonomy and systematics of AM fungi are truly understood. Nonetheless the classification of Schüßler *et al.* (2001b) represents the basis for a new taxonomy for AM fungi, finally acknowledging what has long been obvious, namely that AM fungi are not typical Zygomycetes.

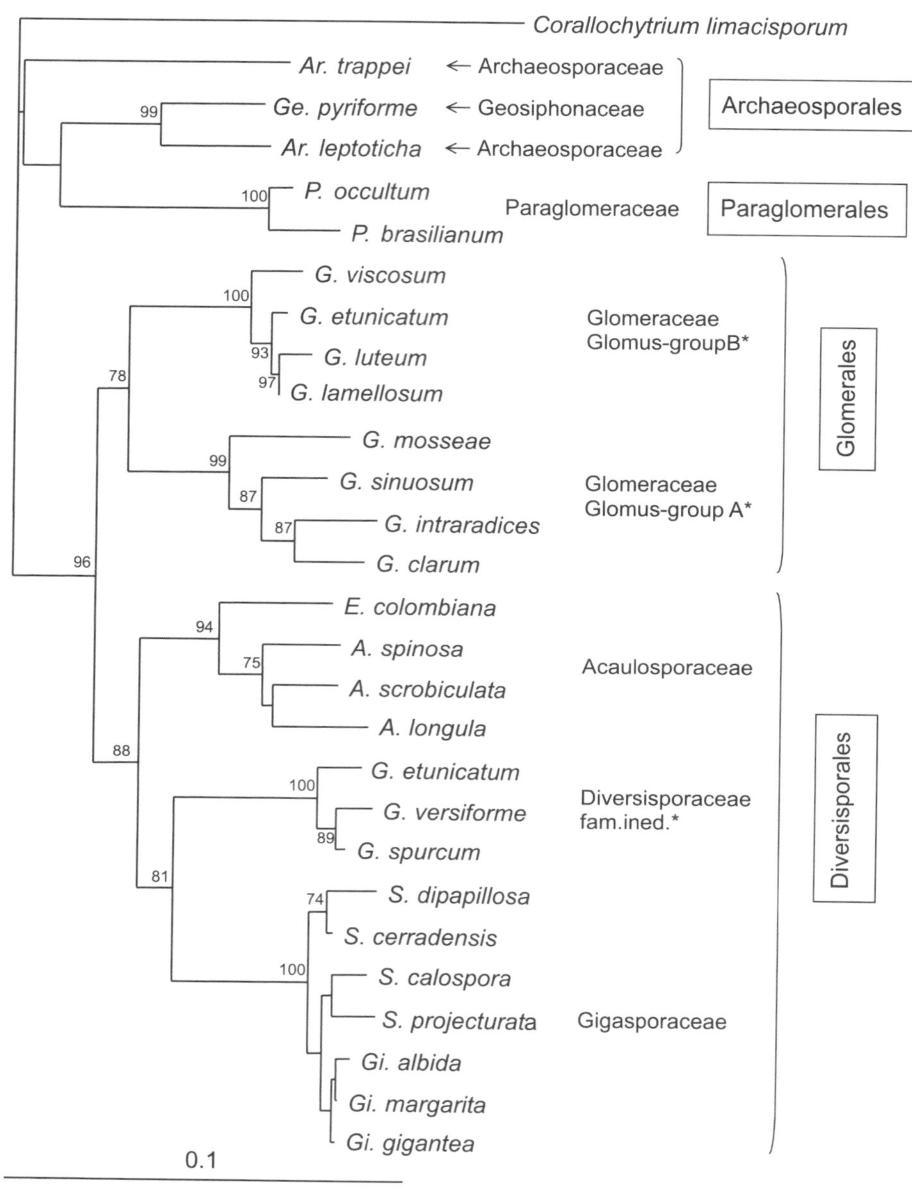

Figure 1.1. Phylogeny of the Glomeromycota (neighbour-joining analysis of SSU rDNA sequences) indicating the families and orders proposed by Schüßler *et al.*, (2001b). Bootstrap values >70% (1000) replicates are shown. Asterisks (*) identify the AM fungal taxa previously classified within the genus *Glomus*. A putative choanozoan, *C. limacisporum* L42528, (Cavalier-Smith and Allsopp, 1996) is used as an outgroup. Accession numbers are as follows: *Ar. trappei* Y17634, *Ge. pyriforme* AJ276074, *Ar. leptoticha* AJ301861, *P. occultum* AJ276082, *P. brasilianum* AJ301862, *G. viscosum* Y17652, *G. etunicatum* Y17639, *G. luteum* AJ276089, *G. lamellosum* AJ276087, *G. mosseae* AJ418853, *G. sinuosum* AJ133706, *G. intraradices* AJ301859 *G. clarum* AJ276084, *E. colombiana* Z14006, *A. spinosa* Z14004, *A. scrobiculata* AJ306442, *A. longula* AJ306439, *G. etunicatum* AJ301860, *G. versiforme* X86687, *G. spurcum* Y17650, *S. dipapillosa* Z14013, *S. cerradensis* AB041344, *S. callospora* AJ306443, *S. projecturata* AJ242729, *Gi. albida* Z14009, *Gi. margarita* X58726, *Gi. Gigantea* Z14010.

1.2 Molecular techniques used to study AM fungi in the field

Molecular techniques have the potential to revolutionise the study of AM fungi in the environment. Previously we have only had access to those AM fungi amenable to trap culture or actively sporulating in the field. Now a variety of nucleic acid-based strategies have been developed that enable the AM fungi to be characterised independently of spore formation. The majority of sequence information for AM fungi is derived from the ribosomal RNA genes (rDNA). In most organisms the ribosomal RNA genes are present in multiple copies arranged in tandem arrays. Each repeat unit consists of genes encoding a small (SSU or 18S) and a large (LSU or 28S) subunit, separated by an internal transcribed spacer (ITS), which includes the 5.8S rRNA gene (*Figure 1.2*). The SSU and 5.8S genes evolve relatively slowly and are useful for studies of distantly related organisms. The LSU and ITS regions evolve more quickly and are useful for fine-scale differentiation between species. In most organisms, the process of concerted evolution ensures that within an individual the multiple copies of rDNA are identical, but early studies looking at the genetic diversity of AM fungi revealed an unexpectedly high degree of ITS sequence variation within single spores (Lloyd-MacGilp *et al.*, 1996; Sanders *et al.*, 1995). Sequence divergence within single spores has since been detected in the ITS of many different species (Antoniolli *et al.*, 2000; Jansa *et al.*, 2002; Lanfranco *et al.*, 1999; Pringle *et al.*, 2000). The diversity is not restricted to the ITS, but has also been detected in the SSU (Clapp *et al.*, 1999) and extensive intrasporal diversity appears to exist in the LSU (Clapp *et al.*, 2001; Rodriguez *et al.*, 2001). Most recently, intrasporal variation has been detected in a gene encoding a binding protein (*BiP*) (Kuhn *et al.*, 2001).

Currently the full implications of the intrasporal variation present within the Glomeromycota are unclear. One obvious problem is that there is no straightforward correlation between sequence identity and species identity, and as a consequence there is no phylogenetic species concept for AM fungi. Furthermore there is no straightforward correlation between the number of sequence variants detected within a root and the number of separate infection events. Thus, ecological studies that use molecular markers to study AM fungal diversity are limited as to the conclusions they can make. Not only is it impossible to ascribe a species name to a sequence type, but it is also impossible to determine how many different AM fungi are colonising each root. As a consequence the majority of studies reviewed in this chapter have simply placed the AM fungi into groups based on sequence similarity, making the assumption that the AM fungi within each grouping share at least some ecological or functional characteristics. Such an assumption is not necessarily unjustified because in general most of the intrasporal sequence variation is relatively minor. Usually phylogenetic analyses reveal that the majority of sequence variants will form a 'core cluster' with a minority of the sequence variants revealing a greater divergence and clustering elsewhere (Clapp *et al.*, 2001, 2002). Even so, more research is clearly needed in order to understand the genetic organisation of AM fungi. For a more detailed discussion of the topic see Clapp *et al.* (2002), and Sanders (2002).

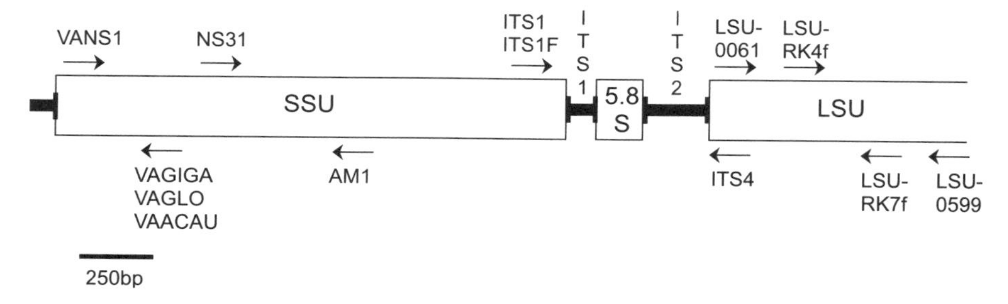

Figure 1.2. A repeat unit of the fungal ribosomal RNA genes, showing the position of primers commonly used for the study of AM fungi.

1.2.1 Community detection

One of the most common goals in arbuscular mycorrhizal research is to determine the role AM fungi play in ecosystem functioning. Previously any such research has been severely limited by the lack of basic information such as the identities and distributions of the fungi. Whilst molecular techniques can help us gain such information, the methods themselves are limited by the available molecular markers. As a rule, in the wider field of molecular ecology, a single set of ribosomal primers is quickly established that is sufficient for all preliminary investigations of the organism(s) in question (see Avise, 1994; Carvalho, 1998). Not so the arbuscular mycorrhizas; as yet there is no single method that can reliably measure *in planta* diversity. Separate from the issue of intraspecies diversity, we have yet to develop markers that can differentiate all AM fungal sequences from non-AM fungal and plant sequences. With the discovery of the highly divergent families of the Archaeosporaceae and Paraglomeraceae the sequence divergence within the Glomeromycota is considerable, so it may never be possible to rely on a single molecular marker. Currently, a compromise must be made between the level of genetic resolution and the number of lineages detected.

The first paper to apply molecular techniques to AM research appeared 10 years ago (Simon *et al.*, 1992). The authors used general eukaryotic primers to amplify SSU sequences from spores and, based on this information, designed the primer VANS1 which they hoped to be a general AM fungal primer. They subsequently designed family-specific primers (VAGIGA, VAGLO, VAACAU) which, when teamed with VANS1, enabled the direct amplification of AM fungi from within plant roots (Simon *et al.*, 1993b). Although these primers appeared to work well on roots from microcosm studies, in the field they were more problematic (Clapp *et al.*, 1995). It was later revealed that the VANS1 site is not well conserved throughout the Glomeromycota (Clapp *et al.*, 1999; Schüßler *et al.*, 2001a).

Helgason *et al.* (1998) also targeted the SSU when they designed the AM1 primer to exclude plant sequences and preferentially amplify AM fungal sequences. Coupled with a general eukaryotic primer, it has been successfully utilised in the field to determine the diversity of AM fungi from many different habitats (Daniell *et al.*, 2001; Helgason *et al.*, 1998, 1999, 2002; Husband *et al.*, 2002a,b; Kowalchuk *et al.*, 2002; Vandenkoornhuyse *et al.*, 2002). However, new sequence data have revealed that the AM1 primer is not well conserved in certain divergent lineages, the Archaeosporaceae and the Paraglomeraceae (Morton and Redecker, 2001). The AM1 primer also contains two mis-matches for sequences belonging to the *Glomus* group B clade defined by Schüßler *et al.* (2001b). In addition, in some habitat types, relatively high proportions (up to 30%) of non-AM fungi, mainly pyrenomycetes, are co-amplified (Daniell *et al.*, 2001). Even so, at this present time AM1 remains the most broadly applicable single primer suitable for field studies, reliably detecting the three traditional families, and having numerous studies that provide useful comparisons.

In contrast, Kjøller and Rosendahl, (2001) have designed LSU primers specific to a subgroup in the Glomeraceae. The primers LSURK4f and LSURK7r are used in a nested PCR following amplification with general eukaryotic primers (LSU0061 and LSU0599; van Tuinen *et al.*, 1998) and are designed to amplify a lineage within the *Glomus* group A clade, including *G. mosseae*, *G. caledonium* and *G. geosporum*. The authors took this approach because preliminary characterisation of the spore populations in their field site determined that the AM fungal community was dominated by many very closely related *Glomus* species that would be very difficult to distinguish using the SSU gene. Therefore they utilised the higher diversity of the LSU to separate the different species, enabling community comparisons to be made within this subgroup. A series of group-specific primers targeting the major lineages within the Glomeromycota has also been designed by Redecker (2000). These primers amplify parts of the SSU, the ITS and the 5.8S gene, although they have yet to be used in the field.

In the field, most plant roots are colonised by more than one AM type; therefore, unless single-taxon primers are used, a way must be found to separate the different types. The majority of the studies using the AM1 primer have used an approach based on cloning, PCR, and restriction fragment length polymorphism (RFLP) to divide the AM fungi into classes. Examples of each class can then be sequenced to give an insight into their identity. If it is assumed each fungal type is amplified and cloned proportionally, then the numbers of each class can be perceived as an approximate estimate of their proportion in the root. Although yielding much valuable information, the cloning step is both expensive and labour-intensive. Recently, Kowalchuk *et al.* (2002) adopted a different technique, denaturing gradient gel electrophoresis (DGGE), Muyzer *et al.*, 1993), to characterise the AM communities. Separation of the different AM types depends on the melting behaviour of the DNA sequence and, in theory, DGGE is sensitive enough to detect differences of a single base pair. A slightly different gel-based technique, single-strand conformation polymorphism (SSCP, Orita *et al.*, 1989) was used by Simon *et al.* (1993b) and Kjøller and Rosendahl (2001). The PCR product is denatured immediately before loading on a non-denaturing gel, and separation is achieved through migrational differences between the sequences as they adopt different conformations within the gel. An alternative approach, terminal restriction fragment length polymorphism (T-RFLP; Liu *et al.*, 1997) was used by Vandenkoornhuyse *et al.* (personal communication). This method uses a PCR in which the primers are fluorescently labelled. After amplification, the PCR product is digested with one or more enzymes generating terminal-labelled fragments that are characteristic in size. In theory, with the appropriate combination of genetic marker and restriction enzyme, terminal fragments can be generated that are diagnostic of individual species.

1.2.2 Specific detection

Frequently it is desirable to focus on the ecology of specific fungal isolates. The method devised by van Tuinen *et al.* (1998) provides a good example of how species-specific primers can track different AM fungal strains in mixed inoculum experiments. The authors designed primers specific for the LSU of each inoculant species, using them in a second round of amplification to gain information on the competitive interactions of the various isolates. This approach has successfully been used and expanded in microcosm experiments testing the effect of sewage sludge treatments on AM fungi (Jacquot *et al.*, 2000; Jacquot-Plumey *et al.*, 2001) and directly in the field studying the effect of heavy-metal polluted soils (Turnau *et al.*, 2001). The study by Turnau *et al.* (2001) further serves to illustrate the discrepancy between spore and root populations of AM fungi. Even though the authors characterised the spore population at their field site and designed primers for all the species isolated, many plant roots did not yield amplified sequences despite being clearly colonised.

The main utility of molecular markers able to differentiate between closely related strains is in the field of molecular taxonomy. The ITS region has been used extensively for such studies and the universal primers ITS1 and ITS4, designed by White *et al.* (1990), have proved especially useful. However, due to the sequence variation present within single spores of AM fungi, the ITS region cannot be used as a taxonomic tool as it has been in other organisms. Nonetheless, the ITS1 and ITS4 primers have been used extensively in AM research to study the nature of this intrasporal sequence variation itself (Antoniolli *et al.*, 2000; Jansa *et al.*, 2002; Lanfranco *et al.*, 1999; Lloyd-MacGilp *et al.*, 1996; Pringle *et al.*, 2000; Sanders *et al.*, 1995).

1.3 The molecular diversity of AM fungi colonising roots in the field

At present we do not know what level of genetic diversity is meaningful in an ecological context, therefore it is not possible to fully interpret the results of ecological studies that use molecular markers. Despite this limitation, molecular techniques have so far provided much valuable information on the diversity of AM fungi across a variety of habitats. The first study (Clapp *et al.*, 1995) that used molecular techniques to analyse the diversity of AM fungi colonising roots in the field used the family-specific primers designed by Simon *et al.* (1992, 1993b). The authors compared molecular data for the presence or absence of each of the families in roots, with counts of spores isolated from the surrounding soil. The morphological and molecular data were largely in agreement for *Acaulospora* and *Scutellospora* types, but there was a large discrepancy for *Glomus* types, whereby *Glomus* spores were rarely found in the soil yet *Glomus* types were frequent colonisers of the roots (Clapp *et al.*, 1995). Thus this study showed conclusively what had long been suspected; spore populations do not accurately reflect the AM fungi colonising roots.

To date the most extensive molecular investigations have all used the AM1 primer, making it possible to draw comparisons between the AM communities in a seminatural woodland (Helgason *et al.*, 1999, 2002), arable sites (Daniell *et al.*, 2001), a seminatural grassland (Vandenkoornhuyse *et al.*, 2002), coastal sand dunes (Kowalchuk *et al.*, 2002) and a tropical forest (Husband *et al.*, 2002a,b). A summary of the levels of AM fungal diversity detected within these habitats is given in *Table 1.1*. Although different degrees of sampling intensity make it difficult to make direct comparisons, collectively these studies appear to reveal an approximate correlation between above- and below-ground diversity. In itself this approximate correlation has an important implication. Van der Heijden *et al.* (1998b) found that plant diversity increased with increasing AM fungal diversity in their experimental microcosm system. They suggested that the increase in plant diversity resulted from the growth of different plants being stimulated by different fungal species and consequently that AM fungal identity and diversity were potential determinants of plant community structure. Although no causal relationship can be drawn from the molecular field data, the demonstration that in the field there is a link between plant and AM fungal diversity is consistent with the hypothesis that AM fungi are potential determinants of ecosystem diversity.

These molecular data also reveal large differences in the AM community composition between the different habitat types. The arable sites, seminatural grassland and tropical forest are all heavily dominated by *Glomus* types, both in terms of the number of types and their abundance. In contrast, the seminatural woodland AM community is more evenly distributed between *Acaulospora* and *Glomus* types, though in terms of abundance, one of the woodland hosts *Hyacinthoides non-scripta* (bluebell), is heavily dominated by a *Scutellospora* type early in the growing season (Helgason *et al.*, 1999). No *Acaulospora* types were detected in the AM community colonising *Ammophila arenaria* in coastal sand dunes; instead the community contained equal numbers of *Glomus* and *Scutellospora* types (Kowalchuk *et al.*, 2002).

Where the experimental design allows it, extensive spatial and temporal heterogeneity is revealed within each habitat. Kowalchuk *et al.* (2002) were able to detect clear differences between the AM communities colonising *Ammophila* in vital and degenerating stands. Not only were the degenerating stands depauperate relative

Table 1.1. A summary of the levels of AM fungal diversity detected across various habitats

Habitat	No. of host species	No. of roots	No. of clones	No. of *Acualo-spora* sp.	No. of *Glomus* sp.	No. of *Giga-spora* sp.	Total no. of types
Sand dune[a]	1	/	n.a.	0	3	3	6
Arable[b]	4	79	303	1	6	1	8
Woodland[c]	6	71	257	5	6	1	13*
Grassland[d]	2	49	2001	2	15	1	18
Tropical forest[e]	2	48	1383	1	21	1	23

Data from [a]Kowlachuk *et al.*, (2002); [b]Daniell *et al.*, (2001); [c]Helgason *et al.*, (1999, 2002); [d]Vandenkoornhuyse *et al.*, (2002); and [e]Husband *et al.*, (2002a).

*An *Archaeospora* type was also detected.

to the vital stands, but the relative signal intensities of the samples from the degenerating stands tended to be substantially reduced. The AM community colonising bluebells in the seminatural woodland shows a clear seasonal succession, initially being dominated by a *Scutellospora* type which later in the season gives way to *Glomus* types if the dominant canopy is *Acer pseudoplatanus*, or *Acaulospora* types if the dominant canopy is *Quercus petraea* (Helgason *et al.*, 1999). These trends match well with the data from morphological analyses of the fungi colonising bluebell roots (Merryweather and Fitter, 1998a,b). Similarly, Husband *et al.* (2002a,b) detected a replacement over time in the presence and abundance of AM fungi colonising cohorts of seedlings in a tropical forest. The grassland mycorrhizal community was also shown to change at each sampling period, and the authors suggested that a shift in field management from grazing to mowing and the subsequent decrease in organic matter might be responsible (Vandenkoornhuyse *et al.*, 2002).

At different spatial scales, the woodland, grassland and tropical forest studies, all detected non-random associations between the plant community and the AM fungal community. The woodland AM community was influenced by the dominant canopy type (Helgason *et al.*, 1999); whereas the grassland AM community (at a single site) was shown to be significantly different between the two host species, *Agrostis capillaris* and *Trifolium repens* (Vandenkoornhuyse *et al.*, 2002). The tropical AM community was also significantly different between the host species, but the environment was found to have a greater influence, such that differences between host species were site-specific (Husband *et al.*, 2002a). Again, although no causal relationship can be made, these non-random patterns of association have an important implication. In recent years, various microcosm studies have shown that the AM fungal community can affect plant diversity or *vice versa* (Bever, 2002; Bever, *et al.*, 1996; Burrows and Pfleger, 2002; Eom *et al.*, 2000; Helgason *et al.*, 2002; Sanders and Fitter, 1992; van der Heijden *et al.*, 1998 a,b). If such processes occur in the environment, some degree of host preference in natural populations is highly likely. Indeed, by selecting morphologically distinct fungal species, McGonigle and Fitter (1990) were the first to demonstrate that non-random associations between different hosts and AM fungi exist in the field. These molecular data support the findings of McGonigle and Fitter (1990) and suggest that in the environment AM fungi may commonly exhibit a host preference.

The study by Helgason *et al.* (2002) is especially relevant because not only did they demonstrate that root colonisation, symbiont compatibility and plant performance varied with each fungus–plant combination in the greenhouse, but they were able to link the functioning of the mycorrhizae with the patterns of association between plants and fungi found in the field. For example, the authors suggested that one of the fungal types, *Glomus* sp. UY1225, appears to be a 'typical' AM fungus. In the field it shows a relatively broad host range and in the laboratory study *Glomus* sp. UY1225 provided some benefit to most of the plant species without greatly benefiting any of them. The only plant species it did not colonise extensively in pots was *Acer*, the only species in the field survey from which it was absent. In fact, the only fungus to colonise and benefit *Acer* in the laboratory study was *Glomus hoi* UY110. In the field the AM type Glo9, which is very closely related to, but distinct from, *G. hoi*, is found almost exclusively in *Acer* roots. The authors suggest that the sequence variation detected within *G. hoi* might represent a fraction of the variation within a single species that would in fact include the field-derived Glo9 sequences. Unfortunately, the low clone numbers generated from the field study make it impossible to test this idea. Even so, the authors argue that given the large impact of *G. hoi* on *Acer* growth, it would be very unexpected to find two functionally unrelated taxa restricted to the roots of *Acer* in the field. If, in the future, it is demonstrated that *G. hoi* and Glo9 are one and the same, the study by Helgason *et al.* (2002) will have provided the first ever evidence of functional selectivity within the arbuscular mycorrhizal symbiosis.

The AM types recognised by these molecular techniques cannot be equated directly with the formal species that are identified on the basis of spore morphology. Even so, many of the sequence types have been recovered repeatedly in different studies, and there are examples of identical sequences being isolated from different habitats (see *Figure 1.3* and * in *Figure 1.4*). It would appear that many of these types represent entities as widespread and stable as those defined by morphology. Furthermore, the SSU rDNA region amplified by the NS31/AM1 primers appears to provide a level of discrimination at approximately the species level (*Figure 1.3*). For example, the *G. mosseae* clade contains numerous sequences from different spores and cultures revealing a level of intraspecies variation comparable to the variation in the field-derived sequence type Glo1b. Similarly, the clade containing the sequence type Glo3 contains numerous sequences that have been isolated repeatedly both from different hosts and time points within a habitat, and from different habitats. A sequence from the *Glomus* sp. isolate UY1225 trapped from the woodland soil (Helgason *et al.*, 2002) matches many of the field-derived Glo3 sequences. Overall this clade reveals a similar level of variation as the *G. mosseae*/Glo1b clade. In contrast the sequence type Glo8 contains three distinct groups, one of which includes sequences from *G. fasiculatum* and *G. vesiculiferum*, another containing a *G. fasiculatum* sequence and the third containing sequences from *G. intraradicies*. There are too few culture-derived sequences to make many comparisons of this nature, and it must be acknowledged that if more sequences per isolate per spore were characterised, the level of intraspecies variation within the SSU could turn out to be much greater than presently recognised. However, the study by Kowalchuk *et al.* (2002) used DGGE to characterise various AM fungal isolates and detected no intraspecies variation, with the exception of *G. clarum* that consistently yielded two bands. Critically, they analysed both single-spore and multispore extracts thus maximising the probability of detecting intraspecies variation if it existed. In theory DGGE is sensitive enough to detect differences of a single basepair, but Kowalchuk *et al.* (2002) were not able to distinguish between two closely related species *Gi. margarita* and *Gi. albida*, the sequences for which differ by approximately five basepairs. Therefore, based on the data currently available, it would seem that small levels of intraspecies variation are present in this region of the SSU, but the variation is not so great as to be prohibitive to community studies.

The phylogenetic tree shown in *Figure 1.4* contains a single example of the different *Glomus* group A sequences isolated from the various habitats. As can be seen, with the exception of the Glo types already discussed, very few of the field-derived sequences group with sequences from AM fungi in culture. This phenomenon is not restricted to studies using the AM1 primer. Kjøller and

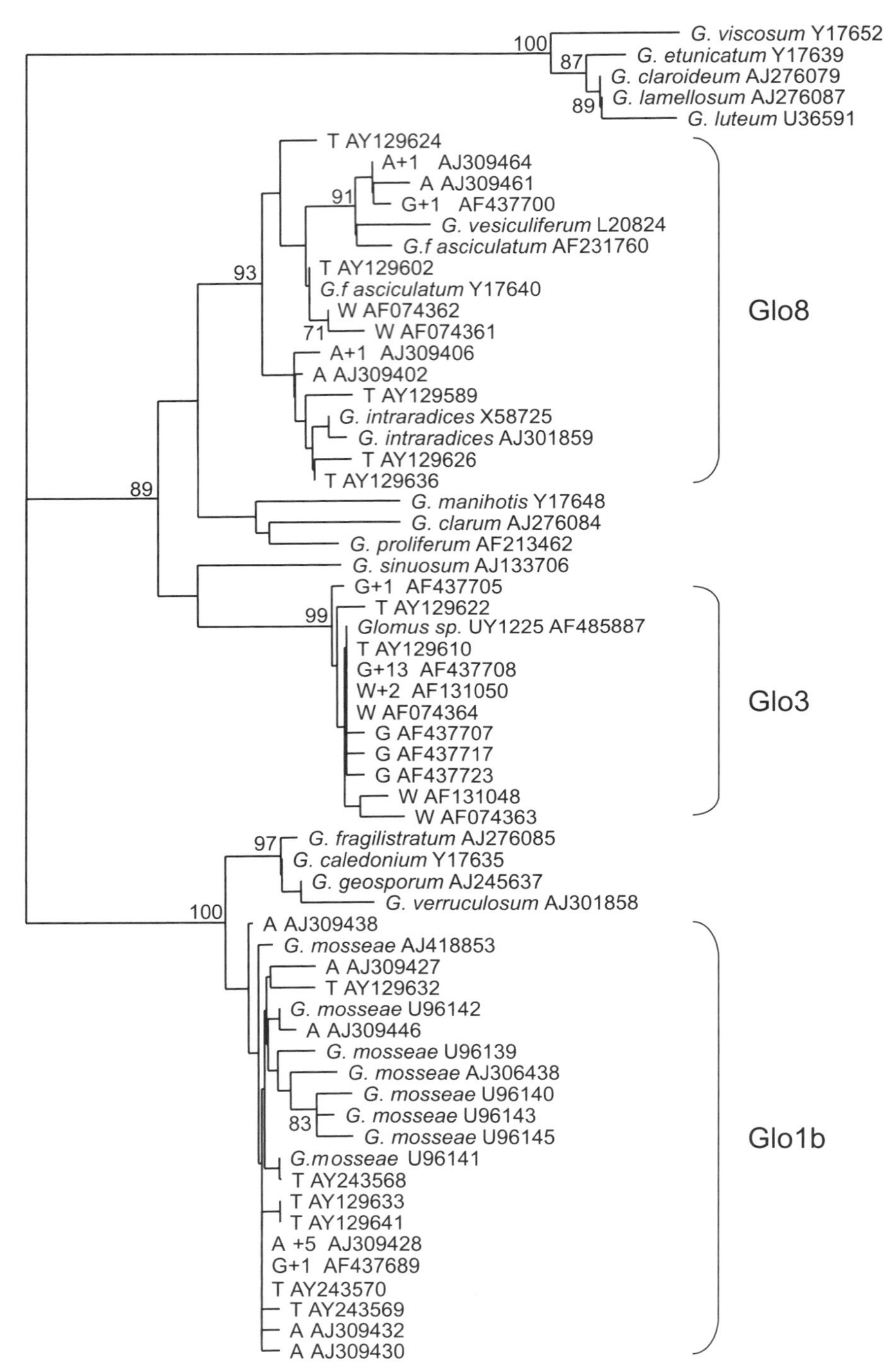

Figure 1.3. Neighbour-joining phylogenetic tree of the Glo1b, Glo3 and Glo8 field-derived sequences recovered from **W** seminatural woodland (Helagson *et al.*, 1999, 2002); **A**, arable sites (Daniell *et al.*, 2001); **G**, seminatural grassland (Vandenkoornhyuse *et al.*, 2002) and **T**, tropical forest (Husband *et al.*, 2002a,b). Bootstrap values >70% are shown (1000 replicates). Multiple identical sequences of the number indicated have been recovered.

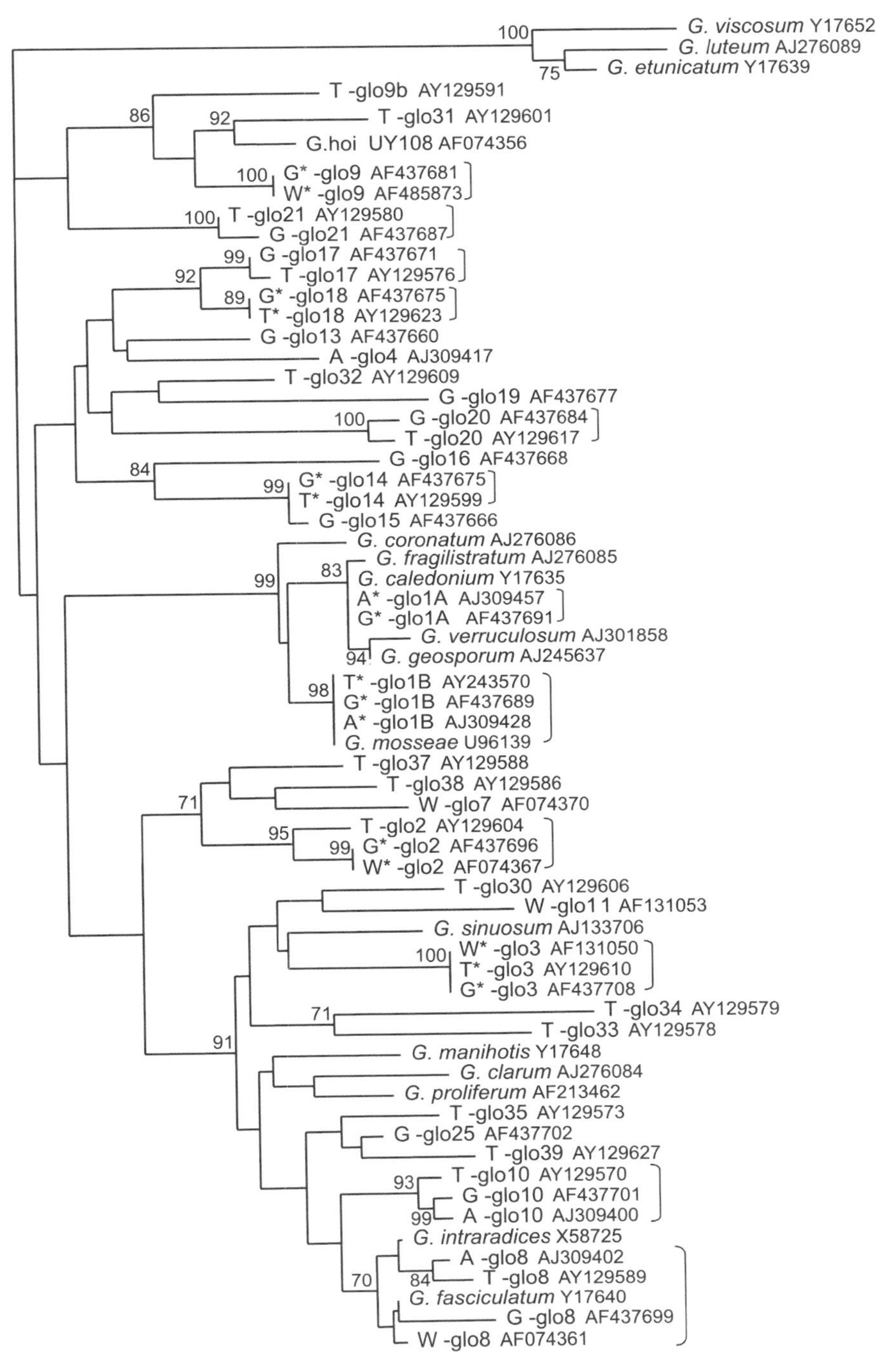

Figure 1.4. Neigbour-joining phylogenetic tree showing examples of the different Glomus-group A sequences isolated from **W** seminatural woodland (Helagson *et al.*, 1999, 2002); **A,** arable sites (Daniell *et al.*, 2001); **G,** seminatural grassland (Vandenkoornhyuse *et al.*, 2002) and **T,** tropical forest (Husband *et al.*, 2002a,b). Bootstrap values >70% are shown (1000 replicates). Asterisks (*) identify identical sequences recovered from different habitats. Brackets indicate identical sequence types recovered from different habitats.

Rosendahl (2001) deliberately designed their LSU primers to focus on a single lineage that includes *G. mosseae*, *G. claroideum* and *G. geosporum*, because these were the species that had been isolated from their field site. Yet despite limiting their study to these groups, they too recovered very few field sequences that matched known isolates. The simplest explanation for these observations is that the number of AM fungi in culture represent but a fraction of the true AM fungal diversity. Recently both Bever *et al.* (2001) and Helgason *et al.* (2002) have put forward arguments to this effect. Helgason *et al.* (2002) also challenge the traditional assumption that AM fungi are not host-specific. They argue such an assumption is based on the fact that: (i) fewer than 200 species have been described; and (ii) the AM fungi in culture tend to have a broad host range. However, they suggest there could be large numbers of as yet undescribed AM fungi that we have not managed to culture precisely because they are more host-selective. The growing number of non-random associations detected between different AM fungi and hosts in the field, plus the minimal overlap between sequences derived from the field and from cultured isolates, would seem to support their claims.

1.4 Conclusions

The application of molecular techniques to the ecological study of AM fungi has led to a number of valuable insights. It has repeatedly been demonstrated that the AM community composition within roots is diverse, changes radically between different habitats, and within habitats between different time points and plant species. This variation itself is proof that the AM fungi in the field are not ecologically equivalent. However, the ecological role of AM fungi will never be fully appreciated until we understand the relationship between morphological, functional and molecular diversity. The challenge for the future is to resolve the genetic structuring of AM fungi so that we may address ecological questions in a determined manner and ultimately establish the link between above- and below-ground ecosystem diversity.

References

Antoniolli, Z.I., Schachtman, D.P., Ophel-Keller, K., Smith, S.E. (2000) Variation in rDNA ITS sequences in *Glomus mosseae* and *Gigaspora margarita* spores from a permanent pasture. *Mycol. Res.* **104**: 708–715.

Avise, J.C. (1994) *Molecular Markers, Natural History and Evolution*. London: Chapman and Hall.

Benny, G.L. (1995) Classical morphology in Zygomycete taxonomy. *Can. J. Bot.-Rev. Can. Botan.* **73**: S725–S730.

Bever, J.D. (2002) Host-specificity of AM fungal population growth rates can generate feedback on plant growth. *Plant Soil* **244**: 281–290.

Bever, J.D., Morton, J.B., Antonovics, J., Schultz, P.A. (1996) Host-dependent sporulation and species diversity of arbuscular mycorrhizal fungi in a mown grassland. *J. Ecol.* **84**: 71–82.

Bever, J.D., Schultz, P.A., Pringle, A., Morton, J.B. (2001) Arbuscular mycorrhizal fungi: More diverse than meets the eye, and the ecological tale of why. *Bioscience* **51**: 923–931.

Burrows, R.L., Pfleger, F.L. (2002) Arbuscular mycorrhizal fungi respond to increasing plant diversity. *Can. J. Bot.-Rev. Can. Botan.* **80**: 120–130.

Carvalho, G.R. (1998) *Advances in Molecular Ecology*. IOS Press, Oxford.

Cavalier-Smith, T., Allsopp, M. (1996) *Corallochytrium*, an enigmatic non-flagellate protozoan related to choanoflagellates. *Euro. J. Protistol.* **32**: 306–310.

Clapp, J.P., Young, J.P.W., Merryweather, J.W., Fitter, A.H. (1995) Diversity of fungal symbionts in arbuscular mycorrhizas from a natural community. *New Phytol.* **130**: 259–265.

Clapp, J.P., Fitter, A.H., Young, J.P.W. (1999) Ribosomal small subunit sequence variation within spores of an arbuscular mycorrhizal fungus, *Scutellospora* sp. *Mol. Ecol.* **8**: 915–921.

Clapp, J.P., Rodriguez, A., Dodd, J.C. (2001) Inter- and intra-isolate rRNA large subunit variation in *Glomus coronatum* spores. *New Phytol.* **149**: 539–554.

Clapp, J.P., Helgason, T., Daniell, T.J., Young, J.P.W. (2002) Genetic studies of the structure and diversity of arbuscular mycorrhizal communities. In: van der Heijden, M.G.A, Sanders, I.R. (eds) *Mycorrhizal Ecology*, pp. 201–224. Heidelberg: Springer.

Daniell, T.J., Husband, R., Fitter, A.H., Young, J.P.W. (2001) Molecular diversity of arbuscular mycorrhizal fungi colonising arable crops. *FEMS Microbiol. Ecol.* **36**: 203–209.

Eom, A.H., Hartnett, D.C., Wilson, G.W.T. (2000) Host plant species effects on arbuscular mycorrhizal fungal communities in tallgrass prairie. *Oecologia* **122**: 435–444.

Gange, A.C., Brown, V.K., Sinclair, G.S. (1993) Vesicular-arbuscular mycorrhiza fungi – a determinant of plant community structure in early succession. *Function. Ecol.* **7**: 616–622.

Gehrig, H., Schüßler, A., Kluge, M. (1996) *Geosiphon pyriforme*, a fungus forming endocytobiosis with *Nostoc* (Cyanobacteria), is an ancestral member of the Glomales: Evidence by SSU rRNA analysis. *J. Mol. Evol.* **43**: 71–81.

Grime, J.P., Mackey, J.M.L., Hillier, S.H., Read, D.J. (1987) Floristic diversity in a model system using experimental microcosms. *Nature* **328**: 420–422.

Hartnett, D.C., Wilson, G.W.T. (1999) Mycorrhizae influence plant community structure and diversity in tallgrass prairie. *Ecology* **80**: 1187–1195.

Helgason, T., Daniell, T.J., Husband, R., Fitter, A.H., Young, J.P.W. (1998) Ploughing up the wood-wide web? *Nature* **394**: 431.

Helgason, T., Fitter, A.H., Young, J.P.W. (1999) Molecular diversity of arbuscular mycorrhizal fungi colonising *Hyacinthoides non-scripta* (bluebell) in a seminatural woodland. *Mol. Ecol.* **8**: 659–666.

Helgason, T., Merryweather, J.W., Denison, J., Wilson, P., Young, J.P.W., Fitter, A.H. (2002) Selectivity and functional diversity in arbuscular mycorrhizas of co-occurring fungi and plants from a temperate deciduous woodland. *J. Ecol.* **90**: 371–384.

Husband, R., Herre, E.A., Turner, S.L., Gallery, R., Young, J.P.W. (2002a) Molecular diversity of arbuscular mycorrhizal fungi and patterns of host association over time and space in a tropical forest. *Mol. Ecol.* **11**: 2669–2678.

Husband, R., Herre, E.A., Young, J.P.W. (2002b) Temporal variation in the arbuscular mycorrhizal communities colonising seedlings in a tropical forest. *Fems Microbiol. Ecol.* **42**: 131–136.

Jacquot, E., van Tuinen, D., Gianinazzi, S., Gianinazzi-Pearson, V. (2000) Monitoring species of arbuscular mycorrhizal fungi in planta and in soil by nested PCR: application to the study of the impact of sewage sludge. *Plant Soil* **226**: 179–188.

Jacquot-Plumey, E., van Tuinen, D., Chatagnier, O., Gianinazzi, S., Gianinazzi-Pearson, V. (2001) 25S rDNA-based molecular monitoring of glomalean fungi in sewage sludge-treated field plots. *Environ. Microbiol.* **3**: 525–531.

Jansa, J., Mozafar, A., Banke, S., McDonald, B.A., Frossard, E. (2002) Intra- and intersporal diversity of ITS rDNA sequences in *Glomus intraradices* assessed by cloning and sequencing, and by SSCP analysis. *Mycol. Res.* **106**: 670–681.

Kjøller, R., Rosendahl, S. (2001) Molecular diversity of glomalean (arbuscular mycorrhizal) fungi determined as distinct *Glomus* specific DNA sequences from roots of field grown peas. *Mycol. Res.* **105**: 1027–1032.

Klironomos, J.N., McCune, J., Hart, M., Neville, J. (2000) The influence of arbuscular mycorrhizae on the relationship between plant diversity and productivity. *Ecol. Lett.* **3**: 137–141.

Kowalchuk, G.A., De Souza, F.A., Van Veen, J.A. (2002) Community analysis of arbuscular mycorrhizal fungi associated with *Ammophila arenaria* in Dutch coastal sand dunes. *Mol. Ecol.* **11**: 571–581.

Kuhn, G., Hijri, M., Sanders, I.R. (2001) Evidence for the evolution of multiple genomes in arbuscular mycorrhizal fungi. *Nature* **414**: 745–748.

Lanfranco, L., Delpero, M., Bonfante, P. (1999) Intrasporal variability of ribosomal sequences in the endomycorrhizal fungus *Gigaspora margarita*. *Mol. Ecol.* **8**: 37–45.

Liu, W.T., Marsh, T.L., Cheng, H., Forney, L.J. (1997) Characterization of microbial diversity by determining terminal restriction fragment length polymorphisms of genes encoding 16S rRNA. *Appl. Environ. Microbiol.* **63**: 4516–4522.

Lloyd-MacGilp, S.A., Chambers, S.M., Dodd, J.C., Fitter, A.H., Walker, C., Young, J.P.W. (1996) Diversity of the ribosomal internal transcribed spacers within and among isolates of *Glomus mosseae* and related mycorrhizal fungi. *New Phytol.* **133**: 103–111.

McGonigle, T.P., Fitter, A.H. (1990) Ecological specificity of vesicular arbuscular mycorrhizal associations. *Mycol. Res.* **94**: 120–122.

Merryweather, J., Fitter, A. (1998a) The arbuscular mycorrhizal fungi of *Hyacinthoides non-scripta* – I. Diversity of fungal taxa. *New Phytol.* **138**: 117–129.

Merryweather, J., Fitter, A. (1998b) The arbuscular mycorrhizal fungi of *Hyacinthoides non-scripta* – II. Seasonal and spatial patterns of fungal populations. *New Phytol.* **138**: 131–142.

Morton, J.B., Benny, G.L. (1990) Revised classification of arbuscular mycorrhizal fungi (Zygomycetes) – a new order, Glomales, 2 new suborders, Glomineae and Gigasporineae, and 2 new families, Acaulosporaceae and Gigasporaceae, with an emendation of Glomaceae. *Mycotaxon* **37**: 471–491.

Morton, J.B., Redecker, D. (2001) Two new families of Glomales, Archaeosporaceae and Paraglomaceae, with two new genera *Archaeospora* and *Paraglomus*, based on concordant molecular and morphological characters. *Mycologia* **93**: 181–195.

Morton, J.B., Bentivenga, S.P., Bever, J.D. (1995) Discovery, measurement, and interpretation of diversity in arbuscular endomycorrhizal fungi (Glomales, Zygomycetes). *Can. J. Bot.-Rev. Can. Bot.* **73**: S25–S32.

Muyzer, G., Dewaal, E.C., Uitterlinden, A.G. (1993) Profiling of complex microbial populations by denaturing gradient gel electrophoresis analysis of polymerase chain reaction amplified genes coding for 16S ribosomal RNA. *Appl. Environ. Microbiol.* **59**: 695–700.

Newsham, K.K., Fitter, A.H., Watkinson, A.R. (1995) Multi-functionality and biodiversity in arbuscular mycorrhizas. *Trends Ecol. Evol.* **10**: 407–411.

O'Connor, P.J., Smith, S.E., Smith, E.A. (2002) Arbuscular mycorrhizas influence plant diversity and community structure in a semiarid herbland. *New Phytol.* **154**: 209–218.

O'Donnell, K., Lutzoni, F.M., Ward, T.J., Benny, G.L. (2001) Evolutionary relationships among mucoralean fungi (*Zygomycota*): Evidence for family polyphyly on a large scale. *Mycologia* **93**: 286–297.

Orita, M., Iwahana, H., Kanazawa, H., Hayashi, K., Sekiya, T. (1989) Detection of polymorphisms of human DNA by gel electrophoresis as single strand conformation polymorphisms. *Proc. Natl Acad. Sci. USA* **86**: 2766–2770.

Pringle, A., Moncalvo, J.M., Vilgalys, R. (2000) High levels of variation in ribosomal DNA sequences within and among spores of a natural population of the arbuscular mycorrhizal fungus *Acaulospora colossica*. *Mycologia* **92**: 259–268.

Redecker, D. (2000) Specific PCR primers to identify arbuscular mycorrhizal fungi within colonized roots. *Mycorrhiza* **10**: 73–80.

Rodriguez, A., Dougall, T., Dodd, J.C., Clapp, J.P. (2001) The large subunit ribosomal RNA genes of *Entrophospora infrequens* comprise sequences related to two different glomalean families. *New Phytol.* **152**: 159–167.

Sanders, I.R. (2002) Ecology and evolution of multigenomic arbuscular mycorrhizal fungi. *Am. Natural.* **160**: S128–S141.

Sanders, I.R., Fitter, A.H. (1992) Evidence for differential responses between host fungus combinations of vesicular arbuscular mycorrhizas from a grassland. *Mycol. Res.* **96**: 415–419.

Sanders, I.R., Alt, M., Groppe, K., Boller, T., Wiemken, A. (1995) Identification of ribosomal DNA polymorphisms among and within spores of the Glomales – application to studies on the genetic diversity of arbuscular mycorrhizal fungal communities. *New Phytol.* **130**: 419–427.

Schüßler, A., Gehrig, H., Schwarzott, D., Walker, C. (2001a) Analysis of partial Glomales SSU rRNA gene sequences: implications for primer design and phylogeny. *Mycol. Res.* **105**: 5–15.

Schüßler, A., Schwarzott, D., Walker, C. (2001b) A new fungal phylum, the Glomeromycota: phylogeny and evolution. *Mycol. Res.* **105**: 1413–1421.

Schwarzott, D., Walker, C., Schüßler, A. (2001) *Glomus*, the largest genus of the arbuscular mycorrhizal fungi (Glomales), is nonmonophyletic. *Mol. Phylogenet. Evol.* **21**: 190–197.

Simon, L., Lalonde, M., Bruns, T.D. (1992) Specific amplification of 18S fungal ribosomal genes from vesicular-arbuscular endomycorrhizal fungi colonizing roots. *Appl. Environ. Microbiol.* **58**: 291–295.

Simon, L., Bousquet, J., Levesque, R.C., Lalonde, M. (1993a) Origin and diversification of endomycorrhizal fungi and coincidence with vascular land plants. *Nature* **363**: 67–69.

Simon, L., Levesque, R.C., Lalonde, M. (1993b) Identification of endomycorrhizal fungi colonizing roots by fluorescent single-strand conformation polymorphism polymerase chain-reaction. *Appl. Environ. Microbiol.* **59**: 4211–4215.

Smith, S.E., Read, D.J. (1997) *Mycorrhizal Symbiosis*, 2nd edn. San Diego: Academic Press.

Streitwolf-Engel, R., Boller, T., Wiemken, A., Sanders, I.R. (1997) Clonal growth traits of two *Prunella* species are determined by co-occurring arbuscular mycorrhizal fungi from a calcareous grassland. *J. Ecol.* **85**: 181–191.

Taylor, T.N., Remy, W., Hass, H., Kerp, H. (1995) Fossil arbuscular mycorrhizae from the Early Devonian. *Mycologia* **87**: 560–573.

Tehler, A., Farris, J.S., Lipscomb, D.L., Källersjö, M. (2000) Phylogenetic analyses of the fungi based on large rDNA data sets. *Mycologia* **92**: 459–474.

Tester, M., Smith, S.E., Smith, F.A. (1987) The phenomenon of nonmycorrhizal plants. *Can. J. Bot.-Rev. Can. Bot.* **65**: 419–431.

Trappe, J.M. (1987) Phylogenetic and ecological aspects of mycotrophy in the angiosperms from an evolutionary standpoint. In: Safir, G.R. (ed) *Ecophysiology of VA Mycorrhizal Plants*, pp. 5–25. Boca Raton, Florida: CRC Press.

Turnau, K., Ryszka, P., Gianinazzi-Pearson, V., van Tuinen, D. (2001) Identification of arbuscular mycorrhizal fungi in soils and roots of plants colonizing zinc wastes in southern Poland. *Mycorrhiza* **10**: 169–174.

van der Heijden, M.G.A., Boller, T., Wiemken, A., Sanders, I.R. (1998a) Different arbuscular mycorrhizal fungal species are potential determinants of plant community structure. *Ecology* **79**: 2082–2091.

van der Heijden, M.G.A., Klironomos, J.N., Ursic, M., Moutoglis, P., Streitwolf-Engel, R., Boller, T., Wiemken, A., Sanders, I.R. (1998b) Mycorrhizal fungal diversity determines plant biodiversity, ecosystem variability and productivity. *Nature* **396**: 69–72.

van Tuinen, D., Jacquot, E., Zhao, B., Gollotte, A., Gianinazzi-Pearson, V. (1998) Characterization of root colonization profiles by a microcosm community of arbuscular mycorrhizal fungi using 25S rDNA-targeted nested PCR. *Mol. Ecol.* **7**: 879–887.

Vandenkoornhuyse, P., Husband, R., Daniell, T.J., Watson, I.J., Duck, J.M., Fitter, A.H., Young, J.P.W. (2002) Arbuscular mycorrhizal community composition associated with two plant species in a grassland ecosystem. *Mol. Ecol.* **11**: 1555–1564.

White, T.J., Bruns, T., Lee, S., Taylor, J. (1990) Amplification and direct sequencing of fungal ribosomal genes for phylogenies. In: innis, M.A., Gelfand, D.H., Sninsky, J.J., White, T.J. (eds) *PCR Protocols: a Guide to Methods and Applications*, pp. 315–322. New York: Academic Press.

2

Rhizobial signals convert pathogens to symbionts at the legume interface

A. Bartsev, H. Kobayashi and W.J. Broughton

2.1 Introduction

Legume root-nodule bacteria (*Rhizobium* and related genera, collectively called rhizobia) initiate, in conjunction with an appropriate legume partner, symbioses of immense global importance in agriculture, biological productivity, plant successions and soil fertility. Establishment of symbioses between host-plants and symbiotic bacteria is a multistep process consisting of signal perception, signal transduction and cellular responses to these signals (Broughton *et al.*, 2000; Perret *et al.*, 2000). Initially, rhizobia in the rhizosphere perceive plant-derived signals (usually flavonoids) by NodD, a LysR-type transcriptional activator. Flavonoids from root exudates accumulate in the rhizobial cytoplasmic membrane (Hubac *et al.*, 1993; Recourt *et al.*, 1989) and probably interact there with NodD. In the presence of compatible flavonoids, NodD triggers transcription of bacterial nodulation genes (*nod*, *noe*, *nol*) from conserved promoter motifs called *nod*-boxes (Broughton *et al.*, 2000). Some of these genes govern the synthesis and excretion of Nod-factors, a family of lipo-chito-oligosaccharides (LCOs), signals that are recognised by the host plant (Geurts and Bisseling, 2002).

Nod-factors induce deformation and curling of the root hairs, the formation of nodule primordia, the expression of early nodulin (ENOD) genes and finally allow rhizobia to penetrate root hairs (Gage and Margolin, 2000; Viprey *et al.* 2000). Rhizobia enter root hairs in a plant-derived tubular structure, called the infection thread. Infection threads grow towards the root inner cortex, and branch on their way. At the same time, rhizobia grow and divide in the infection thread. When the thread reaches the inner cortex, the rhizobia are released into the plant cytoplasm in an endocytotic manner that ensures that derivatives of the infection thread surround them. Finally, the rhizobia differentiate into nitrogen-fixing bacteroids, the metabolism of which is integrated with that of the host. Thus, until nodules begin to senesce, the endosymbionts are maintained either in the infection thread

Plant Microbiology, Michael Gillings and Andrew Holmes

or in the nodule. How do plant hosts distinguish between rhizobia and pathogenic bacteria, which also try to invade? Host plants and their microsymbionts must communicate at each step of recognition, especially to modulate plant defence reactions. While molecular signals conferring host specificity have been well documented in rhizobia (Broughton and Perret, 1999; Perret *et al.*, 2000), the mechanism(s) by which they are perceived is still poorly understood. In this chapter, we describe recent molecular and physiological findings concerning the host plant responses to the signals derived from their endosymbionts.

2.2 Plant responses to Nod-factors: perception and signal transduction

Nod-factors are absolutely required for nodulation. Rhizobia that have been rendered incapable of Nod-factor synthesis, and legume mutants that are defective in Nod-factor perception are incapable of nodulation (i.e. they are Nod$^-$). Nod-factors of various rhizobia share a common core consisting of three to six β-1,4-linked *N*-acetyl-D-glucosamine residues with a fatty acid attached to the nitrogen of the non-reducing sugar moiety (Mergaert *et al.*, 1997). This common backbone is reflected in similarities amongst the *nodABC* genes of the various genera. Most other *nod*-genes are not functionally or structurally conserved however, and are involved in strain-specific modifications of the Nod-factors. Variation in the structure of Nod-factors reflects the rhizobia from which they were isolated and has relatively minor effects on such processes as root-hair deformation, initiation of meristematic activity in the nodules, and the induction of ENODs (Miklashevichs *et al.*, 2001). Recently, Walker *et al.* (2000) showed that a *R. leguminosarum* bv. *vicae nodFEMNTLO* deletion mutant, that produces Nod-factors without host-specific decorations, penetrates root-hairs but cannot induce a functional infection thread, suggesting perhaps that host-specific decorations are not required for entry of rhizobia but are critical for the formation of functional infection threads. Physiological changes caused by Nod-factors are summarized in *Table 2.1*. Nod-factors induce responses not only in root hairs but also in cortex and vascular bundles. How do Nod-factors induce such temporal and spatial changes, including organogenesis?

A primary response of root hairs is the opening of transmembrane channels, causing depolarisation of the root-hair plasma membrane followed by intracellular alkalinisation and periodic oscillations in intracellular calcium levels (Ehrhardt *et al.*, 1992; Felle *et al.*, 1999; Irving *et al.*, 2000). Although all these phenomena are induced by Nod-factors, their biological meaning is not clear. Pharmacological studies give some indication of how this might occur. Mastoparan induces *ENOD12* expression in *Medicago truncatula* and root-hair deformation on *Vicia sativa* (den Hartog *et al.*, 2001; Pingret *et al.*, 1998). Mastoparan activates heterotrimeric G proteins by mimicking the intracellular domain of membrane spanning receptors. In addition, inhibition of phospholipase C by neomycin or by *n*-butyl alcohol blocks root-hair deformation (den Hartog *et al.*, 2001). Kelly and Irving (2001) showed that Nod-factors stimulate membrane-delimited phospholipase C activity in purified plasma membranes of *Vigna unguiculata*. These reports strongly suggest that G protein-induced lipid

Table 2.1. Responses of legume roots to Nod factors (Cullimore *et al.*, 2001)

Tissue	Response	Rapidity of response	Nod-factor concentration	Tested plants
Epidermis	Ion fluxes	Seconds	nm	*Medicago*
	Plasma membrane depolarisation	Seconds	nm	*Medicago*
	Increase in intracellular pH	Seconds	nm	*Medicago*
	Accumulation of Ca^{2+} in root-hair tip	Seconds	nm	*Medicago, Vigna*
	Ca^{2+} spiking	10 mins	nm	*Medicago, Pisum*
	Gene expression (e.g. *ENOD12, RIP1*)	Mins–hours	fm–pm	*Medicago*
	Root-hair deformation	Mins–hours	nm–µm	Many
	Cyto-skeleton modification	Mins–hours	fm–pm	*Phaseolus, Vicia*
Cortex	Gene expression (e.g. *ENOD 20*)	Hours–days	pm	*Medicago*
	Formation of pre-infection threads	Days	nm–µm	*Vicia*
	Cell division leading to nodule primordia formation	Days	nm–µm	Many
Vascular system	Inhibition of polar auxin transport	Mins	mins	*Trifolium*
	Gene expression (e.g. *ENOD 40*)	24 hours–days	nm–µm	*Glycine, Vicia, Medicago*

signalling is part of the Nod-factor signal transduction pathway. Inhibition of phospholipase C also blocks Nod-factor-induced calcium spiking, although how calcium spiking affects Nod-factor signal transduction is still not clear (Engstrom *et al.*, 2002). Perhaps the primary effect of Nod-factors is to activate a G protein-gated Ca^{2+} channel in the plasma membrane of root-hairs. Similarly, activated phospholipase C induces the release of Ca^{2+} from stores within the cell causing spiking in the same time frame as phospholipase C activation. As spiking appears to involve stores around the nucleus (Ehrhardt *et al.*, 1996), both these responses could be causally related. It is not known whether Nod-factor–receptor complex(es) directly interact with the G-protein.

Biochemical approaches led to the characterisation of high-affinity binding sites for Nod-factors on host plant roots. One of these, NFBS2, is located in the plasma membrane and exhibits differential selectivity for Nod-factors in *M. sativa* and *Phaseolus vulgaris* (Gressent *et al.*, 1999). In *Dolichos biflorus*, a lectin (*Db*-LNP) that shows high affinity for Nod-factors, has been characterised (Etzler *et al.*,1999). This *Db*-LNP (*D. biflorus* lectin nucleotide phosphohydrolase) has an apyrase activity and hydrolyses ATP to ADP/AMP. *Db*-LNP showed the highest affinity for Nod-factors from *D. biflorus* symbionts, *B. japonicum* and *Rhizobium* sp. NGR234. Its apyrase activity is stimulated by binding to Nod-factors. Immunofluorescence assays have shown that *Db*-LNP is localised on the surface of the root hairs. GS52, an orthologue of *Db*-LNP from *Glycine max*, is associated with plasma membranes and is transcriptionally activated by rhizobia (Day *et al.*, 2000). In *M. truncatula*, the expression of two of four putative apyrase genes, *Mtapy1* and *Mtapy4*, is induced following inoculation with *R. meliloti* (Cohn *et al.*, 2001). Treatment of roots with antiserum against *Db*-LNP or GS52 inhibited root-hair deformation and nodulation on *D. biflorus* or *G. max*, respectively. Two nodulation deficient mutants of *M. truncatula* lacked the expression of any apyrases. These properties suggest that such LNPs might play a role, perhaps as Nod-factor receptors, in the initiation of the *Rhizobium*-legume symbioses (Kalsi and Etzler, 2000). In animal cells, apyrases play roles in signal transduction by degrading ATP pools. It is thus likely that LNPs modulate the concentration of ATP/ADP/AMP upon binding with Nod-factors. Apyrase activity is also stimulated by Ca^{2+} suggesting that calcium spiking might also be integrated in LNPs-mediated pathways.

Recently, several plant regulatory genes involved in nodule establishment have been discovered. By *Ac* transposon tagging of *Lotus japonicus*, Schauser *et al.* (1999) isolated *Nin*, a transcriptional factor involved in nodule organogenesis. *Nin* mutants are Nod$^-$ on *L. japonicus*, yet inoculation with *Meshorhizboium loti* provokes excessive root-hair deformation but not infection thread formation or cortical cell division. *Nin* is highly transcribed in the nodule primordium and nodule vascular bundles. Since Nin possesses putative membrane-spanning segments and nuclear localisation signals, post-transcriptional regulation including proteolytic cleavage of *Nin* for relocalisation to nuclei, may be part of the signal transduction pathway. It remains unclear which gene(s) is regulated by Nin. Endre *et al.* (2002) cloned a gene called *NORK* (nodule receptor kinase) from a tetraploid *M. sativa* non-nodulation mutant by map-based cloning. This *nork* mutant fails to induce Ca^{2+} spiking and all downstream symbiotic responses to *R. meliloti* or its Nod-factors. Mutations affecting NORK homologues were also found in the non-nodulation mutant *dmi2* of *M. truncatula* and *Pssym19* of *P. sativum*. In parallel, a *NORK* homologue, a symbiosis receptor-like kinase (SymRK), was also cloned from *L. japonicus* (Stracke *et al.*, 2002). The predicted protein possesses a putative extracellular domain containing leucine-rich repeat (LRR) motifs and an intracellular domain with serine/threonine protein kinase signatures (Endre *et al.*, 2002). *NORK* homologues may interact with unknown receptor(s) of Nod-factors by the LRR motifs. Proteins encoded by the *Ljsym1, 5, 70* locus of *L. japonicus* or *Pssym10* of *P. sativum* are candidates for Nod-factor receptors. Other receptor-like kinases that are involved in autoregulation of nodule number, *HAR1* and *GmNARK*, has been cloned from hypernodulation mutants of *L. japonicus* and *G. max* (Krusell *et al.*, 2002 ; Nishimura *et al.*, 2002 ; Searle *et al.*, 2003). *HAR1* and *GmNARK* are highly similar to the *CLAVATA1* of *Arabidopsis thaliana*. Thus, these nodule autoregulation receptor-like kinases may also perceive small peptide signals from the upper parts of plants. None of the protein(s) thought to be phosphorylated by these receptor kinases has been found.

Taken together, these data suggest that there are at least four components to Nod-factor signal transduction pathways. Down-regulation (through mutagenesis or inhibitors) of any of the known components in the Nod-factor perception pathway, severely affects the nodulation process. LNPs can directly interact with Nod-factors and hydrolyse ATP. NORK may affect unknown extracellular protein(s), which perceive Nod-factors and trigger phosphorylation of unknown proteins. *Nin* may be activated transcriptionally and post-transcriptionally as a downstream part of the transduction signal cascade, eventually activating genes involved in nodule development. Direct connections between these steps have yet to be demonstrated however.

2.3 Plant defence responses during the establishment of symbiosis

Similarities between plant responses to symbionts and pathogens exist. Reactive oxygen species (ROS) are produced in early plant defence responses to avirulent pathogens. After a transient and non-specific weak oxidative burst, massive production of ROS, particularly superoxide ($O^{2-\cdot}$) and hydrogen peroxide (H_2O_2), is observed (Baker and Orlandi, 2001; Hammond-Kosack and Jones, 1996; Van Camp *et al.*, 1998). In *M. sativa* plants inoculated with *R. meliloti*, $O^{2-\cdot}$ production was detected in infection threads and in infected cells of young nodules (Santos *et al.*, 2001). H_2O_2 production was also detected in infection threads and in cell walls of infected cells of nodules as an electron-dense deposit stained with cerium chloride. Moreover, Ramu *et al.* (2002) showed that purified *R. meliloti* Nod-factors induced ROS production in the root proximal zone of *M. truncatula*, where rhizobial infection is initiated. ROS induction was not observed in a non-nodulating *M. truncatula* mutant *dmi1* or when non-sulphated Nod-factors were used, suggesting that the oxidative burst is a result of a specific plant response to Nod-factors. Transient expression of *Trprx2*, a peroxidase, was also detected in *Trifolium repens* roots treated with homologous rhizobia, suggesting that oxidative bursts are involved in other symbiotic interactions (Crockard *et al.*,1999).

Plant chitinases are usually induced during pathogen attack suggesting that they play a role in plant defence (Métraux and Boller, 1986). In *G. max* roots, *B. japonicum* Nod-factors induced chitinase CH1 (Xie *et al.*, 1998). *Srchi*13, an early nodulin of *Sesbania rostrata*, is related to acidic class III chitinases and is transiently induced following inoculation with *Azorhizobium caulinodans* (Goormachtig *et al.*, 1998).

What are the roles of these defence-like responses early in symbiosis? Perhaps they are part of the signal transduction pathway. Ramu *et al.* (2002) showed that oxidative bursts are necessary for the induction of *rip1*, a nodulin encoding a putative peroxidase. Rip1 could metabolise H_2O_2, which is harmful to plant cells. In turn, ROS production might affect signalling proteins, including the activity of transcriptional factors and small GTP binding proteins. Chitinases can also degrade Nod-factors. Thus, *srchi*13 may degrade Nod-factors *in vitro* (Goormachtig *et al.*, 1998). A Nod factor-degrading hydrolase of *M. sativa* has also been described (Staehelin *et al.*, 1995). It is also possible that these enzymes participate directly in plant organogenesis. Perhaps, ROS provides oxidant for peroxidase-mediated cell-wall modifications during infection-thread elongation.

2.4 Factors excreted by rhizobia interfere with host defence responses

Since defence-like responses are induced when rhizobia enter roots, rejection of the symbiont does not occur. In contrast to attack by pathogens, most such reactions are transient and local, suggesting the host modulates defence responses to help establishment of symbiosis. Nod-factors are necessary but insufficient to ensure successful nodulation. Perhaps the best example of this is that *R. etli* produces Nod-factors that possess the same structure as those of *M. loti*, but *R. etli* induces nodules that senesce early on *L. japonicus* (Banba *et al.*, 2001), clearly indicating that additional signals are needed for successful symbiosis.

Direct physical contact between the root surface (the rhizoplane) and bacterial cells is mediated by rhizobial exopolysaccharides (EPS) as well as surface polysaccharides, which form a complex macromolecular structure at the bacteria–plant interface. Accumulating evidence suggests that rhizobial polysaccharides can act as signals to suppress plant defence responses (Spaink, 2000).

Thus, *R. meliloti* produces two EPSs – succinoglycan and EPS II. EPS-defective mutants fail to invade nodules because of blockage in infection thread development. Purified low-molecular-weight succinoglycan and EPS II can rescue the nodule invasion defect at picomolar concentrations, suggesting the existence of a specific recognition system for EPS by the plant. Perhaps this recognition system is involved in the suppression of defence responses since EPS mutants are more active in eliciting defence responses (Niehaus *et al.*, 1993; Parniske *et al.*, 1994).

Another polysaccharide family – lipopolysaccharides (LPS) – is essential for bacterial survival under all growth conditions. Symbiotic phenotypes of various LPS-altered mutants indicate that LPS could play an important role during the infection process. Structural LPS mutants of *R. leguminosarum* bv. *viciae* induce ineffective nodules in *P. sativum* (Perotto *et al.*,1994), partly because they are ineffective in colonising the nodule, and partly because they do not form effective bacteroids. Tissue and cell invasion are often associated with host defence. The severity of symbiotic responses is correlated with the degree of LPS structural modifications (Spaink, 2000). These observations suggest an essential role of LPS in the avoidance of host reactions during nodule development. Treatment with purified LPS of *R. meliloti* suppressed the yeast elicitor-induced alkalinisation and oxidative burst reaction in *M. sativa* cell cultures (Albus *et al.*, 2001). Contrasting results were obtained in non-host tobacco cell culture experiments where LPS itself caused alkalinisation and oxidative burst reactions (Albus *et al.*, 2001). These data suggest that *R. meliloti* LPS released from the bacterial surface might function as a specific signal.

Cyclic (1,3)-(1,6)-β-glucans of *B. japonicum* strain USDA110 are osmotically active solutes that play roles during hypo-osmotic adaptation in the periplasmic space. Additionally, evidence suggests the involvement of β-glucans in suppression of defence responses induced by fungal glucans depends on β-glucan structure (Bhagwat *et al.*, 1999).

2.5 Rhizobial type three secretion systems as new elements in symbiotic development

In many Gram-negative bacterial pathogens, specialised type III secretion systems (TTSS) play a critical role during pathogenic interactions with their eukaryotic hosts (Hueck, 1998). TTSS translocate bacterial effector protein(s) directly into the host cytoplasm across the outer and inner membranes. Among the six main groups of secretion systems, TTSS exhibits the most complex architecture (Thanassi and Hultgren, 2000). About 20 proteins are involved in the formation of a membrame-spanning secretion apparatus, which is associated with an extracellular filamentous (pili) structure (Hueck, 1998). The pili are thought to serve as a 'syringe' to help the injection of effector protein(s) into the host cytoplasm. The flagellar assembly apparatus serves as a protein export system and probably represents an evolutionary ancestor of TTSS (Aizawa, 2001; Hueck, 1998; Young and Young, 2002) (*Figure 2.1*). Recently, it has been shown that some of the effector proteins secreted via TTSS possess enzymatic activity, similar to that of kinases or

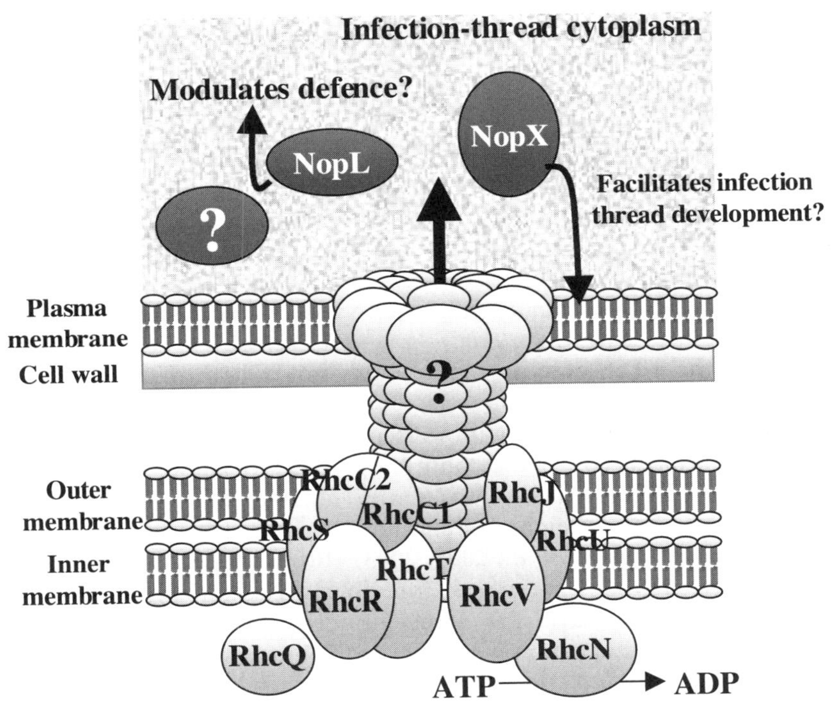

Figure 2.1. Hypothetical model of TTSS functions of *Rhizobium* sp. NGR234 during symbiosis. The structure of the TTSS apparatus was derived from that proposed by Baker *et al.* (1997), with NGR234 gene products replacing their *P. syringae* homologues. TTSS seems to be expressed in the infection-thread upon induction with flavonoids. TTSS injects effectors into the host cytoplasm using energy derived from the hydrolysis of ATP, a reaction that is catalysed by RhcN. Effector proteins may leave rhizobia via pili, but the component(s) of pili are unknown. Results of ectopic expression of NopX and NopL in *L. japonicus* suggest possible roles of the two Nops within this plant (Bartsev *et al.*, 2003; 2004).

phosphatases. These effectors can interfere with phosphorylation of host proteins, resulting in suppression of the defence system thus allowing survival, internalisation and replication of the pathogen. Redirection of transduction of cellular signals may result in disarmament of host immune responses or in cytoskeletal reorganisation. In this way, subcellular niches for bacterial colonisation are formed in a strategy of 'stealth and interdiction' of host defence (Hueck, 1998).

Although TTSSs were previously thought to be unique to pathogenic bacteria, recent surveys of genomes have found TTSSs in *Rhizobium* sp. NGR234 (Freiberg *et al.*, 1997), *M. loti* MAFF303099 (Kaneko *et al.*, 2000) and *B. japonicum* USDA110 (Göttfert *et al.*, 2001). Furthermore, partial sequence data suggest the presence of TTSSs in *R. etli* CFN42 (Gonzalez *et al.*, 2003), as well as *R. fredii* strains USDA257, USDA191 and HH103 (Bellato *et al.*, 1997).

Mutational analyses confirmed that TTSSs are functional in NGR234, *R. fredii* USDA257 and *B. japonicum*. Some proteins are secreted in a TTSS-dependent manner following induction by flavonoids. A NodD-flavonoid-dependent promoter *nod*-box is found in the upstream region of the two-component transcriptional regulator homologue *ttsI* (Marie *et al.*, 2001). TtsI up-regulates parts of the TTSS via a putative promoter motif called the *tts*-box (Krause *et al.*, 2002; Marie *et al.*, 2001; Viprey *et al.*, 1998). Putative-*tts* boxes are found not only upstream of genes located in the TTSS cluster, but also of other genes outside of the cluster in NGR234 (W.J. Deakin, personal communication). Since TTSSs and Nod-genes share regulatory elements, their involvement in symbiotic establishment seems likely.

As shown in *Table 2.2*, rhizobial TTSSs are necessary for optimal nodulation in some symbiotic relationships. Disruption of TTSS-dependent protein secretion affects nodulation in host-specific ways. For example, on *Tephrosia vogelii*, TTSS mutants of NGR234 formed approximately 70% less nodules compared with the wild-type. On the other hand, wild-type NGR234 cannot establish proper, effective symbioses with *Crotalaria juncea* and *Pachyrhizus tuberosus*, but null mutations in the TTSS permit proper nodulation of both plants (Marie *et al.*, 2001, 2003).

Most probably, the responses are caused by the effector protein(s) that are injected into the plant cells. To date, only few proteins are known to be secreted via rhizobial TTSSs (Krishnan *et al.*, 1995; Marie *et al.*, 2001, 2003; Viprey *et al.*, 1998). NGR234 secretes at least eight nodulation outer proteins (Nops) in a TTSS-dependent manner (Marie *et al.*, 2001, 2003). Two of these proteins are NopX (previously called NolX) and NopL (previously y4xL) (Viprey *et al.*, 1998). *R. fredii* appears to have the same proteins (Krishnan *et al.*, 1995). *B. japonicum* does not apparently contain *nopX*, but it possesses an ORF with similarity to *nopL* (Göttfert *et al.*, 2001).

As with pathogens of plant and animals, Nops may be classified into various classes including those that are involved in the formation of the flagellar translocation apparatus and effector proteins that are probably injected into the plant cell. Guttman *et al.* (2002) screened insertion mutants (made using the *avrRpt2*$^{81-225}$ transposon, which can induce hypersensitive responses (HR) on *A. thaliana*) of *Pseudomonas syringae* for effector proteins and found 13 new effectors. All have exceptionally high Ser and low Asp, Leu, Lys contents in their N-termini, suggesting that a specific signal for secretion via TTSS exists.

Table 2.2. Symbiotic phenotype of rhizobia containing a mutated type III secretion gene (after Marie *et al*, 2001).

	No effect	Positive effect	Negative effect
NGR234	*G. max* cv. McCall *G. max* cv. Peking *Leucaena leucocephala* *L. japonicus*	*Flemingia congesta* *Tephrosia vogelii* *V. unguiculata*	*Crotalaria juncea* (Fix⁻ to fix⁺) *Pachyrhizus tuberosus*
fredii HH103	*Cajanus cajan* *Crotalaria juncea* *V. unguiculata*	*G. max* cv. Williams (reduction in competitiveness)	*Erythrina variegata* (Fix⁻ to fix⁺)
fredii USDA257	*G. max* cv. Peking		*Erythrina species* (Fix⁻ to fix⁺) *G. max* cv. McCall (Fix⁻ to fix⁺)
B. japonicum 110*spc*4		*G. max* cv. Williams 10 dpi* *V. unguiculata* 20 dpi* *Macroptilium atropurpureum*	

Responses of various legumes to inoculation with *Rhizobium* sp. NGR234, and derivatives thereof with modified type-three secretion systems. On some plant species, absence of plant secretion had little influence on the symbiotic process (**No effect**). On others, secreted proteins seem to be important for optimal nodulation as their absence leads to a decrease in nodule number or reduction of competitiveness of the secretion mutant, however (**Positive effect**).
*In two cases, obvious differences of nodulation number between mutants and wild-type were observed only at certain periods (dpi; days post inoculation). There are also two types of negative effect exerted by the secreted proteins. TTSS mutants nodulate either more efficiently or convert pseudo-nodules to nitrogen-fixing nodules (Fix⁻ to Fix⁺).

AvrBs2, an effector protein of *X. campestris* pv. *vesicatoria*, was the first protein shown to be injected into plants via TTSS (Casper-Lindley *et al.*, 2002). These workers used AvrBs2 protein fused to an adenylate cyclase gene. The activity of adenylate cyclase depends on the presence of eukaryotic plant calmodulin (thus, is only active after translocation from bacterial cell to plant cytoplasm has occurred). Upon the inoculation of *X. campestris* strain harbouring this fusion, increased cAMP production in *Piper nigrum* cells proved that AvrBs2 had been injected. Szurek *et al.* (2002) showed that AvrBs3 of *X. campestris* pv. *vesicatoria* is injected and localises into the host nucleus by using *in situ* immunocytochemical methods on pepper tissues. These observations raise the possibility that rhizobial Nops are probably injected into the plant cell during nodulation. Moreover, both NopL and NopX have similar N-terminal amino acid compositions as the *P. syringae* effectors.

To assess the functions of Nops inside the host cytoplasm, we ectopically expressed the *nopX* and *nopL* genes of NGR234 within *L. japonicus* using stable *Agrobacterium*-mediated plant transformation techniques. Lines expressing *nopX* grew more rapidly when inoculated with NGR234 than either the wild-type plants or lines transformed with the empty vector (A. Bartsev, unpublished). Thus, the presence of NopX within the plant cells probably helps the establishment of optimal symbiosis. NolX of *R. fredii* USDA257, a close homologue of NopX, is localised in with the membrane of the thread (Krishnan, 2002), where it might facilitate elongation of the thread or help release bacteria from the tips of the infection threads. In turn, this could lead to increased numbers of bacteroids per cell, so explaining the faster growing plants. Interestingly, the expression of *nopX* in *Nicotiana tabacum* plants (non-hosts of NGR234) did not result in the clear phenotype, indicating that action of NopX is specific to symbioses. The same approach was used to elucidate the physiological role of NopL. NopL modulates the plant defence responses following inoculation with rhizobia of *L. japonicus* expressing *nopL* or upon the inoculation with pathogens of *N. tabacum* plants that express *nopL* (Bartsev *et al.*, 2003; 2004). Thus, ectopic gene expression tools are useful in elucidating the function of TTSS effectors during the establishment of symbiotic interactions.

2.6 Conclusions and perspectives

Establishment of symbioses involves overcoming the numerous physical, cellular and molecular barriers presented by the host. Typically, this entails contacting and entering the host, growth and replication of the bacteria using nutrients derived from the plant, avoidance of host defences, and so on. Possible molecular mechanisms by which rhizobia could initiate and maintain symbiotic relationships without triggering plant defence reactions are described here. Many of the molecular mechanisms are still not clear. Nod-factor receptors may or may not have been isolated, but what is the core structure that is necessary for induction of the symbiotic cascade? Many different polysaccharides induce plant responses during symbiosis. That rhizobial TTSSs play a host-specific role in modulation of nodule development raises interesting questions about bacterial evolution and their association with plants. It is also interesting that *R. meliloti* strain 1021 and *M. loti* strain R7A have putative type IV secretion systems (Galibert *et al.*, 2001; Sullivan *et al.*, 2002). In *M. loti*, a *nod*-box probably regulates the secretion system by indirectly up-regulating a two-component regulator VirA. Biochemical, genetic and physiological studies of secreted proteins within plant cells will help to reconstruct the fine-tuning of symbiosis.

We wish to thank Dora Gerber for her unstinting help. This work was supported by the Erna och Victor Hasselblads Stiftelse, the Fonds National de la Recherche Scientifique (Projects 31-30950.91, 31-36454.92, and 31-63893.00), and the Université de Genève.

References

Aizawa, S.I. (2001) Bacterial flagella and type III secretion systems. *FEMS Microbiol. Lett.* **202**: 157–164.

Albus, U., Baier, R., Holst, O., Puhler, A., Niehaus, K. (2001) Suppression of an elicitor-induced oxidative burst reaction in *Medicago sativa* cell cultures by *Sinorhizobium meliloti* lipopolysaccharides. *New Phytol.* **151**: 597–606.

Baker, C.J., Orlandi, E.W. (2001) Active oxygen in plant pathogenesis. *Ann. Rev. Phytopathol.* **33**: 299–321.

Baker, B., Zambryski, P., Staskawicz, B., Dinesh-Kumar, S.P. (1997) Signaling in plant–microbe interactions. *Science* **276**: 726–733.

Banba, M., Siddique, A.-B.M., Kouchi, H., Izui, K., Hata, S. (2001) *Lotus japonicus* forms early senescent root nodules with *Rhizobium etli. Mol. Plant–Microbe Interactions* **2**: 173–180.

Bartsev, A., Boukli, N.M., Deakin, W.J., Staehelin, C and Broughton, W.J. (2003) Purification and phosphorylation of the effector protein NopL from *Rhizobium* sp. NGR234. *FEBS Lett* **554**: 271–274.

Barstev, A.V., Deakin, W.J., Boukli, N.M., Bickley, C., Malnoë, P., Stacey, G., Broughton, W.J and Staehelin, C. (2004) NopL, an effector protein of *Rhizobium* sp. NGR234, thwarts activation of plant defence reactions. *Plant Physiol, in press.*

Bellato, C., Krishnan, H.B., Cubo, T., Temprano, F., Pueppke, S.G. (1997) The soybean cultivar specificity gene *nolX* is present, expressed in a *nodD*-dependent manner, and of symbiotic significance in cultivar-nonspecific strains of *Rhizobium (Sinorhizobium) fredii. Microbiology* **143**: 1381–1388.

Bhagwat, A.A., Mithöfer A., Pfeffer, P.E., Kraus, C., Spickers, N., Hotchkiss, A., Ebel, J., Keister, D.L. (1999) Further studies of the role of cyclic β-glucans in symbiosis. An *ndvC* mutant of *Bradyrhizobium japonicum* synthesizes cyclodecakis-(1→3)-β-Glucosyl. *Plant Physiol.* **119**: 1057–1064.

Broughton, W.J., Perret, X. (1999) Genealogy of legume-*Rhizobium* symbioses. *Current Opin. Plant Biol.* **2**: 305–311.

Broughton, W.J., Jabbouri, S., Perret, X. (2000) Keys to symbiotic harmony. *J. Bacteriol.* **182**: 5641–5652.

Casper-Lindley, C., Dahlbeck, D., Clark, E.T., Staskawicz, B.J. (2002) Direct biochemical evidence for type III secretion-dependent translocation of the AvrBs2 effector protein into plant cells. *Proc. Natl Acad. Sci. USA* **99**: 8336–8341.

Cohn, J.R., Uhm, T., Ramu, S. *et al.* (2001) Differential regulation of a family of apyrase genes from *Medicago truncatula. Plant Physiol.* **125**: 2104–2119.

Crockard, M.A., Bjourson, A.J., Cooper, J.E. (1999) A new peroxidase cDNA from white clover: its characterization and expression in root tissue challenged with homologous Rhizobia, heterologous Rhizobia, or *Pseudomonas syringae. Mol. Plant-Microbe Interactions* **12**: 825–828.

Cullimore, J.V., Ranjeva, R., and Bono, J.J. (2001) Perception of lipo-chitooligosaccharidic Nod fad legumes. *Trends Plant Sci* **6**: 24–30.

Day, R.B., McAlvin, C.B., Loh, J.T., Denny, R.L., Wood, T.C., Young, N.D., Stacey, G. (2000) Differential expression of two soybean apyrases, one of which is an early nodulin. *Mol. Plant-Microbe Interactions* **13**: 1053–1070.

den Hartog, M., Musgrave, A., Munnik, T. (2001) Nod factor-induced phosphatidic acid and diacylglycerol pyrophosphate formation: a role for phospholipase C and D in root hair deformation. *Plant J.* **25**: 55–65.

Ehrhardt, D.W., Atkinson, E.M., Long, S.R. (1992) Depolarization of alfalfa root hair membrane potential by *Rhizobium meliloti* Nod factors. *Science* **256**: 998–1000.

Ehrhardt, D.W., Wais, R., Long, S.R. (1996) Calcium spiking in plant root hairs responding to *Rhizobium* nodulation signals. *Cell* **85**: 673–681.

Endre, G., Kereszt, A., Kevei, Z., Mihacea, S., Kaló, P., Kiss, G.B. (2002) A receptor kinase gene regulating symbiotic nodule development. *Nature* **417**: 962–966.

Engstrom, E.M., Ehrhardt, D.W., Mitra, R.M., Long, S.R. (2002) Pharmacological analysis of Nod factor-induced calcium spiking in *Medicago truncatula*. Evidence for the requirement of type IIA calcium pumps and phosphoinositide signaling. *Plant Physiol.* **128**: 1390–1401.

Etzler, M.E., Kalsi, G., Ewing, N.N., Roberts, N.J., Day, R.B., Murphy, J.B. (1999) A Nod factor binding lectin with apyrase activity from legume roots. *Proc. Natl Acad. Sci. USA* **96**: 5856–5861.

Felle, H.H., Kondorosi, E., Kondorosi, A., Schultze, M. (1999) Elevation of the cytosolic free $[Ca^{2+}]$ is indispensable for the transduction of the Nod factor signal in alfalfa. *Plant Physiol.* **121**: 273–280.

Freiberg, C., Fellay, R., Bairoch, A., Broughton, W.J., Rosenthal, A. and Perret, X. (1997) Molecular basis of symbiosis between *Rhizobium* and legumes. *Nature* **387**: 394–401.

Gage, D.J., Margolin, W. (2000) Hanging by a thread: invasion of legume plants by rhizobia. *Curr. Opin. Microbiol.* **3**: 613–617.

Galibert, F., Finan, T.M., Long, S.R. *et al.* (2001) The composite genome of the legume symbiont *Sinorhizobium meliloti. Science* **293:** 668–672.

Geurts, R., Bisseling, T. (2002) *Rhizobium* Nod factor perception and signalling. *Plant Cell* **14**: S239–249.

Göttfert, M., Rothlisberger, S., Kundig, C., Beck, C., Marty, R., Hennecke, H. (2001) Potential symbiosis-specific genes uncovered by sequencing a 410-kilobase DNA region of the *Bradyrhizobium japonicum* chromosome. *J. Bacteriol.* **183**: 1405–1412.

Gressent, F., Drouillard, S., Mantegazza, N. *et al.* (1999) Ligand specificity of a high-affinity binding site for lipo-chitooligosaccharidic Nod factors in *Medicago* cell suspension cultures. *Proc. Natl Acad. Sci. USA* **96**: 4704–4709.

Gonzalez, V., Bustos, P., Ramirez-Romero, M.A., Medrano-Soto, A., Salgado, H., Hernandez-Gonzalez, I., Hernandez-Celis, J.C., Quintero, V., Moreno-Hagelsieb, G., Girard, L., Rodriguez, O., Flores, M., Cevallos, M.A., Collado-Vides, J., Romero, D., and Davila, G. (2003) The mosaic structure of the symbiotic plasmid of *Rhizobium etli* CFN42 and its relation to other symbiotic genome compartments. *Genome Biol* **4**: R36.

Goormachtig, S., Lievens, S., Van de Velde, W., Van Montagu, M., Holsters, M. (1998) *Srchi13*, a novel early nodulin from *Sesbania rostrata*, is related to acidic class III chitinases. *Plant Cell* **10**: 905–915.

Guttman, D.S., Vinatzer, B.A., Sarkar, S.F., Ranall, M.V., Kettler, G., Greenberg, J.T. (2002) A functional screen for the type III (Hrp) secretome of the plant pathogen *Pseudomonas syringae*. *Science* **295**: 1722–1726.

Hammond-Kosack, K.E., Jones, J.D. (1996) Resistance gene-dependent plant defense responses. *Plant Cell* **8**: 1773–1791.

Hubac, C., Ferran, J., Guerrier, D. (1993) Luteolin absorbtion in *Rhizobium meliloti* wild-type and mutant strains. *J. Gen. Microbiol.* **139**: 1571–1578.

Hueck, C.J. (1998) Type III protein secretion systems in bacterial pathogens of animals and plants. *Microbiol. Molec. Biol. Rev.* **62**: 379–433.

Irving, H.R., Boukli, N.M., Kelly, M.N., Broughton, W.J. (2000) Nod-factors in symbiotic development of root hairs. In: Ridge, R.W., Emons, A.M.C. (eds) *Root Hairs in Cell and Molecular Biology*, pp. 241–265. Springer-Verlag: Tokyo.

Kalsi, G., Etzler, M.E. (2000) Localization of a Nod factor-binding protein in legume roots and factors influencing its distribution and expression. *Plant Physiol.* **124**: 1039–1048.

Kaneko, T., Nakamura, Y., Sato, S. *et al.* (2000) Complete genome structure of the nitrogen-fixing symbiotic bacterium *Mesorhizobium loti* (supplement). *DNA Res.* **7**: 331–338.

Kelly, M.N., Irving, H.R. (2001) Nod factors stimulate plasma membrane delimited phospholipase C activity *in vitro*. *Physiol. Plant.* **113**: 461–468.

Krause, A., Doerfel, A., Göttfert, M. (2002) Mutational and transcriptional analysis of the type III secretion system of *Bradyrhizobium japonicum*. *Mol. Plant-Microbe Interactions* **12**: 1228–1235.

Krishnan, H.B., Kuo, C.-I., and Pueppke, S.g. (1995) Elaboration of flavonoid-induced proteins by the nitrogen-fixing symiont *Rhizobium fredii* is regulated by both *nodD1* and *nodD2*, and is dependent on the cultivar specificity locus, *nolXWBTUV. Microbiol* **141**: 2245–2251.

Krishnan, H.B. (2002) NolX of *Sinorhizobium fredii* USDA257, a type III-secreted protein involved in host range determination, is localized in the infection threads of Cowpea (*Vigna unguiculata* [L.] Walp) and soybean (*Glycine max* [L.] Merr.) nodules. *J. Bacteriol.* **184**: 831–839.

Krusell, L., Madsen, L.H., Sato, S. *et al.* (2002) Shoot control of root development and nodulation is mediated by a receptor-like kinase. *Nature* **420**: 422–426.

Marie, C., Broughton, W.J., Deakin, W.J. (2001) *Rhizobium* type III secretion systems: legume charmers or alarmers? *Curr. Opin. Plant Biol.* **4**: 336–342.

Marie, C. Deakin, W.J., Viprey, V., Kopciñska, J., Golinowski, W., Krishnan, H.B., Perret, X., Broughton, W.J. (2003) Characterisation of Nops, Nodulation Outer Proteins, secreted via the type III secretion system of NGR234. *Mol. Plant–Microbe Interactions* **16**: 743–751.

Mergaert, P., van Montagu, M., Holsters, M. (1997) Molecular mechanisms of Nod factor diversity. *Mol. Microbiol.* **25**: 811–817.

Métraux, J.P., Boller, T. (1986) Local and systemic induction of chitinase in cucumber plants in response to viral, bacterial and fungal infections. *Physiol. Mol. Plant Pathol.* **28**: 161–169.

Miklashevichs, E., Röhrig, H., Schell, J., Schmidt, J. (2001) Perception and signal transduction of rhizobial Nod factors. *Crit. Rev. Plant Sci.* **20**: 373–394.

Niehaus, K., Kapp, D., Pühler, A. (1993) Plant defense and delayed infection of alfalfa pseudonodules induced by an exopolysaccharide (EPS-I)-deficient *Rhizobium meliloti* mutant. *Planta* **190**: 415–425.

Nishimura, R., Hayashi, M., Wu, G.J. *et al.* (2002) *HAR1* mediates systemic regulation of symbiotic organ development. *Nature* **420**: 426–429.

Parniske, M., Schmidt, P.E., Kosch, K., Müller, P. (1994) Plant defense responses of host plants with determinate nodules induced by EPS-defective *exoB* mutants of *Bradyrhizobium japonicum*. *Mol. Plant–Microbe Interactions* **7**: 631–638.

Perotto, S., Brewin, N.J., Kannenberg, E.L. (1994) Cytological evidence for a host-defense response that reduces cell and tissue invasion in pea nodules by lipopolysaccharide-defective mutants of *Rhizobium leguminosarum* strain. *Mol. Plant–Microbe Interactions* **7**: 99–112.

Perret, X., Staehelin, C., Broughton, W.J. (2000) Molecular basis of symbiotic promiscuity. *Microbiol. Mol. Biol. Rev.* **64**: 180–201.

Pingret, J.L., Journet, E.R., Barker, D.G. (1998) *Rhizobium* Nod factor signaling: Evidence for a G protein-mediated transduction mechanism. *Plant Cell* **10**: 659–671.

Ramu, S.K., Peng, H.M., Cook, D.R. (2002) Nod factor induction of reactive oxygen species production is correlated with expression of the early nodulin gene *rip1* in *Medicago truncatula*. *Mol. Plant–Microbe Interactions* **15**: 522–528.

Recourt, K., van Brussel, A.A., Driessen, A.J., Lugtenberg, B.J. (1989) Accumulation of a *nod gene* inducer, the flavonoid naringenin, in the cytoplasmic membrane of *Rhizobium leguminosarum* biovar *viciae* is caused by the pH-dependent hydrophobicity of naringenin. *J. Bacteriol.* **171**: 4370–4377.

Santos, R., Herouart, D., Sigaud, S., Tiuati, D., Puppo, A. (2001) Oxidative burst in alfalfa–*Sinorhizobium meliloti* symbiotic interactions. *Mol. Plant–Microbe Interactions* **14**: 86–89.

Schauser, L., Roussis, A., Stiller, J., Stougaard, J. (1999) A plant regulator controlling development of symbiotic root nodules. *Nature* **11**: 191–195.

Searle, I.R., Men, A.E., Laniya, T.S., Buzas, D.M., Iturbe-Ormaetxe, I., Carroll, B.J., Gresshoff, P.M. (2003) Long-distance signaling in nodulation directed by a CLAVATA1-like receptor kinase. *Science* **299**: 109–112.

Spaink, H.P. (2000) Root nodulation and infection factors produced by rhizobial bacteria. *Ann. Rev. Microbiol.* **54**: 257–288.

Staehelin, C., Schultze, M., Kondorosi, E., Kondorosi, A. (1995) Lipo-chito-oligosaccharide nodulation signals from *Rhizobium meliloti* induce their rapid degradation by the host-plant alfalfa. *Plant Physiol.* **108**: 1607–1614.

Stracke, S., Kistner, C., Yoshida, S. *et al.* (2002) A plant receptor-like kinase required for both bacterial and fungal symbiosis. *Nature* **417**: 959–962.

Sullivan, J.T., Trzebiatowski, J.R., Cruickshank, R.W. *et al.* (2002) Comparative sequence analysis of the symbiosis island of *Mesorhizobium loti* strain R7A. *J. Bacteriol.* **184:** 3086–3095.

Szurek, B., Rossier, O., Hause, G., Bonas, U. (2002) Type III-dependent translocation of the *Xanthomonas* AvrBs3 protein into the plant cell. *Mol. Microbiol.* **46**: 13–23.

Thanassi, D.G., Hultgren, S.J. (2000) Multiple pathways allow protein secretion across the bacterial outer membrane. *Curr. Opin. Cell Biol.* **12**: 420–430.

Van Camp, W., Van Montagu, M., Inze, D. (1998) H_2O_2 and NO: redox signals in disease resistance. *Trends Plant Sci.* **3**: 330–334.

Viprey, V., Del Greco, A., Golinowski, W., Broughton, W.J., Perret, X. (1998) Symbiotic implications of type III protein secretion machinery in *Rhizobium*. *Mol. Microbiol.* **28**: 1381–1389.

Viprey, V., Perret, X., Broughton, W. J. (2000) Host-plant invasion by rhizobia. *Sub-Cell. Biochem.* **33**: 437–456.

Walker, S.A., Viprey, V., Downie, J.A. (2000) Dissection of nodulation signaling using pea mutants defective for calcium spiking induced by Nod factors and chitin oligomers. *Proc. Natl Acad. Sci. USA* **97**: 13413–13418.

Xie, Z.-P., Staehelin, C., Wiemken, A., Broughton, W.J., Müller, J., Boller, T. (1998) Symbiosis-stimulated chitinase isoenzymes of soybean (*Glycine max* (L.) Merr.). *J. Experiment. Bot.* **50**: 327–333.

Young, B.M., Young, G.M. (2002) YplA is exported by the Ysc, Ysa, and flagellar type III secretion systems of *Yersinia enterocolitica*. *J. Bacteriol.* **184**: 1324–1334.

3

The root nodule bacteria of legumes in natural systems

Elizabeth L.J. Watkin

3.1 Introduction

The Leguminosae constitute the third largest family of flowering plants (Sprent, 2001) with approximately 650 genera and 18 000 species (Polhill *et al.*, 1981). It is the most widely distributed family of flowering plants, occupying habitats ranging from rainforest to arid zones throughout the world (Ravin and Polhill, 1981). The Leguminosae consists of three subfamilies, Papilionoideae, Mimosoideae and Caesalpinioideae. Papilionoideae constitute 65% of the legumes and are represented by trees, shrubs and herbs distributed from the tropics to the arctic. The Mimosoideae are the smallest subfamily, comprising 10% of the legumes. Members of this subfamily are often found in the dry areas of the tropics/subtropics and consist mainly of trees and shrubs. The third subfamily, the Caesalpinioideae, comprising 25% of the Leguminosae, are mainly trees growing in the moist tropics. Many legumes play a major role in both natural ecosystems and in agricultural production systems due to their ability to form symbiotic associations (nodules) with Gram-negative, soil-inhabiting bacteria that fix atmospheric N_2. This not only renders the plants independent of soil nitrogen, but also makes them major contributors to soil nitrogen supplies for non-leguminous species. As such they contribute to productivity and sustainability. Approximately 20% of the total legume species have been examined for nodulation, representing all three subfamilies (Sprent, 2001). Nodulation is common within the Papilionoideae and Mimosoideae but only 30% of species within the Caesalpinioideae are nodulated (Allen and Allen, 1981; Sprent, 2001).

The root nodule bacteria (RNB) are currently classified in six genera: *Allorhizobium*, *Azorhizobium*, *Bradyrhizobium*, *Mesorhizobium*, *Rhizobium*, *Sinorhizobium*, with a total of 31 species (Wei *et al.*, 2002). All belong to the alpha subdivision of the proteobacteria, with the exception of two novel taxa, *Ralstonia taiwanensis* (Chen *et al.*, 2001) and *Burkholderia* sp. (Moulin *et al.*, 2001), which are

Plant Microbiology, Michael Gillings and Andrew Holmes

members of the β-*proteobacteria*. The RNB represent great diversity between the genera with some genera being more closely related to non-nodulating bacteria than to each other (Sprent, 2001) the common feature being that they are able to form nitrogen-fixing nodules on some legumes.

While nitrogen fixation in agricultural systems has been widely studied, the area of native legumes has been largely ignored. For the purpose of this chapter native legumes will be defined as legume species that grow in natural systems that have not been subject to cultivation. Native legumes can be of great significance in firewood production, soil stabilisation, mine site rehabilitation and are increasingly important for dealing with the problem of decreasing water quality and salinisation. There have been relatively few studies on host–rhizobial interactions in natural environments, and as such our knowledge of the distribution and importance of the microsymbionts of the many woody and herbaceous legume species in natural ecosystems is limited. Of those studies on native legumes the majority of work has focused on *Acacia* due to the use of many species in agroforestry and rehabilitation, despite this little is known about the specificity of the symbiotic relationship.

In Australia, legumes are a significant and highly diverse component of the native flora comprising 10% of the estimated 18 000 native plant species and they occur in all vegetation types except salt marshes and marine aquatic communities (Davidson and Davidson, 1993). Barnett (1988) summarised reports of nodulation in 52 genera of Australian native legumes. Of these 49 were from the Papilionoideae, and three were from the Mimosoideae. There have been no reports of nodulation of the Australian genera of Caesalpinioideae.

Legumes are often a dominant part of native ecosystems and this may reflect the advantage obtained in low-fertility soils from symbiotic nitrogen fixation. Many natural ecosystems are nitrogen limited, and Australian soils are notoriously nitrogen-deficient. As such, legumes may play a major role in natural ecosystems. For example, it has been demonstrated that *Acacia* species are responsible for substantial levels of nitrogen fixation within natural ecosystems (Hingston *et al.*, 1982; Langkamp *et al.*, 1979, 1982; Monk *et al.*, 1981). In these situations plant–microbial relationships that help circumvent low nutrient levels are likely to be of considerable significance in determining the species composition and structural diversity of plant communities (Allen and Allen, 1981; Read, 1993).

This chapter will detail the function and diversity of RNB associated with native legumes. While the main focus will be the RNB relations with Australian native legumes, examples will also be drawn from other regions of the world.

3.2 The root nodule bacteria–legume symbiosis

The interaction between legume roots and RNB is initiated at the biochemical level. The range of host plants that a species of RNB can nodulate is determined by a 'molecular conversation' between the plant and bacterium. Signals from legume roots (Hungria *et al.*, 1991) activate bacterial nodulation genes (*nod* genes) the products of which, synthesise 'Nod factors', which in turn stimulate nodule development in the plant (Dénarié *et al.*, 1996). Variation in the structure of the

Nod factors results in differing host plant specificity. Specificity in this conversation is displayed in both partners of the legume–RNB symbiosis.
This molecular conversation results in cytological changes within the root enabling the RNB to infect their legume host (*Figure 3.1*). This root infection can be achieved through three possible mechanisms: root hair penetration and infection thread formation as seen in clovers (Hirsch, 1992), entry via wounds or sites of lateral root emergence as occurs in peanuts (Boogerd and van Rossum, 1997), penetration of root primordia found on stems of plants such as *Sesbania* (Boivin *et al.*, 1997).

In the root hair infection mechanism the steps involved in the nodulation process are: recognition by RNB of legume, attachment of RNB to the root, curling of the root hair, root-hair infection by the bacteria, formation of the infection thread, nodule initiation, and transformation of the RNB to bacterioids that fix nitrogen (Allen and Allen, 1981).

Little literature exists on the mechanisms of infection in non-agricultural legumes. Rasanen *et al.* (2001) demonstrated that a range of species from *Acacia* and *Prosopis,* two genera of tree legumes from the Mimosoideae, have been shown to nodulate via root hair penetration. The tree legume *Chamaecytisus proliferus* (tagasaste) has been shown to nodulate through a combination of root hair infection and crack entry (Vega-Hernandez *et al.*, 2001). This area is a rich source of potential research.

3.3 Symbiotic association between native legumes and root nodule bacteria

The degree of specificity of symbiotic associations in agricultural systems has been highly studied but little is known about non-agricultural legumes. Of those native systems studied, with very few exceptions, the root nodule bacteria isolated from native legumes, both in Australia and other parts of the world, demonstrate broad host specificity and varying effectiveness. This is in comparison to agricultural legume species that generally demonstrate a high degree of specificity in their associations with RNB.

When studying the symbiosis between RNB and legumes it is important to consider both the ability of a particular RNB to induce nodules in legume hosts (host range) and the effectiveness of the resultant nitrogen-fixing association.

3.3.1 Host range of isolates from native legumes

Infectiveness can be viewed as the ability of a RNB strain to infect and cause the formation of nodules containing bacteria on the roots or stems of legume hosts. This will vary between different species of legume hosts. Each RNB has the ability to nodulate some but not all legumes and as such RNB can be grouped on the basis of the legume hosts they are able to nodulate (*Figure 3.2A*). Those legumes that are nodulated by a particular RNB are defined as the host range of that RNB. In early studies this phenomenon led to the concept of cross inoculation groups, in which legumes were grouped according to the RNB that would nodulate them. More than 20 groupings were identified (Fred *et al.*, 1932). With further studies of a greater

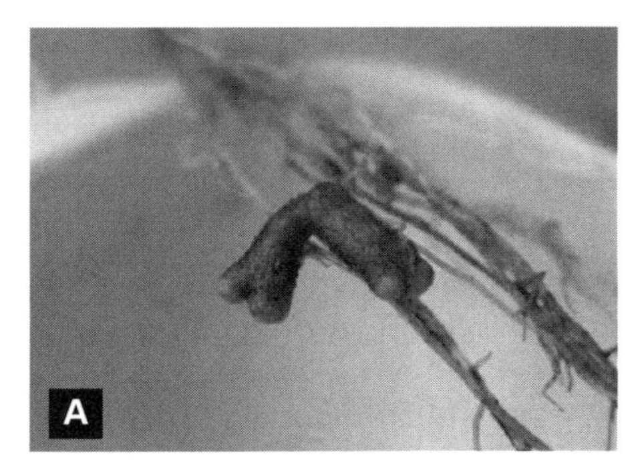

Figure 3.1. Nodules from 12-week-old plants of Papilionoideae and Mimosoideae grown under glasshouse conditions.

A. *Acacia accuminata*
B. *Kennedia prostrata*
C. *Swainsonia formosa*

Photographs courtesy of Ron Yates.

Figure 3.2. Varying host range (A) and effectiveness (B) of single root nodule bacteria isolates on different *Swainsonia* species. Five *Swainsonia* species from the Pilbara region of Western Australia were inoculated with a single RNB strain (different strains for A and B). Left hand pot: *S. pterostylis* (front), two unidentified Swainsonia spp (left and right). Right hand pot: *S. macculochiana* (left) and *S. formosa* (right).
A. *S. pterostylis, S. macculochiana* and one of the unidentified *Swainsonia* sp. (left) *are* unnodulated as indicated by stunted growth and yellow leaves, whereas the unidentified *Swainsonia* sp. (right) and *S. formosa* are nodulated and effectively fixing nitrogen as indicated by dark green leaves.
B. The symbiotic associations with *S. pterostylis, S. macculochiana* and both of the unidentified *Swainsonia* spp are ineffective as indicated by pale green leaves, whereas the symbiosis with *S. formosa* is effective as indicated by dark green leaves. All plants were nodulated. Effectiveness was confirmed by measurement of dry weight of leaves compared to uninoculated control.

Photograph courtesy of Ron Yates.

range of plant species it was found that these groupings became blurred (Wilson, 1944) and other traits are now given more importance. It should be noted that early work on the legume–RNB symbiosis classified the RNB on the basis of growth rates; slow (now the *Bradyrhizobium*) and fast (the *Rhizobium* and *Sinorhizobium*).

In general, it has been demonstrated that RNB isolates from native Australian legumes exhibit a wide host range. In a study of 85 RNB isolates obtained from 83 legume host species from the forests of southwest Western Australia, Lange (1961) found that while all isolates were of the slow-growing bradyrhizobia type they had extremely wide host specificity. Only three isolates nodulated only their host of origin, while the remaining 82 isolates had a broad host range with 53% of isolates nodulating four cross inoculation groups. Similarly, Lawrie (1983) in a study of isolates from 12 native legumes species from southern Victoria (Australia) demonstrated that the majority of isolates showed wide host specificity, being able to nodulate species in both Mimosoideae and Papilionoideae. *Swainsonia lessertiifolia* was the exception with isolates from this host being the only RNB that could nodulated this legume. Both fast- and slow-growing isolates were able to nodulate the remaining 11 species of the legumes investigated. Similarly, Barnet and Catt (1991) found that the RNB isolated from *Acacia* spp. from five climatically diverse widely separated sites in New South Wales (Australia) showed a very wide host range, nodulating members of both Mimosoideae and Papilionoideae regardless of the host of origin. The broad range of hosts of origin for isolates in a study of the native legumes in the open eucalypt forest of south eastern Australia led Lafay and Burdon (1998) to conclude that no clear specificity between RNB genomic species and legume taxa could be determined. A study into the cross inoculation potential of RNB isolated from soil in five sites throughout Western Australia (Watkin, O'Hara, Dilworth and Bennet, 2002, unpublished data) demonstrated a broad host range with RNB isolates able to nodulate across legume families and genera. Yates *et al.* (2004) however demonstrated that isolates from a range of native legume species from the Pilbara region of Western Australia not only nodulated a range of native hosts from both the Papilionoideae and Mimosoideae, but also nodulated a range of exotic legume hosts.

Murray *et al.* (2001) proposed that this broad host specificity has potential links to the distribution of species and that *Acacia* species of restricted geographic distribution may demonstrate a greater specialisation in symbiotic associations (host specificity and effectiveness) than those with a wide distribution. Their work found this was not the case, with no difference in host specificity demonstrated between *Acacia* species of restricted and wide distribution and a wide range in effectiveness demonstrated from isolates obtained from species with varying distribution.

While the previous studies have demonstrated that most RNB isolates from native legumes have a broad host range infecting members from both Mimosoideae and Papilionoideae regardless of their host of origin, examples of greater specialisation in native legume symbioses do exist. Barnet (1988) and Barnet and Catt (1991) isolated very slow-growing strains from *Acacia* species from the alpine region of southeastern Australia that were highly host-specific.

A similar trend in the generally broad host range of isolates from native legumes has been seen in other parts of the world. Parez-Fernandez and Lamont (2003) found a high degree of promiscuity in the legume genera *Cytisus* and

Genista, where species that are native to Spain were able to be nodulated by RNB isolated from a range of seven Australian native legumes from the *Papilionoideae*. Rasanen *et al.* (2001) saw wide host specificity in *Sinorhizobia* isolated from *Acacia senegal* and *Prosopis chilensis* (both from the Mimosoideae), whereas Turk and Keyser (1992) found that the tree legume *Sesbania grandiflora* was highly specific in its RNB requirements. Odee *et al.* (2002) demonstrated a great range in nodulating ability of isolates obtained from soil where a range of *Acacia* species and *Sesbania sesban* (Papilionoideae) grew naturally. *Sesbania sesban* was highly specific in its RNB requirements and was only able to be nodulated by RNB originally isolated from *Sesbania sesban* while isolates from African *Acacias* were able to nodulate widely within their genera but were unable to nodulate *Sesbania sesban.* Santamaria *et al.* (1997) demonstrated varying host specificity in the legume shrubs from the Canary Islands with *Bradyrhizobium* spp. being promiscuous, forming effective nodules on their original hosts as well as *Chamecytisus proliferus* hosts and *Rhizobium* spp. only nodulating their host of origin (*Teline canariensis*).

Generally the RNB of native legumes from all regions of the world demonstrate very wide host range. Not only do they nodulate a wide range of host species within the same subfamily but they also appear to be able to nodulate host species from other subfamilies of the Leguminosae.

3.3.2 The effectiveness of nitrogen fixation in native legume symbiotic associations

The ability of nodulated legumes to fix nitrogen may be important in many natural ecosystems, as well as in agriculture. The effectiveness of a symbiotic association refers to the amount of nitrogen fixed by a particular host (Hansen, 1994). This is affected by two main components – the genetically determined compatibility of the RNB with the host and environmental conditions. A wide range in the ability of isolates of native RNB to form an effective symbiosis with host legumes has been reported. A single isolate may vary in effectiveness across a range of hosts, as well as a range of isolates varying in effectiveness on a single host (Barnet and Catt, 1991; Burdon *et al.*, 1999; Lawrie, 1983; Murray *et al.*, 2001, Thrall *et al.*, 2000; Turk and Keyser, 1992) (*Figure 3.2B*).

In the majority of these studies the effectiveness of the symbiotic associations was assessed in glasshouse experiments by evaluating the increase in dry weight of inoculated plants compared to an uninoculated control. Barnet and Catt (1991) demonstrated, by visual inspection, that 90% of RNB isolated from *Acacia* species had some ability to fix nitrogen, as shown by green plants when grown in seedling agar. However, when tested more stringently, by assessing increased dry matter production in soil when inoculated back onto their host of origin, there was great variation in the effectiveness of RNB isolates with only 36% of strains demonstrating a statistically significant increase in dry weight over their uninoculated controls.

Lawrie (1983) in a study of isolates obtained from 12 legumes from the Mimosoideae and Papilionoideae subfamilies found symbiotic effectiveness, as assessed by increased total nitrogen was usually poor, with only 6% of associations being ranked as effective and 73% as ineffective. Furthermore all combinations of hosts and their own isolates were ineffective while effective associations only formed between plants and isolates from other hosts of origin. No single isolate showed outstanding effectiveness on all hosts. Similarly, Watkin, O'Hara, Dilworth and Bennett (2002, unpublished data) showed that isolates from a range of native legumes in Western Australia from the subfamilies Mimosoideae and Papilionoideae also had a range of effectiveness across hosts of different genera and while nodulating their host of origin well were not necessarily the most effective on that host. Thrall *et al.* (2000) found a range of effectiveness in isolates from several common *Acacia* species but were able to demonstrate that an isolate that was effective on its host of origin was usually effective on other species. This contrasts with work by Burdon *et al.* (1999) who found in *Acacia* species that the performance of an isolate on one host gave little information about its potential performance on plants of a different species of *Acacia* with that strain. Studies conducted on isolates from a range of African native legumes demonstrated a similar situation with isolates obtained from *Acacia* species generally forming the most effective symbiotic associations with species other than their host species of isolation (Odee *et al.*, 2002).

These preceding studies were carried out in glasshouse experiments, under ideal conditions of nutrient and water supply. The question of the extent that native legumes contribute to nitrogen in natural ecosystems would be more realistically demonstrated by measuring nitrogen fixation in the field. Accurate and reliable information on the rates of biological nitrogen fixation in native plant communities is important both for decisions on forest management and on revegetation. Many natural ecosystems are nitrogen-limited, and it is important to understand the constraints to nitrogen fixation within them. However, quantifying this at present is hampered by the absence of good assays, especially for use with woody species (Sprent, 2001). A number of attempts, reviewed by Barnet (1988), have however been made to measure nitrogen fixation in Australian ecosystems where acetylene reduction has been measured in the field to give an estimate of nitrogenase activity. The rates observed are generally low compared to those seen in agricultural systems (463 kg N/ha), compared to 0.005 kg/ha for coastal on low-nutrient sands (Lawrie, 1981), 7.8 kg/ha for some areas of Jarrah forest (Hansen *et al.*, 1987) and 6.4 kg/ha for fertilised *Acacia* (*A. holosericea*) under plantation conditions (Langkamp *et al.*, 1982).

These low rates may be attributed to adverse environmental constraints such as nutrient deficiency and water restriction. Hansen and Pate (1987) demonstrated fixation rates comparable to those seen in agricultural systems when *Acacia* in glasshouse trials were grown under ample moisture and phosphate. Seasonal variation in nitrogen fixation thought to be due to moisture stress has been demonstrated (Barnet *et al.*, 1985; Hansen and Pate, 1987; Hansen *et al.*, 1987; Hingston *et al.*, 1982; Langkamp *et al.*, 1982; Lawrie, 1981; Monk *et al.*, 1981). Gathumbi *et al.* (2002) assessed N_2 fixation in shrub and tree legumes in western Kenya using the ^{15}N natural abundance technique in inoculated pots in the glasshouse. They demonstrated that the nitrogen fixed in non-phosphorous- and -potassium-limiting growth conditions ranged from 24–142 kg N/ha after 9 months. Both native and introduced legumes markedly increase the amount of nitrogen fixed when supplied with phosphate and an adequate supply of water.

In all probability, native legumes in natural ecosystems do not contribute large quantities of fixed nitrogen when compared with agricultural systems. This is due to the combination of a small quantity of nodules, low plant density (Langkamp *et al.*, 1982) and low fixation rates as a result of environmental constraints. In addition, the significance of nitrogen fixation of native legumes on other plants is not clear, as there is no information on the transfer of symbiotically fixed nitrogen to non-leguminous plants.

The RNB of native legumes show a wide host range; being able to nodulate plant species across genera, and a range in symbiotic effectiveness, with the majority of associations being poorly effective. The question then arises as to what benefit the associations have for either partner.

3.3.3 A survival strategy?

A number of authors have suggested that the broad host range and lack of specificity with varying effectiveness seen in native legume species may have evolved as a survival strategy for the microsymbionts. Rasanen *et al.* (2001) compared the nodulation of *Medicago sativa* with that of *Acacia* and *Prosopis*. *M. sativa* is a legume with specific symbiotic associations, being effectively nodulated only by *S. meliloti* and occasionally ineffectively by other RNB. *Acacia* and *Prosopis* are native legumes that are effectively nodulated with a much broader range of fast-growing RNB but also frequently ineffectively nodulated. They propose that *Acacia* and *Prosopis* could not exclude nodulation by unsuitable RNB. Burdon *et al.* (1999) proposed that the advantage gained from any nitrogen-fixing symbiosis is better than none and results in native legumes being non-selective in nodulation. As such an 'anything is better than nothing' tradeoff between symbiotic effectiveness and host range exists, that is, in the bacteria a wider host range is linked with lower levels of effectiveness on any one host. It could be suggested that it is the bacteria driving the nodulation and that the broad host range is a strategy of bacteria to avoid periodically unfavourable growth conditions in the soil (Rasanen *et al.*, 2001). This then leaves the question, is the broad host range demonstrated in the RNB for native legumes a chance trait or a survival mechanism?

3.4 Diversity of organisms that nodulate native legumes

Diversity within the legume–RNB symbiosis can be viewed from either partner's perspective, the diversity of RNB that nodulates a specific host or the diversity of hosts that is nodulated by a specific RNB. For the purposes of this chapter, diversity will be defined as the diversity of RNB that nodulates a particular host. The diversity of the microsymbiont can be assessed at a number of different levels from the genes of the organism through to the isolation host.

Early studies describing diversity of the RNB were based on growth rates, serological response and cross inoculation studies, however they were unable to give precise information on the nature and structure of RNB communities in natural ecosystems. With improvements in the techniques used in fingerprinting there has been an increased interest in the biodiversity of root nodule bacteria including those that nodulate legumes native to Australia and elsewhere.

Caution must be taken when interpreting results on RNB diversity, as it is generally not possible to isolate RNB directly from the soil. With the exception of one study (Zeze *et al.*, 2001) diversity analysis uses isolates trapped using specific host legumes. This can lead to varying pictures of diversity depending on the host used (Bala *et al.*, 2003; Odee *et al.*, 2002).

Similarly, diversity can be affected by environmental stresses (Bala *et al.*, 2003), geography or diversity of isolates in a given location (Barnet and Catt, 1991).

3.4.1 Phenotypic classification

As stated previously each RNB has the ability to nodulate some but not all legumes. Early studies in the diversity of the RNB classified them on the basis of the legumes on which the RNB formed nodules or on cross inoculation groups (Fred *et al.*, 1932). In addition, standard bacteriological methods such as growth rate, carbohydrate utilisation, acid or alkaline production were also used (Lange, 1961).

In the earliest reported study of the characterisation of RNB isolated from native Australian legumes Hannon (1956) isolated 50 strains of slow-growing RNB from endemic species of *Acacia* from Hawkesbury Sandstone soil near Sydney. Norris (1956) reported finding bradyrhizobia-type RNB from a range of native species. Lange (1961) identified RNB from 85 native legume species as slow-growing bacteria of the bradyrhizobia type.

Until relatively recently there was no evidence from isolated bacteria that anything other than slow-growing RNB were indigenous to Australia. Lawrie (1983) was the first author to report fast-growing RNB from temperate Australian woody legumes, as well as slow-growing *Bradyrhizobium*. Classification was once again based solely on cultural characteristics. Fast-growing isolates were only obtained from *Acacia longifolia* var *sophorae* and *Kennedia prostrata*. Isolates demonstrating an intermediate growth rate were isolated from *Swainsonia lessertiifolia*. All other isolates were classified as slow-growing 'bradyrhizobia' RNB. Both *Acacia longifolia* and *Kennedia prostrata* were able to be nodulated by both fast- and slow-growing RNB. Barnet *et al.* (1985) found similar results, obtaining both fast- and slow-growing RNB strains classed as *Rhizobium* and *Bradyrhizobium* from *Acacia* species. An extremely slow-growing *Bradyrhizobium* was also isolated. In these two studies, assignment of fast and slow growers to *Rhizobium* and *Bradyrhizobium* respectively was confirmed by serological testing of antigenic cross-reactions between native root nodule bacteria and representatives of known genera. Barnet *et al.* (1985), comparing protein profiles of the isolates using SDS-PAGE, showed that these groupings contained extremely diverse organisms that were nevertheless more similar to each other than to agricultural species. Comparing the profiles of fast-growing RNB native isolates, they also showed no close relationship with fast-growing control strains from exotic legumes. A similar degree of diversity was observed in the slow-growing isolates and while some similarity of fast-growing isolates was observed between sites, no such similarities were seen in slow-growing isolates. A further study on RNB for Australian *Acacia* from a larger number of sites, identified a few isolates with a intermediate growth rate (perhaps *Mesorhizobium*). Most conformed to *Bradyrhizobium* (Barnet and Catt, 1991).

3.4.2 Phylogenetic characterisation

Chromosomal gene analysis

Recent advances in molecular techniques and interest in native legumes as a potential genomic resource have led to the description of many new genera and species of RNB associated with indigenous legumes (Chen *et al.*, 1997; de Lajudie *et al.*, 1998; Nick *et al.*, 1999; Tan *et al.*, 1999; Wang *et al.*, 1999b) and further demonstrated their diversity (Doignon-Bourcier *et al.*, 2000; Dupuy *et al.*, 1994; Haukka *et al.*, 1996; Wang *et al.*, 1999a). While a wide range of molecular techniques can be employed, the majority of studies conducted on native populations of RNB have used restriction fragment length polymorphism (RFLP) of PCR amplified gene regions (16S rRNA and the 16S–23S ITS) (Khbaya *et al.*, 1998; Lafay and Burdon, 1998; Laguerre *et al.*, 1994; Nick *et al.*, 1999) in conjunction with gene sequencing (particularly 16S rRNA). The majority of these studies having focused on the native legumes of Africa, South and Central America and China. Only a few recent studies have used molecular tools to study diversity in the RNB of native Australian legumes (Lafay and Burdon, 1998, 2001; Marsudi *et al.*, 1999; Yates *et al.*, 2004).

RNB from *Rhizobium*, *Bradyrhizobium* and *Mesorhizobium* have been demonstrated to nodulate native Australian legumes and that within those groupings a high degree of diversity of isolates is demonstrated (Lafay and Burdon, 1998, 2001; Marsudi *et al.*, 1999; Yates *et al.*, 2004). The diversity of RNB isolates from nodules of 32 different legume hosts obtained from 12 locations in southeastern Australia was assessed using PCR-RFLP of 16S rDNA (Lafay and Burdon, 1998). The isolates fell into 21 distinct groupings; the majority of the isolates belonged to the genus *Bradyrhizobium* that contained 16 subgroups (*Figure 3.4*), there were also two *Rhizobium* subgroups (*Figure 3.3A*) and three *Mesorhizobium* subgroups (*Figure 3.3B*). Only one of the subgroups corresponded to a known species (*Rhizobium tropicii*). The distribution of isolates within the groupings was highly unbalanced with 94% of isolates belonging to *Bradyrhizobium* and 58% to genomic species A (*Figure 3.4*). Ninety-seven percent of isolates were contained in eight genomic species with the remaining 13 genomic species containing only 3% of the total isolates. While this indicates that the population is not as diverse as indicated in earlier studies, this disproportionate distribution of isolates may be a result of sampling bias and will be discussed in future sections.

The diversity of RNB isolates from *Acacia saligna* growing in the southwest of Western Australia was characterised on the basis of their growth, physiology and partial 16S rRNA sequencing (Marsudi *et al.*, 1999). Twenty-nine percent of isolates were identified as fast growing and affinity of two groupings to *R. leguminosarum* bv. *phaseoli*, *R. tropicii* was demonstrated. This contrasts with the work of Lafay and Burdon (1998) where only 6% of isolates where *Rhizobium* or *Mesorhizobium*. PCR-RFLP of 16S rDNA was used to assess the diversity of the RNB of 13 different *Acacia* species from 44 different sites in south-eastern Australia (Lafay and Burdon, 2001). Nine genomic species (subgroupings within the same genera) were identified, all similar to those seen in an earlier study (Lafay and Burdon (1998): eight were from *Bradyrhizobium* lineage and four of these genomospecies were related to *B. japonicum* and represent 88% of the total isolates. The remaining genomospecies corresponded to *R. tropicii*. This study demonstrated the dominance of the isolates

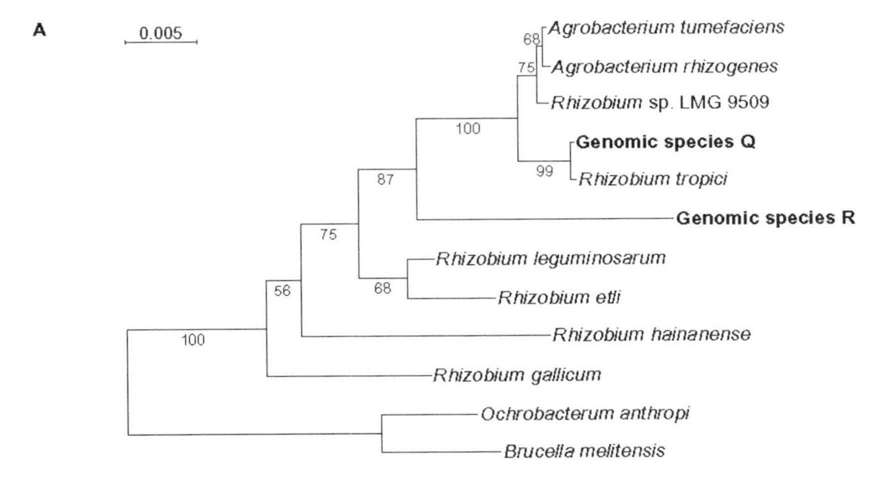

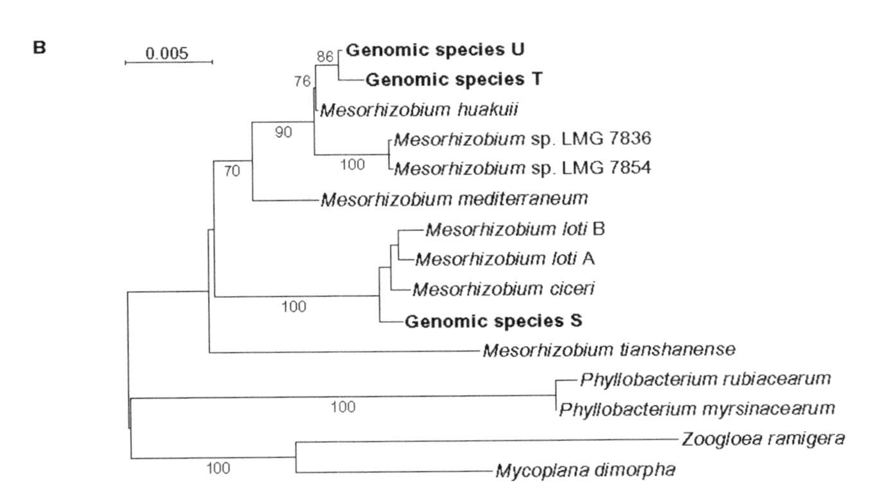

Figure 3.3. Phylogenetic relationships among genomic species belonging to the genera *Rhizobium* (A) and *Mesorhizobium* (B) isolated from Australian native legumes characterised by SSU rDNA PCR-RFLPs. Reproduced with permission from Lafay and Burdon (1998).

by one or two genomic species, but that they were different than those species identified for non-*Acacia* legumes (Lafay and Burdon, 1998), suggesting a difference in nodulation patterns for *Mimosoideae* and *Papilionoideae*. A study of isolates from nodules of legumes growing in the Gascoyne and Pilbara regions of northwest Western Australia identified 65% of the isolates as fast-growing (Yates *et al.*, 2004). On the basis of PCR RAPD analysis, the diversity within the fast-growing isolates was determined to be greater than seen within the slow-growing isolates. The 16S rDNA sequence homology of four isolates to known species was identified, with

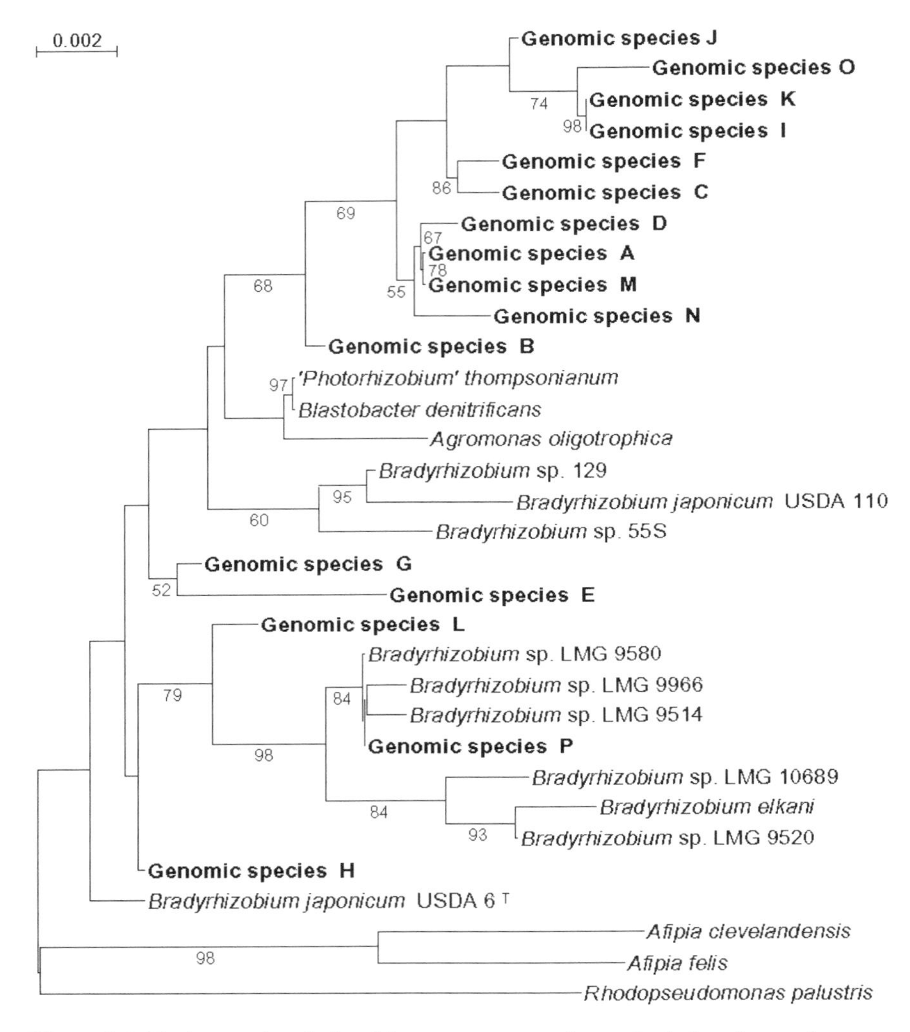

Figure 3.4. Phylogenetic relationships among genomic species belonging to the genera *Bradyrhizobium* isolated from Australian native legumes characterised by SSU rDNA PCR-RFLPs. Reproduced with permission from Lafay and Burdon (1998).

the fast-growing isolates sharing 99% homology with *S. meliloti* and *S. terangae* and the slow-growing isolates sharing 99% homology with *B. elkanii* and *B. japonicum*. Great variability in isolates was also seen by Watkin, Vivas-Marfisi, O'Hara and Dilworth (2003, unpublished data) investigating the diversity of RNB isolates from soil collected at five different sites in Western Australia. While only five percent of the total isolates were fast-growing, compared to the 65% seen in the study of Yates *et al.* (2004), the majority of these were isolated from the soils of a single region (Karijini National Park, in the NW of Western Australia). Based on PCR-RFLP of 16S rDNA, the fast-growing isolates showed great diversity with only two isolates grouping with *S. meliloti* and *Mesorhizobium*. The remaining isolates showed no affinity with known RNB genera/species. The slow-growing isolates showed less diversity with the majority of isolates falling into two genomospecies, which included the reference strains, *B. japonicum* and *B. liaoningense*.

A number of studies on the diversity of the RNB of African legumes have focused on two related plant genera, *Acacia* and *Prosopis* (Mimosoideae), as well as the Papilionoideae *Sesbania sesban*. These tree species naturally occur in arid and semi-arid regions. The range of genomospecies in RNB that nodulate these plants is as diverse as seen in Australia, with *Rhizobium*, *Mesorhizobium*, *Sinorhizobium* and *Bradyrhizobium* being formally described (Haukka *et al.*, 1998; Moreira *et al.*, 1998; Nick *et al.*, 1999; Zhang *et al.*, 1991). Haukka *et al.* (1996) in a study using partial 16S rRNA gene sequencing of fast-growing isolates obtained from *Acacia senegal* and *Prosopis chilensis* determined 12 different sequences, eight of which were novel. Khbaya *et al.* (1998) demonstrated that a high proportion of isolates obtained from four *Acacia* species when analysed using PCR-RFLP of 16S rRNA gene and the 16S–23S rRNA ITS fit within the *Sinorhizobium* lineage. Odee *et al.* (2002) however, demonstrated that eight *Acacia* species in Kenya were nodulated by the five genera of RNB *Agrobacterium*, *Bradyrhizobium*, *Mesorhizobium*, *Rhizobium*, and *Sinorhizobium* using PCR-RFLP of 16S rRNA gene, and that they fell into 12 distinct genotypes. Ba *et al.* (2002), on the basis of whole cell proteins (SDS-PAGE) and 16S rDNA sequence analysis on strains obtained from *Acacia tortilis* demonstrated that most strains were *Mesorhizobium* and *Sinorhizobium* with several different genomospecies. Bala *et al.* (2002) determined that *Sesbania sesban* was nodulated by *Mesorhizobium*, *Sinorhizobium*, *Rhizobium* and *Allorhizobium*, but *Mesorhizobium* accounted for 92% of all isolates. It was previously thought that *Sesbania* was highly specific in RNB requirements (Turk and Keyser, 1992), but this result may have reflected the inability in that particular study to isolate from genera that occurred at a lower frequency. A study of isolates obtained from nodules of 27 legume species native to Senegal (Doignon-Bourcier *et al.*, 1999) produced only slow-growing bacterial strains. Further characterisation of these isolates by PCR-RFLP of 16S rDNA and comparative SDS-PAGE of whole proteins revealed several phylogenetic subgroups of *Bradyrhizobium*. Conversely, McInroy *et al.* (1999) found four genera of RNB (*Bradyrhizobium*, *Mesorhizobium*, *Sinorhizobium* and *Rhizobium*) represented in isolates obtained from African *Acacia* and other tropical woody legumes, with the majority grouped within *Mesorhizobium* and *Sinorhizobium*. Similar diversity of RNB of native legumes is seen in other parts of the world. Tan *et al.* (1999) in a study of the RNB for 11 wild legumes from northwest China, found these legumes were nodulated by *Mesorhizobium*, *Rhizobium* and *Agrobacterium tumefaciens*. Wang *et al.* (1999a) examined the diversity of the RNB from the Mexican legume, *Leucocephal*. PCR-RFLP of 165 rDNA revealed 12 rDNA types that bore similarities to *Mesorhizobium*, *Rhizobium* and *Sinorhizobium*. Seven unique types were identified but most isolates corresponded to *Sinorhizobium*.

The diversity of the RNB that nodulate native legume species worldwide is broad, with plant species able to be nodulated by RNB from a number of genera. While the inherent limitations in studies of diversity that require trap hosts to isolate RNB from the soil must be acknowledged, the use of molecular techniques has shown a picture of much greater diversity than was originally believed, as well as revealing a number of new and novel organisms. The RNB of native legumes provide a rich genetic resource of symbiotic nitrogen-fixing organisms.

Symbiotic gene analysis

The lack of specificity in RNB–host association in native legumes, the broad host range of isolates and the ability of host plants to be nodulated by a diversity of isolates has been noted in Section 3.3.1. Assessing microbial diversity based on 16S rRNA genes enables the allocation of isolates to known groups or the determination of their relatedness to those group's phylogenies. These related groupings have been shown to bear no functional relationship to the host range of the isolates. The following question therefore arises when considering host range, is assessing diversity based on highly conserved chromosomal genes the correct approach? Any set of genes can be used to describe similarities between isolates. Classifications of RNB based on 16S rDNA and the ITS regions indicate there is little phylogenetic correlation between bacteria and their legume hosts (Doyle, 1998). In an attempt to obtain more meaningful phylogenies with respect to the host specificity of RNB, the genes involved in nodulation and symbiosis have received most attention in recent times. It has been demonstrated that phylogenies based on symbiotic genes give better association with those obtained from host range than those based on 16S rRNA (Ba *et al.*, 2002; Doyle, 1998; Haukka *et al.*, 1998; Laguerre *et al.*, 1996; Zhang *et al.*, 2000).

The nodulation (*nod*) and nitrogen fixation (*nif*) genes in RNB are responsible for host specificity and symbiotic nitrogen fixation. All RNB possess a series of nodulation genes (*nod*DABC) termed the 'common nodulation genes'. The product of *nod*D is a protein that regulates the expression of the nodulation genes, whereas the products of the expression of *nod*ABC are responsible for the synthesis of the Nod factor backbone. Some degree in specificity in the production of the Nod factor has been noted for these genes (Downie, 1998) and as such makes these genes ideal to investigate the diversity in host range. By contrast genes homologous to the rhizobial *nif* genes, which are responsible for the synthesis of nitrogenase, the enzyme involved in the conversion of atmospheric nitrogen to ammonia, are found in many bacteria beside RNB. There is some evidence that phylogenies based on the common *nod* genes are closely linked to nodulation groups (Dobert *et al.*, 1994; Ueda *et al.*, 1995a) whereas the phylogeny of *nif*H closely resembles that of phylogenies generated with 16S rRNA gene (Ueda *et al.*, 1995b; Young 1992).

A phylogenetic analysis of *nod*A of isolates from *Acacia tortilis* that represented various genomospecies in *Mesorhizobium* and *Sinorhizobium*, grouped all strains together into the *Acacia–Leucaena–Prosopis* nodulation group and formed a unique phylogenetic cluster (Ba *et al.*, 2002). Although taxonomically diverse, the isolates had all demonstrated similar symbiotic characteristics, and chemical analysis of the Nod factors demonstrated that they were similar within the phylogenetic groupings. Similarly, Tan *et al.* (1999) analysed 35 isolates from 11 wild legumes in northwest China. The isolates where characterised on the basis of PCR-RFLP of 16S rRNA gene and restriction patterns of *nod*DAB and *nif*H genes. Isolates obtained from different plants but grouped in the same clusters based on 16S rDNA RFLP and sequence analysis were found to have similar *nif*H RFLP patterns. The isolates from different hosts however had different *nod*DAB RFLP patterns.

The nodulation genes of a group of diverse RNB isolated from *Astragalus sinicus*, all falling into the genus *Mesorhizobium* but belonging to four different 16S rDNA genotypes were analysed (Zhang *et al.*, 2000). Representatives of each of these groups had the same *nod* gene organisation and identical *nodA* gene sequences. The nodulation genes were conserved while the isolates were chromosomally diverse, indicating phylogenies based on the nodulation genes are closely related to host range.

Haukka *et al.* (1998) in a study of isolates belonging to *Mesorhizobium* and *Sinorhizobium* determined by 16S rRNA gene sequencing analysis with similar host range were analysed on the basis of *nod*A and *nif*H restriction patterns and sequence analysis. The phylogenies based on *nod*A and *nif*H were similar. Groupings obtained from phylogenetic analysis were on taxonomic and geographical divisions. There was no correlation between host range and the phylogeny based on *nod*A, which is contrary to the view that *nif*H phylogeny closely mirrors phylogeny of 16S rRNA while the phylogeny of *nod* genes is closely related to that of the host plants. This may be explained by the isolates initially having a similar host range and as such may produce similar Nod factors.

It has been generally concluded that nodulation genes, which are often located on plasmids for some RNB species can be transferred between strains (Laguerre *et al.*, 1996; Schofield *et al.*, 1987; Sullivan *et al.*, 1995; Urtz and Elkan, 1996; Young and Wexler, 1988). This may explain the observation that RNB, which are taxonomically distinct, but produce similar Nod factors, often have similar host ranges. The *nif* genes, on the other hand, are thought to have similar evolutionary histories to 16S rRNA genes (Ueda *et al.*, 1995b; Young 1992) and as such the phylogenies of *nif*H will closely resemble those of the 16S rRNA gene.

The use of phylogenies based on the nodulation genes of RNB from native legumes may in part explain the apparent broad host range of RNB isolates from native legumes. The initial molecular conversation between the RNB and its host is the critical step in establishing the symbiosis. Therefore, to have similarities in the common nodulation genes is more meaningful in determining a similar host range than the taxonomic classification of these organisms.

3.5 Influences on root nodule bacteria populations and diversity

3.5.1 Soil characteristics

In soils of low fertility, such as Australian soils, plant–microbe interactions are likely to be of considerable significance in determining the species composition and structural diversity of both the plant and microbial communities. Populations of RNB found at a particular site are likely to be determined by either the host plants that are present or the edaphic conditions of that site. It has been stated previously that environmental constraints as well as the method of sampling can influence measures of diversity.

There are conflicting data associating the dominance of particular RNB genera at a site with the climatic, soil and plant characteristics of that site. Some studies have found a positive association (Barnet, 1988; Barnet and Catt, 1991; Barnet *et al.*,

1985, Thrall *et al.*, 2000, Zhang *et al.*, 1991) while a number of other studies have been unable to identify any such association (Lafay and Burdon, 1998; Lawrie, 1983; Marsudi *et al.*, 1999).

In a study of RNB isolated from Australian *Acacia*, Barnet *et al.* (1985) isolated both fast- and slow-growing RNB from two sand dune regions, however, the composition of the RNB populations varied with locality. The coastal site with low organic matter, which was more environmentally extreme, consisted of 21% *Rhizobium* and 79% *Bradyrhizobium*, whereas at the less extreme site, which had higher organic matter, only 12% of isolates were fast growers. The apparent differences in populations from the different localities did not seem to be due to the selection by host plants as the isolates obtained using a common trap host were similar to those obtained from nodules on the local *Acacia*s. As such soil type is therefore implicated.

Lawrie (1983) in a study of native Australian legumes from three sites in southern Victoria found fast-growing isolates in two of the three sites. While these had very different soil types, one slightly acidic at pH 5.5–6.0 and the other alkaline at pH 8.5–9.0, both with differing levels of total nitrogen and available phosphorous, both soils were sands and as such prone to desiccation. The level of organic matter at these sites was not reported but both sites where fast-growing isolates were reported had plant cover to one metre with no upper-storey cover, while the third site, where no fast growers were detected, was a low, open-forest area. This leads to the deduction that organic matter levels would be low at the sites where the fast-growing RNB were isolated and that there would be high soil temperatures in summer. These sites would therefore be subject to rapid desiccation. While the authors hypothesise that host selectivity may be more important than soil type, this conclusion may have been drawn due to incomplete analysis of the soil samples, and could have been confirmed by the use of the same trap host for soil samples from each site.

Barnet and Catt (1991) in a study of RNB isolates from *Acacia* from five climatically diverse and geographically widely spread localities concluded that isolates obtained were more related to soil type than host plant species, with marked geographic localisation noted. The fast-growing isolates they isolated were restricted to sites that were arid with very low organic matter (0.3%) and a neutral pH. Extremely slow-growing isolates were found exclusively in an alpine site with high soil organic matter (24%) and very acidic pH (3.0–4.2). Barnet *et al.* (1992) obtained isolates of native RNB nodulating Australian *Acacia* from a range of habitats in New South Wales (Australia). Fast-growing isolates were uniformly obtained from areas with poor vegetative cover, low soil organic matter, high soil temperatures and low soil water. They demonstrated that abrupt transitions from areas yielding slow-growing RNB to areas yielding fast-growing RNB corresponded to changes in soil type and habitat characteristics. Yates *et al.* (2004) isolated a majority of fast-growing RNB (68% of total) from the Gascoyne and Pilbara regions of NW Western Australia. These soils were subject to high soil temperatures, had sparse vegetation, alkaline pH and low organic matter. Similarly, Watkin (2003, unpublished data) isolated RNB from five sites across Western Australia using trap hosts. Fifty-five percent of fast-growing isolates were obtained from Karijini National Park, in the northwest of the state where soil conditions are similar to those seen by Yates *et al.* (2004).

These studies indicate that sandy sites, with low organic matter and, as such, subject to desiccation, are likely to have fast-growing strains of RNB. Bala *et al.* (2003) in a study of the RNB for a number of leguminous trees isolated from soils from three continents found that soil acidity was highly correlated with genetic diversity among RNB populations. These authors proposed that acid stress could result in selective pressure, the more acid-tolerant genera dominating the population. It was also reported that while the clay content of soil was positively correlated to RNB population numbers and the sand content negatively correlated, neither factor had any correlation with the diversity of isolates. Barnet and Catt (1991) demonstrated a good correlation between the expected severity of the stresses due to heat desiccation and low organic matter and the proportion of fast-growing *Acacia* isolates. The authors questioned if the success of these fast-growing isolates in the hot desert soils was due to an inherent advantage bestowed by a short generation time such that these isolates were physiologically adapted to withstand desert conditions. Low soil organic matter and sparse plant cover therefore limited shading, leading to high soil temperatures with limited water availability. This information would agree with the data presented in this section where fast-growing isolates were seen in sites of neutral to alkaline pH and low organic matter, and as such are likely to be subject to desiccation.

Barnet and Catt (1991) suggested the early concept that Australian native taxa only nodulate with slow-growing strains had arisen because the range of sites included had been too restricted. A sampling bias had therefore limited the isolates obtained to slow-growing RNB. This can in part be confirmed. The work of Lange (1961) in which all isolates were determined to be slow-growing bradyrhizobia-type RNB was limited to the forests of the southwest of Western Australia, with high organic soil matter and shading keeping soil temperatures low. Lafay and Burdon (1998) found no geographic partitioning of RNB isolates in a study of native shrubby legumes in open eucalypt forest in southeast Australia. While only 3% of isolates were identified as fast growers all sample sites were acid or near acid that favour *Bradyrhizobium.* The most abundant of fast-growing genera was *R. tropicii,* which is the most acid tolerant of the *Rhizobium.*

3.5.2 Legume host

The legume host present at the site can influence the diversity of RNB in these soils. It has been suggested by Sadowsky and Graham (1998) that this influence may be exerted via the following mechanisms: relatively non-specific enhancement of RNB because of their ability to metabolise a substance present in root exudates; multiplication and release of rhizobia from nodules;or, the ability of host legumes to select particular groups of RNB from mixed populations.

In agricultural systems it has been demonstrated that population densities of *Rhizobium leguminosarum* bv. *viceae* are influenced by which host is present (Bottomley, 1992; Kucey and Hynes, 1989). Similar results were seen for *Rhizobium leguminosarum* bv. *trifolii* (Bottomley, 1992). While no such studies have been conducted for native legumes Thrall *et al.* (2001) found that native RNB were undetectable in heavily impacted areas where native shrubs had been cleared or there had been continual grazing over a long period of time. The relationship between native legume species present at a particular site and the diversity of RNB at that site is worthy of investigation.

It is difficult to separate the influence of soil type and plant species present at a site on the diversity of RNB present at that site as both are intrinsically linked. The influence of these two factors on the diversity of RNB is an area of potential research.

3.6 Conclusion

Native legumes are a significant and highly diverse component of natural ecosystems due to the nitrogen-fixing symbiosis with soil-inhabiting RNB, nevertheless, this association has remained largely unexplored. Legumes in natural systems are nodulated by a wide diversity of RNB while the RNB isolates from native legumes demonstrate a broad host range and varying effectiveness of the resultant nitrogen-fixing symbiosis. Environmental constraints may result in lower level of nitrogen fixation in natural systems than in agricultural systems, nevertheless, native legumes can be of great significance in firewood production, soil stabilisation, mine site rehabilitation and are important for dealing with the problem of decreasing water quality and salinisation. As novel RNB isolates are increasingly being identified they are also vitally important as a resource for 'exploitable' species. Hence an understanding of the symbiotic relationships in native legumes will be of significance for conservation management, sustainable agriculture and restoring degraded landscapes.

Acknowledgements

I am grateful to Ron Yates for the provision of photographs and Graham O'Hara and Lesley Mutch for constructive comments on this manuscript.

References

Allen, O.N., Allen, E.K. (1981) *The Leguminosae. A Source Book Of Characteristics, Uses, And Nodulation.* University of Wisconsin Press: Madison, WI.

Ba, S., Willems, A., De Lajudie, P. ***et al.*** (2002) Symbiotic and taxonomic diversity of rhizobia isolated from *Acacia tortilis* subsp. *raddiana* in Africa. *System. Appl. Microbiol.* **25**: 130–145.

Bala, A., Murphy, P., Giller, K.E. (2002) Occurrence and genetic diversity of rhizobia nodulating *Sesbania sesban* in African soils. *Soil Biol. Biochem.* **34**: 1759–1768.

Bala, A., Murphy, P.J., Osunde, A.O., Giller, K.E. (2003) Nodulation of tree legumes and the ecology of their native rhizobial populations in tropical soils. *Appl. Soil Ecol.* **22**: 211–223.

Barnet, Y.M. (1988) Nitrogen-fixing symbioses with Australian native legumes. In: Murrell, W.G., Kennedy, I.R. (eds) *Microbiology in Action*, pp. 81–92. Research Studies Press: Chichester.

Barnet, Y.M., Catt, P.C. (1991) Distribution and characteristics of root nodule bacteria isolated from Australian *Acacia* spp. *Plant Soil* **135**: 109–120.

Barnet, Y.M., Catt, P.C., Hearne, D.H. (1985) Biological nitrogen fixation and root nodule bacteria (*Rhizobium* sp. and *Bradyrhizobium* sp.) in two rehabilitating sand dune areas planted with *Acacia* spp. *Aust. J. Botany* **33**: 595–610.

Barnet, Y.M., Catt, P.C., Jenjareontham, R., Mann, K. (1992) Fast-growing root nodule bacteria from Australian *Acacia* spp. In: Palacios, R., Mora, J., Newton, W.E. (eds) *New Horizons in Nitrogen Fixation. Proceedings of the 9th International Congress on Nitrogen Fixation*, p 594. Kluwer Academic Publishers, Cancun, Mexico.

Boivin, C., Ndoye, I., Lajudie, F., Dupuy, N., Dreyfus, B. (1997) Stem nodulation in legumes: Diversity, mechanisms, and unusual characteristics. *Crit. Rev. Plant Sci.* **16**: 1–30.

Boogerd, F.C., van Rossum, D. (1997) Nodulation of groundnut by *Bradyrhizobium* – a simple infection process by crack entry. *FEMS Microbiol. Rev.* **21**: 5–27.

Bottomley, P.J. (1992) Ecology of *Bradyrhizobium* and *Rhizobium*. In: Stacey, G., Burris, R.H., Evans, H.J. (eds) *Biological Nitrogen Fixation*, pp. 293–348. Chapman and Hall: New York, NY.

Burdon, J.J., Gibson, A.H., Searle, S.D., Woods, M.J., Brockwell, J. (1999) Variation in the effectiveness of symbiotic associations between native rhizobia and temperate Australian *Acacia* – within species interactions. *J. Appl. Ecol.* **36**: 398–408.

Chen, W.X., Tan, Z.Y. Gao, J.L., Li, Y., Wang, E.T. (1997) *Rhizobium hainanense* sp. nov, isolated from tropical legumes. *Internat. J. System. Bacteriol.* **47**: 870–873.

Chen, W.-M., Laevens, S., Lee, T.-M., Coenye, T., De Vos, P., Mergaey, M., Vandamme, P. (2001) *Ralstonia taiwanensis* sp. nov., isolated from root nodules of *Mimosa* species and sputum of a cystic fibrosis patient. *Internat. J. System. Evolution. Microbiol.* **51**: 1729–1735.

Davidson, B.R., Davidson, H.F. (1993) Legumes and nitrogen in the Australian vegetation. In: Nutman, P.S. (ed) *Legumes: The Australian Experience*, pp. 131–165. Research Studies Press Ltd: Taunton, Somerset, UK.

de Lajudie, P., Willems, A., Nick, G. *et al.* (1998) Characterization of tropical tree rhizobia and description of *Mesorhizobium plurifarium* sp. nov. *Internat. J. System. Bacteriol.* **48**: 369–382.

Dénarié, J., Debellé, F., Promé, J.-C. (1996) Rhizobium lipo-chito-oligosaccharide nodulation factors: Signalling molecules mediating recognition and morphogenesis. *Ann. Rev. Biochem.* **65**: 503–535.

Dobert, R.C., Breil, B.T., Triplett, E.W. (1994) DNA sequence of the common nodulation genes of *Bradyrhizobium elkanii* and their phylogenetic relationships to those of other nodulating bacteria. *Mol. Plant–Microbe Interactions* 7: 564–572.

Doignon-Bourcier, F., Sy, A., Willems, A., Torck, U., Dreyfus, B., Gillis, M., de Lajudie, P. (1999) Diversity of bradyrhizobia from 27 tropical Leguminosae species native of Senegal. *System. Appl. Microbiol.* **22**: 647–661.

Doignon-Bourcier, F., Willems, A., Coopman, R., Laguerre, G., Gillis, M., de Lajudie, P. (2000) Genotypic characterization of *Bradyrhizobium* strains nodulating small Senegalese legumes by 16S–23S rRNA intergenic gene spacers and amplified fragment length polymorphism fingerprint analyses. *Appl. Environ. Microbiol.* **66**: 3987–3997.

Downie, J.A. (1998) Functions of rhizobial nodulation genes. In: Spaink, H.P., Kondorosi, A., Hooykaas, P.J.J. (eds) *The Rhizobiaceae: Molecular Biology of Model Plant-Associated Bacteria*, pp. 387–402. Kluwer Academic Publishers: Dordrecht, The Netherlands.

Doyle, J.J. (1998) Phylogenetic perspectives on nodulation: evolving views of plants and symbiotic bacteria. *Trends Plant Sci.* **3**: 473–478.

Dupuy, N., Willems, A., Pot, B. *et al.* (1994) Phenotypic and genotypic characterization of bradyrhizobia nodulating the leguminous tree *Acacia albida*. *Internat. J. System. Bacteriol.* **44**: 461–473.

Fred, E.B., Baldwin, I.L., McCoy, E. (1932) *Root Nodule Bacteria and Leguminous Plants*. The University of Wisconsin Press: Madison, WI.

Gathumbi, S.M., Cadisch, G.A., Giller, K.E. (2002) ^{15}N natural abundance as a tool for assessing N_2-fixation of herbaceous, shrub and tree legumes in improved fallows. *Soil Biol. Biochem.* **34**: 1059–1071.

Hannon, N.J. (1956) The status of nitrogen in the Hawkesbury sandstone soils and their plant communities in the Sydney district. I. The significance and level of nitrogen. *Proc. Linneal Soc. NSW* **81**: 119–143.

Hansen, A. (1994) *Symbiotic N_2* Fixation Of Crop Legumes: Achievements And Perspectives. Margraf Verlag: Wilkersheim, Germany.

Hansen, A.P., Pate, J.S. (1987) Comparative growth and symbiotic performance of seedlings of *Acacia* spp. in defined pot culture or as natural understorey components of a eucalypt forest ecosystem in S.W. Australia. *J. Exp. Bot.* **38**: 13.

Hansen, A.P., Pate, J.S., Hansen, A., Bell, D. (1987) Nitrogen economy of post-fire stands of shrub legumes in jarrah (*Eucalyptus marginata* Donn ex Sm.) forest of S.W. Australia. *J. Exp. Bot.* **38**: 26.

Haukka, K., Lindstrom, K., Young, J.P.W. (1996) Diversity of partial 16S rRNA sequences among and within strains of African rhizobia isolated from *Acacia* and *Prosopis*. *System. Appl. Microbiol.* **19**: 352–359.

Haukka, K., Lindstrom, K., Young, J.P.W. (1998) Three phylogenetic groups of *nodA* and *nifH* genes in *Sinorhizobium* and *Mesorhizobium* isolates from leguminous trees growing in Africa and Latin America. *Appl. Environ. Microbiol.* **64**: 419–426.

Hingston, F.J., Malajczuk, N., Grove, T.S. (1982) Acetylene reduction (N_2-fixation) by Jarrah forest legumes following fire and phosphate application. *J. Appl. Ecol.* **19**: 631–645.

Hirsch, A.M. (1992) Developmental biology of legume nodulation. *New Phytol.* **122**: 211–237.

Hungria, M., Johnston, A., Phillips, D.A. (1991) *Rhizobium nod* gene inducers exuded naturally from roots of common bean (*Phaseolus vulgaris* L.). *Plant Physiol.* **97**: 759–764.

Khbaya, B., Neyra, M., Normand, P., Zerhari, K., Filali-Maltouf, A. (1998) Genetic diversity and phylogeny of rhizobia that nodulate *Acacia* spp. in Morocco assessed by analysis of rRNA genes. *Appl. Environ. Microbiol.* **64**: 4912–4917.

Kucey, R.M.N., Hynes, M.F. (1989) Populations of *Rhizobium leguminosarum* biovars *phaseoli* and *viceae* in field after bean or pea in rotation with nonlegumes. *Can. J. Microbiol.* **35**: 661–667.

Lafay, B., Burdon, J.J. (1998) Molecular diversity of rhizobia occurring on native shrubby legumes in southeastern Australia. *Appl. Environ. Microbiol.* **64**: 3989–3997.

Lafay, B., Burdon, J.J. (2001) Small-subunit rRNA genotyping of rhizobia nodulating Australian *Acacia* spp. *Appl. Environ. Microbiol.* **67**: 396–402.

Laguerre, G., Allard, M.-R., Revoy, F., Amarger, N. (1994) Rapid identification of rhizobia by restriction fragment length polymorphism analysis of PCR-amplified 16S rRNA genes. *Appl. Environ. Microbiol.* **60**: 56–63.

Laguerre, G., Mavingui, P., Allard, M., Charnay, M., Louvrier, P., Mazurier, S., Rigottier-Gois, L., Amarger, N. (1996) Typing of Rhizobia by PCR DNA fingerprinting and PCR-Restriction fragment length polymorphism analysis of chromosomal and symbiotic gene regions: application to *Rhizobium leguminosarum* and its different biovars. *Appl. Environ. Microbiol.* **62**: 2029–2036.

Lange, R.T. (1961) Nodule bacteria associated with the indigenous *Leguminosae* of south-western Australia. *J. Gen. Microbiol.* **61**: 351–359.

Langkamp, P.J., Farnell, G.K., Dalling, M.J. (1982) Nutrient cycling in a stand of *Acacia holosericea* A. Cunn ex G. Don. I. Measurements of precipitation interception, seasonal acetylene reduction, plant growth and nitrogen requirement. *Aust. J. Bot.* **30**: 87–106.

Langkamp, P.J., Swinden, L.B., Dalling, M.J. (1979) Nitrogen fixation (acetylene reduction) by *Acacia pellita* on areas restored after mining at Groote Eylandt, Northern Territory. *Aust. J. Bot.* **27**: 353–361.

Lawrie, A.C. (1981) Nitrogen fixation by native Australian legumes. *Aust. J. Bot.* **29**: 143–157.

Lawrie, A.C. (1983) Relationships among rhizobia from native Australian legumes. *Appl. Environ. Microbiol.* **45**: 1822–1828.

Marsudi, N.D.S., Glenn, A.R., Dilworth, M.J. (1999) Identification and characterization of fast- and slow-growing root nodule bacteria from South-Western Australian soils able to nodulate *Acacia saligna*. *Soil Biol. Biochem.* **31**: 1229–1238.

McInroy, S.G., Campbell, C.D., Haukka, K.E., Odee, D.W., Sprent, J.I., Wang, W.J., Young, J.P.W., Sutherland, J.M. (1999) Characterisation of rhizobia from African *Acacias* and other tropical woody legumes using Biolog (TM) and partial 16S rRNA sequencing. *FEMS Microbiol. Lett.* **170**: 111–117.

Monk, D., Pate, J.S., Loneragan, W.A. (1981) Biology of *Acacia pulchella* R. Br. with special reference to symbiotic nitrogen fixation. *Aust. J. Bot.* **29**: 579–592.

Moreira, F.M., Haukka, S.K., Young, J.P.W. (1998) Biodiversity of rhizobia isolated from a wide range of forest legumes in Brazil. *Mol. Ecol.* **7**: 889–895.

Moulin, L., Munive, A., Dreyfus, B., Boivi-Masson, C. (2001) Nodulation of legumes by members of the β-subclass of *Protobacteria*. *Nature* **411**: 948–950.

Murray, B.R., Thrall, P.H., Woods, M.J. (2001) *Acacia* species and rhizobial interactions: Implications for restoration of native vegetation. *Ecol. Manage. Restor.* **2**: 213–219.

Nick, G., de Lajudie, P., Eardly, B.D., Suomalainen, S., Paulin, L., Zhang, X.P., Gillis, M., Lindstrom, K. (1999) *Sinorhizobium arbores* sp. nov and *Sinorhizobium kostiense* sp. nov., isolated from leguminous trees in Sudan and Kenya. *Internat. J. System. Bacteriol.* **49**: 1359–1368.

Norris, D.O. (1956) Legumes and the *Rhizobium* symbiosis. *Empire J. Experiment. Agriculture* **24**: 247–270.

Odee, D.W., Haukka, K., McInroy, S.G., Sprent, J.I., Sutherland, J.M., Young, J.P.W. (2002) Genetic and symbiotic characterization of rhizobia isolated from tree and herbaceous legumes grown in soils from ecologically diverse sites in Kenya. *Soil Biol. Biochem.* **34**: 801–811.

O'Hara, G.W., Howieson, J.G., Graham, P.H. (2002) Nitrogen fixation and agricultural practice. In: Leigh, G.J. (ed) *Nitrogen Fixation at the Millennium*, pp. 391–420. Elsevier Science B.V.

Parez-Fernandez, M.A., Lamont, B.B. (2003) Nodulation and performance of six exotic and six native legumes in six Australian soils. *Aust. J. Bot.* **51**: 543–553.

Polhill, R.M., Raven, P.H., Stirton, C.H. (1981) Evolution and systematics of the Leguminosae In: Polhill R.M., Raven, P.H. (eds) *Advances In Legume Systematics*, pp. 1–26. Royal Botanic Gardens: Kew, UK.

Rasanen, L.A., Sprent, J.I., Lindstrom, K. (2001) Symbiotic properties of sinorhizobia isolated from *Acacia* and *Prosopis* nodules in Sudan and Senegal. *Plant Soil* **235**: 193–210.

Ravin, P.H., Polhill, R.M. (1981) Biogeography of the Leguminosae. In: Polhill R.M., Raven, P.H. (eds) *Advances in Legume Systematics*, pp. 27–34. Royal Botanic Gardens: Kew, UK.

Read, D.J. (1993) Plant–microbe mutualisms and community structure. In: Schultz, E.D., Mooney, H.A. (eds) *Biodiversity and Ecosystem Function*, pp. 181–209. Springer-Verlag: Berlin.

Sadowsky, M.J., Graham, P.H. (1998) Soil biology of the *Rhizobiaceae*. In: Spaink, H.P., Kondorosi, A., Hooykaas, P.J.J. (eds) *The Rhizobiaceae: Molecular Biology of Model Plant-Associated Bacteria*, pp. 155–172. Kluwer Academic Publishers: Dordrecht, The Netherlands.

Santamaria, M., Corzo, J., Leon-Barrios, M.A., Gutierrez-Navarro, A. (1997) Characterisation and differentiation of indigenous rhizobia isolated from Canarian shrub legumes of agricultural and ecological interest. *Plant Soil* **190**: 143–152.

Schofield, P.R., Gibson, A.H., Dudman, W.F., Watson, J.M. (1987) Evidence for genetic exchange and recombination of *Rhizobium* symbiotic plasmids in a soil population. *Appl. Environ. Microbiol.* **53**: 2942–2947.

Sullivan, J.T., Patrick, H.N., Lowther, W.L., Scott, D.B., Ronson, C.W. (1995) Nodulating strains of *Rhizobium loti* arise through chromosomal symbiotic gene transfer in the environment. *Proc. Natl Acad. Sci. USA* **92**: 8985–8989.

Sprent, J.I. (2001) *Nodulation in Legumes*. Royal Botanic Gardens: Kew, UK.

Tan, Z.Y., Wang, E.T., Peng, G.X., Zhu, M.E., Martinez-Romero, E., Chen, W.X. (1999) Characterization of bacteria isolated from wild legumes in the north-western regions of China. *Intern. J. System. Bacteriol.* **49**: 1457–1469.

Thrall, P.H., Burdon, J.J., Woods, M.J. (2000) Variation in the effectiveness of symbiotic associations between native rhizobia and temperate Australian legumes: Within and among genera interactions. *J. Appl. Ecol.* **37**: 52–65.

Thrall, P.H., Murray, B.R., Watkin, E.L.J., Woods, M.J., Baker, K., Burdon, J.J. and Brockwell, J. (2001) Bacterial partnerships enhance the value of native legumes in revegetation and rehabilitation of degraded agricultural lands. *Ecol. Manage. Restor.* **2**: 233–235.

Turk, D., Keyser. H.H. (1992) Rhizobia that nodulate tree legumes: specificity of the host for nodulation and effectiveness. *Can. J. Microbiol.* **38**: 451–460.

Ueda, T., Suga, Y., Yahiro, N., Matsuguchi, T. (1995a) Phylogeny of symplasmids of rhizobia by PCR-based sequencing of a *nod*C segment. *J. Bacteriol.* **177**: 468–472.

Ueda, T., Suga, Y., Yahiro, N., Matsuguchi, T. (1995b) Remarkable N_2-fixing bacterial diversity detected in rice roots by molecular evolutionary analysis of *nif*H gene sequences. *J. Bacteriol.* **177**: 1414–1417.

Urtz, B.E., Elkan, G.H. (1996) Genetic diversity among *Bradyrhizobium* isolates that effectively nodulate peanut (*Arachis hypogaea*). *Can. J. Microbiol.* **42**: 1121–1130.

Vega-Hernandez, M.C., Perez-Galdona, R., Dazzo, F.B., Jarabo-Lorenzo, A., Alfayate, M.C., Leon-Barrios, M. (2001) Novel infection process in the indeterminate root nodule symbiosis between *Chamaecytisus proliferus* (tagasaste) and *Bradyrhizobium* sp. *New Phytol.* **150**: 707–721.

Wang, E.T., Martinez-Romero, J., Martinez-Romero, E. (1999a) Genetic diversity of rhizobia from *Leucaena leucocephala* nodules in Mexican soils. *Mol. Ecol.* **8**: 711–724.

Wang, E.T., Rogel, M.A., Garcia-de los Santos, A., Martinez-Romero, J., Cevallos, M.A., Martinez-Romero, E. (1999b) *Rhizobium etli* bv. *mimosae*, a novel biovar isolated from *Mimosa affinis*. *Intern. J. System. Bacteriol.* **49**: 1479–1491.

Wei, G.H., Wang, E.T., Tan, Z.Y., Zhu, M.E., Chen, W.X. (2002) *Rhizobium indigoferae* sp. nov and *Sinorhizobium kummerowiae* sp. nov., respectively isolated from *Indigofera* spp. and *Kummerowia stipulacea*. *Intern. J. System. Evolution. Microbiol.* **52**: 2231–2239.

Wilson, J.K. (1944) Over five hundred reasons for abandoning the cross inoculation groups of legumes. *Soil Sci.* **58**: 61–69.

Yates, R.J., Howieson, J.G., Nandasena, K.G., O'Hara, G.W. (2004) Root nodule bacteria from indigenous legumes in the north-west of Western Australia and their interaction with exotic legumes. *Soil Biol. Biochem.* (in press).

Young, J.P.W. (1992) Phylogenetic classification of nitrogen-fixing organisms. In: Stacey, G., Burris, R.H., Evans H.J. (eds) *Biological Nitrogen Fixation*, pp. 43-86. Chapman and Hall: New York, N.Y.

Young, J.P.W., Wexler, M. (1988) Symplasmid and chromosomal genotypes are correlated in field populations of *Rhizobium leguminosarum*. *J. Gen. Microbiol.* **134**: 2731–2739.

Zeze, A., Mutch, L.A., Young, J.P.W. (2001) Direct amplification of *nodD* from community DNA reveals the genetic diversity of *Rhizobium leguminosarum* in soil. *Environ. Microbiol.* **3**: 363–370.

Zhang, X., Harper, R., Karsisto, M., Lindstrom, K. (1991) Diversity of *Rhizobium* bacteria isolated from the root nodules of leguminous trees. *Intern. J. System. Bacteriol.* **41**: 104–113.

Zhang, X.X., Turner, S.L., Guo, X.-W., Yang, H.-J., Debellé, F., Yang, G.-P., Dénarié, J., Young, J.P.W., Li, F.-D (2000) The common nodulation genes of *Astragalus sinicus* rhizobia are conserved despite chromosomal diversity. *Appl. Environ. Microbiol.* **66**: 2988–2995.

4

Effects of transgenic plants on soil micro-organisms and nutrient dynamics

Angela Sessitsch, Kornelia Smalla, Ellen Kandeler and Martin H. Gerzabek

4.1 Introduction

Transgenic plants that show herbicide tolerance, resistance to viral, bacterial and fungal diseases, insect resistance, improved product quality and superior agronomic properties are now widely cultivated. However, the possible impact of genetically engineered plants on human health and ecosystem functioning is of increasing concern. Micro-organisms contribute substantially to soil functions as they play an essential role in maintaining soil quality by being involved in nutrient turnover. Furthermore, plant-associated microbes may promote plant growth and health. This chapter reviews research concerning potential effects of transgenic plants on plant-associated microflora by either the synthesis of antimicrobial substances or by unintentional changes due to bacterial transformation. The possible consequences on nutrient turnover in soil due to the cultivation of genetically engineered plants are discussed. Finally, the latest findings regarding the potential for horizontal gene transfer, particularly of antibiotic resistance genes, from transgenic plants to bacteria are presented.

The genetic modification of crops aims at altering, adding or removing a trait in a plant that in many cases cannot be achieved by conventional plant breeding and selection. Desirable properties from other varieties of the plant can be transferred, but the addition of characteristics from unrelated organisms is also possible. Transgenic plants have been developed carrying traits such as herbicide tolerance, resistance to viral, bacterial and fungal diseases, insect resistance, modified plant architecture and development, tolerance to abiotic stresses, production of industrial chemicals and the suitability to be used as a source of fuel. Plant species that

Plant Microbiology, Michael Gillings and Andrew Holmes

have been genetically engineered include mainly maize, tomato, cotton, soybean, oilseed rape and to a lesser extent potato, squash, beet, rice, flax, papaya and cichorium (USDA, 2002). In the year 2000, 36% of all soybean, 16% of cotton, 11% of oilseed rape and 7% of maize grown globally were transgenic (James, 2001). The possible effects of the cultivation and consumption of genetically modified plants (GMPs) on human health and ecosystem functioning is of increasing concern.

Although GMPs are frequently used in agriculture, their impacts on soil micro-organisms and nutrient dynamics are not completely understood. In contrast, it is well known that bacteria belong to the most dominant soil organisms due to their rapid growth and their ability to utilise a wide range of substrates as carbon and nitrogen sources. Many soil micro-organisms are attached to the surface of soil particles and are components of soil aggregates, however, a great number of microbes lives in association with plant roots. Usually, the concentration of bacteria colonizing the soil surrounding roots, i.e. the rhizosphere, is far higher than the number of bacteria living in bulk soil (Lynch, 1990). Plants promote bacterial growth as they provide nutrients due to the exudation of a range of substrates and the decay of senescent roots. Furthermore, the quantity and composition of root exudates determines the microbial community structure. Different populations are found in the rhizospheres of different plant species and at different plant growth stages (Berg *et al.*, 2002; Crowley, 2000; Grayston *et al.*, 1998; Smalla *et al.*, 2001;Yang and Crowley *et al.*, 2000; Gomes *et al.*, 2001). A range of rhizobacteria may also gain entry into the plant, using a number of mechanisms (reviewed by Sturz *et al.*, 2000). Soil micro-organisms determine to a great extent the functioning of terrestrial ecosystems, whereas plant-associated microbes strongly interact with the plant. This interaction may be harmful, neutral or beneficial for the plant. Beneficial effects can be growth stimulation, growth promotion through the enhanced availability of minerals, protection of plants against abiotic stresses and antagonistic effects towards plant pathogens.

The possible impact of GMPs on soil, its organisms and the nutrient/element cycles involved, and soil foodwebs need to be considered. Transgenic plants produce antimicrobial substances that may directly influence soil and plant-associated organisms. Unwanted side effects associated with the cultivation of trangenic plants may include a perturbation of microbial populations in the soil, the rhizosphere or apoplast of plants, leading to an altered function of these organisms. Furthermore, the potential transfer of antibiotic resistance genes that are used as markers in transgenic plants to pathogenic bacteria is of increasing concern. This review addresses the impact of GMPs on plant-associated micro-organisms in relation to environmental and seasonal factors, the likelihood of horizontal gene transfer from transgenic plants to bacteria and possible effects on soil nutrient cycling.

4.2 Rhizosphere communities of plants producing antimicrobial agents

Plant diseases caused by bacterial phytopathogens account worldwide for high production losses, with developing countries being particularly impacted. Resistance traits have often not been introduced into cultivars by conventional breeding and chemical control of bacterial pathogens is not feasible. Genetic transformation offers novel ways to obtain disease resistance by introducing foreign genes into plants of agricultural importance. Transgenic plants including potato (Düring *et al.*, 1993), tobacco (Trudel *et al.*, 1992) and tomato (Stahl *et al.*, 1998) have been developed that produce antimicrobial agents such as the T4-lysozyme. T4-lysozyme is active against Gram-negative as well as Gram-positive bacteria (de Vries *et al.*, 1999) and degrades the murein of the bacterial cell wall by cleaving the β(1-4)-glycosidic bond between *N*-acetylmuramic acid and *N*-acetylglucosamine (Tsugita *et al.*, 1968). Potato expressing the T4-lysozyme gene has been shown to be tolerant towards infection with *Erwinia carotovora*, the cause of blackleg and soft rot (Düring *et al.*, 1993). For the generation of transgenic potato plants a construct was applied, in which the T4-lysozyme gene was fused to the α-amylase leader peptide (Düring, 1993) and therefore the antibacterial agent was secreted from the cytoplasm into the apoplast. Release from the root into the rhizosphere – probably by diffusion – has been reported (Ahrenholtz *et al.*, 2000; de Vries *et al.*, 1999). Furthermore, it has been demonstrated that the T4-lysozyme is still active on the root surface and exhibits bactericidal effects towards root-adsorbed *Bacillus subtilis* cells (Ahrenholtz *et al.*, 2000). Although a range of soil bacteria proved to be sensitive to T4-lysozyme *in vitro* (de Vries *et al.*, 1999), it was postulated that the released enzyme was rapidly degraded in soil under natural conditions or adsorbed to soil particles (Ahrenholtz *et al.*, 2000). Nevertheless, such a release alters the composition of root exudates and may additionally have inhibitory effects on non-target micro-organisms. Therefore, various risk assessment studies have been carried out with T4-lysozyme-producing potato lines.

There is concern that antimicrobial substances produced by transgenic plants may negatively affect the numbers and function of beneficial plant bacteria that colonize the resistant plant or any follow-up crop. Lottmann *et al.* (1999) tested the effect of several transgenic T4-lysozyme-producing potato lines on plant-associated beneficial bacteria under field conditions. Bacterial isolates from the rhizosphere and geocaulosphere (i.e. the tuber surface) were characterised regarding their ability to show antagonistic activity towards the blackleg pathogen *Erwinia carotovora* ssp. *atroseptica* and to produce the phytohormone indole-3-acetic acid (IAA). The genetic modification did not cause any detectable effect on total bacterial counts or on the percentage of potentially beneficial bacteria, although slight differences were found in the species composition of beneficial bacteria (Lottmann *et al.*, 1999). One transgenic potato line, DL4, showed significantly reduced root weight as compared to the parental and other transgenic lines, which has been explained by unintentional changes due to the genetic modification or by somaclonal variation (Lottmann *et al.*, 1999). As the production and release of antimicrobial substances may inhibit the colonisation of beneficial microbial inoculants, Lottmann *et al.* (2000) evaluated the establishment of introduced biocontrol strains in the rhizosphere of transgenic T4-lysozyme-producing potatoes. Two strains with antagonistic activity towards the blackleg pathogen were used for inoculation. The first was characterised as a *Serratia grimesii* strain that was exclusively found in the rhizosphere of non-transgenic control plants (Lottmann *et al.*, 1999) and showed high sensitivity to T4-lysozyme *in vitro*. The second strain, identified as *Pseudomonas putida*, was isolated from transgenic plants and showed high tolerance against T4-lysozyme (Lottmann *et al.*, 2000). Both isolates were able to compete with the indigenous microflora and to colonize roots and tubers of transgenic as well as parental plants. However, during flowering significantly higher numbers of the T4-lysozyme-tolerant *P. putida* strain were found in the rhizosphere of the transgenic line than in that of the control plant. Since this growth stage is also when most T4-lysozyme is produced, it can be concluded that the inoculant strain had a competitive advantage because of its low sensitivity towards the antimicrobial agent (Lottmann *et al.*, 2000). Recently, strains belonging to two bacterial groups that are known for their interaction with plants and plant growth-promoting abilities – pseudomonads and enterics – were isolated from transgenic and non-transgenic potato lines and investigated (Lottmann and Berg, 2001). Bacterial strains were analysed for their biocontrol activities towards bacterial and fungal pathogens, their ability to synthesise IAA and their sensitivity towards T4-lysozyme. Furthermore, strains were investigated by genetic profiling and identified by fatty acid methyl ester analysis. Results showed that the expression of the T4-lysozyme did not affect members of the pseudomonads and enterics and that the distribution of isolates was not influenced by the plant genotype (Lottmann and Berg, 2001).

Most studies regarding the effects of transgenic plants on plant-associated microbial communities are based on the characterisation of isolated strains. However, it is well known that only a minor percentage of natural microbial communities can be cultivated (Amann *et al.*, 1995) due to unknown growth requirements and the fact that bacteria may enter a viable-but-non-culturable state (Troxler *et al.* 1997; van Overbeek *et al.*, 1995). Therefore, Heuer *et al.* (2002a) applied two approaches to analyse bacterial rhizosphere communities of wild-type and transgenic T4-lysozyme-producing potato lines that were grown for 3 years at two distant field sites with different soil types. First, the species composition was determined by cultivation of rhizosphere bacteria and subsequent identification by fatty acid methyl ester analysis. The second approach involved DNA isolation from rhizosphere soil sampled at different plant growth stages, PCR amplification of 16S rRNA genes and their analysis by denaturating gradient gel electrophoresis (DGGE). Both approaches revealed that environmental factors such as plant growth stage, seasonal changes and soil type had a far higher impact on rhizosphere communities than the production of the T4-lysozyme (Heuer *et al.*, 2002a). Similarly, Lukow *et al.* (2000) reported seasonal and spatial shifts in the rhizosphere communities of transgenic GUS/ Barnase/Barstar potato lines and the non-transgenic control plant. The transgenic T4-lysozyme-producing variant DL4 again showed differing bacterial communities as compared to the control line. This deviation was attributed to abnormal growth characteristics of this line as a result of irregular and multiple integration of the transgene or position effects from the insertion site (Heuer *et al.*, 2002a). In a previous study, the effect of T4-lysozyme production on phyllosphere bacteria communities was investigated (Heuer and Smalla, 1999). Although a slightly different species composition was identified in the phyllosphere of transgenic plants as compared to the parental line, the authors concluded that the observed effects were minor relative to the natural variation between field sites (Heuer and Smalla, 1999).

An additional strategy to suppress bacterial pathogens is the addition of genes encoding lytic peptides such as cecropins to the plant genome, as they exhibit significant activity in transgenic tobacco and potato plants (Huang *et al.*, 1997; Jaynes *et al.*, 1993). Cecropins, isolated from the haemolymph of pupae of the giant silk moth, *Hyalaphora cecropia,* (Hultmark *et al.*, 1980), show strong lytic and antimicrobial activity against several Gram-negative and Gram-positive bacteria (Hultmark *et al.*, 1982). In particular, cecropin B proved to be highly toxic against a number of plant-pathogenic bacteria (Jaynes *et al.*, 1987; Nordeen *et al.*, 1992). For the generation of cecropin-expressing potatoes (Keppel, 2000; Kopper, 1999) two modified cecropin B genes were employed; cecropin C38, that lacks the N-terminal signal peptide, and C4, carrying a barley hordothionin signal peptide, which was found to improve post-translational folding (Florack *et al.*, 1995). Recently, culturable *Bacillus* populations colonising the rhizosphere of non-transgenic and transgenic, cecropin-expressing potato lines were compared at different vegetation stages (Sessitsch *et al.*, 2002). The genus *Bacillus* is an important member of rhizosphere communities and several strains have been shown to promote plant growth (Asaka and Shoda, 1996; Wilhelm *et al.*, 1998). At the flowering stage *Bacillus* isolates obtained from cecropin-expressing lines showed significantly reduced diversity as compared to those isolated from the parental plant. However, at the tuber production stage the rhizosphere *Bacillus* populations only showed few differences. Similarly to Lottmann *et al.* (2000) it was demonstrated that strains with a high tolerance of the lytic peptide had a competitive advantage in colonising the rhizosphere of cecropin-producing lines. Besides the different sensitivities of the *Bacillus* community members towards cecropin, unintentionally altered plant characteristics seemed to be responsible for the observed effects (Sessitsch *et al.*, 2002). Currently, rhizosphere and endophytic bacterial communities of field-grown cecropin- and T4-lysozyme-producing potato plants are further investigated by applying cultivation-dependent and -independent approaches.

4.3 Herbicide-tolerant plants and their associated microflora

Currently, tolerance to non-selective broad-spectrum herbicides such as glyphosate or glufosinate is the most important phenotypic trait introduced into transgenic crops and 46% of all released genetically modified plants carry herbicide resistance genes (USDA, 2002). The presence of such genes allows the application of the complementary herbicide at any time killing almost all weeds without damaging the transgenic crop. This leads to a more efficient use of herbicides and a reduction of between 11 and 30% in the total amount of applied herbicides has been reported (AgrEvo *et al.*, 1998). Various trangenic crops tolerate glyphosate, the active ingredient of Roundup. In general, the herbicide is sprayed onto plants and has a systemic effect. It inhibits the enzyme 5-enolpyruvylshikimate-3-phosphate synthetase (EPSPS) that is highly important for the synthesis of aromatic amino acids. Glyphosate-tolerant plants contain an EPSPS gene of the soil bacterium *Agrobacterium tumefaciens* that, because of structural differences, is not inhibited by the herbicide. Several studies indicate that glyphosate is rapidly and completely degraded by soil organisms (Haney *et al.*, 2002; Heinonen-Tanski, 1989). The second herbicide-resistance system involves glufosinate (phosphinothricin), the active ingredient of Basta or Liberty. It inhibits the enzyme glutamine synthetase leading to the accumulation of ammonium within cells and subsequent cell death (Hoerlein, 1994). Glufosinate is naturally produced by *Streptomyces* spp. (Bayer *et al.*, 1972; Wohlleben *et al.*, 1992) and herbicide-resistant crops contain the *pat* gene from this soil bacterium encoding phosphinotricin acetyltranferase that detoxifies glufosinate by acetylation. The herbicide also shows weak antibacterial activity (Bayer *et al.*, 1972) and sensitivity of several soil microbes has been reported (Ahmad and Malloch, 1995). In addition, Kriete and Broer (1996) demonstrated a negative effect due to glufosinate application on the growth of nitrogen-fixing rhizobia, nodule formation and nitrogen fixation. However, other studies demonstrated that many bacteria are resistant to glufosinate or are even able to degrade the herbicide by deamination and decarboxylation (Ahmad and Malloch, 1995; Allen-King *et al.*, 1995; Bartsch and Tebbe, 1989; Tebbe and Reber, 1988, 1991).

Microbial communities associated with the root interior and rhizosphere soil of three oilseed rape cultivars, Parkland (*Brassica rapa*), Excel (*Brassica napus*) and Quest (*Brassica napus*), were analysed by Siciliano *et al.* (1998). Quest has been genetically engineered to tolerate the herbicide glyphosate. Biolog™ plates were used to assess functional diversity, whereas fatty acid methyl ester (FAME) analysis was applied to determine the community composition of plant-associated bacteria. The glyphosate-tolerant cultivar Quest had different endophytic and rhizosphere communities when compared to other lines. Moreover, the microbial population colonising the non-transgenic *Brassica napus* line Excel showed more relatedness to that of a different *Brassica* species (Parkland) than to that associated with the transgenic *Brassica napus* cultivar Quest (Siciliano *et al.*, 1998). However, Excel is not the isogenic, parental line of Quest, and therefore it cannot be excluded that the differences found are due to genotypic differences between these two cultivars rather than due to the genetic modification (Siciliano *et al.*, 1998). In a follow-up study, Siciliano and Germida (1999) assessed the taxonomic diversity of culturable bacteria associated with the transgenic and non-transgenic *Brassica napus* cultivars Quest and Excel. Again, results demonstrated that both lines were colonised by different communities. In addition, this effect was more pronounced with endophytes than with rhizosphere bacteria. As the root exudate composition greatly influences microbial communities in the rhizosphere and rhizoplane (Grayston *et al.*, 1998; Yang and Crowley, 2000), it was postulated that the differences found may be due to a slightly different root exudation of Excel and Quest (Siciliano and Germida, 1999). However, it is not clear whether root exudates also control endophytic communities. Certain plant phenotypic properties may affect the entrance of rhizosphere bacteria into the plant or the proliferation of endophytes. The transgenic line Quest selectively promoted the endophytic growth of certain *Pseudomonas*, *Flavobacter* and *Aureobacter* species (Siciliano and Germida, 1999). The authors claim unintentional changes during the genetic modification for the observed effects, however, verification is needed by testing isogenic lines that differ only in the presence of herbicide tolerance genes. Differences found between the transgenic and non-transgenic *Brassica napus* lines were further confirmed at different field sites and during two different growing seasons (Dunfield *et al.*, 2001). It was postulated that in addition to unintentionally altered plant characteristics, the exudation of the gene product also may be responsible for different plant-associated bacteria.

Recent studies analysed the effects of transgenic, glufosinate-tolerant lines on endophytic and rhizosphere bacterial communities. Dunfield *et al.* (2001) compared microbial populations associated with three transgenic, glufosinate-tolerant *Brassica napus* lines with those of three conventional cultivars of the same species by FAME analysis and community level physiological profiling (CLPP). Results indicated that transgenic plants are more frequently inhabited by Gram-negative bacteria, particularly *Pseudomonas* spp. Furthermore, an indicator fatty acid for certain groups of Gram-negative bacteria including *Chromatium*, *Legionella*, *Rhodospirillum* and *Campylobacter* was found in higher amounts among microbes of herbicide-tolerant lines. Similarly, some Gram-positive bacteria such as *Clostridium* and/or *Bacillus* were also found in higher quantities among root-associated bacteria of genetically modified lines (Dunfield *et al.*, 2001). Recently, Gyamfi *et al.* (2002) compared eubacterial as well as *Pseudomonas* populations in the rhizospheres of glufosinate-tolerant oilseed rape (*Brassica napus*) and its isogenic parental line at different plant growth stages. In addition, the effect of the associated herbicide application on the rhizosphere microflora was assessed. Microbial populations were analysed by a cultivation-independent approach, in which the structural diversity and community composition were determined by denaturating gradient gel electrophoresis (DGGE) analysis of PCR-amplified 16S rRNA genes. Results showed that among the parameters tested, the plant growth stage had the most pronounced effect on the rhizosphere microflora (Gyamfi *et al.*, 2002). Minor differences between the plant-associated micro-organisms of the transgenic and wild-type were detected and it was assumed that unintentionally altered plant characteristics such as a different root exudate composition due to the genetic modification are responsible for this effect. Furthermore, the complementary herbicide Basta had a more pronounced effect on rhizosphere communities than the conventional herbicide Butisan S (Gyamfi *et al.*, 2002). In this experiment, in which transgenic, Basta-resistant oilseed rape and the parental line were grown and treated with Basta and the conventional herbicide Butisan S, respectively, alfalfa was cultivated as follow-up crop in order to assess long-term effects. The number of nodules formed and *Rhizobium* strain diversity were determined. The legume cultivated after the transgenic line in combination with the associated herbicide led to a significantly decreased number of nodules (Gyamfi *et al.*, unpublished results). However, the *Rhizobium* strain diversity was not affected (Gyamfi *et al.*, unpublished results). In Germany, a field experiment was conducted with two glufosinate-tolerant spring oilseed rape hybrids and a glufosinate-tolerant winter oilseed rape as well as with their wild-type counterparts in order to assess the impact of the cultivation of the transgenic lines on the agroecosystem (Becker *et al.*, 2001). The oilseed rape variety and the herbicide application affected the soil microbial biomass and soil basal respiration as well as the soil *Rhizobium leguminosarum* diversity, however, effects due to the genetic modification were not found (Becker *et al.*, 2001). Schmalenberger and Tebbe (2002) compared the bacterial rhizosphere community of a glufosinate-resistant maize with that of the non-transgenic parent line. Plants were grown under conditions common for agricultural practices and the rhizosphere microflora was analysed by PCR-amplification of 16S rRNA genes from isolated DNA and subsequent single-strand conformation polymorphism (SSCP) analysis. Plants hosted different populations at different plant growth stages, but rhizosphere communities associated with transgenic and non-transgenic lines were highly similar. In addition, the herbicide had no detectable effect on the community structure (Schmalenberger and Tebbe, 2002).

4.4 Horizontal transfer of transgenic plant DNA to bacteria

Natural transformation is the most likely mechanism for horizontal transfer of antibiotic resistance genes from transgenic crops to bacteria (Bertolla and Simonet, 1999; Nielsen *et al.*, 1998). In addition, lightning-mediated gene transfer recently shown under laboratory-scale conditions (Demanèche *et al.*, 2001a) could be the potential route for the transfer of transgenic plant DNA to bacteria. The DNA taken up by the bacteria needs to be integrated either into the bacterial genome by homologous recombination, or form an autonomous replicating element. Natural transformation provides a mechanism of gene transfer that enables competent bacteria to generate genetic variability by making use of DNA present in their surroundings (Nielsen *et al.*, 2000a). From laboratory experiments more than 40 bacterial species from different environments are known to be naturally transformable (Lorenz and Wackernagel, 1994; Nielsen *et al.*, 1998). Prerequisites for natural transformation are the availability of free DNA, the development of competence, the uptake and stable integration of the captured DNA. However, there is very limited knowledge of how important natural transformation is in different environmental settings. Two aspects of natural transformation in the environment have been, or are presently studied: the persistence of free DNA and the ability of different bacterial species to take up free DNA under environmental conditions.

4.4.1 Persistence of free DNA in soil

Recent reports have shown that in spite of the ubiquitous occurrence of DNases, high-molecular-weight free DNA could be detected in different environments. It is supposed that free DNA released from micro-organisms or decaying plant material can serve as a nutrient source or as a reservoir of genetic information for autochthonous bacteria. Reports on the persistence of nucleic acids in non-sterile soil have been published (Blum *et al.*, 1997; Nielsen *et al.*, 1997a), and microbial activity was pinpointed as an important biotic factor affecting the persistence of free DNA in soil. Stimulated microbial activity often coincided with an increase in DNase activity in soil (Blum *et al.*, 1997). Nielsen *et al.* (2000a) showed that cell lysates of *Pseudomonas fluorescens*, *Burkholderia cepacia* and *Acinetobacter* spp. were available as a source of transforming DNA for *Acinetobacter* sp. populations in sterile and non-sterile soil for a few days and that cell debris protected DNA from inactivation in soil. Cell walls might play an important role in protecting DNA after cell death (Paget and Simonet, 1997). Long-term persistence of transgenic plant DNA was found by Widmer *et al.* (1996, 1997), Paget and Simonet (1997) and Gebhard and Smalla (1999) in microcosm and field studies. A more rapid breakdown of transgenic DNA was observed at higher soil humidity and temperature. Both factors are supposed to contribute to a higher microbial activity in soil (Blum *et al.*, 1997; Widmer *et al.*, 1996).

Binding of DNA to rather different surfaces such as chemically purified mineral grains of sand (Lorenz and Wackernagel, 1987) and clay (Demanèche *et al.*, 2001b ; Gallori *et al.*, 1994; Khanna and Stotzky, 1992; Pietramellara *et al.*, 1997), non-purified mineral materials (Chamier *et al.*, 1993) as well as humic stubstances (Crecchio and Stotzky, 1998) has been reported (Lorenz and Wackernagel, 1994; Recorbet *et al.*, 1993; Romanowski *et al.*, 1992). For example, clay–DNA complexes have been shown to persist in non-sterile soil up to 15 days after addition to the soil (Gallori *et al.*, 1994) or up to 5 days after introduction of linear duplex DNA into non-sterile soil microcosms (Blum *et al.*, 1997). In the study of Demanèche *et al.* (2001b) plasmid DNA adsorbed on clay particles was found to be not completely degradable even at high nuclease concentrations. There is considerable evidence that nucleic acids released from cells are distributed in the solid as well as the liquid phase depending on physical and chemical properties of the soil. The adsorption of DNA seems to be a charge-dependent process, since the extent of adsorption is affected by the concentration and valencies of cations (Romanowski *et al.*, 1991). In addition, the rate and extent of adsorption of dissolved DNA to minerals depends largely on the type of mineral, the pH of the bulk phase, whereas the conformation and the molecular size of the DNA molecules have a minor effect (for review see Lorenz and Wackernagel, 1994; Paget and Simonet, 1994; Stotzky, 1986). Under field conditions it may well be that transgenic plant DNA is protected by intact plant cells for quite some time. Since plant DNA can persist adsorbed on soil particles or protected in plant cells this DNA could be captured by competent bacteria.

4.4.2 Transfer of marker genes from transgenic plants to soil or rhizosphere bacteria

Long-term persistence even of a small proportion of the released plant DNA is assumed to enhance the likelihood of transformation processes. Furthermore, it was hypothesized that the introduction of bacterial genes, promoter and terminator sequences into the plant genome might lead to an increased probability that the transgenic plant DNA taken up by bacteria can be stably integrated, based on homologous recombination. However, until recently, it was completely unclear whether bacteria could be transformed by plant DNA at all. The high content of non-bacterial DNA and the much higher methylation rate were supposed to prevent a transfer of antibiotic resistance genes from transgenic plant DNA to bacteria. Several groups had failed to detect horizontal gene transfer (HGT) from transgenic plants to bacteria, perhaps because of the absence of homologous sequences in bacteria (Nielsen *et al.*, 1997b) or the use of less efficiently transformable bacteria (Schlüter *et al.*, 1995). The ability of *Acinetobacter* sp. BD413 to capture and integrate transgenic plant DNA based on homologous recombination could be demonstrated under optimised laboratory conditions (de Vries and Wackernagel, 1998; Gebhard and Smalla, 1998). The restoration of a deletion in the *npt*II gene resulted in a kanamycin resistance phenotype, which could be easily detected. This was observed not only with transgenic plant DNA but also with transgenic plant homogenates (Gebhard and Smalla, 1998). However, compared to transformation with chromosomal or plasmid DNA, transformation frequencies with plant DNA or plant homogenates were drastically reduced.

When the experiments initially done by filter transformation were performed in sterile and non-sterile soil, transformation of *Acinetobacter* sp. BD413 pFG4 by transgenic sugar beet DNA could be detected in sterile but not in non-sterile soil (Nielsen *et al.*, 2000b). The authors estimated that numbers of transformants in non-sterile soil would be at 10^{-10} to 10^{-11} and thus below the level of detection. If homologous DNA was present, studies on gene transfer by natural transformation have revealed that additive integration of non-homologous genetic material can occur when flanking homology is present (Gebhard and Smalla, 1998; Nielsen *et al.*, 1998). The restoration of a 10 bp deletion in the *npt*II gene was also observed when *Pseudomonas stutzeri* pMR7 was transformed with transgenic plant DNA (de Vries *et al.*, 2001). However, in this study no transformants were observed in the absence of homologous DNA for both *Acinetobacter* sp. and *P. stutzeri.* This observation confirmed earlier experiments by Nielsen *et al.* (1997b) and suggests that the probability of integration of transgenes in the bacterial genome of the recipient is low if homologous DNA is not present. Although illegitimate recombination was not detected in the absence of homology, its frequency increased by five orders of magnitude when a 1 kb region of homology to recipient DNA was present in the otherwise heterologous donor DNA (de Vries and Wackernagel, 2002). These findings suggest that stretches of homology down to 183 bp served as recombinational anchors facilitating illegitimate recombination. The presence of stretches of homology shared by donor and recipient DNA as requirement for stable integration of DNA taken up is rather well demonstrated. Relatively little is known about the kind of bacteria that become competent under soil or rhizosphere conditions or inside of plants and the biotic and abiotic factors triggering these processes. The major limiting factor for natural transformation remains the presence of competent bacteria and the development of competence. In most studies on transformation, competent bacteria have been inoculated into the soil system studied (Gallori *et al.*, 1994; Nielsen *et al.*, 1997a; Sikorski *et al.*, 1998). Only recently, could Nielsen *et al.* (1997c, 2000b) show that non-competent *Acinetobacter* sp. strain BD413 cells residing in soil could become competent after addition of nutrients. Nutrient solutions used to stimulate competence development in *Acinetobacter* sp. BD413 populations contained inorganic salts and simple compounds corresponding to rhizosphere exudates (Nielsen *et al.*, 2000b). Using the IncQ plasmid derivative pNS1 as transforming DNA, a collection of *P. stutzeri* isolates from soil were analysed for transformability by Sikorski *et al.* (2002). About two-thirds of the isolates were found to be transformable. Interestingly, the transformability differed among the isolates by up to three to four orders of magnitude and thus it appeared that transformability amongst *P. stutzeri* isolates is a rather variable trait (Sikorski *et al.*, 2002). Demanèche *et al.* (2001c) demonstrated that two typical soil bacteria, *Agrobacterium tumefaciens* and *Pseudomonas fluorescens*, can be transformed. Most remarkably, transformants of *P. fluorescens* were obtained in sterile and non-sterile soil but not under various *in vitro* conditions. This finding obviously raises questions about the environmental triggers affecting the transformability of bacteria. *Ralstonia solanacearum*, the causal agent of bacterial wilt, was reported to develop competence *in planta* and to exchange genetic information *in planta* (Bertolla *et al.*, 1997, 1999). However, gene exchange was demonstrated when tomato plants infected with *R. solanacearum* were inoculated with plasmid DNA or during co-infection with *R. solanacearum* carrying different genetic markers, and not during colonisation of transgenic plants. For the first time, Kay *et al.* (2002) could show transformation of *Acinetobacter* BD413 by transplastomic plant DNA *ad planta*. The *aadA* marker gene of transplastomic tobacco plants was captured by *Acinetobacter* sp. BD413 co-colonising *Ralstonia solanacearum*-infected tobacco based on homologous recombination. An RSF1010 derivative containing plastid sequences, including *rbcL* and *aacD* which was introduced into *Acinetobacter* provided the homologous sequences required for homologous recombination. Transformants were detected based on the acquired resistance to spectinomycin. *Ad planta* transformants were obtained only in transplastomic plants but not in nuclear transgenic plants (Kay *et al.*, 2002).

In contrast to transformation, HGT by conjugation or mobilisation under different environmental conditions is much better documented (Thomas and Smalla, 2000). It cannot be excluded that HGT from plants to bacteria may take place in different environmental niches but the ecological significance of such rare events depends upon the selection of the acquired trait and the present dissemination of respective antibiotic resistance genes. The emergence of bacterial antibiotic resistance as a consequence of the widescale use of antibiotics by humans has resulted in a rapid evolution of bacterial genomes. Mobile genetic elements such as transferable plasmids, transposons and integrons have played a key role in the dissemination of antibiotic resistance genes amongst bacterial populations and have contributed to the acquisition and assembly of multiple antibiotic resistance determinants in bacterial pathogens (Heuer *et al.*, 2002b; Levy, 1997; Tschäpe, 1994; Witte, 1998). Since bacteria circulate between different environments and different geographic areas, the global nature of the problem of bacterial antibiotic resistance requires that data on their prevalence, selection and spread are obtained in a more comprehensive way than before. Only few studies have provided data on the prevalence of antibiotic resistance genes used as markers in transgenic plants. Studies on the dissemination of the most widely used marker gene, *npt*II, in bacteria from sewage, manure, river water and soils demonstrated that in a high proportion of kanamycin-resistant enteric bacteria the resistance is encoded by the *npt*II-gene (Smalla *et al.*, 1993).

Bacteria resistant to multiple antibiotics are not restricted to clinical environments but can easily be isolated from different environmental samples and food (Dröge *et al.*, 2000; Heuer *et al.*, 2002b ; Perreten *et al.*, 1997; Smalla *et al.*, 2000). There is substantial movement of antibiotic resistance genes and antibiotic-resistant bacteria between different environments. In assessing the antibiotic resistance problem, a number of factors can be identified that have contributed to the antibiotic resistance problem: the antibiotic itself and the antibiotic resistance trait (Levy, 1997). The genetic plasticity of bacteria has largely contributed to the efficiency by which antibiotic resistance has emerged. However, HGT events have no *a priori* consequence unless there is antibiotic selective pressure (Levy, 1997). Given the fact that antibiotic resistance genes, often located on mobile genetic elements, are already widespread in bacterial populations and that HGT events from transgenic plants to bacteria are supposed to occur at extremely low frequencies, it is unlikely that antibiotic resistance genes used as markers in transgenic crops will contribute significantly to the spread of antibiotic resistance in bacterial populations. However, it cannot be ruled out that hotspots exist, such as the digestive tract of insects, which might promote gene transfer events. There is no doubt that the present problems in human and veterinary medicine due to the selective pressure posed on microbial communities were created by the unrestricted use of antibiotics in medicine and animal husbandries and not by transgenic crops carrying antibiotic resistance markers. Thus the public debate about antibiotic resistance genes in transgenic plants should not diverge the attention from the real causes of bacterial resistance to antibiotics, which is the continued abuse and overuse of antibiotics by physicians and in animal husbandry (Salyers, 1996).

4.5 Impact of genetically modified plants on element dynamics

In contrast to the impact of genetically modified micro-organisms on soil microbes, which has been demonstrated frequently (e.g. Leeflang *et al.*, 2002; Lynch *et al.*, 1994; van Dillewijn *et al.*, 2002), effects of transgenic plants on soil micro-organisms have been rarely reported. In most cases described effects are unclear and their quantification remains difficult. *Figure 4.1* shows some possible impacts of GMPs on carbon and nutrient dynamics. The major aspects are: (i) the excretion of DNA, proteins and other substances through roots; (ii) the possible influence of plant litter, which may decompose differently from non-genetically modified crops due to special substances (toxins) in the tissue or a higher amount of plant material resistant to decomposition; and (iii) an enhanced nutrient and/or water uptake by GMPs.

One aspect of evaluating the environmental impact of GMPs is the release of modified DNA and toxins to soil and the potential subsequent transfer to other

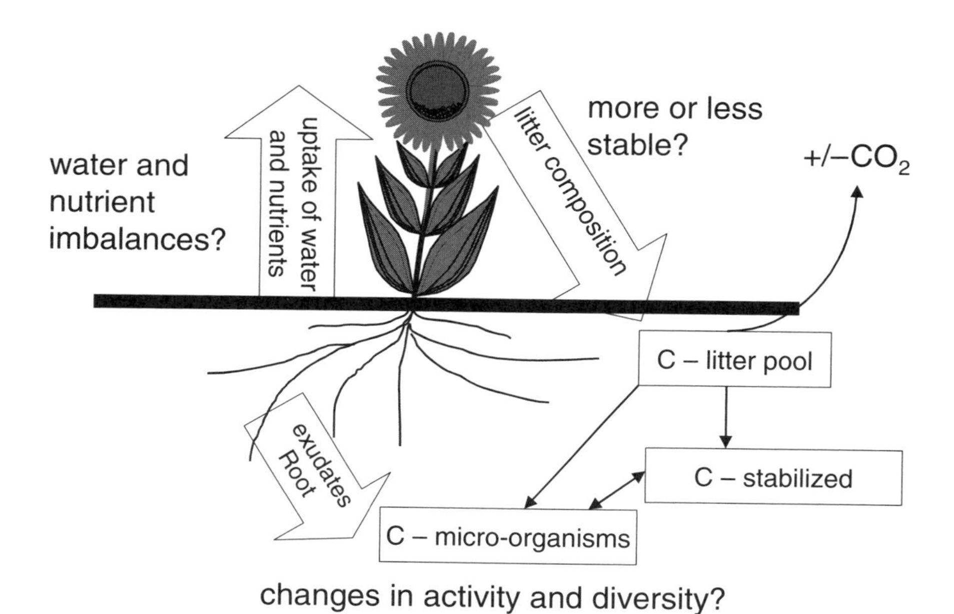

Figure 4.1. Possible effects of genetically modified plants on carbon and nutrient dynamics of the plant–soil–atmosphere system

organisms as well as its impact on the soil foodweb. DNA and toxins produced by GMPs, like plants that produce the Bt endotoxin from *Bacillus thuringiensis* ssp. *kurstaki*, are able to persist in the soil matrix (Stotzky, 2000). This is due to the adsorption properties of soil constituents (Crecchio and Stotzky, 2001; see respective section in this chapter). Effects of the Bt endotoxin on soil microbes, however, could not be clearly shown in experiments (Donegan *et al.*, 1995). Colloidal structures in soil decrease the accessibility of nucleic acids and proteins to enzymes excreted by micro-organisms. This binding effect could decrease both the effect of the excreted substances on other organisms as well as the decomposition by soil microbes. Generally free DNA concentrations in soil are low, ranging from a few to tenths of pg g^{-1} of dry soil, depending on soil management and soil type (Niemeyer and Gessler, 2002).

Transgenic disease resistance is often based on transformation of plants with genes encoding antimicrobial proteins (Cowgill, 2002) and a release of such proteins may have a disadvantageous effect on soil organisms. Griffiths *et al.* (2000) investigated different soil organisms and microbial parameters associated with the growth of transgenic potatoes producing the lectins GNA and ConA in a field experiment. Whereas the protozoan population and microbial activity decreased significantly in comparison to the wild-type plants in a model experiment, these responses could not be detected in a pot experiment. Since protozoa feed on bacteria and contribute to the release and regulation of microbial soil respiration, a long-lasting effect of GMPs on protozoa might therefore influence the turnover of organic carbon in soil. Other effects reported in the literature are, for instance, the smaller arbuscular mycorrhizal infection in transformed alfalfa containing a fungal lignin-peroxidase gene and in tobacco that expresses phosphatases as a means to increase resistance against phytopathogenic fungi (Watrud, 2000). Alteration of the mycorrhizal infection of plants might influence the micronutrient uptake by plants, although this has not been shown in the literature. Another possible effect of GMPs on citrate accumulation and efflux due to the expression of a *Pseudomonas aeruginosa* citrate synthase gene in tobacco, could not be verified in recent experiments (Delhaize *et al.*, 2001). Previously it was reported that an enhanced citrate efflux from genetically modified tobacco roots could improve both the phosphorus uptake and the aluminium tolerance by complexation of free aluminium in the soil solution (de la Fuente *et al.*, 1997).

Transgenic plants might also influence soil processes due to indirect mechanisms such as changes in the composition and quality of the leaf tissue which is decomposed by soil micro-organisms. The suitability of litter of GMPs as a food resource for soil micro-organisms can be influenced by the presence of toxins in the plant tissue as well as by an increase or decrease of compounds resistant to degradation such as lignin. Escher *et al.* (2000) investigated the decomposition of maize litter derived from transgenic plants expressing the Cry1Ab protein from *Bacillus thuringiensis* ssp. *kurstaki*. Nutritional quality of transgenic maize litter for a decomposer (*Porcellio scaber*) was better than that of the non-transgenic variety due to a slightly lower C/N ratio, a lower lignin content and a higher content of soluble carbohydrates. Bacterial growth was equal on leaves of both varieties. Stotzky (2002), on the other hand, showed that microbial decomposition of Bt-maize straw is slower than for non-transgenic varieties. This result was explained by the significantly higher lignin contents of the Bt-maize litter, which was verified for ten Bt maize hybrids. Stotzky

(2002) concluded that the slower degradation of Bt maize hybrids might be beneficial to increase soil organic matter levels. On the other hand, the higher retention time of the toxin in soil might enhance the hazard for non-target organisms.

Effects of plants and soil management on soil microbial communities are evident from the literature. Both plants and fertilizers are drivers of microbial activity and diversity (Gerzabek *et al.*, 2002; O'Donnell *et al.*, 2001). The first authors showed differences in soil enzyme activities of a factor of up to 10.7 (alkaline phosphatase) between peat and animal-manure-treated soil plots and large differences in bacterial diversities due to different soil amendments in the Ultuna long-term field experiment. Simple and frequently used agricultural techniques such as cultivation, mulching and herbicide applications (Wardle *et al.*, 1999) and different crop rotations (Chander *et al.*, 1997) have significant effects on microbial biomass, its activity and soil organic matter turnover. The effects of GMPs in comparison to non-modified plants seem to be less significant in this respect. On the other hand, biotechnology is envisaged to help increasing C inventories in soils. Soil carbon sequestration is one of the key issues in research and politics at the moment and may be improved by: (i) improving net primary production; (ii) manipulating photoassimilate partitioning; (iii) manipulating lignin contents; (iv) engineering C-4 photosynthesis genes into C-3 plants and (v) by engineering N_2-fixation genes into non-leguminous plants (Metting *et al.*, 2001). We tried to evaluate one of the possible effects on the organic carbon inventory of agriculturally used soils (topsoil: 0–20 cm); the alteration in the decomposability of plant litter. The basis for this modelling study was the long-term experiment in Ultuna/Sweden, an agricultural field trial with 14 different treatments, set up in 1956. We used the Ca-nitrate treatment, one of the highly productive variants of the field-experiment. Modelling was performed with the widely used ROTHC-26.3 model developed at the Rothamsted research station (Coleman and Jenkinson, 1999), which has been proven to yield excellent results for the Ultuna site (Falloon and Smith, 2002). This model considers detailed meteorological input, soil clay content, soil depth, soil cover, monthly input of plant residues and farmyard manure (if applicable). Additionally an estimate of the decomposability of the incoming plant material, the DPM/RPM ratio (decomposable plant material vs. resistant plant material) was applied. Based on a weather file created for Ultuna (mean annual temperature: 5.5°C; mean annual precipitation: 660 mm; mean annual evaporation: 537 mm) and a landuse file for the treatment Ca-nitrate with data obtained from Persson and Kirchmann (1994) and Gerzabek *et al.* (1997) plus additional information from a soil sampling in 1998 we calculated the yearly plant residue input to the plots for a measured 'equilibrium' soil organic carbon inventory in 1998. The model suggested a yearly plant residue input of 1.179 t C ha^{-1}. This plant residue input was kept constant for the modelling runs and the basic DPM/RPM value of 1.44 (equivalent to 59% DPM and 41% RPM) was varied. *Table 4.1* shows selected results from this modelling study. An increase of resistant plant material input results in a slightly higher equilibrium organic carbon content in soil, but the effect is smaller than expected. Even a distinct increase in resistant compounds such as a lignin content of 10%, yields only 3.3% higher equilibrium organic carbon level in the topsoil. Changes in soil management have a significantly larger effect (*Table 4.1*). The application of animal manure increases the equilibrium C_{org} inventory to 155% of the Ca-nitrate treatment, in the bare fallow treatment the C_{org} level decreases to 66%.

Table 4.1. Modelled values of equilibrium soil organic carbon stocks in the topsoil of the long-term field experiment at Ultuna/Sweden as influenced by the decomposability of the plant material and measured C-stocks from different treatments of this experiment (year 1998)

Treatment	DPM (%)	RPM (%)	DPM/RPM	t C ha^{-1} modelled	% of Ca-nitrate[a]	t C ha^{-1} measured
Ca-nitrate	59	41	1.44	40.416	100	40.416
	57	43	1.33	40.670	100.6	
	49	51	0.96	41.745	103.3	
Fallow	–	–	–	–	65.8	26.585
NoN	–	–	–	–	81.6	32.960
Animal manure[b]	–	–	–	–	155	62.806
Peat[b]	–	–	–	–	217	87.515

DPM, decomposable plant material; RPM, resistant plant material; NoN, treatment without N-fertilization.
[a] (DPM/RPM = 1.44).
[b] 2000 kg organic carbon ha^{-1} a^{-1} were applied.

We might conclude from the above studies that a significant impact of GMPs on soil microbes exists, but the quantification of these effects with respect to functional microbial diversity, particularly connected with organic matter turnover and nutrient uptake into plants, remains open for further investigation. The improvement of nutrient capture from soil by plants through genetic manipulations might have a more important effect on nutrient dynamics, if methods to improve nutrient uptake strategies of plants by genetic engineering are successful (Hirsch and Sussman, 1999).

References

AgrEvo GmbH, Bund für Lebensmittelrecht und Lebensmittelkunde e.V. (BLL), Monsanto (Germany) GmbH, Novartis (Germany) GmbH (1998) *Kompendium Gentechnologie und Lebensmittel.* Genius GmbH, Darmstadt, Germany.

Ahmad, I., Malloch, D. (1995) Interaction of soil microflora with the bioherbicide phosphinothricin. *Agric. Ecosyst. Environ.* **54**: 165–174.

Allen-King, R.M., Butler, B.J., Reichert, B. (1995) Fate of the herbicide Basta-ammonium in the sandy, low-organic-carbon aquifer at CFB Borden, Ontario, Canada. *J. Cont. Hydrol.* **18**: 161–179.

Amann, R.I., Ludwig, W., Schleifer, K.-H. (1995) Phylogenetic identification and *in situ* detection of individual microbial cells without cultivation. *Microbiol. Rev.* **59**: 143–169.

Ahrenholtz, I., Harms, K., de Vries, J., Wackernagel, W. (2000) Increased killing of *Bacillus subtilis* on the hair roots of transgenic T4 lysozyme-producing potatoes. *Appl. Environ. Microbiol.* **66**: 1862–1865.

Asaka, O., Shoda, M. (1996) Biocontrol of *Rhizoctonia solani* damping-off of tomato with *Bacillus subtilis* RB14. *Appl. Environ. Microbiol.* **62**: 4081–4085.

Bartsch, K., Tebbe, C.C. (1989) Initial steps in the degradation of phosphinothricin (Basta) by soil bacteria. *Appl. Environ. Microbiol.* 55: 711–716.

Bayer, E., Gugel, K.H., Hägele, K., Hagenmaier, H., Jessipow, S., König, W.A., Zähner, H. (1972) Phosphinothricin and phosphinothricyl-alanyl-alanin. *Helv. Chim. Acta* 55: 224–239.

Becker, R., Ulrich, A., Hedtke, C., Honermeier, B. (2001) Einfluss des Anbaus von transgenem herbizidresistentem Raps auf das Agrar-Ökosystem. *Bundesgesundheitsbl-Gesundheitsforsch-Gesundheitsschutz* **44**: 159–167.

Berg, G., Roskot, N., Steidle, A., Eberl, L., Zock, A., Smalla, K. (2002) Plant-dependent genotypic and phenotypic diversity of antagonistic rhizobacteria isolated from different *Verticillium* host plants. *Appl. Environ. Microbiol.* **68**: 3328–3338.

Bertolla, F., Simonet, P. (1999) Horizontal gene transfer in the environment: natural transformation as a putative process for gene transfer between transgenic plants and micro-organisms. *Res. Microbiol.* **150**: 375–384.

Bertolla, F., van Gijsegem, F., Nesme, X., Simonet, P. (1997) Conditions for natural transformation of *Ralstonia solanacearum*. *Appl. Environ. Microbiol.* **63**: 4965–4968.

Bertolla, F., Frostegard, A., Brito, B., Nesme, X., Simonet, P. (1999) During infection of its host, the plant pathogen *Ralstonia solanacearum* naturally develops a state of competence and exchanges genetic material. *MPMI* **12**: 467–472.

Blum, S.E.A., Lorenz, M.G., Wackernagel, W. (1997) Mechanisms of retarded DNA degradation and prokaryotic origin of DNase in nonsterile soils. *Syst. Appl. Microbiol.* **20**: 513–521.

Chamier, B., Lorenz, M.G, and Wackernagel, W. (1993) Natural transformation of *Acinetobacter calcoaceticus* by plasmid DNA adsorbed on sand and groundwater aquifer material. *Appl. Environ. Microbiol.* **59**: 1662–1667.

Chander, K., Goyal, S., Mundra, M.C., Kapoor, K.K. (1997) Organic matter, microbial biomass and enzyme activity of soils under different crop rotations in the tropics. *Biol. Fert. Soils* **24**: 306–310.

Coleman, K., Jenkinson, D.S. (1999) *ROTHC-26.3 A model for the turnover of carbon in soil; model description and windows user guide*. IACR-Rothamsted: Harpenden, UK.

Cowgill, S. (2002) *The effect of transgenic potatoes on non-target organisms.* http://www.bangor.ac.uk/psp/pot_nov00/tech/techhm.htm

Crecchio, C., Stotzky, G. (1998) Binding of DNA on humic acids: Effect on transformation of *Bacillus subtilis* and resistance to DNase. *Soil Biol. Biochem.* **30**: 1061–1067.

Crecchio, C., Stotzky, G. (2001) Biodegradation and insecticidal activity of the toxin from *Bacillus thuringiensis* subsp. *kurstaki* bound on complexes of montmorillonite-humic acids-Al hydroxypolymers. *Soil Biol. Biochem.* **33**: 573–581.

De la Fuente, J.M., Ramirez-Rodriguez, V., Cabrera-Ponce, J.L., Herrera-Estrella, L. (1997) Aluminium tolerance in transgenic plants by alteration of citrate synthesis. *Science* **276**: 1566–1568.

Delhoize, E., Hebb, D.M., Ryan, P.R. (2001). Expression of a Pseudomonas aeruginosa citrate synthase gene in tobacco is not associated with either enhanced citrate accumulation or efflux. *Plant Physiol.* **125**: 2059–2067.

Demanèche, S., Bertolla, F., Buret, F., Nalin, R., Sailand, A., Auriol, P., Vogel, T.M., Simonet, P. (2001a) Laboratory-scale evidence for lightning-mediated gene transfer in soil. *Appl. Environ. Microbiol.* **67**: 3440–3444.

Demanèche, S., Jocteur-Monrozier, L., Quiquampoix, H., Simonet, P. (2001b) Evaluation of biological and physical protection against nuclease degradation of clay-bound plasmid DNA. *Appl. Environ. Microbiol.* **67**: 293–299.

Demanèche, S., Kay, E., Gourbière, F., Simonet, P. (2001c) Natural transformation of *Pseudomonas fluorescens* and *Agrobacterium tumefaciens* in soil. *Appl. Environ. Microbiol.* **67**: 2617–2621.

De Vries, J., Wackernagel, W. (1998) Detection of *npt*II (kanamycin resistance) genes in genomes of transgenic plants by marker-rescue transformation. *Mol. Gen. Genet.* **257**: 606–613.

De Vries, J., Wackernagel, W. (2002) Integration of foreign DNA during natural transformation of *Acinetobacter* sp. by homology-facilitated illegitimate recombination. *Proc. Natl Acad. Sci.* **99**: 2094–2099.

De Vries, J., Harms, K., Broer, I., Kriete, G., Mahn, A., Düring, K., Wackernagel, W. (1999) The bacteriolytic activity in transgenic potatoes expressing a chimeric T4 lysozyme gene and the effect of T4 lysozyme on soil and phytopathogenic bacteria. *Syst. Appl. Microbiol.* **22**: 280–286.

De Vries, J., Meier, P., Wackernagel, W. (2001) The natural transformation of the soil bacteria *Pseudomonas stutzeri* and *Acinetobacter* sp. by transgenic plant DNA strictly depends on homologous sequences in the recipient cells. *FEMS Microbiol. Lett.* **195**: 211–215.

Donegan, K.K., Palm, C.J., Fieland, V.J., Porteous, L.A., Ganio, L.M., Schaller, D.L., Bucao, L.Q., Seidler, R.J. (1995) Changes in levels, species and DNA fingerprints of soil micro-organisms associated with cotton expressing the *Bacillus thuringiensis* var. *kurstaki* endotoxin. *Appl. Soil Ecol.* **2**: 111–124.

Dröge, M., Pühler, A., Selbitschka, W. (2000) Phenotypic and molecular characterization of conjugative antibiotic resistance plasmids isolated from bacterial communities of activated sludge. *Mol. Gen. Genet.* **263**: 471–482.

Dunfield, K.E., Germida, J.J. (2001) Diversity of bacterial communities in the rhizosphere and root interior of field-grown genetically modified *Brassica napus*. *FEMS Microbiol. Ecol.* **38**: 1–9.

Düring, K. (1993) A tightly regulated system for overproduction of bacteriophage T4-lysozyme in *Escherichia coli*. *Protein Expr. Purif.* **4**: 412–416.

Düring, K., Porsch, P., Fladung, M., Lörz, H. (1993) Transgenic potato plants resistant to the phytopathogenic bacterium *Erwinia carotovora*. *Plant J.* **3**: 587–598.

Escher, N., Kach, B., Nentwig, W. (2000) Decomposition of transgenic *Bacillus thuringiensis* maize by micro-organisms and wood lice *Porcellio scaber* (crustacea: isopoda). *Basic Appl. Ecol.* **1**: 161–169.

Falloon, P., Smith, P. (2002) Simulating SOM changes in long-term experiments with RothC and CENTURY: model evaluation for a regional scale application. *Soil Use Management* **18**: 101–111.

Florack, D., Allefs, S., Bollen, R., Bosch, D., Visser, B., Stiekema, W. (1995) Expression of giant silkmoth cecropin B genes in tobacco. *Transgenic Res.* **4**: 132–141.

Gallori, E., Bazzicalupo, M., Dal Canto, L., Fani, R., Nannipieri, P., Vettori, C. (1994) Transformation of *Bacillus subtilis* by DNA bound on clay in non-steile soil. *FEMS Microbiol. Ecol.* **15**: 119–126.

Gebhard, F., Smalla, K. (1998) Transformation of *Acinetobacter* sp. strain BD413 by transgenic sugar beet DNA. *Appl. Environ. Microbiol.* **64**: 1550–1554.

Gebhard, F., Smalla, K. (1999) Monitoring field releases of genetically modified sugar beets for persistence of transgenic plant DNA and horizontal gene transfer. *FEMS Microbiol. Ecol.* **28**: 261–272.

Gerzabek, M.H., Pichlmayer, F., Kirchmann, H., Haberhauer, G. (1997) The response of soil organic matter amendments in a long-term experiment at Ultuna, Sweden. *Europ. J. Soil Sci.* **48**: 273–282.

Gerzabek, M.H., Haberhauer, G., Kandeler, E., Sessitsch, A., Kirchmann, H. (2002) Response of organic matter pools and enzyme activities in particle size fractions to organic amendments in a long-term field experiment. In: Violante, A., Huang, P.M., Bollag, J.-M., Gianfreda, L. (eds) *Soil Mineral-Organic Matter-Micro-organism Interactions and Ecosystem Health, Developments in Soil Science 28B*, pp. 329–344, Elsevier: Amsterdam.

Gomes, N.C.M., Heuer, H., Schönfeld, J., Costa, R., Hagler-Mendonca, L., Smalla, K. (2001) Bacterial diversity of the rhizosphere of maize (*Zea mays*) grown in tropical soil studied by temperature gradient gel electrophoresis. *Plant Soil* **232**: 167–180.

Grayston, S.J., Wang, S.Q., Campbell, C.D., Edwards, A.C. (1998) Selective influence of plant species on microbial diversity in the rhizosphere. *Soil Biol. Biochem.* **30**: 369–378.

Griffiths, B.S., Geoghegan, I.E., Robertson, W.M. (2000) Testing genetically engineered potato, producing the lectins GNA and Con A, on non-target soil organisms and processes. *J. Appl. Ecol.* **37**: 159–170.

Gyamfi, S., Pfeifer, U., Stierschneider, M., Sessitsch, A. (2002) Effects of transgenic glufosinate-tolerant oilseed rape (*Brassica napus*) and the associated herbicide application on eubacterial and *Pseudomonas* communities in the rhizosphere. *FEMS Microbiol. Ecol.* **41**: 181–190.

Haney, R.L., Senseman, S.A., Hons, F.M. (2002) Effect of roundup ultra on microbial activity and biomass from selected soil. *J. Environ. Qual.* **31**: 730–735.

Heinonen-Tanski, H. (1989) The effects of temperature and liming on degradation of glyphosate in two forest soils. *Soil Biol. Biochem.* **21**: 313–317.

Heuer, H., Smalla, K. (1999) Bacterial phyllosphere communities of *Solanum tuberosum* L. and T4-lysozyme-producing transgenic variants. *FEMS Microbiol. Ecol.* **28**: 357–371.

Heuer, H., Kroppenstedt, R.M., Lottmann, J., Berg, G., Smalla, K. (2002a) Effects of T4 lysozyme release from transgenic potato roots on bacterial rhizosphere communities are negligible relative to natural factors. *Appl. Environ. Microbiol.* **68**: 1325–1335.

Heuer, H.H., Krögerrecklenfort, E., Egan, S. *et al.* (2002b) Gentamycin resistance genes in environmental bacteria: Prevalence and transfer. *FEMS Microbiol. Ecol.* **42**: 289–302.

Hirsch, R.E., Sussman, M.R. (1999) Improving nutrient capture from soil by the genetic manipulation of crop plants. *Tibtech* **17**: 356–361.

Hoerlein, G. (1994) Glufosinate (phosphinothricin), a natural amino acid with unexpected herbicidal properties. *Rev. Environ. Contam. Toxicol.* **138**: 73–145.

Huang, Y., Nordeen, R.O., Di, M., Owens, L.D., McBeath, J.H. (1997) Expression of an engineered cecropin gene cassette in transgenic tobacco plants confers disease resistance to *Pseudomonas syringae* pv. *tabaci*. *Phytopathology* **87**: 494–499.

Hultmark, D., Steiner, H., Rasmuson, T., Boman, H.G. (1980) Insect immunity: purification and properties of three inducible bactericidal proteins from hemolymph of immunized pupae of *Hyalophora cecropia*. *Eur. J. Biochem.* **106**: 7–16.

Hultmark, D., Engström, A., Bennich, H., Kapur, R., Boman, H.G. (1982) Insect immunity. Isolation and structure of cecropin D and four minor antibacterial components from cecropia pupae. *Eur. J. Biochem.* **127**: 207–217.

James, C. (2001) Global review of commercialized transgenic crops: 2000. ISAAA Briefs No. 2: Preview. ISAAA, Ithaka, NY. Cited in: Lövei, G.L. (2001) Ecological risks and benefits of transgenic plants. *New Zealand Plant Protect.* **54**: 93–100.

Jaynes, J.M., Xanthopoulos, K.G., Destéfano-Beltrán, L., Dodds, J.H. (1987) Increasing bacterial disease resistance utilizing antibacterial genes from insects. *BioEssays* **6**: 263–270.

Jaynes, J.M., Nagpala, P., Destéfano-Beltrán, L., Huang, J., Kim, J., Denny, T., Cetiner, S. (1993) Expression of a cecropin B lytic peptide analog in transgenic tobacco transfers enhanced resistance to bacterial wilt caused by *Pseudomonas solanacearum*. *Plant Sci.* **89**: 43–53.

Kay, E., Vogel, T.M., Bertolla, F., Nalin, R., Simonet, P. (2002) *In situ* transfer of antibiotic resistance genes from transgenic (transplastomic) tobacco plants to bacteria. *Appl. Environ. Microbiol.* **68**: 3345–3351.

Keppel, M. (2000) *Genetic transformation of potato plants (*Solanum tuberosum *L.) with bacterial resistance enhancing lytic peptides*. Ph.D. thesis, University of Vienna, Austria.

Khanna, M., Stotzky, G. (1992) Transformation of *Bacillus subtilis* by DNA bound on montmorillonite and effect of DNase on the transforming ability of bound DNA. *Appl. Environ. Microbiol.* **58**: 1930–1939.

Kopper, E. (1999) *Transformation of potato (*Solanum tuberosum *L.) in order to obtain increased bacterial resistance*. Ph.D. thesis, Agricultural University Vienna, Austria.

Kriete, G., Broer, I. (1996) Influence of the herbicide phosphinothricin on growth and nodulation capacity of *Rhizobium meliloti*. *Appl. Microbiol. Biotechnol.* **46**: 580–586.

Leeflang, P., Smit, E., Glandorf, D.C.M., van Hannen, E.J., Wernars, K. (2002) Effects of *Pseudomonas putida* WCS358r and its genetically modified phenanzine producing derivative on the *Fusarium* population in a field experiment, as determined by 18S rDNA analysis. *Soil Biol. Biochem.* **34**: 1021–1025.

Levy, S.B. (1997) Antibiotic resistance: an ecological imbalance. In: *Antibiotic Resistance: Origins, Evolution, Selection and Spread*, pp. 1–14, Ciba Foundation Symposium 207, Wiley: Chichester.

Lorenz, M.G., Wackernagel, W. (1987) Adsorption of DNA to sand and variable degradation rates of adsorbed DNA. *Appl. Environ. Microbiol.* **53**: 2948–2952.

Lorenz, M.G., Wackernagel, W. (1994) Bacterial gene transfer by natural genetic transformation in the environment. *Microbiol. Rev.* **58**: 563–602.

Lottmann, J., Berg, G. (2001) Phenotypic and genotypic characterization of antagonistic bacteria associated with roots of transgenic and non-transgenic potato plants. *Microbiol. Res.* **156**: 75–82.

Lottmann, J., Heuer, H., Smalla, K., Berg, G. (1999) Influence of transgenic T4-lysozyme-producing potato plants on potentially beneficial plant-associated bacteria. *FEMS Microbiol. Ecol.* **29**: 365–377.

Lottmann, J., Heuer, H., de Vries, J., Mahn, A., Düring, K., Wackernagel, W., Smalla, K., Berg, G. (2000) Establishment of introduced antagonistic bacteria in the rhizosphere of transgenic potatoes and their effect on the bacterial community. *FEMS Microbiol. Ecol.* **33**: 41–49.

Lövei, G.L. (2001) Ecological risks and benefits of transgenic plants. *New Zealand Plant Protect.* **54**: 93–100.

Lukow, T., Dunfield, P.F., Liesack, W. (2000) Use of the T-RFLP technique to assess spatial and temporal changes in the bacterial community structure within an agricultural soil planted with transgenic and non-transgenic potato plants. *FEMS Microbiol. Ecol.* **32**: 241–247.

Lynch, J.M. (1990) *The Rhizosphere*. Wiley-Interscience: Chichester, England.

Lynch, J.M., de Leij, F.A.A.M., Whipps, J.M., Bailey, M.J. (1994) Impact of GEMMOs on rhizosphere population dynamics. In: O'Gara, F. and Dowling, D.N. (eds) *Molecular Ecology of Rhizosphere Micro-organisms*, pp. 49–55, VCH: Weinheim, Germany.

Metting, F.B., Smith, J.L., Amthor, J.S., Izaurralde, R.C. (2001) Science needs and new technologies for increasing soil carbon sequestration. *Climatic Change* **51**: 11–34.

Nielsen, K.M., van Weerelt, D.M., Berg, T.N., Bones, A.M., Hagler, A.N., van Elsas, J.D. (1997a) Natural transformation and availability of transforming chromosomal DNA to *Acinetobacter calcoaceticus* in soil microcosms. *Appl. Environ. Microbiol.* **63**: 1945–1952.

Nielsen, K.M., Gebhard, F., Smalla, K., Bones, A.M., van Elsas, J.D. (1997b) Evaluation of possible horizontal gene transfer from transgenic plants to the soil bacterium *Acinetobacter calcoaceticus* BD413. *Theor. Appl. Genet.* **95**: 815–821.

Nielsen, K.M., Bones, A.M., van Elsas, J.D. (1997c) Induced natural transformation of *Acinetobacter calcoaceticus* in soil microcosms. *Appl. Environ. Microbiol.* **63**: 3972–3977.

Nielsen, K.M., Bones, A.M., Smalla, K., van Elsas, J.D. (1998) Horizontal gene transfer from transgenic plants to terrestrial bacteria – a rare event? *FEMS Microbiol. Rev.* **22**: 79–103.

Nielsen, K. M., Smalla, K., van Elsas, J.D. (2000a) Natural transformation of *Acinetobacter* sp. strain BD413 with cell lysates of *Acinetobacter* sp., *Pseudomonas fluorescens*, and *Burkholderia cepacia* in soil microcosms. *Appl. Environ. Microbiol.* **66**: 206–212.

Nielsen, K. M., van Elsas, J. D., Smalla, K. (2000b) Transformation of *Acinetobacter* sp. strain BD413 (pFG4D*nptII*) with transgenic plant DNA in soil microcosms and effects of kanamycin on selection of transformants. *Appl. Environ. Microbiol.* **66**: 1237–1242.

Niemeyer, J., Gessler, F. (2002) Determination of free DNA in soils. *J. Plant Nutr. Soil Sci.* **165**: 121–124.

Nordeen, R.O., Sinden, S.L., Jaynes, J.M., Owens, L.D. (1992) Activity of cecropin SB37 against protoplasts from several plant species and their bacterial pathogens. *Plant Sci.* **82**: 101–107.

O'Donell, A.G., Seasman, M., Macrae, A., Waite, I., Davies, J.T. (2001) Plants and fertilizers as drivers of change in microbial community structure and function in soils. *Plant Soil* **232**: 135–145.

Paget, E., Simonet, P. (1994) On the track of natural transformation in soil. *FEMS Microbiol. Ecol.* **15**: 109–118.

Paget, E., Simonet, P. (1997) Development of engineered genomic DNA to monitor the natural transformation of *Pseudomonas stutzeri* in soil-like microcosms. *Can. J. Microbiol.* **43**: 78–84.

Perreten, V., Schwarz, F., Cresta, L., Boeglin, M., Dasen, G., Teuber, M. (1997) Antibiotic resistance spread in food. *Nature* **389**: 801–802.

Persson, J., Kirchmann, H. (1994) Carbon and nitrogen in arable soils as affected by supply of N fertilizers and organic manures. *Agric. Ecosyst. Environ.* **51**: 249–255.

Pietramellara, G., Dal Canto, L., Vettori, C., Gallori, E., Nannipieri, P. (1997) Effects of air-drying and wetting cycles on the transforming ability of DNA bound on clay minerals. *Soil Biol. Biochem.* **29**: 55–61.

Recorbet, G., Picard, C., Normand, P., Simonet, P. (1993) Kinetics of the persistence of chromosomal DNA from genetically engineered *Escherichia coli* introduced into soil. *Appl. Environ. Microbiol.* **59**: 4289–4294.

Romanowski, G., Lorenz, M.G., Wackernagel, W. (1991) Adsorption of plasmid DNA to mineral surfaces and protection against DNase I. *Appl. Environ. Microbiol.* **57**: 1057–1061.

Romanowski, G., Lorenz, M.G., Sayler, G., Wackernagel, W. (1992) Persistence of free plasmid DNA in soil monitored by various methods, including a transformation assay. *Appl. Environ. Microbiol.* **58**: 3012–3019.

Salyers, A. (1996) The real threat from antibiotics. *Nature* **384**: 304.

Schlüter, K., Fütterer, J., Potrykus, I. (1995) Horizontal gene transfer from a transgenic potato line to a bacterial pathogen (*Erwinia chrysanthemi*) occurs – if at all – at an extremely low frequency. *Nature Biotechnol.* **13**: 94–98.

Schmalenberger, A., Tebbe, C.C. (2002) Bacterial community composition in the rhizosphere of a transgenic, herbicide-resistant maize (*Zea mays*) and comparison to its non-transgenic cultivar *Bosphore*. *FEMS Microbiol. Ecol.* **40**: 29–37.

Sessitsch, A., Kan, F.-Y., Pfeifer, U. (2002) Diversity and community structure of culturable *Bacillus* spp. populations in the rhizospheres of transgenic potatoes expressing the lytic peptide cecropin B. *Appl. Soil Ecol.* **22**: 149–158.

Siciliano, S.D., Theoret, C.M., de Freitas, J.R., Hucl, P.J., Germida, J.J. (1998) Differences in the microbial communities associated with the roots of different cultivars of canola and wheat. *Can. J. Microbiol.* **44**: 844–851.

Siciliano, S.D., Germida, J.J. (1999) Taxonomic diversity of bacteria associated with the roots of field-grown transgenic *Brassica napus* cv. Quest, compared to the non-transgenic *B. napus* cv. Excel and *B. rapa* cv. Parkland. *FEMS Microbiol. Ecol.* **29**: 263–272.

Sikorski, J., Graupner, S., Lorenz, M.G., Wackernagel, W. (1998) Natural transformation of *Pseudomonas stutzeri* in a non-sterile soil. *Microbiology* **144**: 569–576.

Sikorski, J., Teschner, N., Wackernagel, W. (2002) Highly different levels of natural transformation are associated with genomic subgroups within a local population of *P. stutzeri* from soil. *Appl. Environ. Microbiol.* **68**: 865–873.

Smalla, K., van Overbeek, L.S., Pukall, R., van Elsas, J.D. (1993) Prevalence of *npt*II and Tn*5* in kanamycin resistant bacteria from different environments. *FEMS Microbiol. Ecol.* **13**: 47–58.

Smalla, K., Heuer, H., Götz, A., Niemeyer, D., Krögerrecklenfort, E., Tietze, E. (2000) Exogenous isolation of antibiotic resistance plasmids from piggery manure slurries reveals a high prevalence and diversity of IncQ-like plasmids. *Appl. Environ. Microbiol.* **66**: 4854–4864.

Smalla, K., Wieland, G., Buchner, A., Zock, A., Parzy, J., Kaiser, S., Roskot, N., Heuer, H., Berg, G. (2001) Bulk and rhizosphere soil bacterial communities studied by denaturing gradient gel electrophoresis: plant-dependent enrichment and seasonal shifts revealed. *Appl. Environ. Microbiol.* **67**: 4742–4751.

Stahl, D.J., Maser, A., Dettendorfer, J., Hornschulte, B., Thomzik, J.E., Hain, R., Nehls, R. (1998) *Increased fungal resistance of transgenic plants by heterologous expression of bacteriophage T4-lysozyme gene*. 7th Int. Congress on Plant Pathology, Abstract: Edinburgh.

Stotzky, G. (1986) Influence of soil mineral colloids on metabolic processes, growth, adhesion, and ecology of microbes and viruses. In: Soil Science Society of America (ed) *Interactions of soil minerals with natural organics and microbes.* SSSA Special Publication. Soil Science Society of America: Madison, WI.

Stotzky, G. (2000) Persistence and biological activity in soil of insecticidal proteins from *Bacillus thuringiensis* and of bacterial DNA bound on clays and humic acids. *J. Env. Qual.* **29**: 691–705.

Stotzky, G. (2002) Clays and humic acids affect the persistence and biological activity of insecticidal proteins from *Bacillus thuringiensis* in soil. In: Violante, A., Huang, P.M., Bollag, J.M., Gianfreda, L. (eds) *Soil mineral–organic matter–micro-organism interactions and ecosystem health, Developments in Soil Science 28B*, pp. 1–16, Elsevier: Amsterdam.

Sturz, A.V., Christie, B.R., Nowak, J. (2000) Bacterial endophytes: potential role in developing sustainable systems of crop production. *Crit. Rev. Pl. Sci.* **19**: 1–30.

Tebbe, C.C., Reber, H.H. (1988) Utilization of the herbicide phosphinothricin as a nitrogen source by soil bacteria. *Appl. Microbiol. Biotechnol.* **29**: 103–105.

Tebbe, C.C., Reber, H.H. (1991). Degradation of [^{14}C]phosphinothricin (Basta) in soil under laboratory conditions: effects of concentration and soil amendments on $^{14}CO_2$ production. *Biol. Fertil. Soils* **11**: 62–67.

Thomas, C.M., Smalla, K. (2000) Trawling the horizontal gene pool. *Microbiology-Today* **27**: 24–27.

Troxler, J., Zala, M., Natsch, A., Moënne-Loccoz, Y., Keel, C., Défago, G. (1997) Predominance of nonculturable cells of the biocontrol strain *Pseudomonas fluorescens* CHAO in the surface horizon of large outdoor lysimeters. *Appl. Environ. Microbiol.* **63**: 3776–3782.

Trudel, J., Potvin, C., Asselin, A. (1992) Expression of active hen egg white lysozyme in transgenic tobacco. *Plant Sci.* **87**: 55–67.

Tschäpe, H. (1994) The spread of plasmids as a function of bacterial adaptability. *FEMS Microbiol. Ecol.* **15**: 23–32.

Tsugita, A., Inouye, M., Terzaghi, E. and Streisinger, G. (1968) Purification of the bacteriophage T4 lysozyme. *J. Biol. Biochem.* **243**: 391–397.

USDA (2002) http://www.aphis.usda.gov/ppq/biotech

van Dillewijn, P., Villadas, P.J., Toro, N. (2002) Effect of a *Sinorhizobium meliloti* strain with a modified *putA* gene on the rhizosphere microbial community of alfalfa. *Appl. Environ. Microbiol.* **68**: 4201–4208.

van Overbeek, L.S., Eberl, L., Giskov, M., Molin, S., van Elsas, J. D. (1995) Survival of, and induced stress resistance in, carbon-starved *Pseudomonas fluorescens* cells residing in soil. *Appl. Environ. Microbiol.* **61**: 4202–4208.

Wardle, D.A., Yeates, G.W., Nicholson, K.S., Bonner, K.I., Watson, R.N. (1999) Response of soil microbial biomass dynamics, activity and plant litter decomposition to agricultural intensification over a seven-year period. *Soil Biol. Biochem.* **31**: 1707–1720.

Watrud, L.S. (2000) Genetically engineered plants in the environment – applications and issues. In: Subba Rao, N.S., Dommergues, Y.R. (eds) *Microbial interactions in agriculture and forestry*, Vol. 2, pp. 61–81. Science Publishers: Enfield/USA.

Widmer, F., Seidler, R.J., Watrud, L.S. (1996) Sensitive detection of transgenic plant marker gene persistence in soil microcosms. *Mol. Ecol.* 5: 603–613.

Widmer, F., Seidler, R.J., Donegan, K.K., Reed, G.L. (1997) Quantification of transgenic plant marker gene persistence in the field. *Mol. Ecol.* **6**: 1–7.

Wilhelm, E., Arthofer, W., Schafleitner, R., Krebs, B. (1998) *Bacillus subtilis* an endophyte of chestnut (*Castanea sativa*) as antagonist against chestnut blight (*Cryphonectria parasitica*). *Plant Cell, Tissue Organ. Cult.* **52**: 105–108.

Witte, W. (1998) Medical consequences of antibiotic use in agriculture. *Science* **279**: 996–997.

Wohlleben, W., Alijah, R., Dorendorf, J., Hillemann, D., Nussbaumer, B., Pelzer, S. (1992) Identification and characterization of phosphinothricin-tripeptide biosynthetic genes in *Streptomyces viridochromogenes*. *Gene* **115**: 127–132.

Yang, C-H., Crowley, D.E. (2000) Rhizosphere microbial community structure in relation to root location and plant iron nutritional status. *Appl. Environ. Microbiol.* **66**: 345–351.

5

Fungal endophytes: hitch-hikers of the green world

K. Saikkonen, M. Helander and S.H. Faeth

5.1 Introduction

By most definitions, fungal endophytes are fungi that live for all, or at least a significant part, of their life cycle asymptomatically and intercellularly within plant tissues (Wilson, 1995). Endophytes are thought to have evolved from parasitic or pathogenic fungi via an extension of latency periods and associated reduction of virulence (e.g., Carroll, 1988). Endophytic fungi are ubiquitous and abundant residents of plants, often more so than pathogens or mycorrhizae (Arnold *et al.*, 2001; Carroll, 1988, 1991). Non-systemic and horizontally transmitted (by spores), endophytes having been found from every plant species examined so far; systemic (growing throughout the host plant) and vertically transmitted (via host seeds) endophytes are less common, but nonetheless have been isolated from the majority of cool-season and some warm-season grass species (Bernstein and Carroll, 1977; Clay, 1988; Clay and Schardl, 2002; Faeth and Hammon, 1997a; Fisher, 1996; Fröhlich *et al.*, 2000; Hawksworth, 1988, 1991; Helander *et al.*, 1994; Lodge *et al.*, 1996; Petrini *et al.*, 1982; Rajagopal and Suryanarayanan, 2000; Rodrigues, 1994, 1996; Saikkonen *et al.*, 2000; Schulz *et al.*, 1993).

Endophytic fungi have attracted increasing attention among biologists and agronomists since observations of toxicoses on livestock grazing on fungally infected forage in the USA and New Zealand in the mid 20th century. Livestock disorders were attributable to alkaloids produced by endophytes belonging to the tribe Balansiae (Ascomycotina), and accumulating evidence has shown that these alkaloids can negatively affect a wide variety of invertebrate and vertebrate herbivores (Bacon *et al.*, 1977; Ball *et al.*, 1993; Breen, 1994; Bush *et al.*, 1997; Durham and Tannenbaum, 1998; Hoveland, 1993; Porter, 1994; Schardl and Phillips, 1997; Siegel *et al.*, 1990; Wilkinson *et al.*, 2000). The majority of these studies have focused on the fungal endophyte *Neotyphodium* (formerly *Acremonium*) and its

Plant Microbiology, Michael Gillings and Andrew Holmes

sexual stage *Epichloë*, which are symbionts of many cool-season grasses of the subfamily Pooideae. In addition to increased herbivore resistance, these endophytes may also increase plant vigor and tolerance to a wide range of environmental conditions when compared to their endophyte-free conspecifics. Because asexual *Neotyphodium* endophytes in cool-season grasses are not known to sporulate in nature and rely upon plant reproduction (hyphae grow into seeds), endophytic fungi are generally considered as strongly mutualistic with their hosts (e.g., Clay, 1990; Clay and Schardl, 2002). The fungus provides a wide range of benefits to the plant while the plant provides nutrients, structural refuge and transmission to the next host plant generation.

Mycologists and ecologists working on endophyte–plant interactions readily accepted these interactions as mutualistic (Breen, 1994; Carroll, 1986, 1991; Clay, 1990; Malinowski and Belesky, 2000; Schardl, 2001). Indeed, the majority of the published studies on endophytes are still based on the conventional wisdom that these fungi are mutualistic symbionts of cool-season grasses (e.g., Clay and Schardl, 2002). However, an increasing number of recent studies, particularly with native grass– and tree–endophyte systems, have shown that endophyte–plant interactions may vary from antagonistic to mutualistic (see e.g. Ahlholm *et al.*, 2002a; Faeth, 2002; Faeth and Sullivan, 2003; Lehtonen *et al.*, unpublished; Saikkonen *et al.*, 1998). The reasons for the strong mutualistic stamp of endophytes are largely historical and system-based. Agronomic or economically important forage grasses with obvious toxic properties drew immediate interest from agronomists, whereas non-toxic infected grasses have only recently been studied (Faeth, 2002; Faeth and Bultman, 2002). Even now, only a minority of studies focus on the ecological importance and evolution of fungal endophytes outside of the agricultural arena of selectively bred, non-native pasture grasses (Faeth, 2002).

5.1.1 Endophyte – constructive or misleading concept?

The term 'endophyte' has been controversial and confusing since it started to appear commonly in the literature (Petrini, 1991; Wennstrom, 1994; Wilson, 1995). A common thread to all notions of endophytic fungi, however, is that these fungi live asymptomatically and internally within host plant tissues. Some arguments about the term resulted from whether latent pathogens or saprophytes, which spend part of their life cycle symptomless, should be considered endophytic (Wennstrom, 1994; Wilson, 1995). However, most now agree that endophytes are fungi that live internally and remain asymptomatic for at least part of their life cycle. More importantly, 'endophyte' became synonymous with 'mutualist', although not originally intended so (De Bary, 1866). More recent evidence suggests that endophytic fungal associations with their host plants encompass the full range of possible ecological interactions, from mutualism through to antagonism (Saikkonen *et al.*, 1998). Asymptomatic fungal infections have been detected from virtually every plant species examined to date, and identical fungal species have been characterized as both endophytic and pathogenic, and asexual (anamorphic) and sexual (teleomorphic) stages of the fungal species often are named differently (Kehr, 1992; Kehr and Wulf, 1993; Paavolainen *et al.*, 2000; Stone, 1987; Stone *et al.*, 1996; Williamson and Sivasithamparam, 1994).

It is increasingly evident that the direction of the interaction is labile in evolutionary time. For example, a mutation of a single locus may convert a fungal plant pathogen to a non-pathogenic endophytic symbiont (Freeman and Rodriguez, 1993). Furthermore, recent empirical evidence suggests that the relative costs and benefits of endophytes (including *Neotyphodium* endophytes in grasses), and hence the direction of the interaction with the host plant are conditional on available resources, life-history characters and genetic combinations of the host and the fungus (Ahlholm *et al.*, 2002 b, c; Cheplick *et al.*, 1989, 2000; Faeth and Bultman, 2002; Faeth and Sullivan, 2003; Faeth *et al.*, 2002; Lehtonen *et al.*, unpublished). Despite the asymptomatic lifestyle of the fungus, these costs are detectable in fitness-correlated plant characters, such as decreased biomass, clonal propagation and sexual reproduction (Ahlholm *et al.*, 2002c; Faeth and Sullivan, 2003). When these costs outweigh associated benefits, the fungus should be considered as a parasite, yet another ecological interaction. Although endophyte is a useful generic term, particularly when referring to completely symptomless *Neotyphodium* endophytes in pooid grasses, caution is advised in presuming the effects on the host. We suggest that endophyte–plant interactions should not be viewed as an entity in their own right, deserving of their own theory, but instead simply represent diverse examples, albeit intriguing ones, of evolutionary and ecological species interactions that vary in time and space.

Several recent papers have reviewed taxonomy, history, chemical ecology, and economic value of endophytes (Ball *et al.*, 1993; Breen, 1994; Clay, 1990; Clay and Schardl, 2002; Hoveland, 1993; Malinowski and Belesky, 2000; Schardl, 2001;Siegel and Bush, 1996). In this review, we address the ecology and the evolutionary strategies of the fungal symbionts. We propose that variation in sexual reproduction and modes of transmission causes variation in the symbiotic character of plant–fungus interaction. These differences among endophytes, in concert with biotic and abiotic environmental factors, are likely to have implications for genotypic diversity, generation time, spatial and temporal distribution of endophytes, and the nature of plant–fungus interactions. By emphasising that endophyte symbiosis is built upon the use and manipulation of other species in ways that increase an individual's fitness (see e.g., Thompson, 1994), our intention is to blur the unnecessary and potentially misleading dichotomy between current theory of the ecology and evolution of plant–endophyte symbiosis and plant–pathogen and plant–parasite interactions. Indeed, we emphasise that endophytes provide a fertile new ground for ecologists and evolutionary biologists interested in evolutionary processes.

5.2 Life history traits of endophytic fungi and host plants

5.2.1 Reproduction and transmission mode of fungi

Reproductive and transmission modes of endophytic fungi are often used synonymously to refer to how fungi spread within and among host plant populations. The key difference is, however, that reproduction mode refers to whether sex occurs or not, whereas mode of transmission describes only those mechanisms by which fungal infections are distributed. Endophytic fungi have two known transmission modes. Fungal hyphae may grow clonally into host seeds and are thereby transmitted to offspring of infected plants, or the fungus may produce spores (Carroll, 1988; Schardl *et al.*, 1994, 1997); the first is commonly termed as vertical and the latter as horizontal transmission of fungi. To fully understand the ecological and evolutionary consequences of these life history strategies, however, it is essential to recognise that fungi may produce either mitotic asexual or meiotic sexual spores. Thus, asexual reproduction of fungi is possible through vertical transmission via host seeds and horizontal transmission by spores, or possibly hyphae (Hamilton, 2002), whereas sexual reproduction requires production of sexual spores and is therefore always horizontal.

The reproductive and transmission mode of the fungus appears to be adapted to the life history of the host, particularly the growth pattern, expected lifetime, and age of sexual maturity of the plant. The vast majority of ecological literature on fungal endophytes associated with grasses has focused on two related fungal genera, *Neotyphodium* and *Epichloë*. Both of them occur as systemic infection (i.e,. growing throughout the host plant to developing inflorescence and seeds), and are transmitted vertically from maternal plants to offspring. *Neotyphodium* endophytes are assumed to be strictly vertically transmitted, and thus, considered 'trapped' in the host plant (see e.g., Clay and Holah, 1999; Wilkinson and Schardl, 1997). In contrast, *Epichloë* endophytes can also be transmitted sexually by spores (e.g., Clay and Schardl, 2002; Schardl, 2001). However, contagious spread should not be ruled out even in *Neotyphodium* endophytes because they produce asexual conidia on growth media (Glenn *et al.*, 1996) and on living plants (White *et al.*, 1996), and recent evidence indicates horizontal transmission in natural grass populations (Hamilton, 2002). Foliar endophytes of woody plants are non-systemic and transmitted horizontally by spores from plant to plant, usually causing highly restricted local infections. Endophytes of woody plants have also been documented in seeds and acorns (Petrini *et al.*, 1992; Wilson and Carroll, 1994), but vertical transmission of woody plant endophytes is probably rare (Saikkonen *et al.*, 1998). Although many tree-endophytes also produce asexual spores, horizontal transmission and sexual reproduction of some fungal species is likely to result in relatively higher genotypic diversity in populations of fungal endophytes in trees than in grasses.

Reproduction and transmission modes are well recognised as important factors related to the epidemiology and evolution of virulence in parasite and pathogen interactions (Bull *et al.*, 1991; Ewald, 1983; Herre, 1993; Herre *et al.*, 1999; Kover *et al.*, 1997; Kover and Clay, 1998; Lipsitch *et al.*, 1996). Mode of transmission, pattern of endophyte infections, architecture and lifespan of the host and the fungus likely affect the probability of endophyte–plant interactions occurring along the continuum from antagonistic to mutualistic interactions (Clay and Schardl, 2002; Saikkonen *et al.*, 1998). Saikkonen *et al.* (1998) suggested that exclusively vertically transmitted asexual grass endophytes are more likely to fall nearer the mutualistic end of the interaction continuum compared with mixed strategy (both vertically and horizontally) or only horizontally transmitted endophytes. However, strict vertical transmission does not guarantee mutualistic interactions with the host (Faeth and Bultman, 2002; Saikkonen *et al.*, 2002), as often assumed (e.g., Clay, 1998; Clay and Schardl, 2002).

5.2.2 Partner fidelity and evolution of virulence

Evolutionary theory predicts that vertical transmission should align the interests of partners toward mutualistic associations, whereas horizontal transmission, with increased opportunities for contagious spread, should promote the evolution of increased virulence (Ewald, 1987; Fine, 1975; Herre, 1993; Kover and Clay, 1998; Lipsitch *et al.*, 1995; Yamamura, 1993). Most empirical literature on endophytes generally supports this theory. Interactions between *Neotyphodium* endophytes and grasses represent an extreme form of partner fidelity, because the fungus spreads only with seeds of infected plants (at least presumed so), and thus the fungus is fully dependent on the host plant for survival and reproduction. *Neotyphodium* interactions are often found as mutualistic, lending support to the theory. In contrast, other grass endophytes, such as some *Epichloë* species, with mixed modes of transmission, may incur severe costs to the host by producing fungal sexual structures (stromata) in the plant inflorescences thereby decreasing seed production of the host plant. In general, endophytes that are transmitted horizontally by spores are only rarely mutualistic and often either neutral or parasitic (see e.g., Ahlholm *et al.*, 2002a; Carroll, 1988; Faeth, 2002; Saikkonen *et al.*, 1996), even though these endophytes too were originally proposed as defensive mutualists against rapidly evolving herbivores (Carroll, 1988).

Although vertically transmitted endophytes appear selected for lowered virulence, their interactions with grasses do not necessarily remain mutualistic and evolutionary stable for several reasons. First, cost and benefits of the partners are not symmetric, even in mutualistic plant–endophyte symbioses. The symbiosis is critical for long-term survival and reproduction of the fungus, which has presumably lost the independent phase of its life cycle. Alternatively, the fungus may only minimally increase plant survival and reproduction. Recent empirical evidence suggests in some environments and for some endophyte–host combinations, the endophyte reduces host growth and reproduction, further skewing the relative cost and benefits of association between partners (Ahlholm *et al.*, 2002c; Cheplick *et al.*, 1989, 2000; Faeth and Sullivan, 2003).

Another important destabilizing factor is the mismatch between genetic diversity of the host grass and asexual endophytes. Asexual, vertically transmitted endophytes, such as *Neotyphodium*, have greatly reduced genetic diversity, and in natural populations, exhibit very low gene flow (Sullivan, 2002; Sullivan and Faeth, 2001). Some genetic diversity is infused by hybridisation events with ancestral *Epichloë* species, but these events are very rare. In a recent study, Sullivan and Faeth (2001) found that three of four natural populations of Arizona fescue harboured only one or two haplotypes of *Neotyphodium*, whereas the fourth was more diverse with seven haplotypes. Thus, at each reproductive episode of the host grass, a more or less genetically uniform endophyte within its maternal plant is embedded in a constantly changing host seed genome, due to sexual recombination and contribution of widely dispersing pollen. Sullivan (2002) argued that this mismatch would select for endophytes that minimise costs to any given host genotype, rather than increased benefits, such that any given endophyte haplotype could generally survive unpredictable host genotypic backgrounds.

Increased benefits of endophyte are typically manifested through increased production or diversity of endophytic alkaloids, nitrogen-rich compounds with associated high costs (Faeth, 2002). The consequence of this strategy is that the majority of vertically transmitted endophytic associations with native grasses may only be weakly mutualistic, such that genetically limited haplotypes can persist over time in an ever-changing (genetically) host background. Endophyte–host associations that are strongly mutualistic (i.e., great benefits) may also be highly costly in terms of high or diverse alkaloid production. Indeed, this is borne out empirically. Faeth (2002) reviewed the literature and found far fewer native grass–endophyte associations that were highly toxic to herbivores than expected based upon estimated species of grasses infected with *Neotyphodium*, contrary to prevailing ideas of endophytic mutualisms. The strategy of many seedborne endophytes may be: do little harm but provide few benefits.

We would predict this scenario to change, however, if genetic diversity of asexual endophytes is more aligned with that of its host grass. In other words, when genetic diversity of the host grass is low, more mutualistic associations are expected because more constant plant genotypic backgrounds appear generation after generation. This appears exactly the case in agronomic grasses such as tall fescue and perennial ryegrass, well known for high and diverse alkaloid production that inhibits herbivores. Cultivars of these agronomic plants are highly inbred and exhibit much lower genetic diversity than their native counterparts (e.g., Saikkonen, 2000). For example, lack of genetic diversity of endophytic fungi inhabiting genetically narrow Kentucky 31 cultivar of tall fescue (Braverman, 1986), a widely used model system in endophyte studies, has been proposed to play a central role in this cultivar's great success in the United States (Ball *et al.*, 1993; Hoveland, 1993; Saikkonen, 2000). Furthermore, the cost of high alkaloid production in agronomic grasses is greatly offset by anthropogenic inputs of fertiliser and water (Faeth, 2002; Faeth and Bultman, 2002; Saikkonen *et al.*, 1998).

5.2.3 Asexual fungi – evolutionary dead ends?

Natural selection operates on heritable properties of individuals, and sexual reproduction promotes genetic variability through outcrossing, permitting rapid response to changing selection pressures (Williams, 1966, 1975). Sexual reproduction also removes accumulating deleterious mutations (Muller, 1964). Thus in theory, although loss of sexual reproduction may provide short-term benefits, it should increase probability of extinction of plant mutualistic fungi. Interestingly, however, in about 20% of all known fungi, including *Neotyphodium* endophytes, sex has never been observed in nature (Carlile *et al.*, 2001), and some may be very old (Blackwell, 2000; Freeman, 1904; Moon *et al.*, 2000). For example, darnel (*Lolium temulentum* L.), known from Roman times for its toxicity, has been found to harbour endophytic fungus *Neotyphodium occultans* (Freeman, 1904; Moon *et al.*, 2000).

There are two hypotheses that may explain how asexual endophytes may be able to cope with changing selection pressures. First, fitness of fungus is intertwined with the fitness of the host plant. Although only one fungal genotype is transmitted vertically to seed progeny, novel genetic combinations of vertically transmitted endophytes and their hosts are formed regularly through sexual reproduction of hosts. Thus, the fungus may be buffered by its outcrossing host that evolves rapidly enough in the face of environmental changes. Recent evidence also indicates the importance of interactive effects of fungal and plant genotypes, which affect the mutual fitness of the fungus and the host plant. Faeth *et al.* (2002) found that plant genotype rather than endophyte haplotype or environmental conditions mostly determined the mycotoxin levels within the examined population of Arizona fescue (*Festuca arizonica*). Second, genetic diversity of asexual, endophytic fungi can increase by means other than sexual reproduction. Molecular evidence suggests that some presumably asexual *Neotyphodium* lineages are hybrids of sexual and asexual endophytes (Clay and Schardl, 2002; Schardl, 2001; Schardl *et al.*, 1994; Tsai *et al.*, 1994; Wilkinson and Schardl, 1997). Molecular techniques are proving to have a revolutionary role in studies examining genetic diversity and specificity of endophytes and host plants. They are providing new insights into how plant resistance to certain species or genotypes of endophytic fungi is correlated (at phenotypic and genotypic levels) to other plant characteristics, such as growth and reproduction; and to what extent genotype–genotype interactions between host plant and fungus determine or constrain performance of partners under variable selection pressures.

5.3 Ecological consequences of endophyte infections

Life history traits, such as the mode of transmission, largely determine the spatial and temporal distribution of endophytes (Saikkonen *et al.*, 1998). Vertically transmitted grass-endophytes usually produce considerable mycelial biomass within the host, sometimes throughout the whole plant and always along the stem to developing flower heads and seeds. The generation time of vertically transmitted grass-endophytes is relatively long, often covering several grass generations. In contrast, abundance and diversity of horizontally transmitted endophytes in plants accumulate throughout the growing season, mostly in foliage (Faeth and Hammon, 1997a; Helander *et al.*, 1994). Individual endophyte infections are localised and the mycelial biomass remains very low relative to plant biomass. Spores are usually dispersed from senescent and abscised leaves, and thus the lifespan of foliage limits the lifespan of most endophytes inhabiting woody plants. Thus, the spatial and temporal patterns of endophytes differ not only between grasses and trees, but also between evergreen and deciduous trees.

5.3.1 Resource allocation among competing plant and fungal functions

Species interactions, even obligate mutualisms, are generally accepted as being based on mutual exploitation rather than reciprocal altruism (Doebeli and Knowlton 1998; Maynard Smith and Szathmáry 1995; Thompson 1994), with sanctions imposed against overexploitation by either partner (e.g., Denison 2000; Pellmyr *et al.*, 1996). Theory predicts that sporulating endophytes should range from negative to positive in their interactions with host plants, and that contagious spreading should favour less-mutualistic interactions (Bull *et al.*, 1991; Ewald, 1994; Saikkonen *et al.*, 1998). Empirical evidence supports this view (Ahlholm *et al.*, 2002a; Faeth and Hammon, 1996, 1997a, b; Faeth and Wilson, 1996; Gange, 1996; Preszler *et al.*, 1996; Saikkonen *et al.*, 1996; Wilson, 1995;

Wilson and Carroll, 1994; Wilson and Faeth, 2001). However, we argue that costs of systemic and vertically transmitted endophytes have been underestimated in past literature, where costs of harbouring endophytes were assumed to be negligible (e.g., Bacon and Hill, 1996). Clearly, systemic *Epichloë* endophytes that form stromata which surround and destroy developing inflorescences (choke disease) during the sexual phase of the fungus, are obviously costly and act parasitically (Breen, 1994; Clay, 1990). *Epichloë* infections, however, may also alter host allocation to reproduction (Meijer and Leuchtmann, 2001; Pan and Clay, 2002) and incur energetic costs to the plant, even when reproducing asexually (Ahlholm *et al.*, 2002c). Furthermore, evidence is accumulating that presumably strictly vertically transmitted and asexual *Neotyphodium* endophytes impose significant costs to the host grass, especially under low-resource conditions, such that costs outweigh benefits, in native grasses (Ahlholm *et al.*, 2002c; Faeth, 2002; Faeth and Sullivan, 2003). Faeth (2002) and Faeth and Bultman (2002) discussed the many costs of harbouring systemic endophytes, especially those that produce alkaloids. Recent empirical evidence confirms these high costs in certain resource environments (Faeth and Sullivan, 2003; Lehtonen *et al.*, unpublished; McCormick *et al.*, 2001). The cost of systemic endophytic infections in native grasses have been overlooked because the vast majority of studies have been conducted under enriched resource environments, either in agronomic environments or greenhouses using fertilised standard potting soil, and agronomic grass cultivars (Ahlholm *et al.*, 2002c; Faeth, 2002).

According to life history theory, competition for limited resources is assumed to result in negative correlations (i.e., trade-offs) between competing functions, such as growth, reproduction, maintenance, and defence (Bazzaz and Grace, 1997; Cody, 1966; Hamilton *et al.*, 2001; Reznick, 1985; Williams, 1966). In other words, when the amount of resources allocated to one function increases, the amount of resources available to other functions should decrease (Bell and Koufopanou, 1986; Stearns, 1989). Perhaps the most commonly studied trade-off is the one between growth and sexual reproduction. If systemic and vertically transmitted endophytes are similar to other inherited properties of host plants, then there may exist trade-offs between the endophyte infection and plant functions (*Figure 5.1*).

Recent evidence has demonstrated that this concept of trade-offs holds for endophyte–plant interactions. Benefits from endophytic fungi do not come without associated costs in terms of resource requirements of the fungus, and its associated alkaloids, even in the assumed mutualism of asexual grass endophytes. Indeed, these costs may outweigh their benefits in resource-limited conditions (*Figure 5.1b*; Ahlholm *et al.*, 2002c; Cheplick *et al.*, 1989, 2000; Faeth and Sullivan, 2003). For example, Ahlholm *et al.* (2002c) detected costs and benefits of endophytes with examined grass species, *Festuca pratensis* and *F. rubra*, in greenhouse experiments. However, costs and benefits are conditional on available resources and differ among the grass species. Costs, in terms of vegetative growth and reproduction, were detected particularly in poor resource conditions, but only in infected *Festuca pratensis*. Under resource limitation, infected *F. pratensis* plants produced fewer tillers and lower root and total biomass relative to uninfected plants. Seed production correlated negatively with vegetative growth of the plant. In contrast, and similar to previous studies, grass endophytes increased host plant

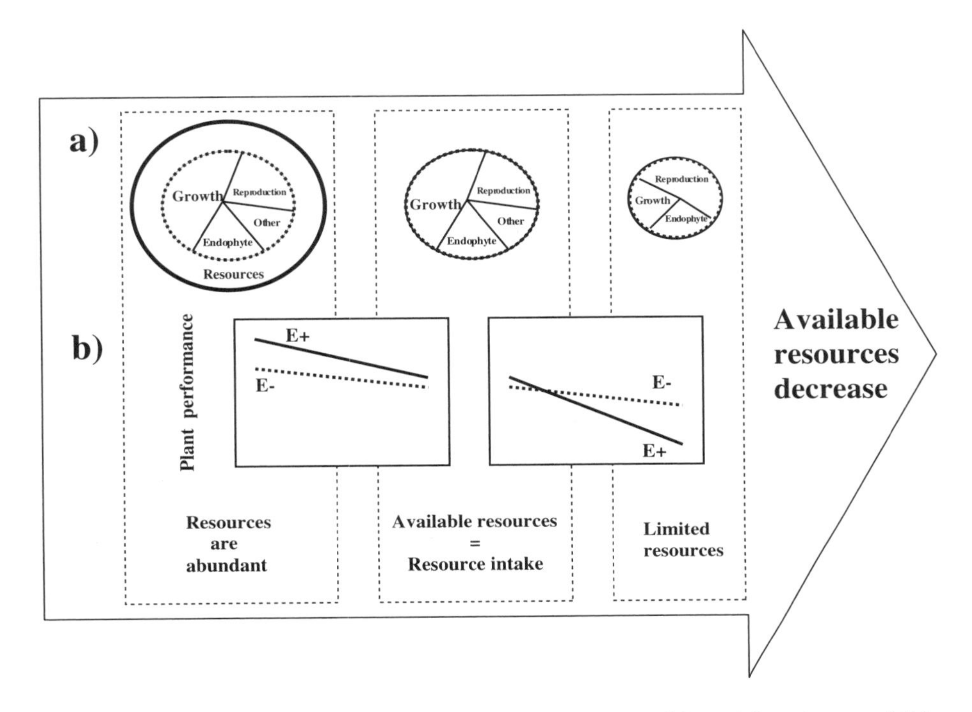

Figure 5.1. (a) Resource allocation among competing plant and fungal functions, and (b) performance of endophyte-free (E-) and endophyte infected (E+) plants in relation to resource availability.

growth in *Festuca rubra* if the resource supply was adequate (Bacon, 1994; Breen, 1994; Cheplick *et al.*, 1989; Clay, 1990; Elmi and West, 1995; Hill, 1994; Marks *et al.*, 1991; West, 1994). Different responses to infection may be related to variation in life history strategies and environmental requirements of the species. *F. pratensis* is more resource-demanding, and occurs typically in agronomic areas and nearby meadows, whereas *F. rubra* often grows in very resource-poor environments. Furthermore, the effects of infection on *F. pratensis* appeared to change over time, emphasising the importance of long-term experiments. Morse *et al.* (2002) found that infected Arizona fescue performed worse than uninfected plants under no or moderate water stress, but showed increased water-use efficiency and growth under prolonged and serve drought. Similarly, Lehtonen *et al.* (unpublished) found that aphid (*Rhopalosiphum padi*) mortality was highest, and reproduction lowest, on *Neotyphodium*-infected *F. pratensis* plants growing in high-nutrient soils, whereas aphid survival on infected plants was comparable to that on endophyte-free plants in low-nutrient soils. Thus, capacity of the fungus to produce nitrogen-based alkaloids probably increased when available nutrients are abundant.

5.3.2 Defensive mutualism or plant resistance to folivorous organisms

The majority of studies on endophyte–plant interactions have emphasised fungal-mediated plant resistance to herbivores. Some seedborne endophytes of grasses have indisputably negative effects on herbivores, especially in agronomic grasses, but these appear far fewer than expected in native grasses (Faeth, 2002). In contrast, most studies with horizontally transmitted endophytes (by spores) of woody plants have shown more variable effects, ranging from beneficial to deleterious on herbivores (Ahlholm *et al.*, 2002a; Faeth and Hammon, 1996, 1997a, b; Faeth and Wilson, 1996; Gange, 1996; Preszler *et al.*, 1996; Saikkonen *et al.*, 1996; Wilson, 1995; Wilson and Faeth, 2001). This variability is likely related to: (i) localized nature of these infections; (ii) absence or at least more variable mycotoxins produced by these fungi (Petrini *et al.*, 1992); or, (iii) that transmission is facilitated by herbivore damage and thus most of these endophytes may have evolved to tolerate or encourage herbivores (Faeth and Hammon, 1997a, b). However, variation in effects on herbivores may also indicate differences in plant quality for fungi and herbivores, without causal association between fungal infection and herbivore performance.

To examine this hypothesis, Ahlholm *et al.* (2002a) compared phenotypic and genetic correlations of fungal frequencies and performance of invertebrate herbivores growing on the same mature half-sib progenies of mountain birches (*Betula pubescens* ssp. *czerepanovii*) in two environments. They found little support for causal association between fungal frequencies and performance of herbivore species. Indeed, only the weak trend between the late-season herbivore, *Dineura pullior*, and the seasonally accumulating fungi suggested direct interactions between these partners, and can be interpreted as a consequence of higher probability of direct encounters between them (Ahlholm *et al.*, 2002a; Faeth and Hammon, 1997a, b; Saikkonen *et al.*, 1996). Direct effects of fungi, fungal-mediated changes in foliage quality, or fungi causing premature senescence and abscission of leaves, could then be expected to negatively impact late-season herbivore species. Instead, genetic correlations between the autumnal moth (*Epirrita autumnata*) and foliar fungi suggest that herbivore performance may be caused by: (i) genetic differences in plant quality for fungi and herbivores; or, (ii) genetic differences in response to environmental conditions.

Genetic analysis (using random amplified microsatellite PCR) of *Venturia ditricha* (teleomorph of *Fusicladium betulae*) revealed that host genotypes, along with environmental conditions, influence the probability of infection by particular endophyte genotypes (Ahlholm *et al.*, 2002b). The most susceptible host genotypes were highly infected with genetically similar endophyte genotypes, whereas the most resistant trees were less infected and were infected by genetically dissimilar endophytes. Additionally, this study showed environment–host genotype interactions, suggesting phenotypic plasticity of host trees; i.e. that the susceptibility of the host to a particular endophyte genotype may change with environment.

Genetic incompatibility also appears to constrain diversity of established genotype–genotype combinations of systemic seedborne endophytes and grasses (Wäli, unpublished data). Creation of novel endophyte–plant combinations in grasses by removal of the fungus from seeds followed with artificial infection is the traditional approach to separate the effects of the fungus from plant responses in endophyte studies (Brem and Leuchtmann, 2001; Clay and Holah, 1999; Faeth *et al.*, 2002; Lehtonen *et al.*, unpublished). However, Wäli found that successful manipulation of infection status depends largely on the compatibility of endophyte and host genotypes selecting for genotype–genotype combinations of fungus and grass (see also Christensen, 1995; Christensen *et al.*, 2001; Leuchtmann, 1992). Studies using artificially created endophyte–host combinations may thus be biased. Field studies that examine the genetic diversity of host grasses and seedborne endophytes in different environments are critical in understanding the full breadth of endophyte–host grass interactions.

Overall, the results described above indicate, first, that performance of heterotrophic organisms, such as herbivores and endophytes, are responses to genetically determined plant qualities rather than interconnected associations between the heterotrophs. Indeed, the seemingly direct interactions between herbivores and horizontally transmitted fungi may actually indicate genetic differences in plant quality for fungi and herbivores or responses to environmental conditions. Second, it is increasingly clear that host plants harbour scores of endophyte species and genotypes, including grasses infected with systemic endophytes (e.g., Schulthess and Faeth, 1998) and these potentially interact with multiple herbivore species. Considering only the effect of a single endophyte and a single herbivore species very likely obscures the complex interactions between endophytes, host plants and herbivores.

5.3.3 Endophytes – rare plant mutualists?

Empirical evidence suggests that interactions between non-systemic and horizontally transmitted endophytes and plants are variable, and range from positive to negative. These fungi are ubiquitous and diverse temporally and spatially. Functionally, they include a wide variety of dormant saprophytes and latent pathogens and their relatives. Every plant studied to date harbours at least one of these endophyte species and many plants, especially woody plants, may contain literally scores or hundreds of species (Arnold *et al.*, 2001; Carroll, 1986; Faeth and Hammon, 1997a; Helander *et al.*, 1994; Petrini, 1991; Preszler *et al.*, 1996). Spores are usually dispersed from senescent and abscised leaves during the season. Thus, seasonal and spatial variation in the incidence of these fungi is largely dependent on the surrounding vegetation, ground topography, host density and abiotic environmental factors such as weather conditions, moisture regime within the microclimate of the plant foliage and plant damage (Ahlholm *et al.*, 2002a; Faeth and Hammon, 1997a; Helander *et al.*, 1994; Saikkonen *et al.*, 1996). Based on the prevalence of sexual reproduction and the mode of endophyte transmission, we predict that systemic, vertically transmitted endophytes in grasses show stronger mutualism with their host plant than non-systemic horizontally transmitted endophytes in woody plants (see e.g., Saikkonen *et al.*, 1998).

Vertically transmitted endophytes form tightly linked and perennial genotype–genotype associations with their host. These endophytes are highly reliant on the host plant for survival and dissemination. Thus, factors that are beneficial or detrimental to the host plant should also be likewise to the endophyte. A vast majority of past studies have predicted that the fungal symbiont has evolved mechanisms to enhance plant growth and survival, thereby resulting in a strong mutualism (e.g., Breen, 1994; Clay, 1990; Clay and Schardl, 2002; Leuchtmann and Clay, 1997). If so, frequencies of infected plants are predicted to increase over time because of endophyte-increased fitness of the host relative to uninfected grasses. This prediction is supported in agronomic grasses (Clay, 1998; Leuchtmann and Clay, 1997). However, accumulating empirical evidence challenges the generality of this prediction, particularly in native grass populations (Faeth, 2002; Faeth and Bultman, 2002; Faeth and Fagan, 2002; Saikkonen *et al.*, 1998, 2000). Recent studies support the idea that plant–endophyte interactions are much more complex and variable than in the agronomic arena of more genetically homogeneous grasses and more uniform abiotic conditions (Ahlholm *et al.*, 2002a, b, c; Faeth and Bultman, 2002; Saikkonen, 2000; Saikkonen *et al.*, 1998, 1999, 2002). Studies clearly show that mutualism is not overwhelming for native grasses (Faeth, 2002; Faeth and Sullivan, 2003; Saikkonen *et al.*, 1998), and infection frequencies are highly variable within and among wild grass populations (Bazely *et al.*, 1997; Clay and Leuchtmann, 1989; Lewis *et al.*, 1997; Saikkonen *et al.*, 2000; Schulthess and Faeth, 1998). Even introduced tall fescue, upon which much of the mutualistic concept of endophyte–plant interactions has been built, apparently loses strong mutualistic effects when naturalised in native plant communities (Spyreas *et al.*, 2001). The recent literature has provided several alternative explanations for observed persistence and variable levels of endophyte infections in natural grass populations that do not necessarily depend on an obligate endophyte–plant mutualism.

First, vertically transmitted fungal endophytes can be maintained within a spatially structured metapopulation of interconnected local grass populations, even if the fungi locally lower the survival or reproductive success of plants (Gyllenberg *et al.*, 2002; Saikkonen *et al.*, 2002). Gyllenberg *et al.* (2002) and Saikkonen *et al.* (2002) questioned the need for mutualism in exclusively seed-transmitted endophytes and, in addition, showed the importance of habitat diversity in relation to endophyte success in vertical transmission. Second, mathematical models also predict that uninfected hosts could be maintained in a population, assuming that loss of infection in seeds from infected plants (due to either hyphae inviability or failure to propagate into seeds), is greater than 10% (Ravel *et al.*, 1997). Third, the costs and benefits of endophyte infection to the host plant may vary spatially and temporally in natural populations, and thus selection and frequency of infected and uninfected hosts should vary accordingly (Ahlholm *et al.*, 2002c; Brem and Leuchtmann, 2001; Cheplick *et al.*, 2000; Lehtonen *et al.*, unpublished); Morse *et al.*, 2002; West *et al.*, 1995). Fourth, asexual endophytes may manipulate host allocation, increasing allocation to female at the expense of male functions (Faeth, 2002; Faeth and Bultman, 2002; Faeth and Sullivan, 2003). Finally, we propose that the assumed strict vertical transmission of asexual endophytes may be erroneous. Although vertical transmission is probably the primary mode of transmission, sporadic horizontal transmission of an endophyte has been proposed (Cabral *et al.*, 1999; White *et al.*, 1996) and recently confirmed in Arizona fescue (Hamilton, 2002).

5.4 Applications

In addition to providing ideal research systems for testing ecological and evolutionary theory, endophytes also have broad economic applications. Because endophytes can affect virtually every type of plant–plant, plant–pathogen, and plant–herbivore interaction (e.g. Barbosa *et al.*, 1991; Clay, 1987; Hammon and Faeth, 1992; Minter, 1981), any human activities (agriculture, deforestation, pollution, etc.), which alter diversity of endophyte–plant interactions, may have unpredictable, indirect effects on population dynamics and community structure of plants, pathogens and herbivores in terrestrial ecosystems. We suggest that better knowledge of endophytic fungi may provide economically measurable deliverables for the principal stakeholders as: (i) deliverables for end-users from the agribusiness; and, (ii) knowledge of how to consider endophytic fungi in sustainable management strategies in the agronomic, forestry and environmental field.

Direct antiherbivore properties of endophytes (particularly in grasses) have already been exploited, for example in:

1. Biocontrol through developing natural pesticides or improvement of herbivore-resistant cultivars by introducing biologically active (e.g., high mycotoxin production) fungal strains into cultivars (Christensen *et al.*, 2001). In addition to economic value, endophytes may lower investments in chemical pest control by providing environmentally friendly and energy-efficient biocontrol, and consumers avoid remnants of chemical pesticides in the crop.
2. Economic value may also arise from understanding harmful effects in agricultural production. Mycotoxins cause decreased weight gain of livestock and animal disorders. For example, use of endophyte-infected tall fescue (particularly variety Kentucky 31) and perennial ryegrass as forage has resulted in poor animal performance causing major economic losses widely in the USA and New Zealand. Economic losses in the USA alone have been estimated at $609 million annually (Hoveland, 1993). In this context, endophytic fungi have been largely ignored in European grass-ecosystems although most pasture grasses used in the northern hemisphere are of Eurasian origin and infected with endophytes (Hartley and Williams, 1956; Saikkonen *et al.*, 2000).
3. Alternative fungal strains which do not produce mycotoxins harmful to vertebrates but increase plant growth, seed production, seed germination rate and stress tolerance can be used to increase productivity when introduced to the cultivars used as forage. This has already been accomplished for some tall fescue and perennial ryegrass cultivars.

Incorporating microbially mediated interactions in ecosystem management may also broaden the scope of conservation biology. Maintenance of species and genetic diversity of microfungi may be important because of antagonistic interactions between some species of endophytic and pathogenic fungi (Minter, 1981). Minter (1981) reported that *Lophodermium seditiosum*, a pathogenic fungus in young pine trees, is excluded from habitats when another congeneric, but non-pathogenic species, *L. conigenum* is present. Endophytes have usually been ignored in these contexts, perhaps because they are invisible, only a few have expertise to collect and culture them, and taxonomy is virtually unknown. In the future, however, endophytes should be considered when developing sustainable management strategies for forestry and agriculture, and restoring damaged terrestrial ecosystems. For example, introduced grass cultivars with their biologically active endophytes can alter community structure (Clay and Holah, 1999), particularly in extreme habitats, such as grasslands subjected to periodic drought or Arctic regions. Alternatively, restoration of native grass species may be unsuccessful unless their endophytes are also considered (Neil *et al.*, 2003). Successful management requires understanding the basic requirements of microbial-mediated interactions across trophic levels, such as: (i) keystone species; (ii) genetic diversity of these species; (iii) mechanisms and dynamics among interacting species; (iv) minimum habitat size and distance of inhabited patches in fragmented habitats for species survival; and, (v) critical threshold levels of these elements for the loss of biodiversity.

Until now most studies on endophytes have focused on northern temperate regions. Despite the increasing interest in endophytes in tropical plants (e.g., Arnold *et al.*, 2001; Dreyfuss and Petrini, 1984; Rodrigues, 1994, 1996), there is still very little known about the endophytes in these habitats which contain more than half of the species in the entire world biota (Wilson, 1988). Endophytes represent one of the largest reservoirs of fungal species (Dreyfuss, 1989). They produce various chemical compounds similarly to higher plants, and some of the bioactive metabolites assumed to be of plant origin, may actually be produced by fungus or plant and fungus together. For instance, several examples suggest that endophytes may be a largely untapped reservoir of new pharmaceutical products. Highlighting only a few of the best-known examples, endophytes have been reported as producers of antibiotics (*Microsphaeropsis* sp.; Tscherter *et al.*, 1988), an important anticancer drug (taxol by *Taxomyces andreanae*; Stierle *et al.*, 1993), and a potent competitive inhibitor of HIV-1 viral protease (L-696,474 (18-dehydroxy cytochalasin H) by *Hypoxylon fragiforme*; Dombrowski *et al.*, 1992; Ondeyka *et al.*, 1992). Thus, as with the recognised importance of higher plants as sources of new crops, new medicines and new industrial products; there may also be an economic justification for conservation of the natural diversity of endophytes.

References

Ahlholm, J., Helander, M.L., Elamo, P., Saloniemi, I., Neuvonen, S., Hanhimäki, S., Saikkonen K. (2002a) Micro-fungi and invertebrate herbivores on birch trees: fungal mediated plant–herbivore interactions or responses to host quality? *Ecol. Letts.* 5: 648–655.

Ahlholm, J.U., Helander, M., Henriksson, J., Metzler, M., Saikkonen, K. (2002b) Environmental conditions and host genotype direct genetic diversity of *Venturia ditricha*, a fungal endophyte of birch trees. *Evolution* **56**: 1566–1573.

Ahlholm, J.U., Helander, M., Lehtimäki, S., Wäli, P., Saikkonen, K. (2002c) Vertically transmitted endophytes: effects of environmental conditions. *Oikos* **99**: 173–183.

Arnold, A.E., Maynard, Z., Gilbert, G.S. (2001) Fungal endophytes in dicotyledonous neotropical trees: patterns of abundance and diversity. *Mycol. Res.* **105**: 1502–1507.

Bacon, C.W. (1994) Toxic endophyte-infected tall fescue and range grasses: historic perspectives. *J. Anim. Sci.* **73**: 861–870.

Bacon, C.W., Hill, N.S. (1996) Symptomless grass endophytes: products of coevolutionary symbioses and their role in ecological adaptations of grasses. In: Redlin, S.C., Carris, L.M. (eds) *Endophytic Fungi in Grasses and Woody Plants. Systematics, Ecology, and Evolution*, pp. 155–178. APS Press: St. Paul, MN.

Bacon, C.W., Porter, J.K., Robbins, J.D., Luttrell, E.S. (1977) *Epichloë typhina* from tall fescue grasses. *Appl. Environ. Microbiol.* **34**: 576–581.

Ball, D.M., Pedersen, J.F., Lacefield, G.D. (1993) The tall fescue endophyte. *Am. Sci.* **81**: 370–381.

Barbosa, P., Kirschik, V.A., Jones, C.L. (eds) (1991) *Microbial Mediation of Plant–herbivore Interactions*. Wiley: New York.

Bazely, D.R., Vicari, M., Emmerich, S., Filip, L., Lin, D., Inman, A. (1997) Interactions between herbivores and endophyte infected *Festuca rubra* from Scottish islands of St. Kilda, Benbecula and Rum. *J. Appl. Ecol.* **34**: 847–860.

Bazzaz, F.A., Grace, J. (eds) (1997) *Plant Resource Allocation*. Academic Press: San Diego, CA.

Bell, G., Koufopanou, V. (1986) The cost of reproduction. In: Dawkins, R., Ridley, M. (eds) *Oxford Surveys in Evolutionary Biology*, Volume 3, pp. 83–131. Oxford University Press: Oxford, UK.

Bernstein, M.E., Carroll, G.C. (1977) Microbial populations on Douglas fir needles. *Microb. Ecol.* **4**: 41–52.

Blackwell, M. (2000) Terrestrial life – fungal from the start. *Science* **289**: 1884–1885.

Braverman, S.W. (1986) Disease resistance in cool-season forage grasses II. *Bot. Rev.* **52**: 1–113.

Breen, J.P. (1994) *Acremonium* endophyte interactions with enhanced plant resistance to insects. *Ann. Rev. Entomol.* **39**: 401–423.

Brem, D., Leuchtmann, A. (2001) *Epichloë* grass endophytes increase herbivore resistance in the woodland grass *Brachypodium sylvaticum*. *Oecologia* **126**: 522–530.

Bull, J.J., Molineux, I.J., Rice, W.R. (1991) Selection of benevolence in host–parasite system. *Evolution* **45**: 875–882.

Bush, L.P., Wilkinson, H.H., Schardl, C.L. (1997) Bioprotective alkaloids of grass–fungal endophyte. *Plant Physiol.* **114**: 1–7.

Cabral, D., Cafaro, M.J., Saidman, B., Lugo, M., Reddy, P.V., White, J.F., Jr. (1999) Evidence supporting the occurrence of a new species of endophyte in some South American grasses. *Mycologia* **91**: 315–325.

Carlile, M.J., Watkinson, S.C., Gooday, G.W. (2001) *The Fungi*. Academic Press: London.

Carroll, G.C. (1986) The biology of endophytism in plants with particular reference to woody perennials. In: Fokkema, N.J., van den Heuvel, J. (eds) *Microbiology of the Phyllosphere*, pp 205–222. Cambridge University Press: Cambridge.

Carroll, G. (1988) Fungal endophytes in stems and leaves: from latent pathogen to mutualistic symbiont. *Ecology* **69**: 2–9.

Carroll, G.C. (1991) Beyond pest deterrence – alternative strategies and hidden cost of endophytic mutualism in vascular plants. In: Andrews, J.H., Hirano, S.S. (eds) *Microbial Ecology of Leaves*, pp. 358–375. Springer-Verlag: New York.

Cheplick, G.P., Clay, K., Marks, S. (1989) Interactions between infection by endophytic fungi and limitation in the grasses *Lolium perenne* and *Festuca arundinaceae*. *New Phytol.* **111**: 89–97.

Cheplick, G.P., Perera, A., Koulouris, K. (2000) Effect of drought on the growth of *Lolium perenne* genotypes with and without endophytes. *Funct. Ecol.* **14**: 657–667.

Christensen, M.J. (1995) Variation in the ability of *Acremonium* endophytes of perennial rye-grass (*Lolium perenne*), tall fescue (*Festuca arundinacea*) and meadow fescue (*F. pratensis*) to form compatible associations in three grasses. *Mycol. Res.* **99**: 466–470.

Christensen, M.J., Bennett, R.J., Schmid, J. (2001) Vascular bundle colonization by *Neotyphodium* endophytes in natural and novel associations with grasses. *Mycol. Res.* **108**: 1239–1245.

Clay, K. (1987) The effect of fungi on the interactions between host plants and their herbivores. *Can. J. Plant Pathol.* **9**: 380–388.

Clay, K. (1988) Fungal endophytes of grasses: a defensive mutualism between plants and fungi. *Ecology* **69**: 10–16.

Clay, K. (1990) Fungal endophytes of grasses. *Ann. Rev. Ecol. Syst.* **21**: 255–297.

Clay, K. (1998) Fungal endophyte infection and the population biology of grasses. In: Cheplick, G. (ed) *Population Biology of Grasses*, pp. 255–285. Cambridge University Press: Cambridge, UK.

Clay, K., Holah, J. (1999) Fungal endophyte symbiosis and plant diversity in successional fields. *Science* **285**: 1742–1744.

Clay, K., Leuchtmann, A. (1989) Infection of woodland grasses by fungal endophytes. *Mycologia* **81**: 805–811.

Clay, K., Schardl, C. (2002) Evolutionary origins and ecological consequences of endophyte symbiosis with grasses. *Am. Nat.* **160**: s99–s127.

Cody, M. (1966) A general theory of clutch size. *Evolution* **20**: 174–184.

De Bary, A. (1866) *Morphologie und Physiologie der Pilze, Flechten, und Myxomyceten. Vol. II.* – Hofmeister's Handbook of Physiological Botany. W. Engelmann: Leipzig.

Denison, R.F. (2000) Legume sanctions and the evolution of symbiotic cooperation by rhizobia. *Am. Nat.* **156**: 567–576.

Doebeli, M., Knowlton, N. (1998) The evolution of interspecific mutualism. *Proc. Natl Acad. Sci. USA* **95**: 8676–8680.

Dombrowski A.W., Bills, G.F., Sabinis, G., Koupal, L.R., Meyer, R., Ondeyka, J.G., Giacobbe, R.A., Monaghan, R.L., Lingham, R.B. (1992) L-696,474, a novel cytochalasin as an inhibitor of HIV-1 protease I. The producing organism and its fermentation. *J. Antibiot.* **45**: 671–678.

Dreyfuss, M.M. (1989) *Microbial Diversity. Microbial Metabolites as Sources for New Drugs*. Princeton Drug Research Symposia: Princeton, NJ.

Dreyfuss, M.M., Petrini, O. (1984) Further investigations on the occurrence and distribution of endophytic fungi in tropical plants. *Bot. Helv.* **94**: 33–40.

Durham, W.F., Tannenbaum, M.G. (1998) Effects of endophyte consumption on food intake, growth, and reproduction in prairie voles. *Can. J. Zool.* **76**: 960–969.

Elmi, A.A., West, C.P. (1995) Endophyte infection effects on stromatal conductance, osmotic adjustment and drought recovery of tall fescue. *New Phytol.* **131**: 61–67.

Ewald, P.W. (1983) Host–parasite relations, vectors, and the evolution of disease severity. *Ann. Rev. Ecol. Syst.* **14**: 465–471.

Ewald, P.W. (1987) Transmission modes and evolution of the parasitism–mutualism continuum. *Ann. N.Y. Acad. Sci.* **503**: 295–306.

Ewald, P.W. (1994) *Evolution of Infectious Disease*. Oxford University Press: New York.

Faeth, S.H. (2002) Are endophytic fungi defensive plant mutualists? *Oikos* **98**: 25–36.

Faeth, S.H., Bultman, T.L. (2002) Endophytic fungi and interactions among host plants, herbivores and natural enemies. In: Tscharntke, T., Hawkins, B.A. (eds) *Multitrophic Level Interactions*, pp. 89–123. Cambridge University Press: Cambridge, UK.

Faeth, S.H., Fagan, W.F. (2002) Fungal endophytes: common host plant symbionts but uncommon mutualists. *Integ. Comp. Biol.* **42**: 360–368.

Faeth, S.H., Hammon, K.E. (1996) Fungal endophytes and phytochemistry of oak foliage: determinants of oviposition preference of leafminers? *Oecologia* **108**: 728–736.

Faeth, S.H., Hammon, K.E. (1997a) Fungal endophytes in oak trees. I. Long-term patterns of abundance and associations with leafminers. *Ecology* **78**: 810–819.

Faeth, S.H., Hammon, K.E. (1997b) Fungal endophytes in oak trees. Experimental analyses of interactions with leafminers. *Ecology* **78**: 820–827.

Faeth, S.H., Sullivan. T.J. (2003) Mutualistic, asexual endophytes in a native grass are usually parasitic. *Am. Nat.* **161**: 310–325.

Faeth, S.H., Wilson, D. (1996) Induced responses in trees: mediators of interactions between macro and micro-herbivores. In: Gange, A.C. (ed) *Multitrophic Interactions in Terrestrial Systems*, pp. 201–215. Blackwell Science: Oxford, UK.

Faeth, S.H., Bush, L.P., Sullivan, T.J. (2002) Peramine alkaloid variation in *Neotyphodium*-infected Arizona fescue: effects of endophyte and host genotype and environments. *J. Chem. Ecol.* **28**: 1511–1525.

Fine, P.E.M. (1975) Vectors and vertical transmission: an epidemiologic perspective. *Ann. N.Y. Acad. Sci.* **266**: 173–194.

Fisher, P.J. (1996) Survival and spread of the endophyte *Stagonospora pteridiicola* in *Pteridium aquilinum*, other ferns and some flowering plants. *New Phytol.* **132**: 119–122.

Freeman, E.M. (1904) The seed fungus of *Lolium tementulum* L., the darnel. *Philos. Trans. R. Soc. Lond., Ser B* **214**: 1–28.

Freeman, S., Rodriguez, R.J. (1993) Genetic conversation of a fungal pathogen to a nonpathogenic, endophytic mutualist. *Science* **260**: 75–78.

Fröhlich, J., Hyde, K.D., Petrini, O. (2000) Endophytic fungi associated with palms. *Mycol. Res.* **104**: 1202–1212.

Gange, A.C. (1996) Positive effects of endophyte infections on sycamore aphids. *Oikos* **75**: 500–510.

Glenn, A.E., Bacon, C.W., Price, R., Hanlin, R.T. (1996) Molecular phylogeny of *Acremonium* and its taxonomic implications. *Mycologia* **88**: 369–383.

Gyllenberg. M., Ion, D., Saikkonen, K. (2002) Vertically transmitted symbionts in structured host metapopulations. *Bull. Math. Biol.* **64**: 959–978.

Hamilton, C.E. (2002) Master's thesis. Arizona State University: Tempe, AZ.

Hamilton, J.G., Zangerl, A.R., DeLuca, E.H., Berenbaum, M.R. (2001) The carbon–nutrient balance hypothesis: its rise and fall. *Ecol. Letts.* **4**: 86–95.

Hammon, K.E., Faeth, S.H. (1992) Ecology of plant-herbivore communities: A fungal component? *Nat. Toxins* **1**: 197–208.

Hartley, W., Williams, R.J. (1956) Centres of distribution of cultivated pasture grasses and their significance for plant introduction. *Proceedings of the 7th International Grassland Congress*, pp. 190–201.

Hawksworth, D.L. (1988) The variety of fungal–algal symbioses, their evolutionary significance, and the nature of lichens. *Bot. J. Linn. Soc.* **96**: 3–20.

Hawksworth, D.L. (1991) The fungal dimension of biodiversity: magnitude, significance, and conservation. *Mycol. Res.* **95**: 641–655.

Helander, M.L., Sieber, T.N., Petrini, O., Neuvonen, S. (1994) Endophytic fungi in Scots pine needles: spatial variation and consequences of simulated acid rain. *Can. J. Bot.* **72**: 1108–1113.

Herre, E.A. (1993) Population structure and the evolution of virulence in nematode parasite of fig wasps. *Science* **259**: 1442–1445.

Herre, E.A., Knowlton, N. Mueller, U.G., Rehner, S.A. (1999) The evolution of mutualism: exploring the paths between conflict and cooperation. *Trends Ecol. Evol.* **14**: 49–53.

Hill, N.S. (1994) Ecological relationships of *Balansia*-infected graminoids. In: Bacon, C.W., White, J.F., Jr. (eds) *Biotechnology of Endophytic Fungi of Grasses*, pp. 59–71. CRC Press: Boca Raton, FL.

Hoveland, C.S. (1993) Importance and economic significance of the *Acremonium* endophytes to performance of animal and grass plant. *Agric. Ecosyst. Environ.* **44**: 3–12.

Kehr, R.D. (1992) Pezicula canker of *quercus-rubra* l, caused by *pezicula-cinnamomea* (dc) sacc .2. morphology and biology of the causal agent. *Eur. J. For. Pathol.* **22**: 29–40.

Kehr, R.D., Wulf, A. (1993) Fungi associated with aboveground portions of declining oaks (*Quercus robur*) in Germany. *Eur. J. For. Pathol.* **23**: 18–27.

Kover, P.X., Clay, K. (1998) Trade-off between virulence and vertical transmission and the maintenance of a virulent pathogen. *Am. Nat.* **152**: 165–175.

Kover, P.X., Dolan, T., Clay, K. (1997) Potential versus actual contribution of vertical transmission to pathogen fitness. *Proc. R. Soc. Lond. B. Biol. Sci.* **264**: 903–909.

Lehtonen, P., Helander, M., Saikkonen, K. (unpublished manuscript) Endophyte infection and soil nutrients interactively affect herbivore performance.

Leuchtmann, A. (1992) Systematics, distribution, and host specificity of grass endophytes. *Nat. Toxins* **1**: 150–162.

Leuchtmann, A., Clay K. (1997) The population biology of grass endophytes. In: Carroll, G., Tudzynski, P. (eds) *The Mycota*, pp. 185–202. Springer: New York.

Lewis, G.C., Ravel, C., Naffaa, W., Astier, C., Charmet, G. (1997) Occurrence of *Acremonium* endophytes in wild populations of *Lolium* spp. in European countries and a relationship between level of infection and climate in France. *Ann. Appl. Biol.* **130**: 227–238.

Lipsitch, M., Nowak, A., Ebert, D., May, R. (1995) The population dynamics of vertically and horizontally transmitted parasites. *Proc. R. Soc. Lond. B. Biol. Sci.* **260**: 321–327.

Lipsitch, M., Siller, S., Nowak, M.A. (1996). The evolution of virulence in pathogens with vertical and horizontal transmission. *Evolution* **50**: 1729–1741.

Lodge, D.J., Fisher, P.J., Sutton, B.C. (1996) Endophytic fungi of *Manilkara bidentata* leaves in Puerto Rico. *Mycologia* **88**: 733–738.

Malinowski, D.P., Belesky, D.P. (2000) Adaptations of endophyte-infected cool-season grasses to environmental stresses: mechanisms of drought and mineral stress tolerance. *Crop Sci.* **40**: 923–940.

Marks, S., Clay, K., Cheplick, C.P. (1991) Effects of fungal endophytes on interspecific and intraspecific competition in the grasses *Festuca arundinaceae* and *Lolium perenne*. *J. Appl. Ecol.* **28**: 194–204.

Maynard Smith, J., Szathmary, E. (1995) *The Major Transitions in Evolution*. W.H. Freeman: Oxford, UK.

McCormick, M.K., Gross, K.L., Smith, R.A. (2001) *Danthonia spicata* (Poaceae) and *Atkinsonella hypoxylon* (Balansiae): Environmental dependence of a symbiosis. *Am. J. Bot.* **88**: 903–909.

Meijer, G., Leuchtmann, A. (2001) Multistrain infections of the grass *Brachypodium sylvaticum* by its fungal endophyte *Epichloë sylvatica*. *New Phytol.* **141**: 355–368.

Minter, D.W. (1981) Possible biological control of *Lophodermium seditiosum*. In: Millar, C.S. (ed) *Current Research on Conifer Needle Diseases*, pp. 67–74. Aberdeen University Press: Aberdeen, UK.

Moon, C.D., Scott, B., Schardl, C.L., Christensen, M.J. (2000) The evolutionary origins of *Epichloë* endophytes from annual ryegrasses. *Mycologia* **92**: 1103–1118.

Morse, L.J., Day, T.A., Faeth, S.H. (2002) Effect of *Neotyphodium* endophyte infection on growth and leaf gas exchange of Arizona fescue under constrasting water availability regimes. *Environ. Exp. Bot.* **48**: 257–268.

Muller, H.J. (1964) The relevance of mutation to mutational advance. *Mut. Res.* **1**: 2–9.

Neil, K.L., Tiller, R.L., Faeth, S.H. (2003) Germination of big Sacaton and endophyte-infected Arizona fescue under water stress. *J. Range Man.* **56**: 616–622.

Ondeyka, J., Hensens, O.D., Zink, D., Ball, R., Lingham, R.B., Bills, G., Dombrowski, A., Goetz, M. (1992) L-696,474, a novel cytochalasin as an inhibitor of HIV-1 protease. II. Isolation and structure. *J. Antibiot.* **45**: 679–685.

Paavolainen, L., Hantula, J., Kurkela, T. (2000) *Pyrenopeziza betulicola* and an anamorphic fungus occurring in leaf spots of birch. *Mycol. Res.* **104**: 611–617.

Pan, J.J., Clay, K. (2002) Infection by the systemic fungus *Epichloë glyceriae* and clonal growth of its host grass *Glyceria striata*. *Oikos* **98**: 37–46.

Pellmyr, O., Leebens-Mack, J., Huth, C.J. (1996) Non-mutualistic yucca moths and their evolutionary consequences. *Nature* **380**: 155–156.

Petrini, O. (1991) Fungal endophytes of tree leaves. In: Andrews, J.H., Hirano, S.S. (eds) *Microbial Ecology of Leaves*, pp. 179–197. Springer-Verlag: New York.

Petrini, O., Stone, J., Carroll, F.E. (1982) Endophytic fungi in evergreen shrubs in western Oregon: a preliminary study. *Can. J. Bot.* **60**: 789–796.

Petrini, O., Sieber T.H., Toti, L., Viret, O. (1992) Ecology, metabolite production, and substrate utilization in endophytic fungi. *Nat. Toxins* **1**: 185–196.

Porter, J.K. (1994) Chemical constituents of grass endophytes. In: Bacon, C.W., White, J.F., Jr. (eds) *Biotechnology of Endophytic Fungi of Grasses*, pp. 103–123. CRC Press: Boca Raton, FL.

Preszler, R.W., Gaylord, E.S., Boecklen, W.J. (1996) Reduced parasitism of a leaf-mining moth on trees with high infection frequencies of an endophytic fungus. *Oecologia* **108**: 159–166.

Rajagopal, K., Suryanarayanan, T.S. (2000) Isolation of endophytic fungi from leaves of neem (*Azadirachta indica* A. Juss.). *Curr. Sci.* **78**: 1375–1377.

Ravel, C., Michalakis, Y., Charmet, G. (1997) The effect of imperfect transmission on the frequency of mutualistic seed-borne endophytes in natural populations of grasses. *Oikos* **80**: 18–24.

Reznick, D. (1985) Cost of reproduction: an evaluation of the empirical evidence. *Oikos* **44**: 257–267.

Rodrigues, K.F. (1994) The foliar endophytes of the Amazonian palm *Euterpe oleracea*. *Mycologia* **86**: 376–385.

Rodrigues, K.F. (1996) Fungal endophytes of palms. In: Redlin, S.C., Carris, L.M. (eds) *Endophytic Fungi in Grasses and Woody Plants*, pp. 121–132. American Phytopathological Society: St Paul, MN.

Saikkonen, K. (2000) Kentucky-31, far from home. *Science* **287**: 1887a.

Saikkonen, K., Helander, M., Ranta, H., Neuvonen, S., Virtanen, T., Suomela, J., Vuorinen, P. (1996) Endophyte-mediated interactions between woody plants and insect herbivores? *Entomol. Exp. Appl.* **80**: 269–171.

Saikkonen, K., Faeth, S.H, Helander, M., Sullivan, T.J. (1998) Fungal endophytes: a continuum of interactions with host plants. *Ann. Rev. Ecol. Syst.* **29**: 319–343.

Saikkonen, K., Helander, M., Faeth, S.H., Schulthess, F., Wilson, D. (1999) Endophyte–grass–herbivore interactions: the case of *Neotyphodium* endophytes in Arizona fescue populations. *Oecologia* **121**: 411–420.

Saikkonen, K., Ahlholm, J., Helander, M., Lehtimäki, S., Niemeläinen, O. (2000) Endophytic fungi in wild and cultivated grasses in Finland. *Ecography* **23**: 360–366.

Saikkonen, K., Gyllenberg, M., Ion, D. (2002) The persistence of fungal endophytes in structured grass metapopulations. *Proc. R. Soc. Lond. B. Biol. Sci.* **269**: 1397–1403.

Schardl, C.L. (2001) *Epichloë festucae* and related mutualistic symbionts of grasses. *Fung. Genet. Biol.* **33**: 69–82.

Schardl, C.L. and Phillips, T.D. (1997) Protective grass endophytes. Where are they from and where are they going? *Plant Dis.* **81**: 430–437.

Schardl, C.L., Leuchtmann, A., Tsai, H.F., Collett, M.A., Watt D.M., Scott D.B. (1994) Origin of fungal symbiont of perennial ryegrass by interspecific hybridization of a mutualist with the ryegrass choke pathogen, *Epichloe typhina*. *Genetics* **136**: 1307–1317.

Schardl, C.L., Leuchtmann, A., Chung, K.R., Penny, D., Siegel, M.R. (1997) Coevolution by common descent of fungal symbionts (*Epichloë* spp) and grass host. *Mol. Biol. Evol.* **14**: 133–143.

Schulthess, F.M., Faeth, S.H. (1998) Distribution, abundances, and associations of the endophytic fungal community of Arizona fescue (*Festuca arizonica*). *Mycologia* **90**: 569–578.

Schulz, B., Wanke, U., Draeger, S., Aust, H.-J. (1993) Endophytes from herbaceous plants and shrubs: effectiveness of surface sterilization methods. *Mycol. Res.* **97**: 1447–1450.

Siegel, M.R., Bush, L.P. (1996) Defensive chemicals in grass–fungal endophyte associations. *Rec. Adv. Phytochem.* **30**: 81–118.

Siegel, M.R., Latch, G.C.M., Bush, L.P., Fannin, F.F., Rowan, D.D., Tapper, B.A., Bacon, C.W., Johnson, M.C. (1990) Fungal endophyte infected grasses: alkaloid accumulation and aphid response. *J. Chem. Ecol.* **16**: 3301–3314.

Spyreas, G., Gibson, D.J., Basinger, M. (2001) Endophyte infection levels of native and naturalized fescues in Illinois and England. *J. Torrey Bot. Soc.* **128**: 25–34.

Stearns, S.C. (1989) Trade-offs in life-history evolution. *Funct. Ecol.* **3**: 259–268.

Stierle, A., Strobel, G., Stierle, D. (1993) Taxol and taxan production by *Taxomyces andreanae*, and endophytic fungus of Pacific yew. *Science* **260**: 214–216.

Stone, J.K. (1987) Initiation and development of latent infections by *Rhabdocline parkeri* on Douglas fir. *Can. J. Bot.* **65**: 2614–2621.

Stone, J.K., Sherwood, M.A., Carroll, G.C. (1996) Canopy microfungi: Function and diversity. *Northwest Sci.* **70**: 37–45.

Sullivan, T.J. (2002) *Geographic Variation in the Interaction between* Neotyphodium endopytes *and Arizona fescue*. PhD dissertation. Arizona State University: Tempe, AZ.

Sullivan T.J., Faeth, S.H. (2001) Genetic variation of *Neotyphodium* in native grass populations. In: Paul, V.H., Dapprich, P.D. (eds) *Proceedings of the 4th International Neotyphodium/Grass Interactions Symposium*, pp. 283–288. Fachbereich Agrarwirtschaft: Soest, Germany.

Thompson, J.N. (1994) *The coevolutionary process*, p. 376. The University of Chicago Press: London, UK.

Tsai, H.-F., Liu, J.-S., Staben, C., Christensen, M.J., Latch, G.C.M., Siegel, M.R., Schardl, C.L. (1994) Evolutionary diversity of fungal endophytes of tall fescue grass by hybridization with *Epichloë* species. *Proc. Natl Acad. Sci. USA* **91**: 2542–2546.

Tscherter, H., Hofmann, H., Ewald, R., Dreyfuss, M.M. (1988) Antibiotic lactone compound. U.S. Patent 4,753,959.

Wennström, A. (1994) Endophyte – the misuse of an old term. *Oikos* **71**: 535–536.

West, C.P. (1994) Physiology and drought tolerance of endophyte-infected grasses. In: Bacon, C.W., White, J.F., Jr. (eds) *Biotechnology of Endophytic Fungi of Grasses*, pp. 87–99. CRC Press: Boca Raton, FL.

West, C.P., Elberson, HW., Elmi, A.A., Buck, G.W. (1995) *Acremonium* effects on tall fescue growth: parasitic or stimulant? *Proceedings of 50th* Southern Pasture & Forage Crop Improvement Conference, pp. 102–111.

White, J.F., Jr., Martin, T.I., Cabral, D. (1996) Endophyte–host associations in grasses. 22. Conidia formation by *Acremonium* endophytes on the phylloplanes of *Agrostis hiemalis* and *Poa rigidifolia*. *Mycologia* **88**: 174–178.

Wilkinson, H.H., Schardl, C.L. (1997) The evolution of mutualism in grass–endophyte associations. In: Bacon, C.W., Hill, N.S. (eds) *Neotyphodium/grass interactions*, pp. 13–26. Plenum Press: New York.

Wilkinson, H.H., Siegel, M.R., Blankenship, J.D., Mallory, A.C., Bush, L.P., Schardl, C.L. (2000) Contribution of fungal loline alkaloids to protection from aphids in a grass–endophyte mutualism. *Mol. Plant–Micro. Interact.* **13**: 1027–1033.

Williams, G.C. (1966) Natural selection, the costs of reproduction, and a refinement of Lack's principle. *Am. Nat.* **100**: 687–690.

Williams, G.C. (1975) *Sex and Evolution*. Princeton University Press: Princeton, NJ.

Williamson, P.M., Sivasithamparam, K. (1994) Factors influencing the establishment of latent infection of narrow-leafed lupins by *Diaporthe toxica*. *Aust. J. Agric. Res.* **45**: 1387–1394.

Wilson. D. (1995) Endophyte – the evolution of a term, and clarification of its use and definition. *Oikos* **73**: 274–276.

Wilson, D., Carroll, G.C. (1994) Infection studies of *Discula quercina*, and endophyte of *Quercus garryana*. *Mycologia* **86**: 635–647.

Wilson, D., Faeth, S.H. (2001) Do fungal endophytes result in selection for leafminer ovipositional preference? *Ecology* **82**: 1097–1111.

Wilson, E.O. (1988) The current state of biological diversity. In: Wilson, E.O. (ed) *Biodiversity*, pp. 3–18. National Academy Press: Washington, DC.

Yamamura, N. (1993) Vertical transmission and evolution of mutualism from parasitism. *Theor. Pop. Biol.* **44**: 95–109.

6

Actinorhizal symbioses: diversity and biogeography

David R. Benson, Brian D. Vanden Heuvel and Daniel Potter

6.1 Introduction

The actinobacterial genus *Frankia* encompasses sporulating filamentous bacteria (actinomycetes) that fix N_2; they are defined by their ability to induce N_2-fixing root nodules on a broad range of 'actinorhizal plants'. Actinorhizal plants, in turn, are defined by their ability to form root nodules when in symbiosis with *Frankia*. Within the root nodule, *Frankia* fixes nitrogen that is transported to the host plant in amounts sufficient to supply most of the plant's nitrogen requirements. This symbiosis allows actinorhizal plants to invade and proliferate in soils that are low in combined nitrogen. Although similar in outcome, the symbiosis differs markedly from the rhizobium–legume symbiosis. The overall nodule architecture more closely resembles a foreshortened lateral root rather than a unique plant organ, and the plants have evolved a variety of mechanisms to modulate the levels of free O_2 that would otherwise inactivate nitrogenase (Benson and Silvester, 1993). In common with legumes, however, the plants belong to the 'nitrogen-fixing Clade' within the Rosid I lineage initially described by Soltis *et al.* (1995).

Since the first successful and confirmed isolation of a *Frankia* strain in 1978 (Callaham *et al.*, 1978), many studies have addressed the diversity and distribution of *Frankia* strains in root nodules, and some have dealt with the biogeographic distribution of strains and plants. It has become clear that the existing biogeographic patterns of *Frankia* strain distribution can be viewed as resulting from adaptation by both plants and *Frankia* strains within a geographic mosaic of environments developed over millions of years. To sort out factors that control the distribution of frankiae, one must know the host ranges of strain groups, the richness (number of unique strains) and evenness (representation of each unique strain) components of strain diversity in nodules in nature and the geographical distribution of both plants and frankiae.

Plant Microbiology, Michael Gillings and Andrew Holmes

This chapter focuses on the broad patterns of *Frankia* strain distribution and diversity as they relate to host plant distribution across a geographical mosaic of environments. It begins with some of the issues that arise in studying the biogeography of the symbiosis, followed by a brief overview of the phylogenetic relationships among actinorhizal plants and among *Frankia* strains. Finally, information will be presented concerning the biogeography of the symbioses, and the diversity of *Frankia* strains that participate in symbiosis in each plant family. The chapter will conclude with a discussion of basic principles that are emerging.

6.2 Practical aspects of studying *Frankia* strain diversity

Several issues must be considered when discussing *Frankia* strain diversity and distribution in natural environments in relation to the plant hosts. These include but are not limited to local patterns of strain distribution in soils, including strain dominance and response to edaphic factors, regional patterns of plant and microbe distribution and global patterns imposed on the plants and micro-organisms by climatic and geological changes. Patterns of symbiotic compatibility between plants and micro-organisms are a function of the natural distribution of both partners across a geographical mosaic of environments (Benson and Clawson, 2000).

While patterns of diversity and distribution do occur, and can be identified with some effort, conceptual difficulties arise when studying the biogeography of actinorhizal symbioses. First is the problem, common to bacteriological studies, of defining a *Frankia* strain. A variety of markers have been used to study the ecological diversity of *Frankia* strains but at different levels of resolution (reviewed in Benson and Silvester, 1993; Schwencke and Caru, 2001). These markers range from simple phenotypic traits like sporulation within nodules to protein pattern and isoenzyme analysis to PCR-RFLP and DNA sequencing. Most recent studies have used the variability of 16S rRNA genes amplified by the PCR from isolates and nodules (for example, Benson *et al.*, 1996; Clawson and Benson, 1999a, b; Clawson *et al.*, 1997, 1998; Ritchie and Myrold, 1999), or PCR-RFLP patterns of variable intergenic regions of *nif* or rRNA genes present in nodules (for example, Jamann *et al.*, 1992, 1993; Lumini *et al.*, 1996; McEwan *et al.*, 1994; Rouvier *et al.*, 1996; Simonet *et al.*, 1991), or repetitive extragenic palindromic-PCR (Rep-PCR) (for example, Jeong and Myrold, 1999; Murry *et al.*, 1997). The resolution of these latter approaches is limited by the variability in the DNA used for analysis. However, it is possible to organise strains into closely related groups that are presumed to share more biological similarity within groups than between more distantly related groups.

A second difficulty is certifying that a compatible organism is absent or even present in a complex soil population. Every strain cannot be everywhere but proving that point can be difficult. Direct detection by isolation is complicated by the slow growth of *Frankia* strains on bacteriological media (10–14 days) and by their low number relative to other bacteria in the soil (Baker and O'Keefe, 1984). Therefore, *Frankia* strains have been detected, and populations assessed, by bioassay and, more rarely, by direct PCR amplification of *Frankia*-specific genes from soil. Much of this work has been reviewed previously (Benson and

Silvester, 1993; Hahn *et al.*, 1999; Lechevalier, 1994; Schwencke and Caru, 2001; Wall, 2000).

Bioassays are performed by diluting soil samples, inoculating plants, and then calculating nodulation units based on the number of nodules formed on plants per gram or cm^3 of soil. The unavoidable difficulty with this type of approach is that it underestimates the number of frankiae present in the soil since only those strains capable of infecting the test plant and that actually encounter an infectible zone on the root and then form a root nodule are counted. In addition, different plant species, even within the same genus, may yield different estimates depending on their susceptibility to the local *Frankia* strains (see, for example, Huss-Danell and Myrold, 1994; Mirza *et al.*, 1994a). Nevertheless, within limits, such an approach allows comparative estimates of the number of strains in soil to be made. Estimates of frankiae populations using PCR methods have yielded some promising results (Myrold and Huss-Danell, 1994; Picard *et al.*, 1992), but the low population levels of frankiae in most soils and the questionable specificity of the primers used for analysis have limited the broad application of this approach (Normand and Chapelon, 1997).

The final problem is one of significance. That is, even if strains are defined with sufficient resolution and their geographic distribution is described, their metabolic contribution to the geographic mosaics in which they live, and their attributes that promote their distribution within the mosaic may not be obvious. The contribution of an individual bacterial strain to the environment under study is difficult to assess unless it is observable and quantifiable. To some extent, the problem of significance is less acute in the case of nitrogen-fixing symbioses where a higher organism chooses bacterial strains that are best suited to enter the symbiosis in the environment under study. Their function is known and at least part of their contribution to the soil economy can be quantified.

6.3 Taxonomy and phylogeny of actinorhizal plants and *Frankia*

6.3.1 Actinorhizal plant phylogeny

According to current taxonomy, actinorhizal plants are classified in eight families (*Table 6.1*). They are widely distributed, found on all continents except for Antarctica, and are a diverse group of mostly woody dicots (*Table 6.1*) (Baker and Schwintzer, 1990). Most members are found in temperate zones, with only a few members being found in tropical environments and a few in Arctic environments (*Table 6.1*). Ecologically, actinorhizal plants are usually pioneers on nitrogen-poor soils, and are frequently found in relatively harsh sites, including glacial till, new volcanic soil, sand dunes, clear cuts, and desert and chaparral (Schwencke and Caru, 2001).

Traditional taxonomic treatments suggested that actinorhizal plant families were at most only distantly related, classified in four of the six major angiosperm subclasses as delimited by Cronquist (1981) (*Table 6.1*). This morphological classification suggested that the actinorhizal symbiosis had evolved many times in angiosperm evolution (Mullin *et al.*, 1990). A dramatic shift in this view occurred

Table 6.1 Classification of actinorhizal plants[a]

Subclass[b]	Order[c]	Family	# nodulated genera/total # of genera[d]	Genus	Distribution of genus[e]
Hamamelidae	Fagales	Betulaceae	1/6	*Alnus*	n. temperate, higher elevations in C. and S. America, n. Africa, Asia
		Casuarinaceae	4/4	*Allocasuarina*	Australia
				Casuarina	Old World tropics
				Ceuthostoma	Philippines, Borneo, New Guinea
				Gymnostoma	Malaysia to W. Pacific
		Myricaceae	2/3	*Comptonia*	e. N. America
				Myrica	nearly cosmopolitan (not Mediterranean, Australia)
Rosidae	Rosales	Elaeagnaceae	3/3	*Elaeagnus*	Europe, Asia, N. America
				Hippophae	temperate Eurasia
				Shepherdia	N. America
		Rhamnaceae[f]	7/55	*Ceanothus*	N. America, esp. California
				Colletia	s. S. America
				Discaria	s. S. America, Australia, New Zealand
				Kentrothamnus	S. America (Bolivia, Argentina)
				Retanilla	S. America (Peru, Chile)
				Trevoa[g]	S. America (Andes)
		Rosaceae	5/100	*Cercocarpus*	w. N. America
				Chamaebatia	California, Baja California
				Dryas	circumboreal, arctic – alpine
				Purshia[h]	w. N. America
Magnoliidae	Cucurbitales	Coriariaceae	1/1	*Coriaria*	Mexico to S. America, w. Mediterranean
Dilleniidae		Datiscaceae	1/1	*Datisca*	w. N. America, s. Asia

[a] Compiled after Baker and Schwintzer (1990), Swensen (1996), Benson and Clawson (2000), and Schwencke and Carú (2001).
[b] According to the classification of Cronquist (1988).
[c] According to the classification of the Angiosperm Phylogeny Group (1998); all of these orders fall in the 'Eurosid I' group of eudicots.
[d] Number of nodulated genera over the total number of described genera in the family
[e] Compiled from Mabberley (1988) and from the International Plant Names Index (www.ipni.org).
[f] *Adolphia* may be actinorhizal, but has not been confirmed (Cruz-Cisneros and Valdés, 1991).
[g] *Talguenea* should be combined under *Trevoa* (Tortosa, 1992).
[h] *Purshia* and *Cowania* have been combined under *Purshia* (Henrickson, 1986).

with the publication in 1993 of the first extensive molecular phylogeny for angiosperms using sequences from the chloroplast gene for the large subunit of ribulose-1,5-bisphosphate carboxylase/oxygenase (*rbcL*) (Chase *et al.*, 1993). This phylogeny placed all actinorhizal angiosperms in the 'Rosid I' Clade, later termed 'Eurosids I' (Angiosperm Phylogeny Group (APG) (1998); *Figure 6.1*). Furthermore, the two families in which symbiotic relationships with *Rhizobium* and related bacteria occur, Fabaceae (containing the legumes) and Cannabaceae

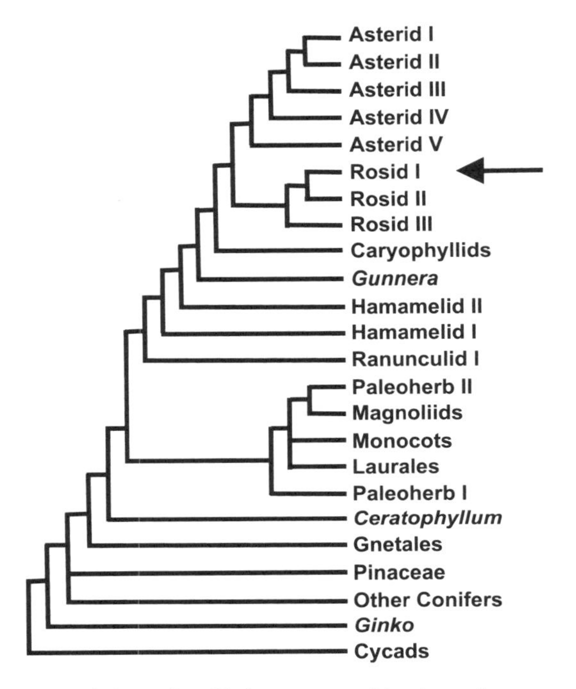

Figure 6.1. A summary of the major Clades recovered in the strict consensus of 3900 most parsimonious trees based on an alignment of 499 rebel sequences representing angiosperm diversity. Adapted and used with permission from the annals of the Missouri Botanical Gardens (Chase *et al.*, 1993). The arrow highlights the Rosid I Clade where all nitrogen-fixing plants, including all actinorhizal and rhizobial plants occur.

(in which only members of the genus *Parasponia* engage in symbiotic nitrogen fixation with rhizobia), are also included in this Clade. This clustering led to the suggestion that the predisposition to form symbiotic nitrogen fixing root nodules may have evolved only once in flowering plants (Soltis *et al.*, 1995).

Later authors examined the phylogenetic relationships of actinorhizal plants in more detail, and provided more rigorous analyses of the origin and evolution of the actinorhizal symbiosis (e.g. Jeong *et al.*, 1999; Roy and Bousquet, 1996; Swensen, 1996; Swensen and Mullin, 1997). In addition, the phylogenetic utility of other markers for resolving relationships among families of angiosperms has been investigated over the last decade. These markers include the 18S ribosomal RNA gene of the nuclear ribosomal DNA repeat and the chloroplast-encoded ATP synthase beta subunit (*atpB*) gene (Soltis and Soltis, 2000). These additional molecular markers have allowed independent assessments of the phylogenetic relationships of the actinorhizal families as well as simultaneous analysis of data from multiple genes representing two cellular compartments.

A maximum parsimony tree of combined data from all three loci is presented in *Figure 6.2*. This treatment places the actinorhizal taxa into three well-supported subClades within Eurosids I. These three subClades have been recognised taxonomically. They are designated as Rosales, Fagales and Cucurbitales (APG, 1998). They contain, respectively, the actinorhizal members of the Rosaceae,

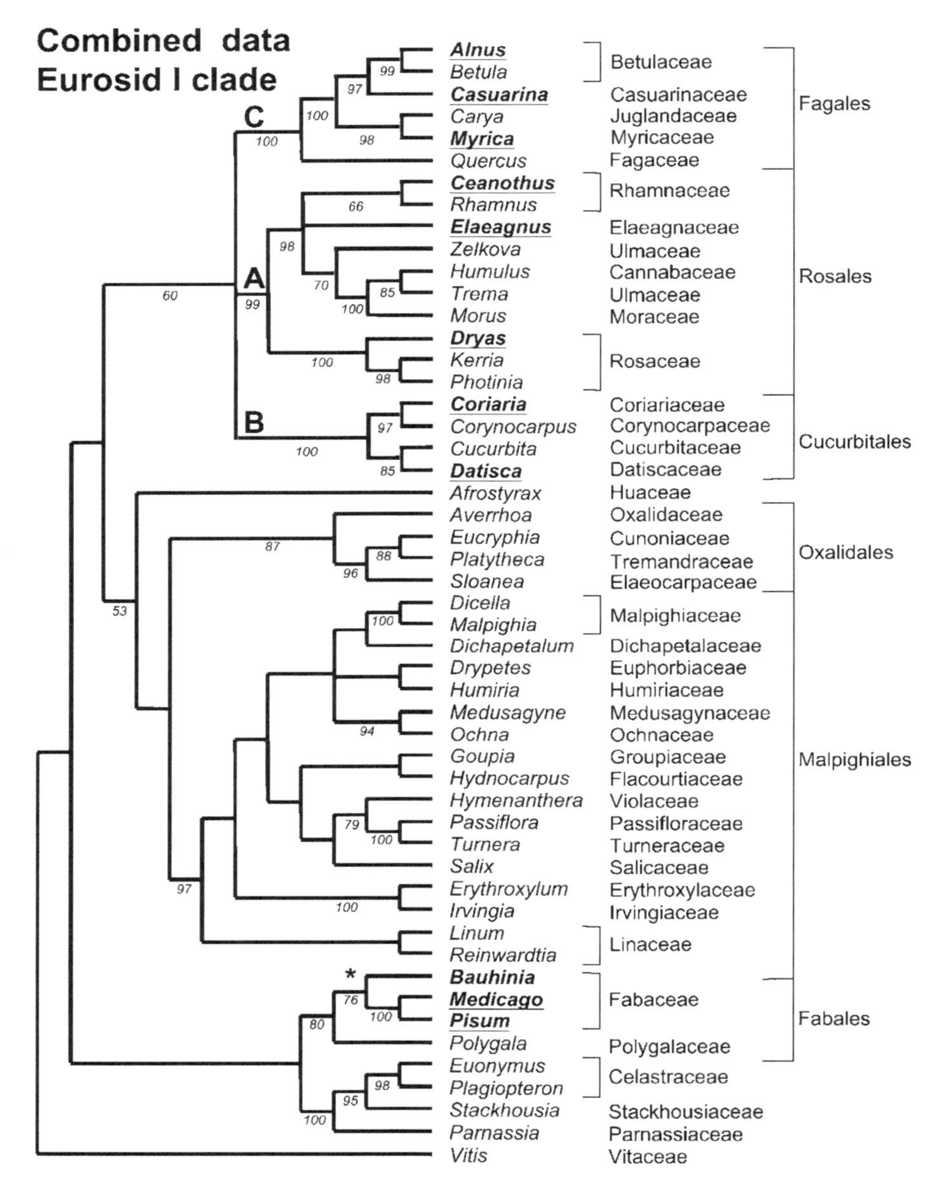

Figure 6.2. A strict consensus of the eight most parsimonious trees based on the combined data from the *rbcL* (1394 aligned basepairs) *atpB* (1520 aligned basepairs), and nuclear ribosomal 18S small subunit (1828 aligned basepairs) for 51 taxa representing all major families in the Rosid I Clade (see *Figure 6.1*). The numbers found below the branches represent bootstrap support from 100 replications. Taxon labels in bold and underlined represent the actinorhizal and rhizobial taxa. The family and order for each taxon in the tree can be seen to the right based on the Angiosperm Phylogeny group (1998) designations. Clades A, B and C highlight the three major Clades that include actinorhizal taxa. The asterisk denotes the rhizobial Clade. Clades A and C correspond to the same Clades identified by Soltis *et al.* (1995) while Clade B was identified as Clade D in their publication.

Rhamnaceae and Elaeagnaceae; those of the Betulaceae, Casuarinaceae and Myricaceae; and those of the Datiscaceae and Coriariaceae (*Table 6.1*, *Figure 6.2*). Maximum parsimony trees based on data from both chloroplast loci yield the same three Clades but trees based on the18S data alone provide weak resolution within Eurosids I and place the actinorhizal taxa in four Clades, resulting in polyphyletic Rosales and Cucurbitales. The discrepancy between topologies may reflect different evolutionary histories for the nuclear and plastid genomes due to past hybridisation or lineage sorting in some lineages, or it may be due to a lack of phylogenetically informative variation within the 18S data set.

Each of the three orders that includes actinorhizal taxa also contains taxa that do not form the symbiosis; indeed, that is also true of several of the actinorhizal families. In some families all members are nodulated (Coriariaceae, Elaeagnaceae, Datiscaceae and Casuarinaceae) whereas in others only a portion of the genera are nodulated (Betulaceae, Myricaceae, Rhamnaceae and the Rosaceae). In at least one case (*Dryas*), nodulation apparently does not extend to all members of a single genus (Kohls *et al.*, 1994). These observations have led to the conclusion that, while the predisposition, or *potential*, to form the nitrogen-fixing symbiosis may have evolved only once, the realization of that potential has occurred and/or been lost multiple times (Benson and Clawson, 2000; Swensen, 1996).

6.3.2 Phylogeny of Frankia

The phylogeny of the genus *Frankia* has been deduced by comparative sequence analysis of the 16S rRNA gene, the genes for nitrogen fixation (*nif* genes) and by other genes (Benson and Clawson, 2000). All analyses agree that the genus is comprised of three major groups or clusters (referred to here as Groups 1, 2 and 3), each having different and sometimes overlapping plant specificity, physiological properties and symbiotic interactions (*Figure 6.3*). Within each group are definable subgroups that constitute 'genospecies' as defined by DNA–DNA homology studies (An *et al.*, 1985; Benson and Clawson, 2000; Dobritsa and Stupar, 1989; Fernandez *et al.*, 1989; Normand *et al.*, 1996).

In general, Group 1 *Frankia* strains form nodules on members of the 'higher' Hamamelidae, now all classified in the order Fagales, including the Betulaceae, Myricaceae and Casuarinaceae. The 'casuarina strains' that primarily infect members of the Casuarinaceae form a subgroup within Group 1. The latter strains also infect members of the Myricaceae as well as *Casuarina* spp., although the extent of their ability to do so in the field is unclear (Simonet *et al.*, 1999). 'Alder strains' generally infect most species of alder tested in greenhouse experiments, with some variability in effectiveness depending on the plant–symbiont combination. They too are generally able to infect members of the Myricaceae.

Group 2 *Frankia* strains are limited to infecting members of the Coriariaceae, Datiscaceae, Rosaceae and *Ceanothus* of the Rhamnaceae. These strains have not been isolated in pure culture despite many attempts to do so by several investigators and may therefore be obligate symbionts. Cross inoculation studies using crushed nodules suggest that symbionts from *Dryas*, *Ceanothus*, *Datisca* and *Coriaria* are in the same cross inoculation group (Kohls, *et al.*, 1994; Mirza *et al.*, 1994b; Torrey, 1990).

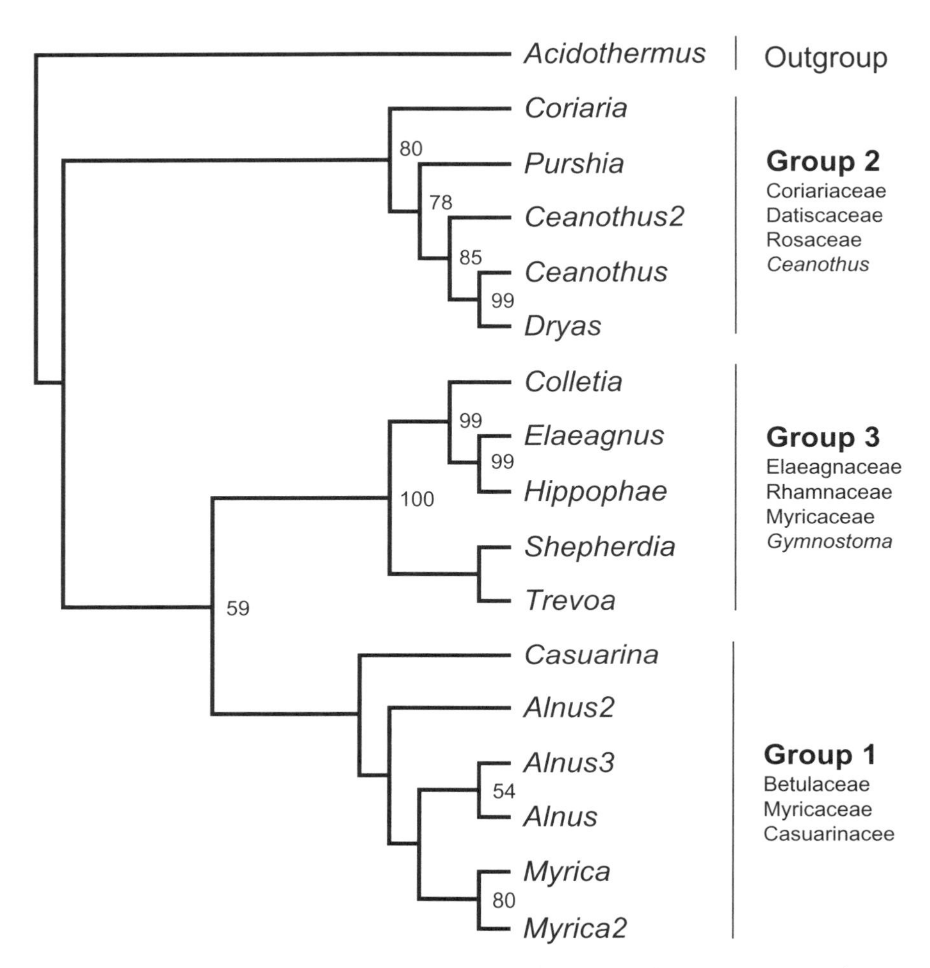

Figure 6.3. Phylogenetic relationships among *Frankia* strains. The major groups of *Frankia* strains are indicated along with the plant families they are known to infect. *Ceanothus* and *Gymnostoma* are listed separately as they are the only members of their families to be infected by the indicated Group of strains. Bootstrap support out of 1000 bootstrap samples is indicated at nodes where it occurred above 50%. Sequences and their accession numbers include: *Acidothermus*, X70635; *Coriaria* nodule, AF063641; *Purshia* nodule AF034776; *Ceanothus1* nodule AF063639; *Ceanothus2* nodule, U69265; *Dryas* nodule, L40616; *Colletia* nodule, AF063640; *Elaeagnus* strain Ea1-2, L40618; *Hippophae* strain HR27-14, L40617; *Shepherdia* strain SCN10a, L40619; *Trevoa* nodule, AF063642; *Casuarina* strain CeD, M55343; *Alnus*, strain AcoN24d, L40610, *Alnus2*, strain AVN17s, L40613; *Alnus3*, strain ACN14a, M88466; *Myrica*, nodule L40622; *Myrica2*, nodule, AF158687.

Group 3 strains form effective nodules on members of the Myricaceae, Rhamnaceae, Elaeagnaceae and *Gymnostoma* of the Casuarinaceae and are sometimes isolated as poorly effective, or non-infective, strains from nodules of the Betulaceae, Rosaceae, other members of the Casuarinaceae and *Ceanothus* of the Rhamnaceae.

Of these three major Clades, most ecological information, including distribution and diversity measurements, is available for Group 1 strains that are commonly known as alder and casuarina strains but that also infect *Myrica* spp. Less is known about members of Group 3 and even less about members of Group 2. Some studies have investigated members of these other groups and these will be mentioned below in the context of their plant families.

6.4 Biogeographic distribution of actinorhizal plants and *Frankia* strains

Actinorhizal plants have a global distribution. They are present on every continent except Antarctica, where their predecessors probably did exist for a time during the late Cretaceous. Each of the eight actinorhizal families has a distinctive native range that varies from very limited to global. The *Frankia* strains that infect these various groups of plants co-exist with the plants and some apparently have an independent life in the soil without the plant.

6.4.1 The Betulaceae

The Betulaceae is composed of six genera and about 130 species (Mabberely, 1988). The family is mostly distributed throughout the temperate regions of the northern hemisphere, with the exception of *Alnus glutinosa* (L.) Gaertn., which is found in Africa, and *A. acuminata* HBK, found throughout Central America south to Argentina. The genus *Alnus* is the only actinorhizal genus within the Betulaceae (*Table 6.1*). The family is well-defined, being held together by the synapomorphies of male and female compound catkins and pollen morphology (Chen *et al.*, 1999). The angiosperm *rbcL* phylogeny of Chase *et al.* (1993) strongly supported the classification of Betulaceae within Fagales (a relationship long recognized based on morphology). More recent studies of this group (Manos and Steele, 1997) have placed the family in a subClade with Casuarinaceae and Ticodendraceae.

Recent molecular phylogenies for the Betulaceae suggest two lineages (*Figure 6.4C*) (Chen *et al.*, 1999). One lineage contains the genera *Corylus*, *Ostryopsis*, *Carpinus* and *Ostrya*, and the other includes *Alnus* and *Betula*. The early divergence of *Alnus* agrees with previous morphological and fossil evidence (Chen *et al.*, 1999). The oldest known fossil infructescence for *Alnus* dates to the mid-Eocene (33–55 MYA), but *Alnus*-like pollen has been reported from much earlier, in the late Cretaceous (83–85 MYA), earlier than any fossils for the other genera in the family (Miki, 1977).

Given the distribution of known fossils and the recent molecular phylogeny, it appears the Betulaceae first originated in a Mediterranean climate in Laurasia during the late Cretaceous (89–65 MYA) (Laurasia was the northern supercontinent formed after Pangaea broke up during the Jurrasic and included what are now North America, Europe, Asia, Greenland, and Iceland). Fossil evidence suggests that all six genera, including *Alnus*, were differentiated by the early Eocene (55 MYA) (Chen *et al.*, 1999). This observation suggests that if the ability to nodulate was ancestral in the Betulaceae, loss of that ability occurred very early on in the evolution of the family.

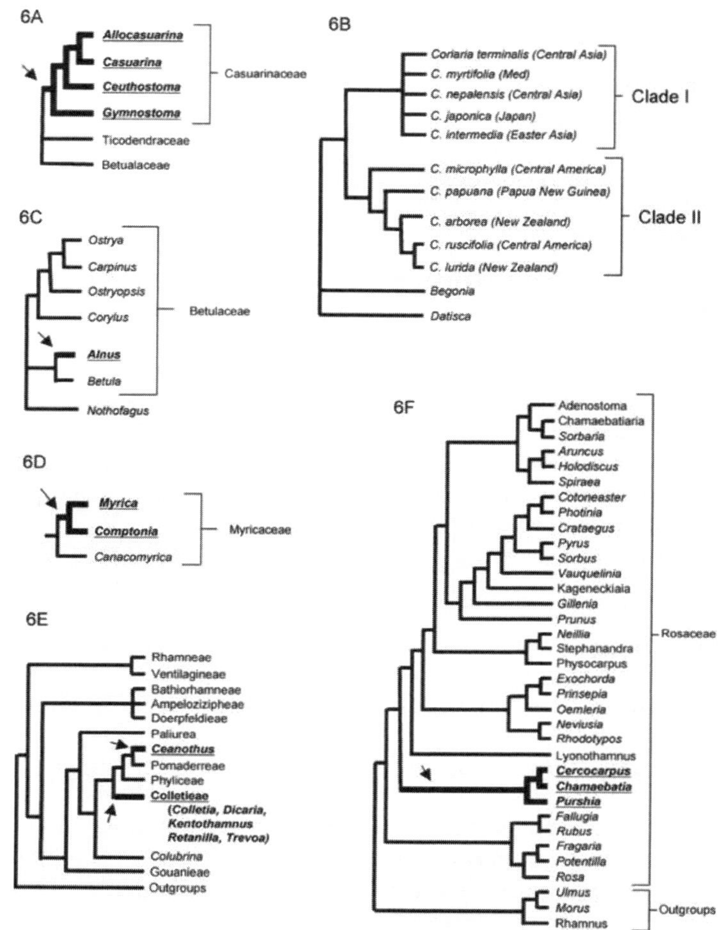

Figure 6.4. Phylogenetic relationships among members of actinorhizal families. **A** – Phylogeny of the Casuarinaceae adapted from Sogo *et al.* (2001). The arrow denotes the possible origin of the actinorhizal symbiosis within the phylogeny, and the thickened branches lead to actinorhizal taxa. Taxon labels in bold and underlined denote the actinorhizal taxa. **B** – A phylogeny for the genus *Coriaria* adapted from Yokoyama *et al.* (2000) based on *rbcL* and *matK* DNA regions. Yokoyama identified two Clades, labelled I and II in the figure. The geographic locations of the taxa in the phylogeny can be found in the parentheses at the end of the taxon name. **C** – A phylogeny for the Betualaceae adapted from Chen *et al.* (1999) based on nuclear ribosomal ITS data and morphology. The taxon labels in bold and italics are actinorhizal. The arrow denotes the possible origin of the actinorhizal symbiosis within the Betulaceae. The thickened branches lead to actinorhizal taxa. **D** – A hypothetical representation of relationships within the Myricaceae adapted from a discussion of morphology and fossil history found in MacDonald (1989). **E** – A representation of a molecular phylogeny of the Rhamnaceae based on *rbcL* and *trnL-F* DNA sequences adapted from Richardson *et al.* (2000). The taxon labels are tribes within the Rhamnaceae. Those taxa with names in bold and underlined are actinorhizal. Arrows denote the two possible origins of the actinorhizal symbiosis. **F** – A representation of a phylogeny for the Rosaceae based on plastid *trnL-F* intergenic spacer and the *matK* DNA regions adapted from Potter *et al.* (2002). Taxon names in both bold and italics are actinorhizal. The arrow denotes the possible origin of the actinorhizal symbiosis within the Rosaceae.

By the early Tertiary (65 MYA), movement between Eurasia and North America was possible, and the range of *Alnus* probably increased. The distribution of *Alnus* to Africa and to Taiwan probably occurred later, during the Pleistocene (1.8 MYA–11 000 years ago) when sea levels were lower (Chen *et al.*, 1999). Beyond the extensive geographic distribution, *Alnus* also grows in a wide range of habitats, from glacial till, sand hills, and bogs to dry volcanic lava, ash alluvium, and water courses (Schwencke and Caru, 2001; Silvester, 1977). To date, all species of *Alnus* examined have been shown to nodulate.

Alnus associates with *Frankia* strains that are similar to those that nodulate the other actinorhizal families in the Fagales, the Casuarinaceae and Myricaceae. These 'alder strains' belong to the diverse Group 1 frankiae. A few members of Group 3 have also been shown to nodulate alders but do so only rarely and are poorly effective (Bosco *et al.*, 1992; Lumini and Bosco, 1996).

Most studies that have focused on the distribution of frankiae in soil have used *Alnus* spp. as the trapping plant, largely because alder seeds are readily available and easily germinated. Except for a few environments such as at the foot of retreating glaciers in Alaska (Kohls, *et al.*, 1994) alder *Frankia* strains are cosmopolitan and seem to persist independently of host plants.

Population estimates of alder frankiae vary from a few per gram to several thousands per gram in soils both with and without actinorhizal plants (Markham and Chanway, 1996; Maunuksela *et al.*, 1999; Myrold *et al.*, 1994; Smolander, 1990; Smolander and Sarsa, 1990; Smolander and Sundman, 1987; Van Dijk, 1979, 1984). Alder strains are commonly detected outside the geographic ranges of their compatible hosts and they persist long after compatible hosts have disappeared from a site (Arveby and Huss-Danell, 1988; Huss-Danell and Frej, 1986; Maunuksela, *et al.*, 1999; Paschke and Dawson, 1992a, 1992b; Smolander and Sundman, 1987; Wollum II *et al.*, 1968). In New Zealand, *Alnus* species nodulate at every site where they are grown at elevations from sea level up to 1700 m, even though the genus is a recent arrival to the islands (Benecke, 1969). Molecular work has shown that the nodules of these 'exotic' plants contain 'typical' *Frankia* strains, that is, those from phylogenetic Group 1 normally associated with the genera (Clawson *et al.*, 1997). Indeed, the diversity of strains in the alder nodules from New Zealand is greater than the diversity of strains infecting the native *Coriaria* sp.

As might be predicted, the nodulation capacity of soils for alders is affected by season (Myrold and Huss-Danell, 1994), acidity (Crannell *et al.*, 1994; Griffiths and McCormick, 1984; Smolander and Sundman, 1987; Zitzer and Dawson, 1992), fertility (Kohls and Baker, 1989; Myrold and Huss-Danell, 1994; Myrold *et al.*, 1994; Sanginga *et al.*, 1989; Thomas and Berry, 1989; Yang, 1995), water availability (Dawson *et al.*, 1986; Nickel *et al.*, 1999; Schwintzer, 1985), the physiological status of *Frankia* strains (Myrold and Huss-Danell, 1994) and by the type of plant cover (Huss-Danell and Frej, 1986; Markham and Chanway, 1996; Myrold and Huss-Danell, 1994; Smolander, 1990; Smolander and Sarsa, 1990; Smolander and Sundman, 1987; Smolander *et al.*, 1988; Zimpfer *et al.*, 1999).

In the case of plant cover, the number of alder nodulation units (NUs) seems to be as high or higher in soil beneath *Betula* (birch) stands than in soils beneath alder (Paschke and Dawson, 1992a; Smolander and Sarsa, 1990; Smolander and Sundman, 1987; Van Dijk *et al.*, 1988) perhaps indicating a rhizosphere relationship between alder frankiae and other members of the Betulaceae. On the other hand, a study on the nodulation capacity of soils beneath birch, pine and spruce indicated similar alder frankiae populations of 3160, 2267 and 2747 NUs g^{-1}, respectively, suggesting that factors other than plant genotype can sustain populations in soil lacking actinorhizal plants (Maunuksela *et al.*, 1999).

Although many studies have been done on enumerating alder frankiae in soil, relatively few have directly examined the correlation between the diversity of the strains trapped and the environmental parameters of the soil examined. Some morphological work has indicated sorting of strains by soil type in alder. For example, the character of sporangium formation in a nodule seems to be stable among strains, thereby enabling geographical studies on sp(+) (containing sporangia) or sp(–) (devoid of sporangia) nodules (Schwintzer, 1990). Thus, in British Columbia, sp(–) nodules of *A. rubra* dominate in submaritime zones with no sp(+) nodules observed. The proportion of sp(+) nodules increased moving inland, up to 53% of the total (Markham and Chanway, 1996). In some studies, more acidic soils appear to select for *Frankia* strains with the sp(+) phenotype (Holman and Schwintzer, 1987; Kashanski and Schwintzer, 1987; Weber, 1986). Other studies have related the development of sp(+) or sp(–) nodules to the age of the stands, to plant selection or to moisture content of the soil (reviewed in (Schwintzer, 1990) and (Markham and Chanway, 1996)).

Alder frankiae can be concluded to be cosmopolitan and quite diverse within the limits of Group 1 strains. Their ubiquity implies that many alder frankiae are soil organisms with an independent existence not requiring continuous symbiotic interaction. It is likely that their wide distribution is related to the ranges of their hosts, both *Alnus* and *Myrica* spp., that extend throughout the northern hemisphere and into South America and Africa. The explanation for their abundance in exotic environments such as New Zealand may lie in their ability to grow as saprophytes in the absence of actinorhizal hosts, although their rapid spread once introduced cannot be ruled out.

6.4.2 The Myricaceae

The Myricaceae is composed of three genera, *Myrica* L., *Comptonia* L'Herit., and *Canacomyrica* Guillaumin. The family is classified in the Higher Hamamelidae, now included in the order Fagales (APG, 1998) and molecular phylogenetic studies have not provided strong support for its relationships within the order (Manos and Steele, 1997). *Myrica* is by far the largest genus, having about 50 described species with a wide distribution in North America, Europe, Africa, and Asia. *Myrica* spp. have been transported to new sites. For example, *M. faya* has been introduced to Hawaii where it has become an invasive exotic pest (Mabberely, 1988). The other two genera are monotypic. *Comptonia peregrina* L. is native to North America and *Canacomyrica monticola* Guillaumin is endemic to New Caledonia. Nodulation has been observed on all species of *Myrica* and *Comptonia*, but has yet to be documented on *Canacomyrica* (Navarro *et al.*, 1999).

To date, there has been no family-wide molecular phylogeny generated for the Myricaceae, yet the fossil history of the family has been well-discussed (see (Macdonald, 1989)). Briefly, there are two different opinions concerning the first appearance of the Myricaceae in the fossil record. One view holds that the

Myricaceae appeared early, during the Santonian period (83–85 MYA), based on *Myrica*-like pollen. The other view holds that the Myricaceae is instead a much later lineage, originating during the Eocene where there is more fossil evidence, and that the previous fossilised pollen was misidentified (Macdonald, 1989).

The geographic origin of the family is also in dispute. Both a Southeast Asian origin during the early Cretaceous (146–65 MYA) and a northern Tethyan origin during the late Cretaceous have been postulated. The enigmatic genus *Canacomyrica* appears to have many morphological similarities to the fossil, ancestral Juglandaceae and may represent the extant relictual ancestor to the Myricaceae (Macdonald, 1989). If *Canacomyrica* does indeed represent the basal lineage to the Myricaceae, than it would appear that the ability to nodulate occurred after the family had diverged or that the lineage leading to *Canacomyrica* lost the ability to nodulate (*Figure 6.4D*). A phylogeny and detailed biogeographic study is needed to further explore this issue.

Myrica species represent an interesting subject for studying the diversity of *Frankia* strains since they are considered to be promiscuous hosts based on the results of greenhouse cross inoculation studies and on ecological evidence (Baker, 1987; Clawson *et al.*, 1998; Torrey, 1990). It has been known for some time that *Myrica* species are effectively nodulated by frankiae from phylogenetic Groups 1 and 3 (Baker, 1987). Group 2 strains have not been reported in *Myrica*.

The natural diversity of *Frankia* strains in the nodules of *Myrica* spp. native to northeast North America has been examined using the variability between 16S rDNA sequences or PCR-RFLP of 16S rDNA as measures of diversity (Clawson and Benson, 1999b; Huguet *et al.*, 2001). In one study, root nodules were collected from 30 sites with *Comptonia peregrina*, 29 with *Myrica pensylvanica* and 37 with *M. gale*, 37 unique sequences were found in 97 nodules analysed. Only two were present in all three plant species and two more were found in both *C. peregrina* and *M. pensylvanica*.

Interestingly, the richness and evenness components of diversity differed markedly between plant species. Nine Group 1 sequences were obtained from 37 *M. gale* nodules but only three dominated, accounting for 81% of the total. *C. peregrina* nodules had 15 Group 1 sequences in 30 nodules with four accounting for 60% of the total. Bayberry (*M. pensylvanica*) nodules yielded the highest diversity with 20 *Frankia* strain sequences in 29 nodules. Of the 20 sequences, 13 from Group 1 were found in 20 nodules, six Group 3 sequences were found in eight nodules, and one nodule yielded a sequence like that of Nod-/Fix- actinomycetes isolated from a variety of actinorhizal plants (Clawson *et al.*, 1998). Identical sequences were commonly found in plants growing at widely dispersed sites indicating that some *Frankia* strains are cosmopolitan. These results should be viewed with the understanding that strains in nodules with identical 16S sequences are not necessarily identical, only very similar.

The northern circumboreal species, *M. gale*, has historically been considered to be a 'promiscuous host' because it nodulated with most *Frankia* isolates in greenhouse trials (Torrey, 1990). It is not, however, overtly promiscuous in nature. This observation has been further confirmed in a separate study using PCR-RFLP of 16S rDNA PCR amplified from root nodules collected from *M. gale* nodules in Canada (Huguet *et al.*, 2001). Low diversity in *M. gale* nodules may be attributed to its preference for growing in water-saturated soils near lakes, swamps or bogs.

Such locations are typically acidic and low in oxygen; such conditions may limit the selection of *Frankia* to those strains capable of tolerating them.

In older studies, *M. gale* exhibited distinctive patterns of distribution of sp(+) and sp(–) (indicating the presence or absence of *Frankia* sporangia in nodules) root nodules with sp(+) strains more common in nodules collected at southern interior and coastal regions and the sp(–) strains more common in northern and western Maine (Schwintzer, 1990). The presence of sp(+) nodules positively correlated with the average number of frostfree days per year, and with the percentage organic matter in the soil, and negatively correlated with pH, with more sp(+) nodules found in more acidic soils. Although the sporulation phenotype has low resolution, its significance is enhanced by the finding that the diversity of frankiae in these nodules is low. A similar sorting by environment has been observed in sp(+) and sp(–) nodules growing on alders (see above). Thus, local edaphic factors are clearly important in determining which *Frankia* strains get into nodules and most likely how strains distribute among environments.

Since many *Myrica* spp. and *C. peregrina* can be nodulated by alder strains, the environmental distribution of Group 1 strains is potentially very wide. The same holds true for the elaeagnus strains from Group 3 that can infect *Myrica* spp. The degree of overlap is not known since the degree of individual strain specificity for different plant species has not been well documented. However, the wide geographical distribution of the plants from these groups most likely accounts for, and reinforces, the similarly wide distribution of homologous symbionts. The distribution of casuarina *Frankia* strains provides a contrast to this ubiquity.

6.4.3 The Casuarinaceae

The Casuarinaceae is a well-characterised family of four genera and roughly 96 species. Based on DNA sequencing, the Casuarinaceae has been placed in the Fagales (APG, 1998; Chase *et al.*, 1993). The family is easily recognised by its slender, wiry branches and highly reduced leaves. It is geographically restricted to Australia and the Melanesian region of the Pacific (Johnson and Wilson, 1989) but *Casuarina* spp. have been naturalised in islands and coastal regions of the Indian Ocean, Africa and the Americas. Recent molecular phylogenies found *Gymnostoma* L.A.S. Johnson sister to the rest of the family and *Allocasuarina* L.A.S. Johnson most derived (Maggia and Bousquet, 1994; Sogo *et al.*, 2001). This topology agrees with the traditional, morphological view based on branchlet and infructescence structure (*Figure 6.4A*) (Sogo *et al.*, 2001). Fossil evidence for the family dates back to the Eocene (55–39 MYA). The family had a much wider distribution in the past, with macrofossils of *Gymnostoma* and *Ceuthostoma* L.A.S. Johnson discovered outside of its present range in New Zealand and South America. The absence of both macrofossils and undisputed fossil pollen in the northern hemisphere has led most researchers to postulate that the family had its origins in Gondwanaland (Johnson and Wilson, 1989).

Members of all four genera nodulate, although the genera differ with respect to the type and diversity of *Frankia* strains with which they associate. *Gymnostoma* spp. interact with diverse Group 3 *Frankia* strains, while more derived members of the family, *Casuarina* L. and *Allocasuarina*, interact with a more reduced set of strains from Group 1. This observation has led authors to postulate that perhaps the family is evolving towards strain-specificity (Maggia and Bousquet, 1994), or the specificity of *Casuarina* and *Allocasuarina* is simply due to the drier habitats of Australia that these genera inhabit, where a smaller subset of *Frankia* strains are able to survive.

The ubiquity of alder and myrica frankiae in soils does not extend to the casuarina frankiae (Diem and Dommergues, 1990), even though both sets of strains belong to phylogenetic Group 1 and the Casuarinaceae is sister to the Betulaceae and Myricaceae among the higher hamamelids. A study done in Jamaica serves to illustrate the status of casuarina strains outside of the native range (Zimpfer *et al.*, 1997). A most probable number approach was used to estimate the relative abundance of frankiae capable of nodulating the native *Myrica cerifera* versus the exotic *Casuarina cunninghamiana* in soils collected at sites lacking actinorhizal plants. Myrica strains occurred in variable abundance at all sites sampled whereas no *C. cunninghamiana* strains were detected. As found elsewhere, the occurrence of myrica strains in Jamaican soils is independent of the host plant whereas casuarina strains seem to depend on the presence of the host. On the other hand, *C. cunninghamiana* is nodulated where it has been established in Jamaica suggesting that a compatible strain had been introduced along with the plant but has not spread. Indeed, in soil collected along transects leading away from *C. cunninghamiana* trees, NUs of casuarina frankiae diminished to undetectable levels at about 20 m away from the trees. Myrica frankiae NUs were present at all distances along the transects despite the nearest host being 25 km distant (Zimpfer *et al.*, 1999). This observation plus the fact that *Casuarina* plants must be deliberately inoculated in areas of the world where they are first planted (Diem and Dommergues, 1990; Simonet *et al.*, 1999) indicates that the presence of the host is critical for maintaining soil populations of casuarina frankiae.

Surprisingly, several lines of evidence indicate that only one group of closely related or identical strains is responsible for nodulating *Casuarina* and *Allocasuarina* spp. outside their normal geographic range (Fernandez *et al.*, 1989; Honerlage *et al.*, 1994; Maggia *et al.*, 1992; Nazaret *et al.*, 1989, 1991; Rouvier *et al.*, 1990, 1996; Simonet *et al.*, 1999). For example, except for one nodule harvested in Kenya, 160 nodules from five *Casuarina* and *Allocasuarina* species sampled outside of Australia from several countries yielded the same PCR-RFLP group that dominated culture collections. The same group was identified in other work using DNA–DNA reassociation (Fernandez *et al.*, 1989), and by PCR-RFLP of the intergenic regions between the *rrs* and *rrl* genes and between the *nifH* and *nifD* genes (Honerlage *et al.*, 1994; Maggia *et al.*, 1992; Rouvier *et al.*, 1992; Simonet, *et al.*, 1999). That group may be the one best adapted to a saprophytic lifestyle in an exotic environment and is the one most likely to be cultured from root nodules (Simonet *et al.*, 1999).

More diversity prevails within the native range of the plants. Twenty-two nodules collected from two *Casuarina* spp. and two *Allocasuarina* spp. in Australia yielded a total of five *rrn* and *nif* PCR-RFLP groups (Rouvier *et al.*, 1996). Nodules from *C. equisetifolia* contained one group in six of eight nodules and another group was present in the remaining two nodules. The dominant strain belonged to the same PCR-RFLP group found in nodules collected from regions of the world where *Casuarina* has been introduced. An additional group was found in six *C. cunninghamiana* nodules, another was found in four *Allocasuarina lehmaniana* nodules, and the final group was found in four *A. torulosa* nodules. Despite the relatively small number of nodules used in the study some degree of plant species–*Frankia* strain specificity was noted. In another study using a similar approach, seven groups were found among 110 nodules sampled from five Casuarinaceae species in Australia (Simonet *et al.*, 1999). Each IGS group was found in only one or two of the plant species. This specificity could be due to the host plant selecting specific strains in the rhizosphere from among a population of strains, or to environmental selection limiting the types of frankiae that are available in a particular soil.

Another member of the Casuarinaceae, *Gymnostoma*, provides a contrast to its nearest relatives. *Gymnostoma* spp. are nodulated by members of phylogenetic Group 3 *Frankia* strains rather than by the casuarina strains from Group 1. The reason for this symbiont shift is not obvious but may be related to an early allopatric distribution of the plants and presumably symbionts (*Gymnostoma* in the north and east of Gondwana islands; *Casuarina*/*Allocasuarina* in the drier Australia) (Simonet *et al.*, 1999). Consistent with this hypothesis is the observation that the only *Casuarina* species present in New Caledonia, *C. collina*, commonly contains both Group 1 casuarina strains and Group 3 strains; the latter are similar or identical to those nodulating *Gymnostoma* (Gauthier *et al.*, 1999).

Also in contrast to other members of the Casuarinaceae is a relatively higher diversity of *Frankia* strains found in *Gymnostoma* nodules. A study in New Caledonia using PCR-RFLP of the ribosomal *rrs-rrl* intergenic spacer as a probe detected 17 different RFLP patterns in 358 nodules from eight *Gymnostoma* species (Navarro *et al.*, 1999). No sharp species specificity was noted among the 17 patterns but a correlation was found between soil type, host species and RFLP pattern. One pattern predominated and accounted for 56% of the total from all species. It was, however, absent from 45 nodules obtained from two *Gymnostoma* species growing at acidic sites and was thus deemed specific for ultramafic soils. Another pattern belonged to a more cosmopolitan strain group found in all species on all soils and was represented in 14.5% of the total. Strains that nodulate *Gymnostoma* are widespread in New Caledonia with some indication, based on trapping experiments, that populations are amplified in soils beneath *Gymnostoma* but also persist without a covering host plant (Gauthier *et al.*, 2000). Populations are also amplified in the rhizosphere of a member of a non-nodulating member of the Rhamnaceae (Gauthier *et al.*, 2000).

The conclusion from these studies and others (Jaffre *et al.*, 2001) is that different populations of Group 3 frankiae colonize different *Gymnostoma* spp. more as a function of soil type rather than species specificity. This work, together with the *M. gale* work noted above, provide the strongest examples of strain sorting by environment.

6.4.4 The Elaeagnaceae

The Elaeagnaceae is a well-defined family of three genera *Elaeagnus* L., *Hippophae* L. and *Shepherdia* Nutt. *Elaeagnus* is distributed across North America and Eurasia, although the range of this genus has greatly increased due to cultivation and use in land reclamation (Baker and Schwintzer, 1990). In Australia and western North America, *Elaeagnus* is often considered an invasive exotic

(Mabberely, 1988). *Elaeagnus* has about 45 described species, most of which have been shown to nodulate (*Table 6.1*). *Hippophae* is native to central Asia, distributed from the North Sea to the Black Sea and east to the Himalayas. The number of species within *Hippophae* has been in dispute, with numbers ranging from one to seven with numerous subspecies. This discrepancy in classification is primarily due to hybridisation and intergradation of morphology (Bartish *et al.*, 2002). *Shepherdia* is composed of three species, restricted to North America (Mabberely, 1988).

The phylogenetic placement of the Elaeagnaceae within angiosperm diversity has been in dispute. The *rbcL* angiosperm phylogeny placed the family close to the Rhamnaceae within the Rosales (Chase *et al.*, 1993), the order in which it is currently classified (APG, 1998). Further phylogenetic studies have placed the Elaeagnaceae sister to the Rhamnaceae, within the Rhamnaceae, or in a loose alliance with the Barbeyaceae, Ulmaceae and Cannabaceae (Richardson *et al.*, 2000). A comprehensive phylogeny for the Elaeagnaceae has not been constructed, yet members of all three genera were included in an *rbcL* phylogeny constructed by Swensen (1996). In that tree, *Hippophae* was sister to *Elaeagnus* and *Shepherdia*. The Elaeagnaceae has a late appearance in the fossil record (Oligocene 22–39 MYA), and based on present distribution, the family most likely originated in Laurasia before the breakup of the continents in the northern hemisphere (Bartish *et al.*, 2002).

Elaeagnus, *Shepherdia* and *Hippophae* spp. are generally well nodulated even in geographical regions where they are not native or where alternate hosts are absent. The *Frankia* strains present in root nodules seem to be shared among the three genera and all belong to Group 3 (Benecke, 1969; Clawson *et al.*, 1998; Huguet *et al.*, 2001; Jamann *et al.*, 1992). Thus, elaeagnus strains from Group 3 can be considered to be cosmopolitan. In part, their wide distribution may stem from their roles as potentially effective symbionts in four of the actinorhizal plant families (Elaeagnaceae, Rhamnaceae, Casuarinaceae (*Gymnostoma*), Myricaceae) and their less well-characterized roles as occasional symbionts in the Betulaceae or as associative strains not clearly involved in nitrogen fixation in the Rosaceae, *Ceanothus* and members of the Casuarinaceae other than *Gymnostoma* (Benson and Clawson, 2000). Like alder strains, the specificity exhibited by individual Group 3 strains is not well characterized.

In Europe, *Elaeagnus* spp. have been recently introduced but *Hippophae rhamnoides* was present throughout Europe during all stages of the late glaciation and probably helped maintain the populations of Group 3 frankiae (Jamann *et al.*, 1992). The same situation holds for North America where most *Elaeagnus* species are introduced and universally nodulated. The native *Shepherdia* is infected by a wide variety of Group 3 frankiae that can also be presumed to infect introduced *Elaeagnus* spp. (Huguet *et al.*, 2001).

Few reports have directly addressed the diversity and distribution of *Frankia* strains that infect members of the Elaeagnaceae. However, all species in the family examined thus far are effectively nodulated only by a set of *Frankia* strains belonging to phylogenetic Group 3. Molecular studies using sequencing of 16S rRNA genes and DNA–DNA hybridisation have indicated that isolated strains are diverse within the confines of Group 3 (Fernandez *et al.*, 1989; Huguet *et al.*, 2001; Nazaret *et al.*, 1989). However, a survey of published Clade 3 sequences from the field reveals that some are cosmopolitan (Clawson *et al.*, 1998; Nalin *et al.*, 1997). For example, an identical partial 16S rDNA sequence has been reported from an *E. angustifolia* and *Myrica pensylvanica* growing in Connecticut, an *E. pungens* in Hamilton, New Zealand, and *Discaria trinervis, Talguenea quinquenervia, Trevoa trinervis* and an unidentified *Elaeagnus* in Chile (Clawson *et al.*, 1998). The same sequence was also reported as belonging to a major group of strains in France (Ea1-2, HR27-14) (Jamann *et al.*, 1992). While this finding may partly reflect the DNA region that was sequenced, it does support the notion that a group of elaeagnus strains (Genomic group 4 (Fernandez *et al.*, 1989)) is widely distributed in nature. A study focused on the distribution of elaeagnus strains through a soil column collected from an area lacking host plants yielded seven PCR-RFLP profiles for DNA obtained from nodules induced by trapping experiments. Six of the profiles corresponded to previously identified genomic species in France and the seventh, collected from the deepest layers, was unique. Thus, a relatively high diversity was found in the samples but it was within the bounds of the diversity of strains known to infect the plants (Nalin *et al.*, 1997).

It would be interesting to determine the patterns of richness and evenness of elaeagnus strains across the native distribution zones of the various species. For example, Russian olive, *E. angustifolia* L., has been widely transplanted as a windbreak or ornamental throughout the world. A useful study might be to compare the diversity of strains found in root nodules within its native range to that of strains found outside its native ranges where it has been transplanted.

6.4.5 The Rhamnaceae

The Rhamnaceae is distributed worldwide, containing 50 genera and about 900 species (Richardson *et al.*, 2000). Traditional taxonomic treatments have placed the Rhamnaceae with the Vitaceae based on shared floral characters (Takhtajan, 1980) or with the Elaeagnaceae based on shared vegetative features (Thorne, 1992). The angiosperm *rbcL* phylogeny placed the Rhamnaceae within the Eurosid I Clade and indicated a close relationship between the Rhamnaceae and the Elaeagnaceae in the Rosales. Quite surprisingly, in past *rbcL* reconstructions, the Rhamnaceae is paraphyletic with Barbeyaceae, Dirachmaceae, and the Elaeagnaceae (Richardson *et al.*, 2000). Further molecular data have not supported this topology, but instead favour a monophyletic Rhamnaceae (Richardson *et al.*, 2000).

Six genera within the Rhamnaceae have been identified as nodulating with *Frankia* strains. Except for *Ceanothus* L., all belong to the tribe Colletieae Reis. Ex. Endl. These genera are: *Colletia* Comm. ex Juss. (17 species found in South America), *Discaria* Hook. (15 described species found in South America, Australia, and New Zealand), *Kentrothamnus* Susseng. and Overk. (one species restricted to Bolivia), *Retanilla* (DC) Brongn. (four species found in Peru and Chile), and *Trevoa* Mires ex. Hook. (one species found in South America). *Trevoa* was recently revised to include the previously separate, actinorhizal genus *Talguenea* (Tortosa, 1992). The one member of the tribe Colletieae whose actinorhizal nature is unconfirmed is *Adolphia* Meisner., located in southwestern North America (Cruz-Cisneros and Valdés, 1991). The other actinorhizal genus in the Rhamnaceae is *Ceanothus* L., a strictly North American genus of approximately 55 species (Mabberely, 1988).

Most nodulated members of the Rhamnaceae grow in dry matorral or chaparral regions.

The recent molecular phylogenies constructed for the Rhamnaceae by Richardson *et al.* (2000) found that the five genera within the tribe Colletieae were indeed monophyletic (*Figure 6.4E*). However, the genus *Ceanothus* did not cluster with the Colletieae, giving rise to the possibility that the actinorhizal symbiosis may have evolved twice within the Rhamnaceae, although the authors indicate that the inclusion of more data may unite *Ceanothus* as sister to the Colletieae.

The Rhamnaceae appears to be a very old lineage, with a rhamnaceous fossilised flower and pollen dating to 94–96 MYA to give a minimum age for the family (Basinger and Dilcher, 1984). Both *Ceanothus* and the tribe Colletieae belong to a large Clade within the family termed the ziziphoid group, which is mostly distributed in the southern hemisphere, suggesting that this branch of the family may be of Gondwanan origin. The major exception to this distributional hypothesis is the genus *Ceanothus*, that Richardson *et al.* (2000) have suggested may have been part of the ziziphoid group with a Laurasian distribution before the Gondwanan split and has had a relictual distribution in North America, primarily California. This hypothesis requires that the genus *Ceanothus* be quite old (65 MYA). An ancient split between *Ceanothus* and the tribe Colletieae may explain why the two groups differ in the *Frankia* strains with which they associate.

Members of the tribe Colletieae in the southern hemisphere associate with ubiquitous Group 3 *Frankia* strains that potentially also associate with the Elaeagnaceae, Myricaceae and *Gymnostoma*. Although several *Frankia* strains have been isolated and characterized from the root nodules of the South American Colletieae (Carú, 1993; Schwencke and Caru, 2001), studies have not yet been done on the ecological diversity patterns of strains in the nodules from different species or environments.

The North American *Ceanothus* spp., on the other hand, associate primarily with Group 2 *Frankia* strains similar to those that nodulate *Datisca*, *Coriaria* and the actinorhizal Rosaceae (see *Figure 6.3*) (Benson and Clawson, 2000). The approximately 55 species of this genus are limited to western parts of North America with the range of one, *C. americanus*, extending to the east coast. Some work has addressed the diversity of symbionts in North American *Ceanothus* root nodules.

An initial study on *C. americanus* found a relatively high level of diversity of *Frankia* strains in root nodules as assessed using RFLP of total DNAs probed with *nifDH* genes or with random probes (Baker and Mullin, 1994). In a separate study, repetitive extragenic palindromic PCR (Rep-PCR) was used as a measure of diversity in six *Ceanothus* spp. taken from seven sites in a 10 mile radius along coastal southern California (Murry *et al.*, 1997). Overall, 54 nodules yielded 11 different Rep-PCR patterns, some of which were very similar to others. Subsequent sequencing of a region of the 16S rRNA gene from a few nodules indicated habitation by Group 3 *Frankia* strains, that is, elaeagnus strains.

This finding is at odds with other studies that have detected Group 2 *Frankia* strains in *Ceanothus* nodules. The picture is further clouded by the finding that some isolates from *Ceanothus* nodules can infect *Elaeagnus* spp. while others belong to a group of Nod-/Fix- strains that sometimes occupy actinorhizal nodules. None of the isolates, however, can reinfect *Ceanothus* plants (Lechevalier and Ruan, 1984; Ramirez-Saad *et al.*, 1998; Torrey, 1990). California is considered to be the centre of *Ceanothus* distribution and might be expected to support a diverse population of ceanothus frankiae, by analogy with the situation for casuarina strains in their native Australia. On the other hand, cohabitation of *Ceanothus* nodules by more than one organism might explain some of the diversity observed by molecular techniques. Additional work needs to be done to sort out the relationship of the different lineages of bacteria that inhabit *Ceanothus* root nodules.

A study on *Ceanothus* in Oregon suggested a relationship between strains and the soil conditions from where nodules were harvested (Ritchie and Myrold, 1999). This work relied on a PCR-RFLP analysis of the ribosomal *rrs-rrl* region. Four RFLP groups were identified with one predominating in mountainous regions and two others limited to the Willamette Valley. The fourth group was limited to *C. americanus* collected from Tennessee. Overall, the diversity of strains reported was less than that reported using other methods. In a similar manner, sampling of nodules from co-populations of different *Ceanothus* species indicated that *Frankia* strain PCR-RFLP patterns were more likely to be related to the environment from which the nodules came than to the plant species infected (Jeong, 2001; Jeong and Myrold, 1999).

In their native range, *Ceanothus* strains have been enumerated by trapping experiments from soils with and without hosts. Populations have been found to be amplified beneath *Ceanothus* stands although sites lacking host plants retained a small population (Jeong, 2001; Wollum II *et al.*, 1968). Low levels of *Ceanothus* nodulation by soils beneath old-growth (300 years) Douglas Fir stands has been noted (Wollum II *et al.*, 1968). However, even in soil beneath host plants, the nodulation capacity is low; in one study nodulation units were estimated at 3.6 to 5.2 NUs g^{-1} soil, which is at the low end of estimates for alder-type frankiae in soils lacking alders (Jeong, 2001). This low population density seems to be characteristic of *Ceanothus* strains and may reflect an actual low population or an inherent difficulty in nodulating *Ceanothus* plants in the greenhouse (Rojas *et al.*, 2001). *Frankia* strains in trapping experiments were found to have similar levels of diversity in both forest soil and *Ceanothus* stands albeit at different population densities. No strong correlation has yet been found with strain type (as determined by rep-PCR or PCR-RFLP) and *Ceanothus* species.

Ribosomal RNA gene sequences amplified from *Ceanothus* nodules are generally very similar (99–100%) to each other and to some amplified from nodules in the Rosaceae, Datiscaceae and Coriariaceae, suggesting that some Group 2 strains are globally dominant (Benson *et al.*, 1996; Ramirez-Saad *et al.*, 1998; Ritchie and Myrold, 1999). This low diversity may also reflect the fact that relatively few 16S rDNA sequences have been obtained from Group 2 *Frankia* strains. Plants from these families share an overlapping range in western North America although *Coriaria* and *Datisca* are more widespread with disjunct populations in several parts of the world (Benson and Clawson, 2000). It is possible that *Ceanothus* became geographically isolated from other Rhamnaceae and subsequently specialised in the Clade 2 *Frankia* strains that may have been more adapted to the environment or simply more numerous because of their proximity to other actinorhizal plants.

6.4.6 The Coriariaceae

The Coriariaceae is a monotypic family whose taxonomic placement has varied considerably in different past treatments; molecular data firmly place it within the Cucurbitales (APG, 1998). The only genus, *Coriaria*, consists of between five and 20 species. Such a wide range in the number of described species, depending on the particular classification, is due to the large, shared morphological variation displayed by members of this genus (Yokoyama *et al.*, 2000). *Coriaria* L. has one of most spectacular native geographic distributions of any genus of its size, being found in four areas worldwide, the Mediterranean, Southeast Asia, Central and South America, and the Pacific islands of New Zealand and Papua New Guinea (Skog, 1972). Such a conspicuous geographic disjunction has attracted many previous authors to hypothesise about the origin and diversification of *Coriaria* (see review in Yokoyama *et al.* (2000)). In a recent molecular phylogeny, Yokoyama *et al.* (2000) tested these previous hypotheses and found that the most basal diverging members of the genus are present in Asia and Central America, leading to the conclusion that the genus originated in either Eurasia or North America. In addition, application of a molecular clock hypothesis led the authors to suggest that the genus had an origin some 60 MYA, far older than an estimate of 5–11 MYA based on fossil evidence (Yokoyama *et al.*, 2000). Based on the present distribution of *Coriaria*, an older date for the origin and diversification of the family may indeed be correct.

The molecular phylogeny for *Coriaria* produced two main Clades. Clade I consisted of taxa from the Mediterranean and Asia and Clade II consisted of taxa from Central and South America. The authors concluded that simple vicariance and dispersal caused by glaciation and drying during the Cenozoic may account for the distribution of *Coriaria* in Clade I, but could not be used to explain the distribution of the *Coriaria* diversity present in Clade II. The topology presented in Clade II favours the interesting hypothesis of long-distance dispersal from Central America to the Pacific islands, followed by another migration back to South America (Chile) (*Figure 6.4B*).

Nodules have been observed on *Coriaria* species from New Zealand (*C. arborea*, *C. plumosa*), from Central America (*C. microphylla*), Europe (*C. myrtifolia*) and Central Asia (*C. nepalensis*) (Mirza *et al.*, 1994a; Nick *et al.*, 1992; Silvester, 1977). The total number of *Coriaria* spp. able to nodulate has yet to be determined. However, at least one species from all four major zones of diversity has been shown to nodulate, and known nodulating species are present in both Clade I and Clade II, indicating that the association with *Frankia* strains appears to be widespread throughout the genus. The distribution of *Coriaria* strains in soils devoid of *Coriaria* hosts has not been addressed. Some studies in New Zealand indicate that *Coriaria arborea* plants are nodulated wherever planted and will readily nodulate in new volcanic soils.

The *Frankia* strains associating with *Coriaria* are closely related to the unisolated Group 2 strains that associate with *Ceanothus*, to strains that associate with members of the actinorhizal Rosaceae and to strains associating with *Datisca*, (Benson and Clawson, 2000). Available information suggests that the richness of strains is low in the Rosaceae, Datiscaceae and Coriariaceae. For example, *Coriaria arborea* nodules in New Zealand yielded only two 16S rRNA gene sequences, differing by a single nucleotide, from 12 nodules collected at distant locales on the North Island (Clawson *et al.*, 1997). Additional sequences from a total of 30 nodules from *C. arborea* and four more from *C. plumosa* collected in New Zealand yielded the same sequences (DRB, unpublished).

Similarly, a collection of short 16S rDNA sequences spanning another 16S region (V6) PCR-amplified from *Coriaria* nodules collected in New Zealand, France and Mexico had only one mismatch in 274 bp analysed (Nick *et al.*, 1992). A further study in Pakistan used the V2 16S rDNA region and found some diversity in both *Coriaria nepalensis* and *Datisca cannabina* that would have been missed using the region analysed by Nick *et al.* (1992). Nevertheless, the number of differences among the sequences was still low suggesting low overall diversity of frankiae within the Coriariaceae and Datiscaceae. No studies have been done to date on the distribution of these strains in soils from areas that lack *Coriaria* spp., so their ubiquity remains unknown.

6.4.7 The Datiscaceae

As traditionally circumscribed, the Datiscaceae *sensu lato* includes three genera, *Datisca* L. (including two species), *Tetrameles* R. Br. (one species), and *Octomeles* Miq. (one species). The family is classified in the order Cucurbitales (APG, 1998). Recent molecular phylogenetic work within the family has shown the Datiscaceae *sensu lato* to be paraphyletic with respect to the Begoniaceae. This result has supported the classification, as originally proposed by Airy Shaw (1964) based on morphology, of *Tetrameles* and *Octomeles* in Tetramelaceae, leaving only the genus *Datisca* in the Datiscaceae. Therefore, the revised Datiscaceae no longer contains non-nodulating genera (Swensen *et al.*, 1994, 1998).

The two species of *Datisca*, *D. cannabina* L. and *D. glomerata* (Presl.) Baill., are adapted to Mediterranean climates and have an interesting distribution. *D. cannabina* is found in the Mediterranean basin and *D. glomerata* is found on the western slope of the Sierra Nevada from northern California to Baja California (Swensen *et al.*, 1994). Plants in California and the Mediterranean basin are known to have some taxonomic affinities (North America and Europe were only separated since the Tertiary) (Solbrig *et al.*, 1977). Detailed phylogenies for *Datisca* indicate that geographic subdivision rather than long-distance dispersal accounts for the present day distribution (Swensen *et al.*, 1998). Since the Mediterranean climate is relatively new, established only since the Pleistocene, it is more likely that after the vicariance both species converged on the Mediterranean climate instead of an ancestor to the two species being preadapted to the Mediterranean climate (Solbrig *et al.*, 1977). Fossil wood from India suggests that the Datiscaceae may have arisen in the Eocene (55–39 MYA), although it is important to point out that there is some question whether the fossil remains are correctly identified as Datiscaceae (Cronquist, 1981). Both species of *Datisca* are actinorhizal (Swensen *et al.*, 1994).

As noted above, *Frankia* strains that inhabit *Datisca* nodules appear to be closely related to those found in *Coriaria*, *Ceanothus* and the actinorhizal Rosaceae (Benson and Clawson, 2000; Benson *et al.*, 1996; Mirza *et al.*, 1994a). In fact, crushed nodule inoculations indicate that *Dryas*, *Ceanothus*, *Datisca* and *Coriaria* are in the same cross inoculation group (Kohls *et al.*, 1994; Mirza *et al.*, 1994b;

Torrey, 1990). The distribution of datisca frankiae in soils has not been extensively studied. Some work indicates that the distribution of strains parallels the distribution of plants on a regional scale. For example, in Pakistan, all soils tested yielded nodules on *Datisca* except one from an eroded area (Mirza *et al.*, 1994a). Companion experiments testing for the nodulation of *Coriaria* with the same soils yielded less nodulation with some soils failing to nodulate, indicating that *Coriaria nepalensis* was more difficult to nodulate, in agreement with previous observations (Bond, 1962). The distribution of datisca *Frankia* strains outside the native range of the plants is unknown.

6.4.8 The Rosaceae

The Rosaceae is a large, economically important family with roughly 122 genera and 3000 species (Heywood, 1993). The family is distributed worldwide, but is found especially in north temperate regions. The Rosaceae has traditionally been subdivided into four subfamilies; the Rosoideae, the Spiraeoideae, the Maloideae and the Amygdaloideae on the basis of fruit type (Schulze-Menz, 1964). Due to the family's economic importance, it has been subject to many evolutionary and phylogenetic studies (Evans *et al.*, 2000; Kalkman, 1988; Morgan *et al.*, 1994; Potter, 1997; Potter *et al.*, 2002; Rohrer *et al.*, 1991). The first *rbcL* phylogeny (Morgan *et al.*, 1994) for the Rosaceae found that the four traditional subfamilies were not natural, and instead Clades appeared to correspond to base chromosome number and not fruit type. The *rbcL* phylogeny and later phylogenetic studies using other molecular markers, have found a strongly supported Clade consisting of the four actinorhizal genera of the Rosaceae (*Figure 6.4F*). These genera include *Cercocarpus* HBK (six to ten species restricted to southwestern North America), *Purshia* (eight species also restricted to southwestern North America), *Chamaebatia* (two species found in California), and *Dryas* (two species found circumpolar in alpine and Arctic habitats) (Evans *et al.*, 2000; Morgan *et al.*, 1994; Potter, 1997; Potter *et al.*, 2002). *Cowania* was recently combined with *Purshia* under the name *Purshia* (Henrickson, 1986).

The relationships near the base of the Rosaceae phylogenetic tree have not been resolved, but studies based on the chloroplast *matK* and *trnL-F* regions suggest that there are three main lineages in the family: the traditional Rosoideae (with some modifications), the actinorhizal Clade, and the rest of the family (Potter *et al.*, 2002). This orientation suggests that either the ability to nodulate evolved once as the family was beginning to diverge, or that nodulation was present in the common ancestor of the family and was lost twice in its diversification.

Aside from a few sequences of 16S rDNA that have been obtained by PCR amplification from root nodules (Benson *et al.*, 1996; Bosco *et al.*, 1994), very little is known about the ecology or diversity of Clade 2 frankiae that inhabit nodules in the Rosaceae. As noted above, the actinorhizal Rosaceae appear to associate only with *Frankia* strains related to those that nodulate the genera *Ceanothus*, *Coriaria* and *Datisca*. These four groups of plants share, at least in part, a range in western North America, although *Coriaria* and *Datisca* are more widespread (see above). Interestingly, an identical partial 16S rDNA sequence has been reported in *Purshia tridentata*, *P. glandulosa*, *Cowania stansburiana*, *Chamaebatia foliosa*, *Ceanothus velutinus*, *C. griseus*, *C. ceruleus* and *Dryas dummondii* all originating in

North America (DRB, unpublished). It is tempting to speculate that the presence of Clade 2 *Frankia* strains in these plants is related to their overlapping biogeography during the breakup of Laurasia and Gondwana in the late Cretaceous.

Nodulation in the rosaceous actinorhizal plants is sporadic (Klemmedson, 1979). One study reported nodulation rates of 8.3–32.2% of field plants of *Cercocarpus, Cowania* and *Purshia* (Nelson, 1983). Some species of *Dryas* have not been observed to nodulate (Kohls *et al.*, 1994). Both *D. octapetala* and *D. integrifolia* have been reported to bear nodules in the older literature but the observations are in need of verification (Baker and Schwintzer, 1990). A putative hybrid between *D. drummondii* and *D. integrifolia* found in Glacier Bay National Park, *D. drummondii*, var. *eglandulosa*, apparently does not nodulate even when deliberately inoculated in the greenhouse (Kohls *et al.*, 1994). When *Dryas* or other actinorhizal rosaceous plants are inoculated in the greenhouse with either soil or crushed nodules, nodules develop beginning 6–8 weeks after inoculation. This slow development contrasts with the 2–3 weeks normally required for nodules to appear on inoculated *Alnus* or *Myrica*.

Few studies have focused on the presence of rosaceous-infective frankiae in soils. What little information is available seems to suggest that strains are distributed in areas where the plants grow but are not abundant outside those areas. Kohls *et al.* (1994) found that soils from Glacier Bay, Alaska, where *Dryas* is abundant, failed to induce nodules on *Cercocarpus betuloides* but did contain *Frankia* strains that nodulated *Dryas drummondii* and *Purshia tridentata*. Crushed nodules from *Dryas* also nodulated *Dryas* and *Purshia* but not *Cercocarpus* suggesting that the cercocarpus strains may differ in some manner from the dryas strains. In the same study, ineffective (unable to fix nitrogen) nodules were formed on *Cercocarpus ledifolius* by CcI3, Cms13 and EuI1b. These strains are from *Casuarina cunninghamiana, Cowania mexicana* and *Elaeagnus umbellata*, respectively, suggesting that these strains may participate in forming ineffective nodules in the field. Other work has shown that *Ceanothus, Cercocarpus, Cowania* (now *Purshia*), *Chamaebatia* and *Purshia* can be nodulated by crushed nodules or soil from beneath *Chamaebatia* and *Cowania* (Nelson and Lopez, 1989; D. Nelson, personal communication).

6.5 Summary

The present patterns of distribution of actinorhizal plants and *Frankia* strains have been formed by the evolutionary histories of the plants, the movement of continents and adaptation of both symbionts to new environments as they have emerged over the past 120 million years. The eight actinorhizal plant families have very different distributions, estimated times of origin and fossil histories. The Casuarinaceae and Rhamnaceae appear to have a Gondwanan origin and the remaining actinorhizal families appear to have originated in Laurasia. The oldest fossil evidence provides a minimum age for some actinorhizal lineages, the Rhamnaceae and Myricaceae, in the Cretaceous (94 MYA). Molecular evidence suggests that the various lineages that eventually gave rise to present day actinorhizal plants were established shortly after the Mid- to Late Cretaceous appearance of eudicots about 125 MYA (Crane *et al.*, 1995; Magallon *et al.*, 1999).

This was a time period dominated by the separation of Gondwana from Laurasia. The major Groups of *Frankia* strains may have emerged at about the same time (Benson and Clawson, 2000).

Beyond distributions and origin dates, the actinorhizal families differ in the degree of nodulation within each family. In the Casuarinaceae, Coriariaceae, Datiscaceae and Elaeagnaceae, all genera nodulate. In the Betulaceae, Myricaceae, Rhamnaceae and Rosaceae, a variable number of the lineages nodulate ranging from three of four genera in the Myricaceae to five of 122 genera in the Rosaceae. Molecular phylogenies have demonstrated that the actinorhizal plant families have a common ancestor that was predisposed to nodulation (Soltis *et al.*, 1995). The number of times this predisposition became reality will never be known with any certainty. It is clear however that the symbiosis has been lost on many occasions as illustrated by the sporadic distribution of nodulating plants between and within orders, families and genera (Benson and Clawson, 2000).

At the local level, patterns of *Frankia* strain distribution are generally characterized by dominance of one particular strain depending on edaphic factors present in the soil (Clawson and Benson, 1999b; Huguet *et al.*, 2001; McEwan *et al.*, 1999). Soil conditions appear at least as important as, if not determinative, in the strain of *Frankia* that succeeds in nodulating appropriate hosts. This conclusion is supported by direct demonstrations of dominance in alder and myrica stands (Clawson and Benson, 1999a, 1999b; McEwan *et al.*, 1999; Van Dijk, 1984), and the observation that casuarinas are necessary for the persistence of casuarina strains when the plants are introduced outside their native range. This dominance effect forms the local pieces of the greater geographical mosaic.

A broader view of the patterns of symbiont associations provides some interesting observations related to vicariance of plant distributions. For example, there exist at least two cases where geographic separation has apparently led to a sorting of frankiae within a plant family. The cases include the South American Rhamnaceae versus the North American *Ceanothus* which interact with Clade 3 and Clade 2 frankiae respectively, and the Australian *Casuarina* versus the Pacific island species of *Gymnostoma* that interact mainly with Clade 1 and Clade 3 frankiae, respectively. It is possible that ancestors of these genera were infected by a greater range of *Frankia* strains that narrowed as the plants radiated into new environments. The mechanism of specialisation is unknown but might include differing abilities of *Frankia* strains to adapt to particular soils or climates, co-speciation of the plant and symbionts, or bottleneck effects on bacterial and plant diversity during climate fluctuations.

Another pattern that emerges indicates that the more widely distributed plants, such as *Alnus* and *Myrica* are infected by strains that are also widely distributed in soil, whereas the geographically limited plants *Casuarina* and *Allocasuarina* are infected by strains that are also geographically limited. Similarly, *Elaeagnus* species are globally distributed and strains that infect (Group 3) them also appear to be cosmopolitan. Frankiae that infect *Elaeagnus* species are also capable of infecting most nodulated members of the Rhamnaceae, plus *Gymnostoma* of the Casuarinaceae, and, to a lesser degree, some alders and many myricas. In this regard, less is known about the distribution of Group 2 frankiae. Those strains form the basal group of *Frankia*, and seem to be, as far as is known, obligate symbionts, although some evidence suggests that they can persist without the continued presence of a host plant (Jeong, 2001). At present they are considered to have less diversity than strains in Groups 1 and 3. This lack of diversity may be an artifact of the few sequences that have been obtained or it may reflect the lack of a soil existence and increased reliance on the host. For that reason, one might anticipate that their distribution in soil parallels the patchy distribution of their hosts. This hypothesis remains to be tested.

References

Airy Shaw, H.K. (1964) Diagnoses of new families, new names, etc. for the seventh edition of Willis's "Dictionary". *Kew Bull.* **18**: 249–273.

An, C.S., Riggsby, W.S., Mullin, B.C. (1985) Relationships of *Frankia* isolates based on deoxyribonucleic acid homology studies. *Int. J. Syst. Bacteriol.* **35**: 140–146.

Angiosperm Phylogeny Group (1998) An ordinal classification for the families of flowering plants. *Ann. Missouri Bot. Gard.* **85**: 531–553.

Arveby, A.S., Huss-Danell, K. (1988) Presence and dispersal of infective *Frankia* in peat and meadow soils in Sweden. *Biol. Fert. Soils* **6**: 39–44.

Baker, D.D. (1987) Relationships among pure cultured strains of *Frankia* based on host specificity. *Physiol. Plant.* **70**: 245–248.

Baker, D., O'Keefe, D. (1984) A modified sucrose fractionation procedure for the isolation of frankiae from actinorhizal root nodules and soil samples. *Plant Soil* **78**: 23–28.

Baker, D.D., Mullin, B.C. (1994) Diversity of *Frankia* nodule endophytes of the actinorhizal shrub *Ceanothus* as assessed by RFLP patterns from single nodule lobes. *Soil. Biol. Biochem.* **26**: 547–552.

Baker, D.D., Schwintzer, C.R. (1990) Introduction. In: Schwintzer, C.R., Tjepkema, J.D. (eds) *The Biology of Frankia and Actinorhizal Plants*, pp. 3–13. Academic Press: New York.

Bartish, I.V., Jeppsson, N., Nybom, H., Swenson, U. (2002) Phylogeny of *Hippophae* (Elaeagnaceae) inferred from Parsimony Analysis of chloroplast DNA and morphology. *Syst. Bot.* **27**: 41–54.

Basinger, J., Dilcher, D. (1984) Ancient bisexual flowers. *Science* **224**: 511–513.

Benecke, U. (1969) Symbionts of alder nodules in New Zealand. *Pl. Soil* **30**: 145–149.

Benson, D.R., Clawson, M.L. (2000) Evolution of the actinorhizal plant symbioses. In: Triplett, E.W. (ed) *Prokaryotic Nitrogen Fixation: A Model System for Analysis of Biological Process*, pp. 207–224. Horizon Scientific Press: Wymondham, UK.

Benson, D.R., Silvester, W.B. (1993) Biology of *Frankia* strains, actinomycete symbionts of actinorhizal plants. *Microbiol. Rev.* **57**: 293–319.

Benson, D.R., Stephens, D.W., Clawson, M.L., Silvester, W.B. (1996) Amplification of 16S rRNA genes from *Frankia* strains in root nodules of *Ceanothus griseus, Coriaria arborea, Coriaria plumosa, Discaria toumatou*, and *Purshia tridentata*. *Appl. Environ. Microbiol.* **62**: 2904–2909.

Bond, G. (1962) Fixation of nitrogen in *Coriaria myrtifolia*. *Nature (London)* **193**: 1103–1104.

Bosco, M., Fernandez, M.P., Simonet, P., Materassi, R., Normand, P. (1992) Evidence that some *Frankia* sp. strains are able to cross boundaries between *Alnus* and *Elaeagnus* host specificity groups. *Appl. Environ. Microbiol.* **58**: 1569–1576.

Bosco, M.S., Jamann, S., Chapelon, C., Simonet, P., Normand, P. (1994) *Frankia* microsymbiont in *Dryas drummondii* nodules is closely related to the microsymbiont of *Coriaria* and genetically distinct from other characterized *Frankia* strains. In: Hegazi, H.A., Fayez, M., Monib, M. (ed) *Nitrogen Fixation with Non-legumes*, pp. 173–183. The American University in Cairo Press: Cairo, Egypt.

Callaham, D., DelTredici, P., Torrey, J.G. (1978) Isolation and cultivation in vitro of the actinomycete causing root nodulation in *Comptonia*. *Science* **199**: 899–902.

Carú, M. (1993) Characterization of native *Frankia* strains isolated from Chilean shrubs (Rhamnaceae). *Plant and Soil* **157**: 137–145.

Chase, M.W., Soltis, D.E., Olmstead, R.G. *et al.* (1993) Phylogenetics of seed plants: An analysis of nucleotide sequences from the plastid gene *rbc*L. *Ann. Missouri Bot. Gard.* **80**: 528–580.

Chen, Z., Manchester, S.R., Sun, H. (1999) Phylogeny and evolution of the Betulaceae inferred from DNA sequences, morphology, and paleobotany. *Am. J. Bot.* **86**: 1168–1181.

Clawson, M.L., Benson, D.R. (1999a) Dominance of *Frankia* strains in stands of *Alnus incana* subsp. *rugosa* and *Myrica pensylvanica*. *Can. J. Bot.* **77**: 1203–1207.

Clawson, M.L. and Benson, D.R. (1999b) Natural diversity of Frankia strains in actinorhizal root nodules from promiscuous hosts in the family Myricaceae. *Appl. Environ. Microbiol.* **65**: 4521–4527.

Clawson, M.L., Benson, D.R., Stephens, D.W., Resch, S.C., Silvester, W.B. (1997) Typical *Frankia* infect actinorhizal plants exotic to New Zealand. *New Zealand J. Bot.* **35**: 361–367.

Clawson, M.L., Caru, M., Benson, D.R. (1998) Diversity of frankia strains in root nodules of plants from the families Elaeagnaceae and Rhamnaceae. *Appl. Environ. Microbiol.* **64**: 3539–3543.

Crane, P.R., Friis, E.M., Pedersen, K.R. (1995) The origin and early diversification of angiosperms. *Nature* **374**: 27–33.

Crannell, W.K., Tanaka, Y., Myrold, D.D. (1994) Calcium and pH interaction on root nodulation of nursery-grown red alder (*Alnus rubra* Bong.) seedlings by *Frankia*. *Soil Biol. Biochem.* **26**: 607–614.

Cronquist, A. (1981) *An Integrated System of Classification of Flowering Plants.* Columbia University Press: New York.

Cronquist, A. (1988) *The Evolution and Classification of Flowering Plants.* The New York Botanical Garden: New York.

Cruz-Cisneros, R., Valdés, M. (1991) Actinorhizal root nodules on *Adolphia infesta* (H.B.K.) Meissner (Rhamnaceae). *Nitrogen Fixing Tree Res. Rep.* **9**: 87–89.

Dawson, J.O., Kowalski, D.G., Dart, P.J. (1986) Variation with soil depth, topographic position and host species in the capacity of soils from an Australian locale to nodulate *Casuarina* and *Allocasuarina* seedlings. *Plant Soil* **118**: 1–11.

Diem, H.G., Dommergues, Y.R. (1990) Current and potential uses and management of Casuarinaceae in the tropics and subtropics. In: Schwintzer, C.R. and Tjepkema, J.D. (eds) *The Biology of Frankia and Actinorhizal Plants*, pp. 317–342. Academic Press: San Diego.

Dobritsa, S.V., Stupar, O.S. (1989) Genetic heterogeneity among *Frankia* isolates from root nodules of individual actinorhizal plants. *FEMS Microbiol. Letts.* **58**: 287–292.

Evans, R.C., Alice, L.A., Campbell, C.S., Kellogg, E.A., Dickinson, T.A. (2000) The granule-bound starch synthase (GBSSI) gene in Rosaceae: multiple loci and phylogenetic utility. *Mol. Phylogen. Evol.* **17**: 388–400.

Fernandez, M.P., Meugnier, H., Grimont, P.A.D., Bardin, R. (1989) Deoxyribonucleic acid relatedness among members of the genus *Frankia*. *Int. J. Syst. Bacteriol.* **39**: 424–429.

Gauthier, D., Jaffre, T., Prin, Y. (1999) Occurrence of both *Casuarina*-infective and *Elaeagnus*-infective *Frankia* strains within actinorhizae of *Casuarina collina*, endemic to New Caledonia. *European J. Soil. Biol.* **35**: 9–15.

Gauthier, D., Jaffre, T., Prin, Y. (2000) Abundance of *Frankia* from *Gymnostoma* spp. in the rhizosphere of *Alphitonia neocaledonica*, a non-nodulated Rhamnaceae endemic to New Caledonia. *Eur. J. Soil Biol.* **36**: 169–175.

Griffiths, A.P., McCormick, L.H. (1984) Effects of soil acidity on nodulation of *Alnus glutinosa* and viability of *Frankia*. *Plant Soil* **79**: 429–434.

Hahn, D., Nickel, A., Dawson, J. (1999) Assessing *Frankia* populations in plants and soil using molecular methods. *FEMS Microbiol. Ecol.* **29**: 215–227.

Henrickson, J. (1986) Notes on Rosaceae. *Phytologia* **60**: 468.

Heywood, V.H. (1993) *Flowering Plants of the World*, Oxford University Press: New York.

Holman, R.M., Schwintzer, C.R. (1987) Distribution of spore-positive and spore-negative nodules in nodules of *Alnus incana* spp. *rugosa* in Maine, USA. *Plant Soil* **104**: 103–111.

Honerlage, W., Hahn, D., Zepp, K., Zyer, J., Normand, P. (1994) A hypervariable region provides a discriminative target for specific characterization of uncultured and cultured *Frankia*. *Syst. Appl. Microbiol.* **17**: 433–443.

Huguet, V., McCray Batzli, J., Zimpfer, J.F., Normand, P., Dawson, J.O., Fernandez, M.P. (2001) Diversity and specificity of *Frankia* strains in nodules of sympatric *Myrica gale*, *Alnus incana*, and *Shepherdia canadensis* determined by *rrs* gene polymorphism. *Appl. Environ. Microbiol.* **67**: 2116–2122.

Huss-Danell, K., Frej, A.-K. (1986) Distribution of *Frankia* in soil from forest and afforestation sites in northern Sweden. *Plant Soil* **90**: 407–418.

Huss-Danell, K., Myrold, D.D. (1994) Intrageneric variation in nodulation of *Alnus*: consequences for quantifying *Frankia* nodulation units in soil. *Soil Biol. Biochem.* **26**: 525–532.

Jaffre, T., McCoy, S., Rigault, F., Navarro, E. (2001) A comparative study of flora and symbiotic microflora diversity in two *Gymnostoma* formations on ultramafic rocks in New Caledonia. *S. African J. Sci.* **97**: 599–603.

Jamann, S., Fernandez, M.P., Moiroud, A. (1992) Genetic diversity of Elaeagnaceae-infective *Frankia* strains isolated from various soils. *Acta Oecologica* **13**: 395–405.

Jamann, S., Fernandez, M.P., Normand, P. (1993) Typing method for N_2-fixing bacteria based on PCR-RFLP – application to the characterization of *Frankia* strains. *Mol. Ecol.* **2**: 17–26.

Jeong, S.-C. (2001) Population size and diversity of *Frankia* in soils of *Ceanothus velutinus* and Douglas-fir stands. *Soil Biol. Biochem.* **33**: 931–941.

Jeong, S.-C., Myrold, D.D. (1999) Genomic fingerprinting of *Frankia* microsymbionts from *Ceanothus* copopulations using repetitive sequences and polymerase chain reactions. *Can. J. Bot.* **77**: 1220–1230.

Jeong, S.C., Ritchie, N.J., Myrold, D.D. (1999) Molecular phylogenies of plants and *Frankia* support multiple origins of actinorhizal symbioses. *Mol. Phylogen. Evol.* **13**: 493–503.

Johnson, L.A.S., Wilson, K.L. (1989) Casuarinaceae: a synopsis. In: Crane, P.R., Blackmore, S. (eds) *Evolution, systematics, and fossil history of the Hamamelidae, Vol. II*, pp. 167–188. Clarendon Press: Oxford, UK.

Kalkman, C. (1988) The phylogeny of the Rosaceae. *Bot. J. Linnean Soc.* **98**: 35–53.

Kashanski, C.R., Schwintzer, C.R. (1987) Distribution of spore-positive and spore-negative nodules of *Myrica gale* in Maine, USA. *Plant Soil* **104**: 113–120.

Klemmedson, J.O. (1979) Ecological importance of actinomycete-nodulated plants in the Western United States. *Bot. Gaz.* **140**(Suppl.): S91–S96.

Kohls, S.J., Baker, D.D. (1989) Effects of substrate nitrate concentration on symbiotic nodule formation in actinorhizal plants. *Plant Soil* **118**: 171–179.

Kohls, S.J., Thimmapuram, J., Buschena, C.A., Paschke, M.W., Dawson, J.O. (1994) Nodulation patterns of actinorhizal plants in the family Rosaceae. *Plant and Soil* **162**: 229–239.

Lechevalier, M.P. (1994) Taxonomy of the genus *Frankia* (*Actinomycetales*). *Int. J. Syst. Bacteriol.* **44**: 1–8.

Lechevalier, M.P., Ruan, J.S. (1984) Physiology and chemical diversity of *Frankia* spp. isolated from nodules of *Comptonia peregrina* (L.) Coult. and *Ceanothus americanus* L. *Plant Soil* **78**: 15–22.

Lumini, E., Bosco, M. (1996) PCR-restriction fragment length polymorphism identification and host range of single-spore isolates of the flexible *Frankia* sp. strain UFI 132715. *Appl. Environ. Microbiol.* **62**: 3026–3029.

Lumini, E., Bosco, M., Fernandez, M.P. (1996) PCR-RFLP and total DNA homology revealed three related genomic species among broad-host-range *Frankia* strains. *FEMS Microbiol. Ecol.* **21**: 303–311.

Mabberely, D.J. (1988) *The Plant-book*. Cambridge University Press: Cambridge, UK.

Macdonald, A.D. (1989) The morphology and relationships of the Myricaceae. In: Crane, P.R., Blackmore, S. (eds) *Evolution, Systematics, and Fossil History of the Hamamelidae*, pp. 147–165. Clarendon Press: Oxford, UK.

Magallon, S., Crane, P.R., Herendeen, P.S. (1999) Phylogenetic pattern, diversity, and diversification of eudicots. *Ann. Missouri Bot. Gard.* **86**: 297–372.

Maggia, L., Bousquet, J. (1994) Molecular phylogeny of the actinorhizal Hamamelidae and relationships with host promiscuity towards *Frankia*. *Mol. Ecol.* **3**: 459–467.

Maggia, L., Nazaret, S., Simonet, P. (1992) Molecular characterization of *Frankia* isolates from *Casuarina equisetifolia* root nodules harvested in West Africa (Senegal and Gambia). *Acta Oecol.* **13**: 453–461.

Manos, P.S., Steele, K.P. (1997) Phylogenetic analysis of "higher" Hamamelididae based on plastid sequence data. *American J. Bot.* **84**: 1407–1419.

Markham, J.H., Chanway, C.P. (1996) *Alnus rubra* nodulation capacity of soil under five species from harvested forest sites in coastal British Columbia. *Plant Soil* **178**: 283–286.

Maunuksela, L., Zepp, K., Koivula, T., Zeyer, J., Haahtela, K., Hahn, D. (1999) Analysis of *Frankia* populations in three soils devoid of actinorhizal plants. *FEMS Microbiol. Ecol.* **28**: 11–21.

McEwan, N.R., Wheeler, C.T., Milner, J.J. (1994) Strain discrimination of cultured and symbiotic *Frankia* by PCR-RFLP. *Soil Biol. Biochem.* **26**: 241–245.

McEwan, N.R., Gould, E.M.O., Wheeler, C.T. (1999) The competitivity, persistence and dispersal of *Frankia* strains in mine spoil planted with inoculated *Alnus rubra*. *Symbiosis* **26**: 165–177.

Miki, A. (1977) Late Cretaceous pollen and spore floras of northern Japan: composition and interpretation. *J. Fac. Sci., Hokkaido Univ., Ser. IV* **17**: 399–436.

Mirza, M.S., Hameed, S., Akkermans, A.D.L. (1994a) Genetic diversity of *Datisca cannabina*-compatible *Frankia* strains as determined by sequence analysis of the PCR-amplified 16S rRNA gene. *Appl. Environ. Microbiol.* **60**: 2371–2376.

Mirza, S.M., Akkermans, W.M., Akkermans, A.D.L. (1994b) PCR-amplified 16S rRNA sequence analysis to confirm nodulation of *Datisca cannabina* L. by the endophyte of *Coriaria nepalensis* Wall. *Plant and Soil* **160**: 147–152.

Morgan, D.R., Soltis, D.E., Robertson, K.R. (1994) Systematic and evolutionary implications of *rbc*L sequence variation in Rosaceae. *American J. Bot.* **81**: 890–903.

Mullin, B.C., Swensen, S.M., Goetting-Minesky, P. (1990) Hypotheses for the evolution of actinorhizal symbioses. In: Gresshoff, P.M., Roth, L.E., Stacey, G., Newton, W.E. (eds) *Nitrogen Fixation: Achievements and Objectives*, pp. 781–787. Chapman & Hall: New York.

Murry, M.A., Konopka, A.S., Pratt, S.D., Vandergon, T.L. (1997) The use of PCR-based typing methods to assess the diversity of *Frankia* nodule endophytes of the actinorhizal shrub *Ceanothus*. *Physiol. Plant.* **99**: 714–721.

Myrold, D.D., Hilger, A.B., Huss-Danell, K., Martin, K.J. (1994) Use of molecular methods to enumerate *Frankia* in soil. In: Ritz, K., Dighton, J., Giller, K.E. (eds) *Beyond the Biomass*, pp. 127–136. John Wiley: Chichester, UK.

Myrold, D.D., Huss-Danell, K. (1994) Population dynamics of *Alnus*-infective *Frankia* in a forest soil with and without host trees. *Soil Biol. Biochem.* **26**: 533–540.

Nalin, R., Normand, P., Domenach, A.-M. (1997) Distribution and N_2-fixing activity of *Frankia* strains in relation with soil depth. *Physiol. Plant.* **99**: 732–738.

Navarro, E., Jaffre, T., Gauthier, D., Gourbiere, F., Rinaudo, G., Simonet, P., Normand, P. (1999) Distribution of *Gymnostoma* spp. microsymbiotic *Frankia* strains in New Caledonia is related to soil type and to host-plant species. *Mol. Ecol.* **8**: 1781–1788.

Nazaret, S., Simonet, P., Normand, P., Bardin, R. (1989) Genetic diversity among *Frankia* strains isolated from *Casuarina* nodules. *Plant Soil* **118**: 241–247.

Nazaret, S., Cournoyer, B., Normand, P., Simonet, P. (1991) Phylogenetic relationships among *Frankia* genomic species determined by use of amplified 16S rDNA sequences. *J. Bacteriol.* **173**: 4072–4078.

Nelson, D., Lopez, C.F. (1989) Variation in nitrogen fixation among populations of *Frankia* sp. and *Ceanothus* sp. in actinorhizal association. *Biol. Fert. Soils* **7**: 269–274.

Nelson, D.L. (1983) Occurrence and nature of actinorhizae on *Cowania stansburiana* and other Rosaceae. In: Tiedmann, A.R., Johnson, K.L. (eds) *Proceedings – Research and Management of Bitterbrush and Cliffrose in Western North America*, pp. 225–239. Forest Service, USDA: Ogden, UT.

Nick, G., Paget, E., Simonet, P., Moiroud, A., Normand, P. (1992) The nodular endophytes of *Coriaria* sp. form a distinct lineage within the genus *Frankia*. *Mol. Ecol.* **1**: 175–181.

Nickel, A., Hahn, D., Zepp, K., Zeyer, J. (1999) In situ analysis of introduced *Frankia* populations in root nodules of *Alnus glutinosa* grown under different water availability. *Can. J. Bot.* **77**: 1231–1238.

Normand, P., Chapelon, C. (1997) Direct characterization of *Frankia* and of close phylogenetic neighbors from an *Alnus viridis* rhizosphere. *Physiol. Plant.* **99**: 722–731.

Normand, P., Orso, S., Cournoyer, B., Jeannin, P., Chapelon, C., Dawson, J., Evtushenko, L., Misra, A.K. (1996) Molecular phylogeny of the genus *Frankia* and related genera and emendation of family *Frankiaceae*. *Int. J. Syst. Bacteriol.* **46**: 1–9.

Paschke, M.W., Dawson, J.O. (1992a) *Frankia* abundance in soils beneath *Betula nigra* and other non-actinorhizal woody plants. *Acta Oecologica* **13**: 407–415.

Paschke, M.W., Dawson, J.O. (1992b) The occurrence of *Frankia* in tropical forest soils of Costa Rica. *Plant Soil* **142**: 63–67.

Picard, C., Ponsonnet, C., Paget, E., Nesme, X., Simonet, P. (1992) Detection and enumeration of bacteria in soil by direct DNA extraction and polymerase chain reaction. *Appl. Environ. Microbiol.* **58**: 2717–2722.

Potter, D. (1997) Variation and phylogenetic relationships of nucleotide sequences for putative genes encoding polygalacturonase inhibitor protein (PGIPs) among taxa of Rosaceae. *American J. Bot.* **84**(6 Suppl.): 224.

Potter, D., Gao, F., Botiri, P.E., Oh, S.-H., Baggett, S. (2002) Phylogenetic relationships in Rosaceae inferred from chloroplast *matK* and *trnL-F* nucleotide sequence data. *Plant Syst. Evol.* **231**: 77–89.

Ramirez-Saad, H., Janse, J.D., Akkermans, A.D.L. (1998) Root nodules of *Ceanothus caeruleus* contain both the N_2-fixing *Frankia* endophyte and a phylogenetically related Nod-/Fix- actinomycete. *Can. J. Microbiol.* **44**: 140–148.

Richardson, J.E., Fay, M.F., Cronk, Q.C.B., Bowman, D., Chase, M. (2000) A phylogenetic analysis of Rhamnaceae using *rbcL* and *trnL-F* plastid DNA sequences. *American J. Bot.* **87**: 1309–1324.

Ritchie, N.J., Myrold, D.D. (1999) Geographic distribution and genetic diversity of *Ceanothus*-infective *Frankia* strains. *Appl. Environ. Microbiol.* **65**: 1378–1383.

Rohrer, J.R., Robertson, K.R., Phipps, J.B. (1991) Variation in structure among fruits of Maloideae (Rosaceae). *American J. Bot.* **78**: 1617–1635.

Rojas, N.S., Li, C.Y., Perry, D.A., Ganio, L.M. (2001) *Frankia* and the nodulation of red alder and snowbrush grown on soils from Douglas-fir forests in the H. J. Andrews experimental forest of Oregon. *Appl. Soil Ecol.* **17**: 141–149.

Rouvier, C., Nazaret, S., Fernandez, M.P., Picard, B., Simonet, P., Normand, P. (1990) *rrn* and *nnif* intergenic spacers and isoenzyme patterns as tools to characterize *Casuarina*-infective *Frankia* strains. *Acta Oecol.* **13**: 487–495.

Rouvier, C., Nazaret, S., Fernandez, M.P., Picard, B., Simonet, P., Normand, P. (1992) *rrn* and *nif* intergenic spacers and isoenzyme patterns as tools to characterize *Casuarina*-infective *Frankia* strains. *Acta Oecol.* **13**: 487–495.

Rouvier, C., Prin, Y., Reddell, P., Normand, P., Simonet, P. (1996) Genetic diversity among *Frankia* strains nodulating members of the family Casuarinaceae in Australia revealed by PCR and restriction fragment length polymorphism analysis with crushed root nodules. *Appl. Environ. Microbiol.* **62**: 979–985.

Roy, A., Bousquet, J. (1996) The evolution of the actinorhizal symbiosis through phylogenetic analysis of host plants. *Acta Bot. Gallica* **143**: 635–650.

Sanginga, N., Danso, S.K.A., Bowen, G.D. (1989) Nodulation and growth response of *Allocasuarina* and *Casuarina* species to phosphorus fertilization. *Plant Soil* **118**: 125–132.

Schulze-Menz, G.K. (1964) Rosaceae. In: Melchior, H. (ed) *Engler's Syllabus der Pflanzenfamilien II 12th ed.*, pp. 209–218. Gebruder Boerntraeger: Berlin, Germany.

Schwencke, J., Caru, M. (2001) Advances in actinorhizal symbiosis: Host plant-*Frankia* interactions, biology, and application in arid land reclamation. A review. *Arid Land Res. Manag.* **15**: 285–327.

Schwintzer, C.R. (1985) Effect of spring flooding on endophyte differentiation, nitrogenase activity, root growth and shoot growth in *Myrica gale*. *Plant Soil* **87**: 109–124.

Schwintzer, C.R. (1990) Spore-positive and spore-negative nodules. In: Schwintzer, C.R., Tjepkema, J.D. (eds) *The Biology of Frankia and Actinorhizal Plants*, pp. 177–193. Academic Press: New York.

Silvester, W.B. (1977) Dinitrogen fixation by plant associations excluding legumes. In: Hardy, R.W.F., Gibson, A.H. (eds) *A Treatise on Dinitrogen Fixation*, pp. 141–190. John Wiley: New York.

Simonet, P., Grosjean, M.C., Misra, A.K., Nazaret, S., Cournoyer, B., Normand, P. (1991) *Frankia* genus-specific characterization by polymerase chain reaction. *Appl. Environ. Microbiol.* **57**: 3278–3286.

Simonet, P., Navarro, E., Rouvier, C. *et al.* (1999) Co-evolution between *Frankia* populations and host plants in the family Casuarinaceae and consequent patterns of global dispersal. *Environ. Microbiol.* **1**: 525–533.

Skog, L.E. (1972) The genus *Coriaria* (Coriariaceae) in the western Hemisphere. *Rhodera* **74**: 242–253.

Smolander, A. (1990) *Frankia* populations in soils under different tree species with special emphasis on soils under *Betula pendula*. *Plant Soil* **121**: 1–10.

Smolander, A., Sarsa, M.-L. (1990) *Frankia* strains in soil under *Betula pendula:*: behavior in soil and in pure culture. *Plant Soil* **122**: 129–136.

Smolander, A., Sundman, V. (1987) *Frankia* in acid soils of forests devoid of actinorhizal plants. *Physiol. Plant.* **70**: 297–303.

Smolander, A., Van Dijk, C., Sundman, V. (1988) Survival of *Frankia* strains introduced into soil. *Plant Soil* **106**: 65–72.

Sogo, A., Setoguchi, H., Noguchi, J., Jaffre, T., Tobe, H. (2001) Molecular phylogeny of Casuarinaceae based on *rbcL* and *matK* gene sequences. *J. Plant Res.* **114**: 459–464.

Solbrig, O.T., Cody, M.L., Fuentes, E.R., Glanz, W., Hunt, J.H., Moldenke, A.R. (1977) The origin of the biota. In: Mooney, H.A. (ed) *Convergent Evolution in Chile and California: Mediterranean Climate Ecosystems*, pp. 8–20. Dowden, Hutchinson, and Ross: Stroudsburg, PA.

Soltis, D.E., Soltis, P.S. (2000) Contributions of plant molecular systematics to studies of molecular evolution. *Plant Mol. Biol.* **42**: 45–75.

Soltis, D.E., Soltis, P.S., Morgan, D.R., Swensen, S.M., Mullin, B.C., Dowd, J.M., Martin, P.G. (1995) Chloroplast gene sequence data suggest a single origin of the predisposition for symbiotic nitrogen fixation in angiosperms. *Proc. Natl Acad. Sci. USA* **92**: 2647–2651.

Swensen, S.M. (1996) The evolution of actinorhizal symbioses: Evidence for multiple origins of the symbiotic association. *Am. J. Bot.* **83**: 1503–1512.

Swensen, S., Mullin, B. (1997) Phylogenetic relationships among actinorhizal plants. The impact of molecular systematics and implications for the evolution of actinorhizal symbiosis. *Physiol. Plant.* **99**: 565–573.

Swensen, S.M., Mullin, B.C., Chase, M.W. (1994) Phylogenetic affinities of the Datiscaceae based on an analysis of nucleotide sequences from the plastid *rbcL* gene. *Syst. Bot.* **19**: 157–168.

Swensen, S.M., Luthi, J.N., Rieseberg, L.H. (1998) Datiscaceae revisited: Monophyly and the sequence of breeding system evolution. *Systematic Bot.* **23**: 157–169.

Takhtajan, A. (1980) Outline of the classification of flowering plants (Magnoliophyta). *Bot. Rev.* **46**: 226–359.

Thomas, K.A. and Berry, A.M. (1989) Effects of continuous nitrogen application and nitrogen preconditioning on nodulation and growth of *Ceanothus griseus* var. *horizontalis*. *Plant Soil* **118**: 181–187.

Thorne, R.F. (1992) Classification and geography of flowering plants. *Bot. Rev.* **46**: 225–348.

Torrey, J.G. (1990) Cross-inoculation groups within *Frankia*. In: Schwintzer, C.R., Tjepkema, J.D. (eds) *The Biology of Frankia and Actinorhizal Plants*, pp. 83–106. Academic Press: New York.

Tortosa, R.D. (1992) El complejo *Retanilla-Talguenea-Trevoa* (Rhamnaceae). *Darwiniana* **31**: 223–252.

Van Dijk, C. (1979) Endophyte distribution in the soil. In: Gordon, J.C., Wheeler, C.T., Perry, D.A. (eds) *Symbiotic Nitrogen Fixation in the Management of Temperate Forests*, pp. 84–94. Oregon State Univ. Press: Corvallis, OR.

Van Dijk, C. (1984) Ecological aspects of spore formation in the *Frankia-Alnus* symbiosis, Ph.D. Thesis, Univ Leiden.

Van Dijk, C., Sluimer, A., Weber, A. (1988) Host range differentiation of spore-positive and spore-negative strain types of *Frankia* in stands of *Alnus glutinosa* and *Alnus incana* in Finland. *Physiol. Plant.* **72**: 349–358.

Wall, L.G. (2000) The actinorhizal symbiosis. *J. Plant Growth Regul.* **19**: 167–182.

Weber, A. (1986) Distribution of spore-positive and spore-negative nodules in stands of *Alnus glutinosa* and *Alnus incana* in Finland. *Plant Soil* **96**: 205–213.

Wollum II, A.G., Youngberg, C.T., Chichester, F.W. (1968) Relation of previous timber stand age to nodulation of *Ceanothus velutinus*. *Forest Sci.* **14**: 114–118.

Yang, Y. (1995) The effect of phosphorus on nodule formation and function in the *Casuarina-Frankia* symbiosis. *Plant Soil* **176**: 161–169.

Yokoyama, J., Suzuki, M., Iwatsuki, K., Hasebe, M. (2000) Molecular phylogeny of *Coriaria*, with special emphasis on the disjunct distribution. *Mol. Phylogen. Evol.* **14**: 11–19.

Zimpfer, J.F., Smyth, C.A., Dawson, J.O. (1997) The capacity of Jamaican mine spoils, agricultural and forest soils to nodulate *Myrica cerifera, Leucaena leucocephala* and *Casuarina cunninghamiana*. *Physiol. Plant.* 5: 664–672.

Zimpfer, J.F., Kennedy, G.J., Smyth, C.A., Hamelin, J., Navarro, E., Dawson, J.O. (1999) Localization of *Casuarina*-infective *Frankia* near *Casuarina cunninghamiana* trees in Jamaica. *Can. J. Bot.* **77**: 1248–1256.

Zitzer, S.F., Dawson, J.O. (1992) Soil properties and actinorhizal vegetation influence nodulation of *Alnus glutinosa* and *Elaeagnus angustifolia* by *Frankia*. *Plant Soil* **140**: 197–204.

7

Chemical signalling by bacterial plant pathogens

Clare L. Pemberton, Holly Slater and George P.C. Salmond

7.1 Introduction

The regulation of virulence in phytopathogenic bacteria is essential for a successful infection. Pathogens must sense their surroundings and determine when, and equally importantly when not, to attack a host. Unnecessary virulence factor synthesis leads to an unrewarded cell metabolic load and may result in elimination of bacteria by the host defence systems. Sensing the presence of a susceptible host can occur through environmental factors such as osmolarity and nutrient availability. This sensing is achieved through a number of two-component phosphorelay systems and intracellular regulatory networks. Bacteria have also been shown to work together as communities, rather than discrete units, coordinating physiological functions, including the production of disease. This coordination occurs via a phenomenon called quorum sensing.

Quorum sensing is a signalling mechanism that allows organisms to control physiological functions in response to population size. By elaborating signals that can be detected by other species members, coordinated responses to environment changes can be mounted. The nature of these signals varies, from small peptides used by Gram-positive bacteria (Kleerebezem *et al.*, 1997; Lazazzera and Grossman, 1998) and acyl homoserine lactones (acyl HSLs) used by many Gram-negative bacteria (reviewed in Whitehead *et al.*, 2001), to species-specific systems including 3-hydroxypalmitic acid methyl ester signals in *Ralstonia solanacearum*, the diffusible factors of *Xanthomonas campestris* and opine signalling in *Agrobacterium tumefaciens*.

The advantages of virulence regulation based on population density are dictated by the necessary components of a successful invasion. Pathogens must identify a susceptible host and then attack it in sufficient numbers to cause disease. Invasion by a large pathogen population should overwhelm a host, allowing successful colonisation of the infection site. Where small numbers of

Plant Microbiology, Michael Gillings and Andrew Holmes

bacteria are used, minimal damage will be caused and the host alerted to the presence of invaders that can then be eliminated by its defence systems. One example of this phenomenon is seen with the maceration of plant tissue by *Erwinia carotovora* subspecies *carotovora* (*Ecc*). The damage of plant tissue by extracellular enzymes from *Ecc* releases plant cell contents into the surrounding environment. The released contents are detected by the plant, which can then mount a defence response and eliminate the pathogens if they are present in sufficiently small numbers (Palva *et al.*, 1993).

Another advantage of quorum-sensing regulation in bacterial systems is the ability to coordinate the production of multiple virulence factors with one external signal. In *Ecc* the synthesis of multiple plant-macerating enzymes and the antibiotic 1-carbapen-2-em-3-carboxylic acid (carbapenem) are regulated by population density. This means that when the bacterium degrades plant cells, releasing vital nutrients, it also begins production of the carbapenem which may play a role in eliminating competitors, leaving available nutrients for use by *Ecc* cells (Axelrood *et al.*, 1988; Salmond *et al.*, 1995).

This review summarises the quorum-sensing regulation of virulence in phytopathogenic bacteria. It begins with a description of the archetypal quorum-sensing system and its major components and then illustrates the adaptations of this system for virulence regulation in various phytopathogens.

7.2 Acyl HSL-based regulation of virulence factors

7.2.1 Vibrio fischeri lux system – archetypal quorum-sensing regulation

The first discovered, and best-studied, example of an acyl HSL-regulated system in bacteria is the *lux* system in *Vibrio fischeri*. This organism lives symbiotically in the light organs of the squid, *Euprymna scolopes* (Ruby 1999). At high cell densities *V. fischeri* cells produce a blue-green light by the action of the luciferase enzyme encoded by the *lux* gene cluster. This bioluminescence is exploited by the squid to perform counter-illumination, a form of camouflage where shadows cast in the moonlight are removed by projection of the squid's own light supply, protecting it from nocturnal predators (Visick and McFall-Ngai, 2000).

The *lux* gene cluster consists of two bidirectionally transcribed operons (Engebrecht *et al.*, 1983; Swartzman *et al.*, 1990). In one direction the transcriptional regulator *luxR* is transcribed, with the rest of the cluster transcribed in the other direction (*Figure 7.1*). This second operon consists of *luxI*, the product of which is responsible for the synthesis of acyl HSL; *luxAB* which encode subunits of the luciferase enzyme responsible for light generation; *luxCDE* which encode products that form a multienzyme complex to synthesise the aldehyde substrate of the luciferase and *luxG* which encodes a probable flavin reductase, generating another luciferase substrate (Zenno and Saigo, 1994).

The acyl HSL produced by LuxI has been identified as *N*-(3-oxohexanoyl)-L-homoserine lactone (OHHL), a small diffusible molecule derived from fatty acids (Eberhard *et al.*, 1981; Engebrecht and Silverman, 1984; Kaplan and Greenberg, 1985). At low cell density *luxI* is expressed at a basal level so OHHL concentrations remain low. When the *V. fischeri* population size increases, so does the environmental

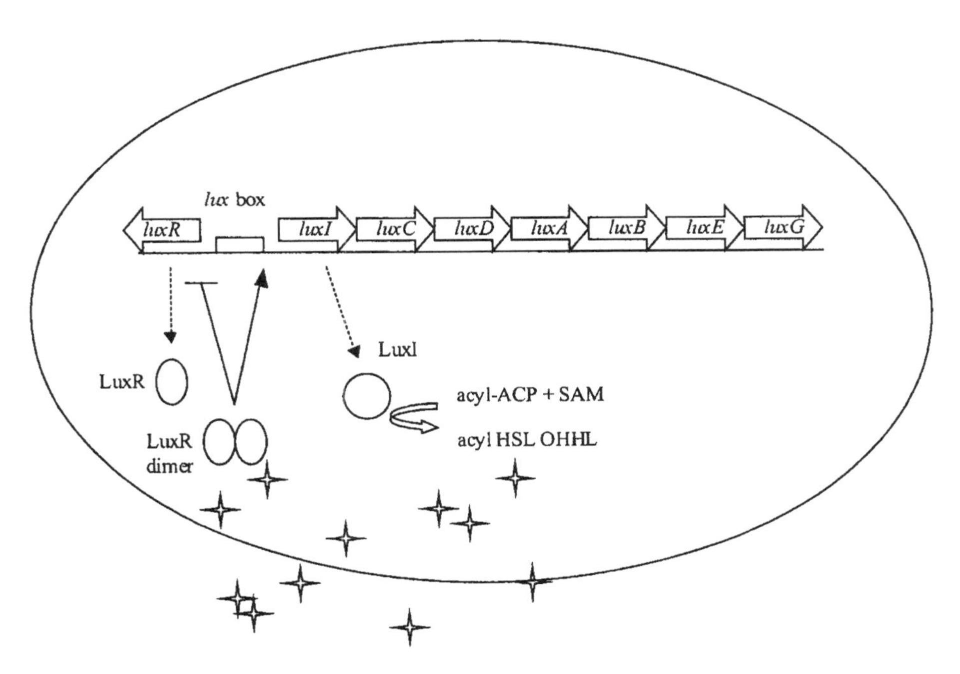

Figure 7.1. Regulation of the *lux* operon in *Vibrio fischeri.* At low cell density the *lux* operon and *luxR* are expressed at basal levels. This means that the concentration of the signalling molecule OHHL (✧) is also low as it is synthesised by the product of *luxI*, a member of the *lux* operon. At high cell density, the levels of OHHL are proportionally higher as the molecule can freely diffuse between *Vibrio* cells. At this higher concentration, OHHL promotes LuxR dimerisation and activation, which stimulates transcription of *luxICDABEG*. This operon encodes the components for bacterial bioluminescence enzyme, luciferase, (*luxAB*) and the synthesis of its substrate (*luxCDE*). An increase in OHHL synthesis is also observed as levels of LuxI increase. Once it has bound OHHL, LuxR represses its own synthesis by an unknown mechanism. Positive regulation is indicated by unbroken arrows and repression by —| symbols.

concentration of OHHL. Once this concentration passes a threshold level, the acyl HSL is believed to bind the transcriptional regulator LuxR, activating the protein, possibly by a conformational change. Activated LuxR is capable of binding DNA, in particular a 20 bp region upstream from the *luxI* transcriptional start site (Devine *et al.*, 1988, 1989; Egland and Greenberg, 1999). This region possesses dyad symmetry and is known as the *lux* box. LuxR binding activates the transcription of the *luxICDABEG* operon, producing both bioluminescence and increased OHHL synthesis.

At low cell density LuxR activates its own transcription by a currently unknown mechanism (Shadel and Baldwin, 1991, 1992a). Once OHHL has bound and activated the regulator, LuxR begins to repress its own synthesis (Dunlap and Ray, 1989). The mechanism for this is also currently unclear although it appears to involve the presence of a *lux* box within *luxD* and may serve to limit the autoinduction of bioluminescence (Shadel and Baldwin, 1992b).

7.2.2 The LuxR family of transcriptional regulators

A family of transcriptional regulators, similar to LuxR and involved in quorum-sensing signalling, have been identified (reviewed in Whitehead *et al.*, 2001). Each consists of a C-terminal DNA-binding domain with a helix-turn-helix motif, a short linker domain and an N-terminal acyl HSL-binding domain (Choi and Greenberg, 1991, 1992a; Hanzelka and Greenberg, 1995; Shadel *et al.*, 1990; Slock *et al.*, 1990). Typically these proteins exist as monomers that dimerize in the presence of their cognate acyl HSL. This generates an active form of the regulator which is then capable of transcriptional activation by binding DNA at conserved *lux* boxes in the promoters of target genes (Choi and Greenberg, 1992b). Both the bioluminescence regulator from *V. fischeri*, LuxR, and TraR, the regulator of plasmid transfer in *Agrobacterium tumefaciens*, function in this way (Qin *et al.*, 2000; Zhu and Winans, 2001). Upon activation by OHHL, the C-terminal portion of LuxR interacts with the σ^{70} subunit of RNA polymerase (Finney *et al.*, 2002). This promotes binding of the complex at the *lux* box where it then activates target gene transcription. The conversion of LuxR to its inactive state, once it is no longer required, is believed to involve the N-terminus of the protein. Work carried out using truncated LuxR proteins showed that the deletion of its N-terminus enables LuxR to activate transcription of the *lux* operon independently of OHHL (Choi and Greenberg, 1991). It is believed that the N-terminus may inhibit LuxR by some structural occlusion of the multimerization or DNA binding domains. This conformation is altered by the binding of OHHL, relieving the inhibition and activating the regulator.

There are also some LuxR homologues that do not follow the typical pattern of activation. Some proteins are able to dimerise in the absence of their cognate acyl HSL. These include CarR, the regulator of carbapenem in *Ecc*, and EsaR, the regulator of exopolysaccharide production in *Pantoea stewartii* (Qin *et al.*, 2000; Welch *et al.*, 2000). CarR is also unusual as it is able to bind target DNA in the absence of acyl HSL although it is only activated once it forms multimeric complexes with the signalling molecule. A similar phenomenon is observed with ExpR, a LuxR homologue from *Erwinia chrysanthemi* (Nasser *et al.*, 1998; Reverchon *et al.*, 1998).

The crystal structure of TraR from *A. tumefaciens* has been solved at 1.66 Å resolution (Vannini *et al.*, 2002; Zhang *et al.*, 2002). This structure, of the regulator bound to its acyl HSL and target DNA, was found to be an asymmetrical dimer. The protein, whose cognate synthase is encoded by the gene *traI*, is activated by *N*-(3-oxooctanoyl)-L-homoserine lactone (OOHL). The binding of OOHL at its N terminus is necessary for the dimerization of TraR as it is believed to act as a scaffold around which the necessary folding can take place. OOHL becomes embedded in the central region of the dimer with virtually no solvent contacts possible. TraR monomers bind one molecule of OOHL and half a *tra* (*lux*) box each. The binding of OOHL may stabilise the protein, saving it from proteolytic degradation.

Analysis of the structure of TraR has provided evidence for the theory that the LuxR proteins are formed from the fusion of two ancestral proteins. The TraR N terminus contains a GAF/PAS domain found in several proteins that function as small molecule-binding modules (Vannini *et al.*, 2002). Fusion between a GAF/PAS molecule adapted for OOHL binding and a small HTH-binding domain may have been responsible for the generation of an ancestral form of this regulator. Formation by the fusion of two other separate proteins may explain the asymmetry observed in TraR.

7.2.3 The LuxI family of acyl HSL synthases

This family of proteins, which has more than 40 members currently, is responsible for the production of the acyl HSL signalling molecules used in Gram-negative quorum sensing. The LuxI family shows a higher degree of conservation than the LuxR proteins, an average of 37% identity, and possesses eight entirely conserved residues (Watson *et al.*, 2002). Each protein is approximately 200 amino acids in length. Acyl HSLs are formed from the substrates S-adenosylmethionine (SAM) and acylated acyl carrier proteins (acyl ACP) (Eberhard *et al.*, 1991; Parsek *et al.*, 1999). The synthase first catalyses the acylation of SAM by acyl ACP and then the methionine moiety of SAM is lactonised, producing acyl HSL.

The crystal structure of EsaI from *Pantoea stewartii* (Section 7.2.6) has been determined to 1.8 Å resolution and was found to exhibit considerable structural similarity to the GNAT family of *N*-acetyltransferases (Watson *et al.*, 2002). In EsaI all eight of the LuxI family-conserved residues are present on the same face of the enzyme. Most were found within the active site cleft, a V-shaped region formed from nine α helices surrounding a highly twisted eight-stranded β sheet structure. The remaining three residues were found in the disordered N terminus which forms a highly mobile region. It is believed that this structure may undergo a conformational change or become more stable once the enzyme substrates have bound. Analysis of the conserved residues in this region has supported the theory that this is where SAM and acyl-ACP interact. The proposed model for HSL synthesis is that acyl-ACP binds the synthase resulting in a conformational change of the N-terminal domain of the protein. SAM then binds and the reaction proceeds (Watson *et al.*, 2002).

Members of the LuxI acyl synthase family produce a wide range of different acyl HSLs. This variability largely results from the specificity of the enzyme for the acyl chain. In EsaI the 3-oxo-hexanoyl portion of acyl-ACP fits neatly into the binding pocket (Watson *et al.* 2002). By altering the peptide sequence the size and specificity of synthase active sites can be altered, allowing different acyl ACPs to be used and a variety of acyl HSLs to be generated.

Another family of acyl HSL synthases has also been identified (Gilson *et al.*, 1995; Hanzelka *et al.*, 1999; Kuo *et al.*, 1994). This includes the enzymes AinS, which directs the synthesis of *N*-octanoyl-L-homoserine lactone (OHL) in *V. fischeri*, and LuxLM in *Vibrio harveyi*. The lactone synthesised by the former enzyme is believed to bind LuxR at low cell density. These proteins share no sequence identity with the LuxI family.

7.2.4 Regulation of exoenzyme production in Erwinia spp.

E. carotovora subspecies *carotovora* (*Ecc*) produces an array of plant-macerating enzymes, causing soft rotting disease in a number of economically important crops including potato and carrot (reviewed in Pérombelon 2002). These enzymes include cellulases, proteases and pectinases and are controlled by quorum sensing

(Jones *et al.*, 1993; Pirhonen *et al.*, 1993). At high cell density, elevated concentrations of the acyl HSL OHHL up-regulate synthesis of exoenzymes.

OHHL production in *Ecc* is directed by CarI synthase (also known as ExpI, OhlI and HslI), a homologue of LuxI (Bainton *et al.*, 1992a, b; Chatterjee *et al.*, 1995; Jones *et al.*, 1993;Pirhonen *et al.*, 1993). Deletion of *carI* has been found to reduce exoenzyme synthesis and lead to reduced virulence of *Ecc in planta* (Swift *et al.*, 1993). The predicted cognate LuxR homologue of CarI involved in the control of exoenzymes has not yet been identified. Two LuxR homologues already identified in *Ecc*, CarR and ExpR (RexR or EccR), have been found to have little effect on enzyme phenotypes (McGowan *et al.*, 1995; Rivet 1998). The overexpression of *carR* from a multicopy plasmid was found to produce a slight decrease in enzyme synthesis, although this is believed to result from the sequestration of OHHL, as deletion of *carR* has no effect. In *Ecc* SCCI 3193, however, an apparently strain-specific increase in pectate lyase enzymes and OHHL levels was observed upon deletion of *expR* (Andersson *et al.*, 2000).

The quorum-sensing regulation of extracellular enzymes in *Ecc* is part of an extensive network of regulators that act upon the virulence factors (*Figure 7.2*). Many of these have global effects, including the regulators RpoS, HexA and KdgR (reviewed by Whitehead *et al.*, 2002). Possibly the most important of these systems is the RsmA/*rsmB* system (Chatterjee *et al.*, 1995; Cui *et al.*, 1995). This system involves the RNA-binding protein, RsmA and an untranslated RNA molecule, *rsmB*, which together form a post-transcriptional control system that is believed to form a link between quorum-sensing signals and the intracellular control network (Liu *et al.*, 1998). RsmA is a negative regulator of virulence factors, including exoenzymes, which acts by degrading target gene mRNA. The *rsmB* RNA neutralises the effect of this regulator by apparently binding to and sequestering the protein. Deletion strains of *rsmA* have been found to produce exoenzymes and other quorum-sensing-regulated pathogenicity determinants independently of the presence of OHHL (Chatterjee *et al.*, 1995; Cui *et al.*, 1996). With the observation that *rsmB* deletion mutants produced reduced virulence factors with or without OHHL, these findings suggest that acyl HSL may control exoenzyme synthesis via the RsmA/*rsmB* network (Chatterjee *et al.*, 2002). OHHL is believed to manifest this effect via repression of *rsmA* transcription although the absence of a *lux* box in the *rsmA* promoter suggests that this occurs through another, possibly unidentified, regulator.

Acyl HSLs may also be involved in the regulation of exoenzymes in *E. chrysanthemi* (*Echr*), which produces soft rots and vascular wilts in several plant hosts. The bacterium produces three lactones – OHHL, *N*-hexanoyl-L-homoserine lactone (HHL) and *N*-decanoyl-L-homoserine lactone (DHL). OHHL and HHL are synthesised by the product of *expI* (Nasser *et al.*, 1998). This gene is found next to the convergently transcribed *expR*. Deletion of either gene has been found to have no effect on exoenzyme production. This observation is unexpected as ExpR binds the upstream region of various *Echr* pectinases genes, suggesting a role for the protein in the regulation of these genes (Nasser *et al.*, 1998; Reverchon *et al.*, 1998). ExpR also binds a *lux* box present in its own promoter, repressing transcription. This is relieved in the presence of OHHL, which causes bound ExpR to be released from its target promoter.

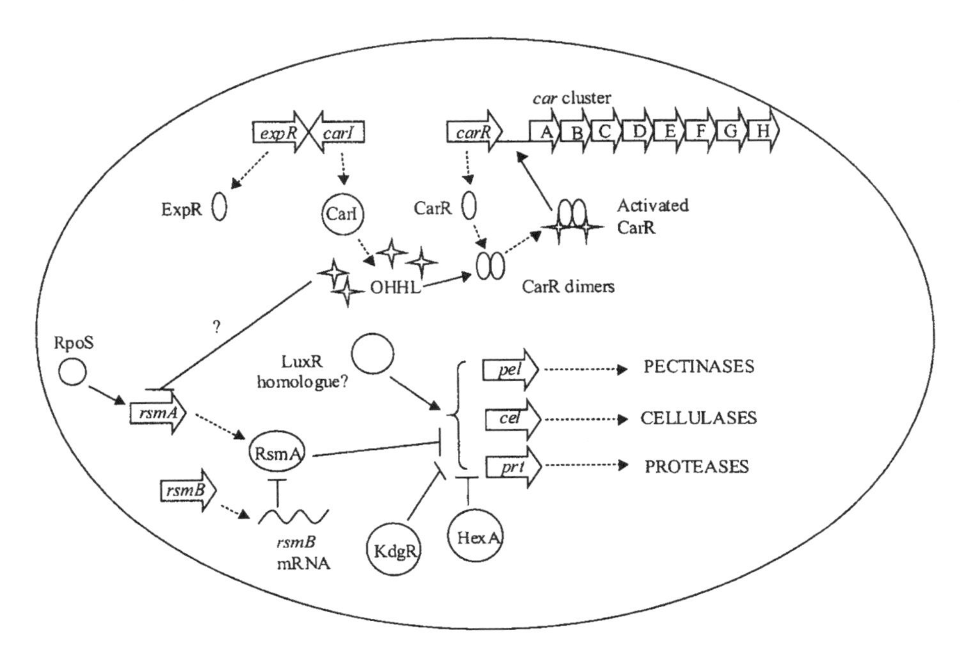

Figure 7.2. The regulation of carbapenem and exoenzymes in *Ecc. Transcriptional activation of the* car cluster in *Ecc* is mediated by OHHL-activated CarR, with OHHL molecules represented by ✦ symbols. The cluster encodes the synthesis of carbapenem (*carA-E*), resistance to the antibiotic (*carFG*) and a protein of currently unknown function (*carH*). Although CarR can dimerise and bind DNA in the absence of OHHL, it requires the signal molecule, synthesised by CarI, to become a transcriptional activator. Such positive regulation is indicated by unbroken arrows and repression by —| symbols.

The *carI* gene is linked to a gene encoding another LuxR homologue, ExpR. Neither this regulator nor CarR have been found to have any significant effect on exoenzyme synthesis. Exoenzyme levels are positively regulated by OHHL although this may occur via another, as yet unidentified, LuxR homologue. Several repressors of enzyme synthesis have been characterised, including HexA, KdgR and the Rsm system (for a review of these and other regulators see Whitehead *et al.*, 2002). The latter involves the RNA-binding protein RsmA which is itself repressed by untranslated *rsmB* RNA. *rsmB* RNA is believed to sequester RsmA before it can bind and degrade target RNA. The transcription of *rsmA* is activated by RpoS and may be repressed by OHHL. It is possible that the effects of OHHL on enzymes are manifested through the Rsm system.

7.2.5 Regulation of antibiotic synthesis in Erwinia spp.

Certain strains of *Ecc* produce the simple β-lactam antibiotic 1-carbapen-2-em-3-carboxylic acid (carbapenem) (Parker *et al.*, 1982). Production of this broad-spectrum antibiotic is under quorum-sensing control via OHHL synthesised by CarI (Bainton *et al.*, 1992a, b; Chatterjee *et al.*, 1995; Jones *et al.*, 1993; Pirhonen *et al.*, 1993) (*Figure 7.2*). This regulation is mediated by the LuxR homologue CarR, which activates expression of the *carA-H* genes encoding carbapenem production (McGowan *et al.*, 1995). The biosynthetic enzymes are encoded by the genes *carA–E*, with *carFG* encoding a resistance mechanism to the effects of the antibiotic (McGowan *et al.*, 1996, 1997). The function of CarH is currently unknown. The *car* cluster is positioned ~ 150 bp downstream of *carR*, with the transcriptional start of *carA* located in this intergenic region. The *carI* gene is unlinked to the *car* cluster. It is found in a separate location with the convergently transcribed *expR*, encoding another LuxR homologue with no function in carbapenem regulation.

CarR-mediated activation of the *car* cluster requires concentrations of OHHL above 0.1 $\mu g.ml^{-1}$. This induction of carbapenem synthesis normally occurs during the late log or early stationary phases of growth although precocious induction can be achieved by the addition of exogenous OHHL. CarR exists as a dimer and can bind the *carR–carA* intergenic region even in the absence of OHHL (Welch *et al.*, 2000). By binding two molecules of OHHL per dimer, CarR is activated and can then induce carbapenem synthesis. The mode of this regulation remains undetermined and any need for the OHHL signalling molecule can be circumvented by the overexpression of CarR. It may be that, by binding the protein, OHHL makes CarR more resistant to proteolytic degradation.

A similar quorum-sensing system, designated EcbRI, has been identified in *E. carotovora* subsp. *betavasculorum* (*Ecb*), the causative agent of soft rot in sugar beet (Costa and Loper, 1997). This is believed to regulate the production of another, as yet unidentified, antibiotic. Homologues of the *Ecc car* genes appear to be widespread amongst *Erwinia* spp. (Holden *et al.*, 1998). These clusters are believed to be cryptic, however, due to the absence of functional *carR* genes in these species because the provision of *carR in trans* has been found to restore antibiotic production in many strains. The absence of functional *carR* could be due to the fitness cost incurred by producing antibiotics in certain ecological niches where they confer no advantage to their host.

7.2.6 Quorum sensing in *Pantoea stewartii*

P. stewartii (formerly *Erwinia stewartii*) is the causative agent of Stewart's wilt in sweetcorn. The bacterium causes wilting by production of large amounts of capsular polysaccharide (stewartan), encoded by the *cps* cluster, which can block xylem vessels and limit plant water transport (Braun, 1982). Two levels of control for the *cps* genes have been identified so far. Primary regulation occurs in a cell density-dependent manner through EsaR and EsaI, homologues of LuxR and LuxI, respectively (Beck von Bodman and Farrand, 1995). A secondary system also exists involving the RcsAB proteins, a system which alters capsule synthesis in response to environmental factors in a number of different bacteria including *E. coli* (Gottesman and Stout, 1991; Wehland *et al.*, 1999).

EsaR is an unusual LuxR homologue as, unlike most of the LuxR proteins identified so far, it is a negative regulator, the effects of which are relieved rather than induced by the presence of OHHL (Beck von Bodman *et al.*, 1998). Deletion mutants of *esaR* were found to be hypermucoid, a phenotype which could not be restored by the addition of exogenous OHHL. The genes *esaR* and *esaI*, encoding the regulator and its cognate OHHL synthase, are located next to each other and are convergently transcribed (Beck von Bodman and Farrand, 1995). Unlike *esaI*, the promoter region of *esaR* contains a *lux* box sequence through which EsaR is believed to repress its own synthesis. Recent studies have indicated that EsaR is unlikely to play any part in the regulation of *esaI* expression, as *esaR* deletion mutants produce the same levels of OHHL as wild-type strains (Minogue *et al.*, 2002).

At low cell density, EsaR is believed to dimerise and repress *esaR* and *cps* gene expression. This dimerised, active, DNA-binding form is only found in the absence of OHHL (Qin *et al.*, 2000). Once OHHL concentrations increase at high cell density, the acyl HSL binds and inactivates available EsaR, relieving *cps* repression. EsaR binds one molecule OHHL per protein monomer (Minogue *et al.*, 2002). This is believed to induce structural changes in the protein, removing its ability to repress *cps* expression. The loss of repression may be because EsaR is no longer able to bind DNA or may result from conformational changes rendering the protein susceptible to proteolytic degradation. OHHL synthesis by EsaI, a protein that shows structural similarity to *N*-acetyltransferases, appears to be constitutive (Watson *et al.*, 2002). It is possible that, at low cell densities, EsaR sequesters the cellular pool of OHHL preventing premature expression of the *cps* genes by another quorum-sensing system (Minogue *et al.*, 2002). Once the levels of OHHL exceed maximum levels of EsaR, this other system would be activated and the *cps* genes would be expressed.

7.2.7 Quorum sensing in Agrobacterium tumefaciens – the regulation of the Ti plasmid

A. tumefaciens is the causative agent of crown gall disease. These crown gall tumours are produced by deregulated cell division induced by plant hormones encoded on T DNA, bacterial genetic material incorporated in the plant nucleus (reviewed in Zhu *et al.*, 2000). This oncogenic DNA, which also codes for the synthesis of small carbon compounds called opines, is introduced into the plant cell in a conjugation-like process. T DNA originates from the Ti (*t*umour-*i*nducing) plasmid in *A. tumefaciens* and is transported via a system of Vir proteins, including the VirB pili which deliver it to the plant cell cytoplasm.

The main function of this incorporation of *A. tumefaciens* DNA into host nuclei is believed to be the production of opines. These act as a nutrient source for the surrounding bacterial population as well as functioning as signal molecules (Dessaux *et al.*, 1998). As part of a two-tier network, involving both plant-derived and bacterial signals, opines act to induce their own catabolic systems and transfer of the Ti plasmid within the *A. tumefaciens* population (*Figure 7.3*). This conjugal transfer is thought to ensure that all bacteria surrounding the plant cell can catabolise the opines available (Piper *et al.*, 1999). Bacterial quorum-sensing signals are also employed to control this transfer as this ensures that the population of donor cells present is sufficiently high to achieve maximum conjugation efficiency (Piper and Farrand, 2000).

The Ti plasmid encodes several components — the T DNA transferred into plant cells, the *vir* genes encoding the apparatus for this movement, a *rep* region coding for Ti replication, *tra* and *trb* genes for conjugal Ti transfer and finally the genes coding for opine catabolism and uptake (Zhu *et al.*, 2000). The latter are induced by their cognate substrate, one of two classes of opine. These different opines separate Ti plasmids into two classes, octopine-type induced by octopine and nopaline-type plasmids induced by agrocinopines A and B (Ellis *et al.*, 1982;

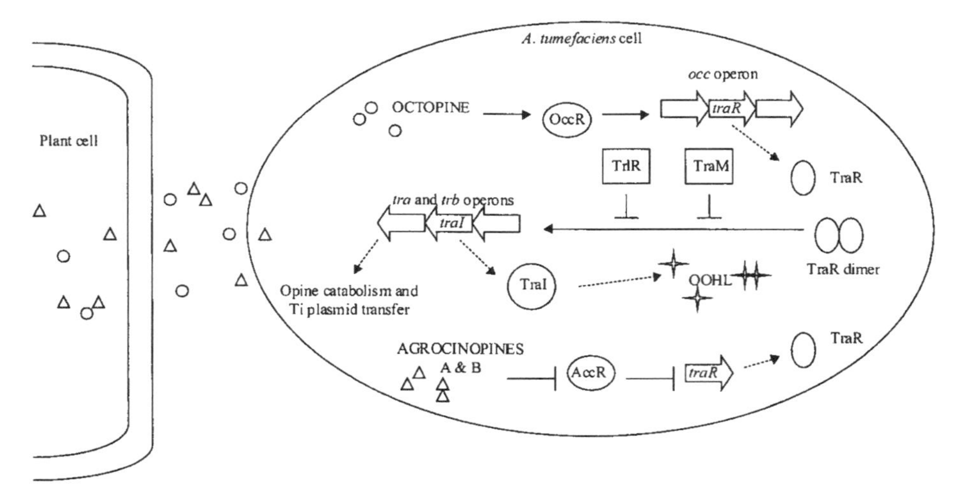

Figure 7.3. The Tra system of *A. tumefaciens.* Opine production by plant cells is encoded by bacterial T DNA which is incorporated into the plant nucleus. Opines are secreted and imported into bacterial cells where they activate their own catabolic systems and transfer of the Ti plasmid carrying T DNA. Octopine (○) activates OccR which induces the transcription of the *occ* operon, containing *traR*. TraR dimerises in the presence of OOHL (✦) and activates the *tra* and *trb* operons. The OOHL synthase, TraI, is encoded by the cognate gene within the *trb* operon. A similar system is seen with agrocinopines (△) which inhibit the regulator AccR, relieving its repression of *traR* transcription. The TraM and TrlR proteins inhibit TraR-activated expression of the *trb* and *tra* operons. Such negative regulation is indicated by —| symbols while positive regulation is shown by unbroken arrows.

Klapwijk *et al.,* 1978). Two different quorum-sensing-based mechanisms are used to control the transfer of these different plasmids.

In octopine-inducible Ti plasmids, the presence of octopine activates the LysR-type regulator OccR (Habeeb *et al.,* 1991; Wang *et al.,* 1992). This induces the transcription of *traR,* a member of the 14 gene *occ* operon encoding products for the uptake and catabolism of octopine (Fuqua and Winans, 1996; Piper *et al.,* 1999). TraR is a LuxR homologue responsible for the transcriptional activation of the *tra* and *trb* genes, coding for the transfer components of the Ti plasmid (Fuqua and Winans, 1994). This activation can only occur once a threshold concentration of the acyl HSL signalling molecule OOHL has been reached, as this molecule is required for the dimerisation and activation of TraR (Qin *et al.,* 2000; see also Section 7.2.2). A negative autoregulatory system exists for the regulator, as the *traI* gene, encoding OOHL synthase, is located in the *trb* operon whose transcription is activated by TraR.

A similar system is present in nopaline-type plasmids although the initial opine induction differs slightly. In this system, the FucR-type regulator AccR represses *traR* transcription until the presence of agrocinopine blocks this control, allowing production of the transcriptional activator (Beck von Bodman *et al.,* 1992; Kim and Farrand, 1997; Piper *et al.,* 1999). The function of TraR then falls under the control of quorum sensing as in octopine-type plasmids. Recently Ti plasmid pAtK84b was identified in strain *A. radiobacter* K84 which contains inducible operons for two different opines (Oger and Farrand, 2002). This plasmid also contains two copies of *traR* and was the first plasmid to be identified that is controlled by more than one opine.

TraR is subject to a further level of regulation in the form of two proteins that bind to the activator, preventing it from carrying out its function. The 11 kDa protein TraM is found on both types of Ti plasmid (Hwang *et al.*, 1999). It is believed to inhibit the function of TraR by sequestering the protein before it can activate transcription (Swiderska *et al.*, 2001). In order to achieve activation of the *tra* and *trb* genes, including *traM* itself, the level of TraR produced must exceed that of TraM (Piper and Farrand, 2000). This system is thought to function in order to stop the transfer of Ti plasmids once conditions become unfavourable, a theory which is supported by the constitutive transfer phenotype observed in *traM* mutants. The second TraR regulatory protein is TrlR (TraS) (Zhu and Winans, 1998). This protein, whose transcription is induced by mannopine, is homologous to TraR. TrlR is believed to originate from a TraR protein that contained a frameshift mutation in its C terminal DNA-binding domain. This complementarity means that each TrlR monomer can form heterodimers by binding one molecule of TraR, inhibiting the ability of TraR to bind the promoters of its target genes (Chai *et al.*, 2001). The activation of TrlR transcription by mannopine is inhibited by the presence of more favourable catabolites including succinate and tryptone (Zhu and Winans, 1999).

A further DNA transfer system has been identified in *A. tumefaciens* C58 (Chen *et al.*, 2002). This was designated AvhB (*Agrobacterium virulence homologue* Vir*B*), based on the homology of seven of the ten genes in this operon to the VirB proteins encoded on the Ti plasmid. This system has been found to mediate conjugal transfer yet it was not found to be essential for virulence and may therefore be expressed in different environments from the Ti-based VirB system. This suggests that further, currently unidentified, factors may also play a role in the regulation of transfer in *A. tumefaciens*.

7.2.8 Quorum sensing in phytopathogenic Pseudomonas spp.

Although relatively little work has been carried out on quorum sensing in plant pathogenic pseudomonads, a number of species have been found to produce acyl HSLs. Over 100 soilborne and plant-associated strains were tested for acyl HSL production using a crossfeeding assay of violacein production in a lactone-deficient strain of *Chromobacterium violaceum* (Elasri *et al.*, 2001). All of those positively identified were plant-associated strains, with 49% of these being phytopathogens. LuxI homologues have been located in several *P. syringae* pathovars including PsyI in *P. syringae* pv *tabaci*, AhlI in *P. syringae* pv *syringae* and PsmI in *P. syringae* pv *maculicola*.

In *P. syringae* pv *maculicola*, open reading frames *psmI*, encoding an acyl HSL synthase, and *psmR*, encoding a LuxR homologue, are convergently transcribed and possess a small region of overlap (Elasri *et al.*, 2001). Expression of PsmI has been found to confer the ability to produce acyl HSL on a non-producing strain of *P. fluorescens*. The expression of both PsmI and PsmR, however, removes this ability, possibly due to negative regulation of *psmI* by PsmR. This theory is supported by the existence of a *lux* box in the promoter region of *psmI*.

P. syringae pv *syringae*, the causative agent of brown spot in beans, possesses the LuxI homologue AhlI (Kinscherf and Willis, 1999). The main acyl HSL produced by AhlI in this bacterium is OHHL. Deletion of *ahlI* was found to eliminate all acyl HSL production, whilst reducing bacterial viability on plant surfaces and leaving levels of protease and syringomycin antibiotic unaffected. The removal of the two-component system GacAS from the bacterium was found to eliminate the characteristic swarming motility of *P. syringae* pv *syringae* whilst leaving the bacteria deficient in acyl HSL levels. The removal of acyl HSL production alone, however, was found to have no effect on motility. This suggests that acyl HSL regulation in this bacterium plays a part in a larger network of intracellular regulators.

7.3 Non-acyl HSL systems

7.3.1 Quorum sensing in Ralstonia solanacearum

R. solanacearum causes vascular wilt in over 200 plant species with worldwide distribution (reviewed in Schell, 2000). The bacteria invade via the plant root system and then disseminate through host xylem vessels. It is here that they employ their primary virulence factor, exopolysaccharide I (EPS I), an acidic polymer that contributes to wilting by blocking water transport within the xylem (Schell, 1996). Other pathogenicity determinants of *R. solanacearum* include plant cell wall-degrading enzymes, siderophores for iron acquisition, and flagella. Together these factors are subject to a complex regulatory network controlled by bacterial cell density.

A central factor in this control is the LysR-type regulator PhcA (Brumbley *et al.*, 1993) (*Figure 7.4*). This has been shown to activate exopolysaccharide, endoglucanase and pectin methyl esterase production, whilst reducing expression of polygalacturonase and motility via the PehRS two component system. Mutants defective in *phcA* have therefore been found to be almost avirulent.

PhcA is itself indirectly regulated by the quorum-sensing signal 3-hydroxypalmitic acid methyl ester (3-OH PAME) (Flavier *et al.*, 1997). This diffusible and volatile compound is synthesised by the product of *phcB*, a gene found in the *phcBSR* operon (Clough *et al.*, 1997). The enzyme PhcB contains a motif typical of SAM-dependent methyltransferases and is therefore believed to convert a fatty acid to its volatile methyl ester, 3-OH PAME. The control of *phcA* expression by 3-OH PAME occurs via a two-component system consisting of the response regulator PhcR and PhcS, its cognate sensor histidine kinase. At low cell density, and therefore low 3-OH PAME concentration, PhcS is believed to phosphorylate PhcR which then represses the expression of PhcA. The exact mechanism of PhcR action is not yet known as it appears to contain no DNA-binding domain. It may therefore act via alternative regulators or directly on PhcA to regulate its activity. At higher cell densities, a critical 3-OH PAME concentration of more than 5 nM is reached. The signal molecule then acts to reduce the ability of PhcS to phosphorylate PhcR, inhibiting *phcA* repression and leading to the transcriptional activation of certain virulence factors.

The Phc system may therefore effect a phenotypic switch between the early and late phases of virulence (Genin and Boucher, 2002). At low cell density, early in

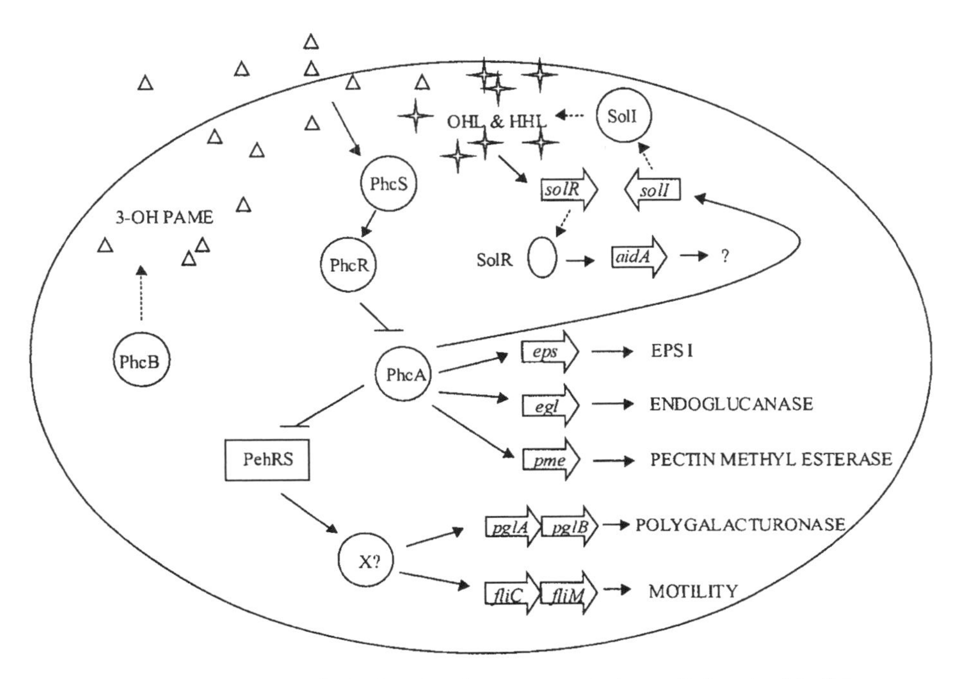

Figure 7.4. Quorum sensing in *Ralstonia solanacearum*. At low cell density PhcR is phosphorylated by PhcS, and can then repress PhcA by a currently unknown mechanism. This allows the PhcA-inhibited systems, including PehRS, to function, activating transcription of genes involved in motility and polygalacturonase synthesis, possibly via another regulator. Once the population size increases, a higher concentration of the PhcB-synthesised signalling molecule 3-OH PAME (△) is observed. This inhibits the action of PhcS and PhcA levels increase. PhcA represses the PehRS system whilst inducing the production of EPS I, endoglucanase and pectin methyl esterase. It also activates the transcription of the *solIR* genes. SolI synthesises the acyl HSLs OHL and HHL (✧), which are capable of activating SolR, a transcriptional regulator and LuxR homologue. The targets of SolR activation remain undetermined, with the exceptions of *solI* and *aidA*, a gene which encodes a protein of unknown function. Positive regulation is indicated by unbroken arrows and negative regulation by —| symbols.

infection, low levels of 3-OH PAME ensure that PhcA-repressed functions such as motility are expressed (Allen *et al.*, 1997; Garg *et al.*, 2000a). This is believed to be an important factor in the initial colonisation of the host by *R. solanacearum*, while the bacterium is essentially non-motile throughout the remainder of the infection (Tans-Kersten *et al.*, 2001). At higher cell densities, 3-OH PAME levels relieve the PhcRS-mediated repression of PhcA. This is then able to activate the expression of virulence factors such as EPS I and exoenzymes which play a role in infection once initial colonisation has taken place. A functional Phc-like system has been identified in the non-pathogenic strain *Ralstonia eutropha* where it serves to control motility and siderophore synthesis (Garg *et al.*, 2000b).

A secondary quorum-sensing mechanism also exists in *R. solanacearum* (Flavier *et al.*, 1997). The *solI* and *solR* genes, encoding *luxI* and *luxR* homologues respectively, form a typical acyl-HSL-based system. The expression of *solIR* requires

RpoS and PhcA activation, thereby forming a hierarchy within *R. solanacearum* quorum-sensing networks. Two signal molecules are believed to be employed by the *solIR* system, *N*-octanoyl-L-homoserine lactone (OHL) and HHL, production of which is eliminated in SolI mutants. The target genes of SolR remain largely unidentified – *solIR* mutations have no clear effect on virulence and the only regulated gene identified so far is the gene *aidA*, of currently unknown function. It is possible that this system may play a role in the regulation of factors involved in the terminal stages of wilting disease.

Sequencing of the genome of *R. solanacearum* strain GMI1000 has revealed a third putative quorum-sensing locus (Genin and Boucher 2002; Salanoubat *et al.*, 2002). A pair of open reading frames with homology to *solI* and *solR* have been identified on the 2.1 Mb megaplasmid of the *R. solanacearum* bipartite genome. This suggests that the complex regulatory network already revealed in *R. solanacearum* may be even more intricate than previously believed.

7.3.2 Quorum sensing in Xanthomonas campestris

X. campestris pv. *campestris* (*Xcc*) is a pathogen of crucifers where it causes black rot disease by producing extensive tissue damage (Onsando, 1992). The bacterium possesses a number of virulence factors including a range of plant cell wall-degrading enzymes and exopolysaccharide (EPS) composed of xanthan gums (Chun *et al.*, 1997; Dow and Daniels, 1994). The regulation of these virulence factors is, in part, controlled by two quorum-sensing systems. These rely on separate small signalling molecules called DF (*d*iffusible *f*actor) and DSF (*d*iffu*s*ible extracellular *f*actor) (Barber *et al.*, 1997; Poplawsky and Chun, 1997). The two systems have been shown by crossfeeding studies to be entirely independent as over-production of either factor is unable to complement a mutation in the other (Poplawsky *et al.*, 1998). Both systems work to control the transcription of their target genes via a series of other regulators.

The DF signalling molecule, which is possibly a butyrolactone, is produced by all xanthomonads (Poplawsky and Chun, 1997). It is synthesised by the product of the *pigB* gene which, in *Xcc* B24, lies in the 25 kb *pigABCDEFG* cluster (Poplawsky and Chun, 1997). DF activates expression of the yellow pigment xanthomonadin and the *gum* operon encoding the synthesis and export of EPS (Chun *et al.*, 1997). The reduced production of these factors in *pigB* mutants can be restored by the addition of exogenous DF by crossfeeding from a wild-type strain. The presence of *pigB* in *Xcc* has been shown to be essential for the epiphytic survival of the bacterium (Poplawsky and Chun, 1998).

The regulation of EPS is also under the control of the DSF signalling molecule (Barber *et al.*, 1997). This factor has been shown to upregulate both EPS and exoenzyme expression through the *rpf* (*r*egulation of *p*athogenicity *f*actors) locus in *Xcc* 8004 (Tang *et al.*, 1991; Dow *et al.*, 2000). The nine genes, *rpfA–I*, present in this operon have all been shown to play a role in DSF-mediated regulation. The genes *rpfB* and *rpfF* encode enzymes required for synthesis of DSF (Barber *et al.*, 1997). These homologues of a long chain fatty acid CoA ligase and enoyl CoA hydratase respectively may function to divert lipid metabolism intermediates to the synthesis of DSF. The signalling molecule is probably a fatty acid derivative, although it is not believed to be an acyl HSL. The deletion of either gene results in the loss of DSF production as well as DSF-regulated phenotypes, but only those in *rpfF* can be restored by the addition of exogenous signalling molecule.

The other genes of the *rpf* operon are regulators. The products of *rpfC* and *rpfG* are thought to form a two-component system involved in the sensing and controlling of DSF levels (Slater *et al.*, 2000). RpfC contains both sensor kinase and response regulator-like domains and may be a hybrid formed from the fusion of two separate proteins (Tang *et al.*, 1991). Mutants defective in *rpfC* have increased DSF production and down-regulated levels of EPS and exoenzymes and it may therefore repress the transcription of DSF whilst activating that of virulence factors. RpfG is thought to be the cognate response regulator of RpfC and is phosphorylated by the sensor kinase following its own autophosphorylation. RpfG possesses a typical receiver domain attached to a HD-GYP domain of the HD superfamily (Galperin *et al.*, 2001). This may have phosphodiesterase activity and could be involved in diguanylate signalling, although this role has not yet been demonstrated. The way in which RpfG regulates gene expression is not yet known and may take place through some secondary regulator, possibly in response to sensing environmental cues.

The RpfA protein is a homologue of bacterial aconitase enzymes (Wilson *et al.*, 1998). In *rpfA* mutants the major *Xcc* aconitase is absent and intracellular iron levels are reduced. The protein may regulate virulence factor expression in response to changes in intracellular iron concentration.

RpfH is structurally homologous to the membrane-spanning region of RpfC, although it does not appear to have a sensor kinase domain. The function of this protein has not yet been determined although it appears to be non-essential for virulence as *rpfH* mutants only have slightly reduced levels of enzymes and EPS (Slater *et al.*, 2000).

While DSF and the *rpf* systems play an important part in the regulation of exoenzyme expression they are not the only system present in *Xcc* to control this virulence determinant. Evidence of this was provided when the addition of exogenous DSF to *Xcc* cultures was found to be insufficient for precocious induction of enzymes. Other factors implicated in their regulation include nutrient availability and a protein homologous to the cAMP receptor protein (Hsaio and Tseng, 2002). The conservation of the *rpf* genes within other *Xanthomonas* spp. and closely related bacteria suggests that, although their contribution is not the sole regulation acting upon the exoenzyme genes, it is of great importance. *rpf* gene clusters have been identified in *X. axonopodis* pv. *citri* (*Xac*), the causal agent of citrus canker, *X. oryzae* pv. *oryzae* which produces bacterial leaf blight in rice and *Xylella fastidiosa* causing diseases such as citrus variegated chlorosis (Chatterjee and Sonti, 2002; da Silva *et al.*, 2001, 2002). Both *Xac* and *X. fastidiosa* contain partial *rpf* clusters with *rpfD* and *rpfH* missing from *X. fastidiosa* and *rpfI* and *rpfH* from *Xac*. The cluster differences between *Xac* and *Xcc* are thought to originate in the different levels of damage inflicted on their hosts (da Silva *et al.*, 2002). In *Xac* tissue is macerated to a lesser extent than in *Xcc* infections and it may be that RpfI, which is missing in the former, regulates this extensive damage. In *X. oryzae* pv. *oryzae* the RpfF protein has been found to have a slightly different function from its *Xcc* counterpart (Chatterjee and Sonti, 2002). In *X. oryzae* pv. *oryzae rpfF* mutants virulence and DSF levels are decreased while EPS and enzyme levels are unaffected and siderophore production is increased. Further study of the role of this protein has lead to the hypothesis that it may be involved in controlling an iron uptake system. The counterpart protein in *Xcc* is involved in DSF synthesis and is not known to be involved in iron uptake. The only iron-related *rpf* gene product identified in *Xcc* is RpfA, although deletion of the gene is not known to have any effect on siderophore levels. RpfA has not yet been investigated in *X. oryzae* pv. *oryzae*.

7.3.3 Quorum sensing and nodulation in Rhizobium spp.

Although not pathogenic to plants, reference should also be made to the Rhizobiaceae which utilise multiple signalling pathways to alter leguminous plant physiology, producing nodules in which nitrogen fixation can take place. *Rhizobium* spp. can communicate with host plants by sensing secreted flavonoid signals. These induce the production of bacterial Nod factors, lipo-chitin oligosaccharides which induce nodulation in the plant root (reviewed in Spaink, 2000). Several rhizobia have been identified that also possess LuxIR-type quorum-sensing systems and in numerous cases were found to produce multiple acyl HSLs (Cha *et al.*, 1998; Daniels *et al.*, 2002). *Rhizobium leguminosarum* has been found to contain a network of quorum-sensing systems including the CinIR and RaiIR systems (Wilkinson *et al.*, 2002; Wisniewski-Dye *et al.*, 2002). Although the exact cellular processes regulated by this network have not yet been fully characterised, the systems are implicated in the inhibition of nodulation and the conjugation of the symbiotic (Sym) plasmid. Quorum-sensing systems have also been identified in *Sinorhizobium meliloti*. The SinIR system in this bacterium is required for maximal nodule formation and SinI was found to synthesise novel acyl HSLs (Marketon *et al.*, 2002). In *S. meliloti* Rm1021 a LuxR homologue, designated ExpR, was found to activate the production of exopolysaccharide II required for root nodule invasion, and homologues of the TraR and TraM proteins of *A. tumefaciens* have been identified (Marketon *et al.*, 2002; Pellock *et al.*, 2002).

7.4 Concluding remarks

7.4.1 Further possibilities for quorum sensing in phytopathogens

The discovery of the furanosyl borate diester signalling molecule designated AI-2 (*a*uto*i*nducer-2) has highlighted further possibilities for bacterial communication (Bassler *et al.*, 1994; Chen *et al.*, 2002). Although the exact function of this molecule in bacterial interactions has not yet been determined, it is possible that it may be involved in intercellular communication, distinguishing it from the intracellular AI-1 (acyl HSL) molecules. AI-2 is produced from SAM, like AI-1, although its synthesis involves three enzymatic steps and the final product bears no structural resemblance to an acyl HSL (Schauder *et al.*, 2001).

The *luxS* gene encoding AI-2 synthase, the final enzyme in the synthesis pathway, has been identified in a large number of bacterial species. The precise physiological function of AI-2 remains unknown although it has been implicated in the regulation of virulence in the human pathogens *E. coli* EPEC and EHEC (Sperandio *et al.*, 1999). As *luxS* homologues have now been identified in *Ecc* (S. Coulthurst, personal commununication), and seems likely to be found in *Erwinia carotovora* subsp. *atroseptica* (The Pathogen sequencing Group Sanger

Institute) and *Echr* (The Institute for Genomic Research), it is possible that some link between AI-2 and the regulation of phytopathogenicity may be found.

There is also evidence suggesting that further bacterial quorum-sensing systems remain currently undiscovered. In *Vibrio cholerae* three parallel quorum-sensing systems have been identified and implicated in the regulation of virulence (Miller *et al.*, 2002). Two of these systems have been characterised and neither uses the typical LuxIR system. These systems, designated systems one and two, utilise AI-2 and CsqA-dependent autoinducer signals respectively.

7.4.2 Why study phytopathogenic bacterial signalling?

A key reason for the study of signalling in bacterial plant pathogens is the potential of these systems as targets for the control of disease. Several possible methods of disrupting quorum-sensing signalling pathways have been investigated although, so far, none has been adapted for large-scale implementation in plants. One possible strategy involves the use of molecular mimicry by molecules such as the furanones secreted by the marine alga *Delisea pulchra* (Kjelleberg *et al.*, 1997). These signals, which show some structural similarity to acyl HSLs, have been shown to inhibit quorum-sensing-dependent phenotypes such as the production of carbapenem in *Ecc* (Manefield *et al.*, 2000). Such AHL mimics have also been identified in plants including pea and soybean (Teplitski *et al.*, 2000).

Quorum-sensing-regulated infections can also be limited by the use of acyl HSL degradases. By degrading the signalling molecules, bacterial populations would be unable to determine their size and quorum-sensing-regulated phenotypes would never be induced. The presence of these degradase enzymes has been demonstrated in *Variovax paradoxus* and *Bacillus* sp. 240B1, with sequence homologues identified in other bacterial species including *Agrobacterium tumefaciens* (Dong *et al.*, 2000; Leadbetter and Greenberg, 2000). The degradase isolated from *Bacillus* sp. 240B1, AiiA, has been expressed in *Ecc* SCG1 where it was shown to reduce the levels of OHHL production and cause a reduction in virulence (Dong *et al.*, 2000). AiiA has also been expressed from tobacco where a similar reduction of virulence in *Ecc* was observed (Dong *et al.*, 2001; see also Chapter 8).

A further strategy is the production of plant species that express acyl HSL synthases. By producing its own source of the signalling molecule, which can be sensed by bacteria, the plant may induce precocious expression of virulence determinants in invading pathogens. This could lead to their detection by the plant immune system, before there are sufficient numbers to successfully produce disease, which can then eliminate the infection. Transgenic tobacco plants containing the *expI* gene of *Ecc* have been created and were shown to produce their own OHHL supply (Mäe *et al.*, 2001). When infected with *Ecc* SCC3193 these plants demonstrated increased resistance to the pathogen. However, success in this system depends on precocious exoenzyme induction via acyl HSLs, and this is not a universal, or indeed even common, response in *Ecc* strains (unpublished). This control method was also investigated using transgenic potato plants containing the *yenI* gene of *Yersinia enterolitica* (Fray *et al.*, 1999). These potatoes, which were shown to produce their own OHHL and HHL, were then infected with *Erwinia carotovora* subspecies *atroseptica*. Precocious induction of virulence at low innoculum levels (10^2 cells) was observed although, instead of resulting in pathogen elimination, this was found to produce disease. This system, therefore, produced disease at smaller pathogen population sizes than in untransformed plants (Fray 2002).

Quorum sensing plays a vital role in the regulation of virulence in plant pathogenic bacteria. Although much is already known about these regulation networks, evidence of additional complexity within these systems is constantly being revealed. The study of this regulation has already revealed several possible targets for the treatment of plant disease caused by quorum-sensing-dependent bacteria. Further study in this area, however, is required in order to fully comprehend bacterial signalling and how it could be harnessed to prevent disease.

Acknowledgements

We would like to thank Martin Welch, Neil Whitehead and all the members of the Salmond group for helpful discussions. Clare Pemberton and Holly Slater are supported by grants from the Biotechnology and Biological Sciences Research Council (UK).

References

Allen, C., Gay, J., Simon-Buela, L. (1997) A regulatory locus, *pehSR*, controls polygalacturonase production and other virulence factors in *Ralstonia solanacearum*. *Mol. Plant–Microbe Interact.* **10**: 1054–1064.

Andersson, R.A., Eriksson, A.R.B., Heikinheimo, R., Mae, A., Pirhonen, M., Koiv, V., Hyytiainen, H., Tuikkala, A., Palva, E.T. (2000) Quorum sensing in the plant pathogen *Erwinia carotovora* subsp. *carotovora*: the role of $expR_{Ecc}$. Mol. *Plant–Microbe Interact.* **13**: 384–393.

Axelrood, P.E., Rella, M., Schroth, M.N. (1988) Role of antibiosis in competition of *Erwinia* strains in potato infection courts. *Appl. Environ. Microbiol.* **54**: 1222–1229.

Bainton, N.J., Stead, P., Chhabra, S.R., Bycroft, B.W., Salmond, G.P.C., Stewart, G.S.A.B., Williams, P. (1992a) *N*-(3-oxohexanoyl)-L-homoserine lactone regulates carbapenem antibiotic production in *Erwinia carotovora*. *Biochem. J.* **288**: 997–1004.

Bainton, N.J., Bycroft, B.W., Chhabra, S.R. *et al.* (1992b) A general role for the *lux* autoinducer in bacterial cell signalling: control of antibiotic biosynthesis in *Erwinia*. *Gene* **116**: 87–91.

Barber, C.E., Tang, J.L., Feng, J.X., Pan, M.Q., Wilson, T.J.G., Slater, H., Dow, J.M., Williams, P., Daniels, M.J. (1997) A novel regulatory system required for pathogenicity of *Xanthomonas campestris* is mediated by a small diffusible signal molecule. *Mol. Microbiol.* **24**: 555–566.

Bassler, B.L., Wright, M., Silverman, M.R. (1994) Multiple signaling systems controlling expression of luminescence in *Vibrio harveyi*: sequence and function of genes encoding a second sensory pathway. *Mol. Microbiol.* **13**: 273–286.

Beck von Bodman, S., Farrand, S.K. (1995) Capsular polysaccharide biosynthesis and pathogenicity in *Erwinia stewartii* require induction by an *N*-acylhomoserine lactone autoinducer. *J. Bacteriol.* **177**: 5000–5008.

Beck von Bodman, S., Hayman, G.T., Farrand, S.K. (1992) Opine catabolism and conjugal transfer of the nopaline Ti plasmid pTiC58 are coordinately regulated by a single repressor. *Proc. Natl Acad. Sci. USA* **89**: 643–647.

Beck von Bodman, S., Majerczak, D.R., Coplin, D.L. (1998) A negative regulator mediates quorum sensing control of exopolysaccharide production in *Pantoea stewartii* subsp. *stewartii*. *Proc. Natl Acad. Sci. USA* **95**: 7687–7692.

Braun, E.J. (1982) Ultrastructural investigation of resistant and susceptible maize inbreds infected with *Erwinia stewartii*. *Phytopathol.* **72**: 159–166.

Brumbley, S.M., Carney, B.F., Denny, T.P. (1993) Phenotype conversion in *Pseudomonas solanacearum* due to spontaneous inactivation of PhcA, a putative LysR transcriptional regulator. *J. Bacteriol.* **175**: 5477–5487.

Cha, C., Gao, P., Chen, Y.C., Shaw, P.D., Farrand, S.K. (1998) Production of acyl-homoserine lactone quorum sensing signals by Gram negative plant-associated bacteria. *Mol. Plant–Microbe Interact.* **11**: 1119–1129.

Chai, Y., Zhu, J., Winans, S.C. (2001) TrlR, a defective TraR-like protein of *Agrobacterium tumefaciens*, blocks TraR function in vitro by forming inactive TrlR:TraR dimers. *Mol. Microbiol.* **40**: 414–421.

Chatterjee, S., Sonti, R.V. (2002) *rpfF* mutants of *Xanthomonas oryzae* pv. *oryzae* are deficient for virulence and growth under low iron conditions. *Mol. Plant–Microbe Interact.* **15**: 463–471.

Chatterjee, A., Cui, Y., Liu, Y., Dumenyo, C.K., Chatterjee, A.K. (1995) Inactivation of *rsmA* leads to overproduction of extracellular pectinases, cellulases and proteases in *Erwinia carotovora* subsp. *carotovora* in the absence of the starvation/cell density sensing signal, *N*-(3-oxohexanoyl)-L-homoserine lactone. *Appl. Environ. Microbiol.* **61**: 1959–1967.

Chatterjee, A., Cui, Y., Chatterjee, A.K. (2002) RsmA and the quorum sensing signal, *N*-(3-oxohexanoyl)-L-homoserine lactone, control the levels of *rsmB* RNA in *Erwinia carotovora* subsp. *carotovora* by affecting its stability. *J Bacteriol.* **184**: 4089–4095.

Chen, L., Chen, Y., Wood, D.W., Nester, E. (2002a) A new type IV secretion system promotes conjugal transfer in *Agrobacterium tumefaciens. J. Bacteriol.* **184**: 4838–4845.

Chen, X., Schauder, S., Potier, N., Van Dorsselaer, A., Pelczer, I., Bassler, B.L. and Hughson, F.M. (2002b) Structural identification of a bacterial quorum sensing signal containing boron. *Nature* **415**: 545–549.

Choi, S.H., Greenberg E.P. (1991) The C-terminal region of the *Vibrio fischeri* LuxR protein contains an autoinducer-independent *lux* gene activating domain. *Proc. Natl Acad. Sci. USA* **88**: 11115–11119.

Choi, S.H., Greenberg E.P. (1992a) Genetic dissection of DNA binding and luminescence gene activation by the *Vibrio fischeri* LuxR protein. *J. Bacteriol.* **174**: 4064–4069.

Choi, S.H., Greenberg E.P. (1992b) Genetic evidence for multimerization of LuxR, the transcriptional activator of *Vibrio fischeri* luminescence. *Mol. Mar. Biol. Biotechnol.* **1**: 408–413.

Clough, S.J., Lee, K.E., Schell, M.A., Denny, T.P. (1997) A two-component system in *Ralstonia (Pseudomonas) solanacearum* modulates production of *phcA*-regulated virulence factors in response to 3-hydroxypalmitic acid methyl ester. *J. Bacteriol.* **179**: 3639–3648.

Costa, J.M., Loper, J.E. (1997) Ecbl and EcbR: homologs of LuxI and LuxR affecting antibiotic and exoenzyme production by *Erwinia carotovora* subsp. *betavasculorum. Can. J. Microbiol.* **43**: 1164–1171.

Cui, Y., Chatterjee, A., Liu, Y., Dumenyo, C.K., Chatterjee, A.K. (1995) Identification of a global repressor gene, *rsmA*, of *Erwinia carotovora* subsp. *carotovora* that controls extracellular enzymes, *N*-(3-oxohexanoyl)-L-homoserine lactone and pathogenicity in soft rotting *Erwinia* spp. *J. Bacteriol.* **177**: 5108–5115.

Cui, Y., Madi, L., Mukherjee, A., Dumenyo, C.K., Chatterjee, A.K. (1996) The RsmA$^-$ mutants of *Erwinia carotovora* subsp. *carotovora* strain Ecc 71 overexpress $hrpN_{Ecc}$ and elicit a hypersensitive reaction-like response in tobacco leaves. *Mol. Plant–Microbe Interact.* **9**: 565–573.

Daniels, R., De Vos, D.E., Desair, J. *et al.* (2002) The *cin* quorum sensing locus of *Rhizobium etli* CNPAF512 affects growth and symbiotic nitrogen fixation. *J. Biol. Chem.* **277**: 462–468.

da Silva, F.R., Vettore, A.L., Kemper, E.L., Leite, A., Arruda, P. (2001) Fastidian gum: the *Xylella fastidiosa* Exopolysaccharide possibly involved in bacterial pathogenicity. *FEMS Micro. Lett.* **203**: 165–171.

da Silva, A.C.R., Ferro, J.A., Reinach, F.C. *et al.* (2002) Comparison of the genomes of two *Xanthomonas* pathogens with differing host specificities. *Nature* **417**: 459–463.

Dessaux, Y., Petit, A., Farrand, S.K., Murphy, P.J. (1998) Opines and opine-like molecules involved in plant-Rhizobiaceae interactions. In: Spaink, H.P., Kondrossi, A., Hooykaas, P.J.J. (eds) *The Rhizobiaceae: Molecular Biology of Modern Plant Associated Bacteria*, pp. 199–233. Kluwer: Dordrecht.

Devine, J.H., Countryman, C., Baldwin, T.O. (1988) Nucleotide sequence of the *luxR* and *luxI* genes and structure of the primary regulatory region of the *lux* regulon of *Vibrio fischeri* ATCC 7744. *Biochemistry* **27**: 837–842.

Devine, J.H., Shadel, G.S., Baldwin, T.O. (1989) Identification of the operator of the *lux* regulon from the *Vibrio fischeri* strain ATCC7744. *Proc. Natl Acad. Sci. USA* **86:** 5688–5692.

Dong, Y.H., Xu, J.L., Li, X.Z., Zhang, L.H. (2000) AiiA, an enzyme that inactivates the acylhomoserine lactone quorum sensing signal and attenuates the virulence of *Erwinia carotovora*. *Proc. Natl Acad. Sci. USA* **97**: 3526–3531.

Dong, Y.H., Wang, L.H., Xu, J.L., Zhang, H.B., Zhang, H.F., Zhang, L.H. (2001) Quenching quorum sensing-dependent bacterial infection by an *N*-acyl homoserine lactone. *Nature* **411**: 813–817.

Dow, J.M., Daniels, M.J. (1994) Pathogenicity determinants and global regulation of pathogenicity of *Xanthomonas campestris* pv. *campestris*. In: Dangl, J.L. (ed) *Bacterial Pathogenesis of Plants and Animals*, pp. 29–41. Springer-Verlag: Heidelberg, Germany.

Dow, J.M., Feng, J-X., Barber, C.E., Tang, J-L., Daniels, M.J. (2000) Novel genes involved in the regulation of pathogenicity factor production within the *rpf* gene cluster of *Xanthomonas campestris*. *Microbiology* **146**: 885-891.

Dunlap, P.V., Ray, J.M. (1989) Requirement for autoinducer in transcriptional negative autoregulation of the *Vibrio fischeri luxR* gene in *Escherichia coli*. *J. Bacteriol.* **171**: 3549–3552.

Eberhard, A., Burlingame, A.L., Eberhard, C., Kenyon, G.L., Nealson, K.H., Oppenheimer, N.J. (1981) Structural identification of autoinducer of *Photobacterium fischeri* luciferase. *Biochemistry* **20**: 2444–2449.

Eberhard, A., Longin, T., Widrig, C.A., Stranick, S.J. (1991) Synthesis of the *lux* gene autoinducer in *Vibrio fischeri* is positively autoregulated. *Arch. Microbiol.* **155**: 294–297.

Egland, K.A., Greenberg, E.P. (1999) Quorum sensing in *Vibrio fischeri*: elements of the *luxI* promoter. *Mol. Microbiol.* **31**: 1197–1204.

Elasri, M., Delorme, S., Lemanceau, P., Stewart, G., Laue, B., Glickmann, E., Oger, P.M., Dessaux, Y. (2001) Acyl-homoserine lactone production is more common among plant-associated *Pseudomonas* spp. than among soilborne *Pseudomonas* spp. *Appl. Environ. Microbiol.* **67**: 1198–1209.

Ellis, J.G., Kerr, A., Petit, A., Tempé, J. (1982) Conjugal transfer of nopaline and agropine Ti-plasmids – the role of agrocinopines. *Mol. Gen. Genet.* **186**: 269–273.

Engebrecht, J., Nealson, K., Silverman, M. (1983) Bacterial bioluminescence: isolation and genetic analysis of functions from *Vibrio fischeri*. *Cell* **32**: 773–781.

Engebrecht, J., Silverman, M. (1984) Identification of genes and gene products necessary for bacterial bioluminescence. *Proc. Natl Acad. Sci. USA* **81**: 4154–4158.

Finney, A.H., Blick, R.J., Murakami, K., Ishihama, A., Stevens, A.M. (2002) Role of the C terminal domain of the alpha subunit of RNA polymerase in LuxR-dependent transcriptional activation of the *lux* operon during quorum sensing. *J. Bacteriol.* **184**: 4520–4528.

Flavier, A.B., Ganove-Raeva, L.M., Schell, M.A., Denny, T.P. (1997) Hierarchical autoinduction in *Ralstonia solanacearum*: control of acyl-homoserine lactone production by a novel autoregulatory system responsive to 3-hydroxypalmitic acid methyl ester. *J. Bacteriol.* **179**: 7089–7097.

Fray R.G. (2002) Altering plant–microbe interaction through artificially manipulating bacterial quorum sensing. *Ann. Bot.* **89**: 245–253.

Fray, R.G., Throup, J.P., Wallace, A., Daykin, M., Williams, P., Stewart, G.S.A.B., Grierson, D. (1999) Plants genetically modified to produce *N*-acylhomoserine lactones communicate with bacteria. *Nature Biotechnol.* **17**: 1017–1020.

Fuqua, W.C., Winans, S.C. (1994) A LuxR-LuxI type regulatory system activates *Agrobacterium* Ti plasmid conjugal transfer in the presence of a plant tumour metabolite. *J. Bacteriol.* **176**: 2796–3806.

Fuqua, C., Winans, S.C. (1996) Localization of OccR-activated and TraR-activated promoters that express two ABC-type permeases and the *traR* gene of Ti plasmid pTiR10. *Mol. Microbiol.* **20**: 1199–1210.

Galperin, M.Y., Nikolskaya, A.N., Koonin, E.V. (2001) Novel domains of the prokaryotic two-component signal transduction systems. *FEMS Micro. Lett.* **203**: 11–21.

Garg, R.P., Huang, J., Yindeeyoungyeon, W., Denny, T.P., Schell, M.A. (2000a) Multicomponent transcriptional regulation at the complex promoter of the exopolysaccharide I biosynthetic operon of *Ralstonia solanacearum*. *J. Bacteriol.* **182**: 6659–6666.

Garg, R.P., Yindeeyoungyeon, W., Gilis, A., Denny, T.P., van der Lelle, D., Schell, M.A. (2000b) Evidence that *Ralstonia eutropha* (*Alcaligenes eutrophus*) contains a functional homologue of the *Ralstonia solanacearum* Phc cell density sensing system. *Mol. Microbiol.* **38**: 359–367.

Genin, S., Boucher, C. (2002) *Ralstonia solanacearum*: secrets of a major pathogen unveiled by analysis of its genome. *Mol. Plant. Pathol.* **3**: 111–118.

Gilson, L., Kuo, A., Dunlap, P.V. (1995) AinS and a new family of autoinducer synthesis proteins. *J. Bacteriol.* **177**: 6946–6951.

Gottesman, S., Stout, V. (1991) Regulation of capsular polysaccharide synthesis in *Escherichia coli* K12. *Mol. Microbiol.* 5: 1599-1606.

Habeeb, L.F., Wang, L., Winans, S.C. (1991) Transcription of the octopine catabolism operon of the *Agrobacterium* tumor-inducing plasmid pTiA6 is activated by a LysR-type regulatory protein. *Mol. Plant–Microbe Interact.* **4**: 379–385.

Hanzelka, B.L., Greenberg, E.P. (1995) Evidence that the N-terminal region of the *Vibrio fischeri* LuxR protein constitutes an autoinducer-binding domain. *J. Bacteriol.* **177**: 815–817.

Hanzelka, B.L., Parsek, M.R., Val, D.L., Dunlap, P.V., Cronan Jr., J.E., Greenberg, E.P. (1999) Acylhomoserine lactone synthase activity of the *Vibrio fischeri* AinS protein. *J. Bacteriol.* **181**: 5766–5770.

Holden, M.T.G., McGowan, S.J., Bycroft, B.W., Stewart, G.S.A.B., Williams, P., Salmond, G.P.C. (1998) Cryptic carbapenem antibiotic production genes are widespread in *Erwinia carotovora*: facile *trans* activation by the *carR* transcriptional regulator. *Microbiology* **144**: 1495–1508.

Hsaio, Y.M., Tseng, Y.H. (2002) Transcription of *Xanthomonas campestris prt1* gene encoding protease 1 increases during stationary phase and requires global transcription factor Clp. *Biochem. Biophys. Res. Commun.* **295**: 43–49.

Hwang, I., Smyth, A.J., Luo, Z., Farrand, S.K. (1999) Modulating quorum sensing by antiactivation: TraM interacts with TraR to inhibit activation of Ti plasmid conjugal transfer genes. *Mol Microbiol.* **34**: 282–294.

Jones, S., Yu, B., Bainton, N.J. *et al.* (1993) The *lux* autoinducer regulates the production of exoenzyme virulence determinants in *Erwinia carotovora* and *Pseudomonas aeruginosa*. *EMBO J.* **12**: 2477–2482.

Kaplan, H.B., Greenberg, E.P. (1985) Diffusion of autoinducer is involved in regulation of the *Vibrio fischeri* luminescence system. *J. Bacteriol.* **163**: 1210–1214.

Kim, H., Farrand, S.K. (1997) Characterization of the *acc* operon from the nopaline-type Ti plasmid pTiC58, which encodes utilization of agrocinopines A and B and susceptibility to agrocin 84. *J. Bacteriol.* **179**: 7559–7572.

Kinscherf, T.G., Willis, D.K. (1999) Swarming by *Pseudomonas syringae* B728a requires *gacS* (*lemA*) and *gacA* but not the acyl-homoserine lactone biosynthetic gene *ahlI*. *J. Bacteriol.* **181**: 4133–4136.

Kjelleberg, S., Steinberg, P., Givskov, M., Gram, L., Manefield, M., de Nys, R. (1997) Do marine natural products interfere with prokaryotic AHL regulatory systems? *Aquat. Microb. Ecol.* **13**: 85–93.

Klapwijk, P.M., Scheulderman, T., Schilperoort, R.A. (1978) Coordinated regulation of octopine degradation and conjugative transfer of Ti plasmids in *Agrobacterium tumefaciens*: evidence for a common regulatory gene and separate operons. *J. Bacteriol.* **136**: 775–785.

Kleerebezem, M., Quadri, L.E.N., Kuipers, O.P., de Vos, W.M. (1997) Quorum sensing by peptide pheromones and two-component signal-transduction systems in Gram-positive bacteria. *Mol. Microbiol.* **24**: 895–904.

Kuo, A., Blough, N.V., Dunlap, P.V. (1994) Multiple *N*-acyl-L-homoserine lactone autoinducers of luminescence in the marine symbiotic bacterium *Vibrio fischeri*. *J. Bacteriol.* **176**: 7558–7565.

Lazazzera, B.A., Grossman, A.D. (1998) The ins and outs of peptide signaling. *Trends Microbiol.* 7: 288–294.

Leadbetter, J.R., Greenberg, E.P. (2000) Metabolism of acyl-homoserine lactone quorum sensing signals by *Variovorax paradoxus*. *J. Bacteriol.* **182**: 6921–6926.

Liu, Y., Cui, Y., Mukherjee, A., Chatterjee, A.K. (1998) Characterisation of a novel RNA regulator of *Erwinia carotovora* subsp. *carotovora* that controls production of extracellular enzymes and secondary metabolites. *Mol. Microbiol.* **29**: 219–234.

Mäe, A., Montesano, M., Koiv, V., Palva, E.T. (2001) Transgenic plants producing the bacterial pheromone *N*-acyl homoserine lactone exhibit enhanced resistance to the bacterial phytopathogen *Erwinia carotovora*. *Mol. Plant–Microbe Interact.* **14**: 1035–1042.

Manefield, M., Harris, L., Rice, S.A., de Nys, R., Kjelleberg, S. (2000) Inhibition of luminescence and virulence in the black tiger prawn (*Penaeus monodon*) pathogen *Vibrio harveyi* by intercellular signal antagonists. *Appl. Environ. Microbiol.* **66**: 2079–2084.

Marketon, M.M., Gonzalez, J.E. (2002) Identification of two quorum sensing systems in *Sinorhizobium meliloti*. *J. Bacteriol.* **184**: 3466–3475.

Marketon, M.M., Gronquist, M.R., Eberhard, A., Gonzalez, J.E. (2002) Characterization of the *Sinorhizobium meliloti sinR/sinI* locus and the production of novel *N*-acyl homoserine lactones. *J. Bacteriol.* **184**: 5686–5695.

McGowan, S., Sebaiha, M., Jones, S. *et al.* (1995) Carbapenem antibiotic production in *Erwinia carotovora* is regulated by CarR, a homologue of the LuxR transcriptional activator. *Microbiology* **141**: 541–550.

McGowan, S.J., Sebaihia, M., Porter, L.E., Stewart, G.S.A.B., Williams, P., Bycroft, B.W., Salmond, G.P.C. (1996) Analysis of bacterial carbapenem antibiotic production genes reveals a novel b-lactam biosynthesis pathway. *Mol. Microbiol.* **22**: 415–426.

McGowan, S.J., Sebaihia, M., O'Leary, S., Hardie, K.R., Williams, P., Stewart, G.S.A.B., Bycroft, B.W., Salmond, G.P.C. (1997) Analysis of the carbapenem gene cluster of *Erwinia carotovora*: definition of the antibiotic biosynthetic genes and evidence for a novel b-lactam resistance mechanism. *Mol. Microbiol.* **26**: 545–556.

Miller, M.B., Skorupski, K., Lenz, D.H., Taylor, R.K., Bassler, B.L. (2002) Parallel quorum sensing systems converge to regulate virulence in *Vibrio cholerae*. *Cell* **110**: 303–314.

Minogue, T.D., Wehland-von Trebra, M., Bernhard, F., von Bodman, S.B. (2002) The autoregulatory role of EsaR, a quorum-sensing regulator in *Pantoea stewartii* ssp. *stewartii*: evidence for a repressor function. *Mol. Microbiol.* **44**: 1625–1635.

Nasser, W., Bouillant, M.L., Salmond, G., Reverchon, S. (1998) Characterization of the *Erwinia chrysanthemi expI – expR* locus directing the synthesis of two *N*-acyl-homoserine lactone signal molecules. *Mol. Microbiol.* **29**: 1391–1405.

Oger, P., Farrand, S.K. (2002) Two opines control conjugal transfer of an *Agrobacterium* plasmid by regulating expression of separate copies of the quorum-sensing activator gene *traR*. *J. Bacteriol.* **184**: 1121–1131.

Onsando, J.M. (1992) Black rot of crucifers. In: Chaube, H.S., Kumar, J., Mukhopadhyay, A.N., Singh, U.S. (eds) *Diseases of Vegetables and Oil Seed Crops*, pp. 243–252. Prentice Hall: New Jersey, USA.

Palva, T.K., Holmstrom, K.O., Heino, P., Palva E.T. (1993) Induction of plant defense response by exoenzymes of *Erwinia carotovora* subsp. *carotovora*. *Mol. Plant–Microbe Interact.* **6**: 190–196.

Parker, W.L., Rathnum, M.L., Wells, J.S., Trejo, W.H., Principe, P.A., Sykes, R.B. (1982) SQ27850, a simple carbapenem produced by species of *Serratia* and *Erwinia*. *J. Antibiot.* **35**: 653–660.

Parsek, M.R., Val, D.L., Hanzelka, B.L., Cronan, J.E., Greenberg, E.P. (1999) Acyl homoserine-lactone quorum-sensing signal generation. *Proc. Natl. Acad. Sci. USA* **96**: 4360–4365.

Pellock, B.J., Teplitski, M., Boinay, R.P., Bauer, W.D., Walker, G.C. (2002) A LuxR homolog controls production of symbiotically active extracellular polysaccharide II by *Sinorhizobium meliloti*. *J. Bacteriol.* **184**: 5067–5076.

Pérombelon, M.C.M. (2002) Potato diseases caused by soft rot erwinias: an overview of pathogenesis. *Plant Pathol.* **51**: 1–12.

Piper, K.R., Farrand, S.K. (2000) Quorum sensing but not autoinduction of Ti plasmid conjugal transfer requires control by the opine regulon and the antiactivator TraM. *J. Bacteriol.* **182**: 1080–1088.

Piper, K.R., Beck von Bodman, S., Hwang, I., Farrand, S.K. (1999) Hierarchical gene regulatory systems arising from fortuitous gene associations: controlling quorum sensing by the opine regulon in *Agrobacterium*. *Mol. Microbiol.* **32**: 1077–1089.

Pirhonen, M., Flego, D., Heikinheimo, R., Palva, E.T. (1993) A small diffusible signal molecule is responsible for the global control of virulence and exoenzyme production in *Erwinia carotovora*. *EMBO J.* **12**: 2467–2476.

Poplawsky, A.R., Chun, W. (1997) *pigB* determines a diffusible factor needed for extracellular polysaccharide slime and xanthomonadin production in *Xanthomonas campestris* pv. *campestris*. *J. Bacteriol.* **179**: 439–444.

Poplawsky, A.R., Chun, W. (1998) *Xanthomonas campestris* pv. *campestris* requires a functional *pigB* for epiphytic survival and host infection. *Mol. Plant–Microbe Interact.* **11**: 466–475.

Poplawsky A.R., Chun, W., Slater, H., Daniels, M.J., Dow, J.M. (1998) Synthesis of extracellular polysaccaharide, extracellular enzymes, and xanthomonadin in *Xanthomonas campestris*: evidence for the involvement of two intercellular regulatory signals. *Mol. Plant–Microbe Interact.* **11**: 68–70.

Qin, Y., Luo, Z., Smyth, A.J., Gao, P., Beck von Bodman, S., Farrand S.K. (2000) Quorum-sensing signal binding results in dimerization of TraR and its release from membranes into the cytoplasm. *EMBO J.* **19**: 5212–5221.

Reverchon, S., Bouillant, M.L., Salmond, G., Nasser, W. (1998) Integration of the quorum-sensing system in the regulatory networks controlling virulence factor synthesis in *Erwinia chrysanthemi*. *Mol. Microbiol.* **29**: 1407–1418.

Rivet, M.M. (1998) Investigation into the regulation of exoenzyme production in *Erwinia carotovora* subspecies *carotovora*. PhD Thesis, University of Warwick: Warwick, UK.

Ruby, E.G. (1999) The *Euprymna scolopes–Vibrio fischeri* symbiosis: a biomedical model for the study of bacterial colonization of animal tissue. *J. Mol. Microbiol. Biotechnol.* **1**: 13–21.

Salanoubat, M., Genin, S., Artiguenave, F. *et al.* (2002) Genome sequence of the plant pathogen *Ralstonia solanacearum*. *Nature* **415**: 497–502.

Salmond, G.P.C., Bycroft, B.W., Stewart, G.S.A.B., Williams, P. (1995) The bacterial 'enigma': cracking the code of cell–cell communication. *Mol. Microbiol.* **16**: 615–624.

Schauder, S., Shokat, K., Surette, M.G., Bassler, B.L. (2001) The LuxS family of bacterial autoinducers: biosynthesis of a novel quorum sensing signal molecule. *Mol. Microbiol.* **41**: 463–476.

Schell, M.A. (1996) To be or not to be: how *Pseudomonas solanacearum* decides whether or not to express virulence genes. *Eur. J. Plant Pathol.* **102**: 459–469.

Schell, M.A. (2000) Control of virulence and pathogenicity genes of *Ralstonia solanacearum* by an elaborate sensory network. *Annu. Rev. Phytopathol.* **38**: 263–292.

Shadel, G.S., Baldwin, T.O. (1991) The *Vibrio fischeri* LuxR protein is capable of bidirectional stimulation of transcription and both positive and negative regulation of the *luxR* gene. *J. Bacteriol.* **173**: 568–574.

Shadel, G.S., Baldwin, T.O. (1992a) Positive autoregulation of the *Vibrio fischeri luxR* gene. *J. Biol. Chem.* **267**: 7696–7702.

Shadel, G.S., Baldwin, T.O. (1992b) Identification of a distantly located regulatory element in the *luxCD* gene required for negative autoregulation of the *Vibrio fischeri luxR* gene. *J. Biol. Chem.* **267**: 7690–7695.

Shadel, G.S., Young, R., Baldwin, T.O. (1990) Use of regulated cell lysis in a lethal genetic selection in *Escherichia coli*: identification of the autoinducer-binding region of the LuxR protein from *Vibrio fischeri* ATCC7744. *J. Bacteriol.* **172**: 3980–3987.

Slater, H., Alvarez-Morales, A., Barber C.E., Daniels, M.J., Dow, J.M. (2000) A two-component system involving an HD-GYP domain protein links cell-cell signalling to pathogenicity gene expression in *Xanthomonas campestris*. *Mol. Microbiol.* **38**: 986–1003.

Slock, J., van Riet, D., Kolibachuk, D., Greenberg, E.P. (1990) Critical regions of the *Vibrio fischeri* LuxR protein defined by mutational analysis. *J. Bacteriol.* **172**: 3974–3979.

Spaink, H.P. (2000) Root nodulation and infection factors produced by Rhizobial bacteria. *Annu. Rev. Microbiol.* **54**: 257–288.

Sperandio, V., Mellies, J.L., Nguyen, W., Shin, S., Kaper, J.B. (1999) Quorum sensing controls expression of the type III secretion gene transcription and protein secretion in enterohemorrhagic and enteropathogenic *Escherichia coli*. *Proc. Natl Acad. Sci. USA* **96**: 15196–15201.

Swartzman, E., Kapoor, S., Graham, A.F., Meighen, E.A. (1990) A new *Vibrio fischeri lux* gene precedes a bidirectional termination site for the *lux* operon. *J. Bacteriol.* **172**: 6797–6802.

Swiderska, A., Berndtson, A.K., Cha, M., Li, L. Beaudoin III, G.M.J., Zhu, J., Fuqua, C. (2001) Inhibition of the *Agrobacterium tumefaciens* TraR quorum sensing regulator. *J. Biol. Chem.* **276**: 49449–49458.

Swift, S., Winson, M.K., Chan, P.F. *et al.* (1993) A novel strategy for the isolation of *luxI* homologues: evidence for the widespread distribution of a LuxR: LuxI superfamily in enteric bacteria. *Mol. Microbiol.* **10**: 511–520.

Tang, J.L., Liu, Y.N., Barber, C.E., Dow, J.M., Wootton, J.C., Daniels, M.J. (1991) Genetic and molecular analysis of a cluster of *rpf* genes involved in positive regulation of synthesis of extracellular enzymes and polysaccharide in *Xanthomonas campestris* pathovar *campestris*. *Mol. Gen. Genet.* **226**: 409–417.

Tans-Kersten, J., Huang, H., Allen, C. (2001) *Ralstonia solanacearum* needs motility for invasive virulence on tomato. *J. Bacteriol.* **183**: 3597–3605.

Teplitski, M., Robinson, J.B., Bauer, W.D. (2000) Plants secrete substances that mimic bacterial *N*-acyl homoserine lactone signal activities and affect population density-dependent behaviours in associated bacteria. *Mol. Plant–Microbe Interact.* **13**: 637–648.

Vannini, A., Volpari, C., Gargiolo, C., Muraglia, E., Cortese, R., De Francesco, R., Nedderman, P., Di Marco, S. (2002) The crystal structure of the quorum sensing protein TraR bound to its autoinducer and target DNA. *EMBO J.* **21**: 4393–4401.

Visick, K.L., McFall-Ngai, M.J. (2000) An exclusive contract: specificity in the *Vibrio fischeri–Euprymna scolopes* partnership. *J. Bacteriol.* **182**: 1779–1787.

Wang, L., Helmann, J.D., Winans, S.C. (1992) The *A. tumefaciens* transcriptional activator OccR causes a bend at a target promoter, which is partially relaxed by a plant tumor metabolite. *Cell* **69**: 659–667.

Watson, W.T., Minogue, T.D., Val, D.L., von Bodman, S.B., Churchill, M.E.A. (2002) Structural basis and specificity of acyl-homoserine lactone signal production in bacterial quorum sensing. *Mol. Cell.* **9**: 685–694.

Wehland, M., Kiecker, C., Coplin, D.L., Kelm, O., Saenger, W., Bernhard, F. (1999) Identification of an RcsA/RcsB recognition motif in the promoters of exopolysaccharide biosynthetic operons from *Erwinia amylovora and Pantoea stewartii* subspecies stewartii. *J. Biol. Chem.* **274**: 3300–3307.

Welch, M., Todd, D.E., Whitehead, N.A., McGowan, S.J., Bycroft, B.W., Salmond, G.P.C. (2000) *N*-acyl homoserine lactone binding to the CarR receptor determines quorum-sensing specificity in *Erwinia*. *EMBO J.* **19**: 631–641.

Whitehead, N.A., Barnard, A.M.L., Slater, H., Simpson, N.J.L., Salmond G.P.C. (2001) Quorum-sensing in Gram-negative bacteria. *FEMS Micro. Rev.* **25**: 365–404.

Whitehead, N.W., Byers, J.T., Commander, P. *et al.* (2002) The regulation of virulence in phytopathogenic *Erwinia* species: quorum sensing, antibiotics and ecological considerations. *Anton. van Leeuwen.* **81**: 223–231.

Wilkinson, A., Danino, V., Wisniewski-Dye, F., Lithgow, J.K., Downie, J.A. (2002) *N*-acyl homoserine lactone inhibition of rhizobial growth is mediated by two quorum sensing genes that regulate plasmid transfer. *J. Bacteriol.* **184**: 4510–4519.

Wilson, T.J.G., Bertrand, N., Tang, J.L., Feng, J.X., Pan, M.Q., Barber, C.E., Dow, J.M., Daniels, M.J. (1998) The *rpfA* gene of *Xanthomonas campestris* pathovar *campestris*, which is involved in the regulation of pathogenicity factor production, encodes an aconitase. *Mol. Microbiol.* **28**: 961–970.

Wisniewski-Dye, F., Jones, J., Chhabra, S.R., Downie, J.A. (2002) *raiIR* genes are part of a quorum sensing network controlled by *cinI* and *cinR* in *Rhizobium leguminosarum*. *J. Bacteriol.* **184**: 1597–1606.

Zenno, S., Saigo, K. (1994) Identification of the genes encoding NAD(P)H-flavin oxidoreductases that are similar in sequence to *Escherichia coli* Fre in four species of luminous bacteria: *Photorhabdus luminescens*, *Vibrio fischeri*, *Vibrio harveyi*, and *Vibrio orientalis*. *J. Bacteriol.* **176**: 3544–3551.

Zhang, R., Pappas, T., Brace, J.L. *et al.* (2002) Structure of a bacterial quorum sensing transcription factor complexed with pheromone and DNA. *Nature* **417**: 971–974.

Zhu, J., Winans, S.C. (1998) Activity of the quorum sensing regulator TraR of *Agrobacterium tumefaciens* is inhibited by a truncated, dominant defective TraR-like protein. *Mol Microbiol.* **27**: 289–297.

Zhu, J., Winans, S.C. (1999) Autoinducer binding by the quorum sensing regulator TraR increases affinity for target promoters *in vitro* and decreases TraR turnover rates in whole cells. *Proc. Natl Acad. Sci. USA* **96**: 4832–4837.

Zhu, J., Winans, S.C. (2001) The quorum-sensing transcriptional regulator TraR requires its cognate signaling ligand for protein folding, protease resistance, and dimerization. *Proc. Natl Acad. Sci. USA* **98**: 1507–1512.

Zhu, J., Oger, P.M., Schrammeijer, B., Hooykaas, P.J.J., Farrand, S.K., Winans, S.C. (2000) The bases of crown gall tumorigenesis. *J. Bacteriol.* **182**: 3885–3895.

8

Quorum quenching – manipulating quorum sensing for disease control

Lian-Hui Zhang

8.1 Introduction

Host plant resistance has been used extensively for disease control in diverse crop species. It is governed in many cases by the 'gene-for-gene' system, i.e., the specific recognition between pathogen *avr* (avirulence) gene and its cognate plant disease resistance (R) gene. A plant displays a resistance phenotype when corresponding *avr* and R genes are present in the pathogen and the plant, respectively, or becomes susceptible if either is absent or inactive (for review see Dangl and Jones, 2001). However, in many cases, host plant resistance is not durable as a result of constant genetic evolution in pathogens, in particular, loss of avirulence genes (for review see Leach *et al.*, 2001).

Many efforts have been made to sustain plant resistance and to identify novel strategies for the prevention and control of microbial diseases. Research progress in recent years has shown that population control of bacterial virulence is a very promising target for prevention of infectious disease. For many bacterial pathogens the outcome of host–pathogen interactions strongly depends on bacterial population density, that is, a threshold cell population of each pathogen is required to establish a successful infection. It has been established in recent years that many bacterial pathogens, if not all, have sophisticated genetic control networks to enable coordination of production of virulence factors with cell population size, thus ensuring a concerted attack to overcome the host defence responses. This mechanism is widely known as quorum sensing (Fuqua *et al.*, 1996).

The quorum-sensing bacteria produce, detect and respond to small signal molecules known as quorum-sensing signals or autoinducers. Several families of

Plant Microbiology, Michael Gillings and Andrew Holmes

quorum-sensing signals are involved in the regulation of bacterial virulence (for review see Whitehead *et al.*, 2001b). Among them, acyl homoserine lactones (AHLs) are one family of the most characterised quorum-sensing signals found in many Gram-negative bacterial species. AHLs are involved in regulation of a range of biological activities including pathogenesis-related processes, such as conjugal transfer of Ti Plasmid, expression of virulence genes and formation of biofilms (Allison *et al.*, 1998; Beck von Bodman and Farrand, 1995; Davies *et al.*, 1998; Jones *et al.*, 1993; Passador *et al.*, 1993; Pirhonen *et al.*, 1993; Zhang *et al.*, 1993). Chapter 7. provides an excellent review on quorum-sensing signalling in plant bacteria.

Although the target genes regulated by AHLs are extremely varied in different bacterial species, the key components of all AHL quorum-sensing systems are AHL signals, which are produced by AHL synthases, and the LuxR-type transcription factors. Three types of AHL synthases have been identified; LuxI (Schaefer *et al.*, 1996), AinS (Hanzelka *et al.*, 1999), and HdtS (Laue *et al.*, 2000). These enzymes do not share significant homologies, although LuxI and AinS appear to use the same substrates in the synthesis of AHLs (Hanzelka *et al.*, 1999; More *et al.*, 1996). Of these three, the LuxI-type enzymes appear to be the most common among the AHL-producing bacteria.

Different bacterial species usually produce different AHLs. About ten AHL molecules have been structurally characterized (for review see Miller and Bassler, 2001; Whitehead *et al.*, 2001b). The AHL derivatives share identical homoserine lactone moieties but differ in the length and structure of their acyl groups. The structural diversity of AHLs may underpin the specificity of quorum-sensing signalling systems and thus prevent cross talking between different bacterial species (Welch *et al.*, 2000).

The majority of LuxR-type proteins are AHL-dependent positive transcription factors. In the absence of AHLs, these proteins are very unstable (Zhu and Winans, 1999), and functionally inactive (Welch *et al.*, 2000). Binding of AHLs stabilizes LuxR-type proteins and induces formation of dimers, or even polymers, that can bind to DNA and initiate transcription of the target genes (Qin *et al.*, 2000; Welch *et al.*, 2000; Zhu and Winans, 1999). The exception is the EsaR of *Pantoea stewartii*, it acts as a repressor of exopolysaccharide synthesis and this suppression is released by 3-oxo-C6-HSL (Beck von Bodman *et al.*, 1998).

Several promising strategies targeting AHL signals and the LuxR-type transcription factors have been reported over the last few years. Some approaches intend to confuse the invading bacterial pathogens by producing high levels of AHLs in transgenic plants (Fray, 2002; Fray *et al.*, 1999; Mäe *et al.*, 2001); while the others aim to block bacterial quorum-sensing signalling using either chemical inhibitors or AHL-degrading enzymes (Dong *et al.*, 2001; Givskov *et al.*, 1996). The latter approaches were conveniently termed as quorum quenching as a vivid contrast to quorum sensing (Dong *et al.*, 2000, 2001). This chapter provides an overview of these promising strategies but emphasises the quorum-quenching approaches and underlying mechanisms. The review is not confined to plant bacterial pathogens since quorum sensing is a generic mechanism conserved in both plant and human bacterial pathogens. Rather than present a comprehensive summary of all related works, this review focuses on the specific experimental approaches that illustrate the general concepts of quorum quenching.

8.2 The enzymes inactivating AHL signals

Although the target genes regulated by AHLs are extremely varied and regulatory mechanisms are likely diversified, the general mechanism of AHL-mediated quorum-sensing signalling is very much conserved. When cell population is low, the concentration of the quorum-sensing signal is too low to be detected. When a sufficiently high bacterial population is present, the signals reach a threshold level that triggers the bacteria to respond by activating or repressing specific target gene(s). This drives bacterial cells to switch on new sets of biological functions such as production of pathogenic factors (for review see Fuqua *et al.*, 1996, 2001). It is apparent that the concentration of AHLs is a key factor in bacterial virulence gene expression. As AHL production is autoregulated by itself (for review see Fuqua *et al.*, 1996), a simple strategy to keep AHL production in check, and thus suppress the expression of virulence genes, is to inactivate the AHL signals produced by pathogenic bacteria.

The first AHL-inactivation enzyme, encoded by the *aiiA* gene, was identified in *Bacillus* sp. 240B1, a Gram-positive bacterium (Dong *et al.*, 2000). $AiiA_{240B1}$ is a novel enzyme with no obvious homologues in public databases. Sequence alignments with known proteins and the subsequent site-directed mutagenesis indicated that AiiA contains a motif, which resembles the zinc-binding motif of several enzymes in the metallohydrolase superfamily (Dong *et al.*, 2000). Chemical and biochemical analyses showed that AiiA opened up the homoserine lactone ring of AHLs and decreased their biological activity more than 1000 times (see *Figure 8.1*; Dong *et al.*, 2001). These data unequivocally established that AiiA is an AHL-lactonase.

AiiA homologues were later found in many subspecies of *Bacillus thuringiensis* and closely related *Bacillus* species, including *B. cerus* and *B. mycoides* (Dong *et al.*, 2002; Lee *et al.*, 2002; Reimmann *et al.*, 2002). These proteins share high identities, ranging from 89–96%, with the $AiiA_{240B1}$ lactonase. Interestingly, some Gram-negative bacteria also produce AHL-lactonase. The *attM* gene of *Agrobacterium tumefaciens*, encodes an AiiA homologue, which controls AHL-signal turnover in a

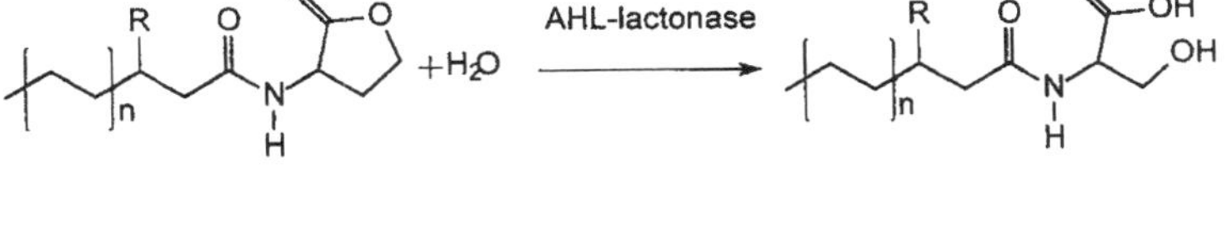

Figure 8.1. AHL-lactonase hydrolyses acylhomoserine lactones in the presence of water to produce acyl homoserines that does not have biological activity at physiologically relevant concentrations.

growth-phase-dependent pattern (Zhang *et al.*, 2002). AttM shares only 35% identity with $AiiA_{240B1}$ but does contain a 'HxDH~59aa~H~21aa~D' motif that is conserved across all the *Bacillus* homologues (Dong *et al.*, 2002). Known AHL-lactonases are all small proteins consisting of 250–264 amino acid residues (Dong *et al.*, 2000, 2002; Zhang *et al.*, 2002).

AHL-lactonase was shown to be a very potent enzyme, capable of degradation of AHL signals produced by bacterial pathogens at physiologically relevant rates and concentrations. Expression of *aiiA* in *Erwinia carotovora* abolished the release of AHL signals to the culture fluid, significantly reduced production of extracellular pectate lyase, pectin lyase, and polygalacturonase, and attenuated pathogenicity on host plants such as Chinese cabbage and eggplant (aubergine) (see *Figure 8.2*; Dong *et al.*, 2000). *E. carotovora* is an important plant pathogen internationally. It produces and secretes exoenzymes that act as virulence determinants of soft rot diseases of many vegetables and plants (Frederic *et al.*, 1994).

For practical applications of AHL-lactonase in disease control, the key question is whether exogenous AHL-lactonase can effectively quench bacterial quorum-sensing signalling. The impact of AHL-lactonase on quorum-sensing bacterial pathogens was tested in transgenic plants. The $aiiA_{240B1}$ gene was cloned in a plant expression vector pBI121 and introduced into tobacco and potato by *Agrobacterium*-mediated transformation. The AHL-lactonase protein contents in transgenic tobacco leaves and potato tubers were estimated to be 2–7 ng and 20–110 ng per mg of soluble proteins, respectively. Transgenic plants expressing AHL-lactonase showed significantly enhanced resistance to *E. carotovora* infection (Dong *et al.*, 2001). There was a strong correlation between the AHL-lactonase activity in transgenic plants and the disease severity. The plants expressing higher amounts of AHL-lactonase showed less maceration symptoms than those that produce lower levels of the enzyme (see *Figure 8.3*). The transgenic potato, which produced higher concentrations of AHL-lactonase than transgenic tobacco, showed resistance to significantly higher doses of *E. carotovora* pv. *carotovora* inoculum than the transgenic tobacco plants (Dong *et al.*, 2002). These data suggest the possibility for further enhancing disease resistance by increasing AHL-lactonase expression level, e.g., through promoter manipulation or codon usage optimisation.

Another significant finding is that when plants expressing AHL-lactonase were challenged with a high dose of bacterial inoculum, even though the soft rot

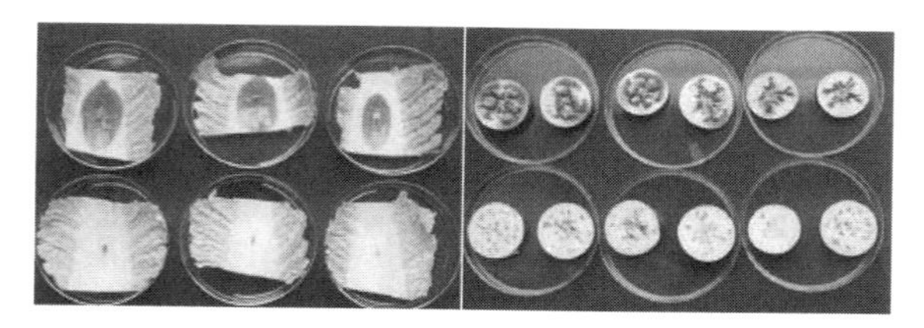

Figure 8.2. Expression of the *aiiA* gene in *E. carotovora* SCG1 attenuates bacterial virulence. Three microlitres of fresh cultures (2×10^9 colony forming units per litre) of SCG1 (top row) and SCG1 (*aiiA*) (bottom row) were inoculated on Chinese cabbage (left) and eggplant (right). The photographs were taken 2 days after inoculation.

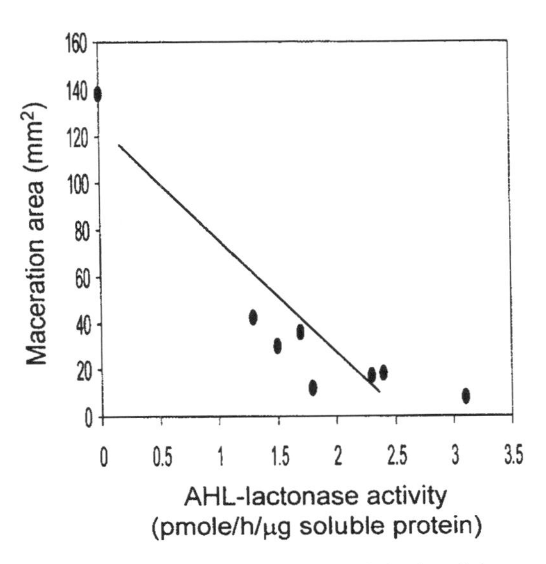

Figure 8.3. The relationship of AHL-lactonase activity in *aiiA* transgenic and control tobacco plants and the severity of the soft rot disease caused by *E. carotovora*. The plants were inoculated as described in the *Figure 8.2* legend.

symptoms were initiated, the symptom development was stopped a few hours after inoculation. In contrast the untransformed control plants experienced progressive maceration (Dong *et al.*, 2001). The logical explanation is that the quorum-quenching enzyme slows down production of virulence factors, allowing the host plants time to enact their defence mechanisms and stop the pathogenic invaders.

Another mechanism by which AHLs are degraded by bacteria has been reported (Leadbetter, 2001; Leadbetter and Greenberg, 2000). An isolate of *Variovorax paradoxus* (*Betaproteobacteria*), was shown to utilise AHL signals as the sole carbon source. During growth on AHL, homoserine lactone was released into the medium as a major degradation product, whereas the fatty acid was metabolised as an energy source. These data indicate the existence of an AHL-acylase, which hydrolyses the amide linkage between the acyl chain and the homoserine moiety of AHL molecules. We have recently cloned a gene encoding a novel and potent AHL-acylase from another betaproteobacterial species, *Ralstonia* sp XJ12B. The enzyme belongs to the family of N-terminal hydrolases and appears to be widely conserved in different bacterial species based on a sequence homology search (Lin *et al.*, 2003).

8.3 The chemicals accelerating LuxR-type protein turn over

The seaweed *Delisea pulchra* produces a number of halogenated furanones which showed potent antifouling activities (de Nys *et al.*, 1993, 1995). These furanone compounds are structurally similar to AHLs. These properties prompted Kjelleberg and colleagues to test whether furanones could block bacterial quorum sensing (Givskov *et al.*, 1996). The subsequent studies showed that the halogenated furanones inhibit several biological activities controlled by AHL-dependent quorum-sensing systems, such as swarming motility of *Serratia liquefaciens* (Givskov *et al.*, 1996), luminescence and virulence of *Vibrio harveyi* (Manefield *et al.*, 2000), antibiotics and exoenzyme production in *E. carotovora* pv. *carotovora* (Manefield *et al.*, 2001), and biofilm development by *Pseudomonas aeruginosa* (Hentzer *et al.*, 2002). (5*Z*)-4-bromo-5(bromomethylene)-3-butyl-2(5*H*)-furanone, one of the halogenated furanone derivatives, was found to inhibit biofilm formation and swarming motility in *E. coli* and *Bacillus subtilis* (Ren *et al.*, 2001, 2002). These two bacteria do not produce AHL signals, but contain AI-2-dependent quorum-sensing systems (Bassler *et al.*, 1997; Hilgers and Ludwig, 2001). Hence, it appears that the halogenated furanones are non- specific intercellular signal antagonists and may have considerable potential in controlling biofilm formation and bacterial virulence.

Although halogenated furanones are structurally similar to AHLs, they did not form a stable complex with the LuxR protein of *Vibrio fischeri* or the CarR protein of *E. carotovora* pv. *carotovora* (Manefield *et al.*, 2001, 2002). Rather they appear to cause the accelerated turnover of the AHL-dependent transcription factors such as LuxR by an unknown mechanism (Manefield *et al.*, 2002). Western analysis showed that the half-life of the LuxR protein overproduced in *E. coli* was reduced up to 100-fold in the presence of halogenated furanones. This is significant, as the primitive role of AHLs appears to be maintenance of the cellular concentration of LuxR-type protein by binding to LuxR-type protein and stabilising the protein against proteolytic degradation. Zhu and Winans (1999) showed that TraR is very unstable with a half-life of about 2 min, whereas binding of 3-oxo-C8-HSL to the protein increased its half-life up to 35 min.

8.4 Quorum-quenching substances in terrestrial plants

Besides the seaweed *Delisea pulchra,* terrestrial plants also produce chemicals that inhibit AHL-dependent quorum-sensing signalling. Bauer and colleagues (Teplitski *et al.*, 2000) showed that crude exudates from Pea (*Pisum sativum*) and Crown vetch (*Coronilla varia*) strongly inhibited the AHL-induced synthesis of violacein in *Chromobacterium violaceum*. Though the chemical nature of the inhibitory substances and the mode of action are not clear, the finding is significant since it shows that blocking pathogen cell–cell signaling could also be a natural plant defence mechanism against pathogenic invaders.

8.5 Overproduction of AHL signals in transgenic plants

Why do bacterial pathogens need quorum-sensing systems and produce pathogenic factors only at high cell density? A likely possibility is to prevent premature production of pathogenic factors that may trigger local or systemic plant defence responses. Pathogens may mount their concerted attack only when the cell population around the infection site is high, so as to overcome plant defences and establish infection. Based on this reasoning, two groups have tested whether transgenic plants producing high levels of AHL could lure bacterial pathogens to mount a pathogenic attack prematurely, and thus win the competition with the pathogen. The *yenI* and *expI* genes from *Yersinia enterocolitica* and *E. carotovora* pv. *carotovora*, respectively, were introduced separately into potato and tobacco. The two genes encode the same function, synthesis of 3-oxo-C6-HSL, which regulates production of virulence factors in *E. carotovora*. While the *expI* transgenic tobacco showed enhanced resistance (Mäe *et al.*, 2001), the *yenI* transgenic potatoes were more susceptible than the untransformed control plants (Fray, 2002). As the intensity and speed of defence responses of different host plants could differ, and the quorum-sensing threshold set by different *E. carotovora* isolates may vary; this approach may require a subtle fine-tuning to suit specific host–pathogen combinations.

8.6 Conclusions and future prospects

The promising outcomes of the above-described proof-of-concept approaches represent a considerable advance in bacterial disease control. It has been clearly established now that quorum quenching is a feasible approach to control bacterial infections. However, we should also be aware that our understanding about quorum-sensing regulation of bacterial virulence is still fragmentary, with most information coming from *in vitro* experiments. Host and environmental factors could also play significant roles in modulation of bacterial quorum-sensing systems. Good examples are plants as well as other bacterial species that could produce AHL mimic compounds that activate bacterial quorum-sensing systems (Pierson *et al.*, 1998; Teplitski *et al.*, 2000). Further investigation on bacterial quorum-sensing systems, especially in the context of host–pathogen interaction, would be essential to maximise the potential of the quorum-quenching strategy in our fight against bacterial plagues.

Despite that challenges remain (Whitehead *et al.*, 2001a), one of the most attractive features apparently exploited by the quorum-quenching approach is that it allows the host valuable time to activate defence mechanisms to stop and eliminate pathogenic invaders (Dong *et al.*, 2001). Such an integration of quorum-quenching mechanisms with host defence systems could be the most economical way to tap the natural self-protection capability of host plants, because constitutive expression of host resistance genes often causes severe yield and biomass penalties. As quorum-sensing regulation of virulence appears to be one of the common strategies that many bacterial pathogens, if not all, have adopted during evolution to ensure their survival in host–pathogen competition, the quorum-quenching concept could have fundamental implications in our future formulation of practical approaches to control various bacterial pathogens.

References

Allison, D.G., Ruiz, B., Sanjose, C., Jaspe, A., Gillbert, P. (1998) Extracellular products as mediators of the formation and detachment of *Pseudomonas fluorescens* biofilms. *FEMS Microbiol. Lett.* **167**: 179–184.

Bassler, B.L., Greenberg, E.P., Stevens, A.M. (1997) Cross-species induction of luminescence in the quorum-sensing bacterium *Vibrio harveyi*. *J. Bacteriol.* **179**: 4043–4045.

Beck von Bodman, S., Farrand, S.K. (1995) Capsular polysaccharide biosynthesis and pathogenicity in *Erwinia stewartii* require induction by an *N*-acylhomoserine lactone autoinducer. *J. Bacteriol.* **177**: 5000–5008.

Beck von Bodman, S.B., Majerczak, D.R., Coplin, D.L. (1998) A negative regulator mediates quorum-sensing control of exopolysaccharide production in *Pantoea stewartii* subsp. *stewartii. Proc. Natl Acad. Sci. USA* **95**: 7687–7692.

Dangl, J.L., Jones, J.D.G. (2001) Plant pathogens and integrated defense responses to infection. *Nature* **411**: 826–833.

Davies, D.G., Parsek, M.R., Pearson, J.P., Iglewski, B.H., Costerton, J.W., Greenberg, E.P. (1998) The involvement of cell-to-cell signals in the development of a bacterial biofilm. *Science* **280**: 295–298.

de Nys, R., Wright, A.D., Konig, G.M., Sticher, O. (1993). New halogenated furanones from the marine alga *Delisea pulchra* (cf. Fimbriata). *Tetrahedron* **49**: 11213–11220.

de Nys, R., Steinberg, P., Willemsen, P., Dworjanyn, S., Gabelish, C., King, R. (1995) Broad spectrum effects of secondary metabolites from the red alga *Delisea pulchra* in antifouling assays. *Biofouling* **8**: 259–271.

Dong, Y.H., Xu, J.L., Li, X.Z., Zhang, L.H. (2000) AiiA, an enzyme that inactivates the acylhomoserine lactone quorum-sensing signal and attenuates the virulence of *Erwinia carotovora. Proc. Natl Acad. Sci. USA* **97**: 3526–3531.

Dong, Y.H., Wang, L.H., Xu, J.L., Zhang, H.B., Zhang, X.F., Zhang, L.H. (2001) Quenching quorum-sensing-dependent bacterial infection by an *N*-acyl homoserine lactonase. *Nature* **411**: 813–817.

Dong, Y.H., Gusti, A.R., Zhang, Q., Xu, J.L., Zhang, L.H. (2002). Identification of quorum-quenching *N*-acyl homoserine lactonases from *Bacillus* species. *Appl. Environ. Microbiol.* **68**: 1754–1759.

Fray, R.G. (2002) Altering plant–microbe interaction through artificially manipulating bacterial quorum sensing. *Annals Botany* **89**: 245–253.

Fray, R.G., Throup, J.P., Wallace, A., Daykin, M., Williams, P., Srewart, G.S.A.B., Grierson, D. (1999) Plants genetically modified to produce *N*-acylhomoserine lactones communicate with bacteria. *Nature Biotech.* **17**: 1017–1020.

Frederic, B., Van Gijsegem, F., Chatterjee, A.K. (1994). Extracellular enzymes and pathogenesis of soft-rot Erwinia. *Ann. Rev. Phytopathol.* **32**: 201–234.

Fuqua, C., Winans, S.C., Greenberg, E.P. (1996) Census and consensus in bacterial ecosystems: the LuxR-LuxI family of quorum-sensing transcriptional regulators. *Ann. Rev. Microbiol.* **50**: 727–751.

Fuqua, C., Parsek, M.R., Greenberg, E.P. (2001) Regulation of gene expression by cell-to-cell communication: acyl-homoserine lactone quorum sensing. *Annu. Rev. Genet.* **35**: 439–468.

Givskov, M., de Nys, R., Manefield, M., Gram, L., Maximilien, R., Eberl, L., Molin, S., Steinberg, P.D., Kjelleberg, S. (1996) Eukaryotic interference with homoserine lactone-mediated prokaryotic signaling. *J. Bacteriol.* **178**: 6618–6622.

Hanzelka, B.L., Parsek, M.R., Va., D.L., Dunlap, P.V., Cronan Jr., J.E., Greenberg, E.P. (1999) Acylhomoserine lactone synthase activity of the *Vibrio fischeri* Ains protein. *J. Bacteriol.* **178**: 372–376.

Hentzer, M., Riedel, K., Rasmussen, T.B. *et al.* (2002) Inhibition of quorum sensing in *Pseudomonas aeruginosa* biofilm bacteria by a halogenated furanone compound. *Microbiology* **148**: 87–102.

Hilgers, M.T., Ludwig, M.L. (2001) Crystal structure of the quorum-sensing protein LuxS reveals a catalytic metal site. *Proc. Natl Acad. Sci. USA* **98**: 11169–11174.

Jones, S.M., Yu, B., Bainton, N.J. *et al.* (1993) The *Lux* autoinducer regulates the production of exoenzyme virulence determination in *Erwinia carotovora* and *Pseudomonas aeruginosa*. *EMBO J.* **12**: 2477–2482.

Laue, R.E., Jiang, Y., Chhabra, S.R., Jacob, S., Steward, G.S.A.B., Hardman, A., Downie, J.A., O'Gara, F., Williams, P. (2000) The biocontrol strain *Pseudomonas fluorescens* F113 produces the *Rhizobium* small bacteriocin, *N*-(3-hydroxy-7-cis- tetradecenoyl)homoserine lactone, via HdtS, a putative novel *N*-acylhomoserine lactone synthase. *Microbiology* **146**: 2469–2480.

Leach, J.E., Vera Cruz, G.M., Bai, J., Leung, H. (2001) Pathogen fitness penalty as a predictor of durability of disease resistance genes. *Annu. Rev. Phytopathol.* **39**: 187–224.

Leadbetter, J.R. (2001) News and views: plant microbiology – quieting the raucous crowd. *Nature* **411**: 748–749.

Leadbetter, J.R., Greenberg, E.P. (2000) Metabolism of acyl-homoserine lactone quorum-sensing signals by *Variovorax paradoxus*. *J. Bacteriol.* **182**: 6921–6926.

Lee, S.J., Park, S.Y., Lee, J.J., Yum, D.Y., Koo, B.T., Lee, J.K. (2002) Genes encoding the *N*-acyl homoserine lactone-degrading enzyme are widespread in many subspecies of *Bacillus thuringiensis*. *Appl. Environ. Microbiol.* **68**: 3919–3924.

Lin, Y.H., Xu, J.L., Hu, J., Wang, L.H., Ong, S.L., Leadbetter, J.R., Zhang L.H. (2003) Acyl-homoserine lactone acylase from *Ralstonia* str. XJ12B represents a novel and potent class of quorum quenching enzymes. *Mol. Microbiol.* **47**: 849–860.

Mäe, A., Montesano, M., Koiv, V., Palva, E.T. (2001) Transgenic plants producing the bacterial pheromone *N*-acyl-homoserine lactone exhibit enhanced resistance to the bacterial phytopathogen *Erwinia carotovora*. *Mol. Plant–Microbe Interact.* **14**: 1035–1041.

Manefield, M., Harris, L., Rice, S.A., de Nys, R., Kjelleberg, S. (2000) Inhibition of luminescence and virulence in the black tiger prawn (*Penaeus monodon*) pathogen *Vibrio harveyi* by intercellular signal antagonists. *Appl. Environ. Microbiol.* **66**: 2079–2084.

Manefield, M., Welch, M., Givskov, M., Salmond, G.P.C., Kjelleberg, S. (2001) Halogenated furanones from the red alga, *Delisea pulchra*, inhibit carbapenem antibiotics synthesis and exoenzyme virulence factor production in the phytopathogen *Erwinia carotovora*. *FEMS Microbiol. Lett.* **205**: 131–138.

Manefield, M., Rasmussen, T.B., Henzter, M., Andersen, J.B., Steinberg, P., Kjelleberg, S., Givskov, M. (2002) Halogenated furanones inhibit quorum sensing through accerated LuxR turnover. *Microbiology* **148**: 1119–1127.

Miller, M.B., Bassler, B.L. (2001) Quorum sensing in bacteria. *Annu. Rev. Microbiol.* **55**: 165–199.

Moré, M.I., Finger, L.D., Stryker, J.L., Fuqua, C., Eberhard, A., Winan, S.C. (1996) Enzymatic synthesis of a quorum-sensing autoinducer through use of defined substrates. *Science* **272**: 1655–1658.

Passador, L., Cook, J.M., Gambello, M.J., Rust, L., Iglewski, B.H. (1993) Expression of *Pseudomonas aeruginosa* virulence genes requires cell-to-cell communication. *Science* **260**: 1127–1130.

Pierson, E.A., Wood, D.W., Cannon, J.A., Blachere, F.M., Pierson III, L.S. (1998) Interpopulation signaling via *N*-acyl-homoserine lactones among bacteria in the wheat rhizosphere. *Mol. Plant–Microbe Interact.* **11**: 1078–1084.

Pirhonen, M., Flego, D., Heikinheimo, R., Palva, E. (1993) A small diffusible signal molecule is responsible for the global control of virulence and exoenzyme production in the plant pathogen *Erwinia carotovora*. *EMBO J.* **12**: 2467–2476.

Qin, Y., Luo, Z., Smyth, A.J., Gao, P., Beck von Bodman, S., Farrand, S.K. (2000) Quorum-sensing signal binding results in dimerization of TraR and its release from membranes into the cytoplasm. *EMBO J.* **19**: 5212–5221.

Reimmann, C., Ginet, N., Michel, L. *et al.* (2002) Genetically programmed autoinducer destruction reduces virulence gene expression and swarming motility in *Pseudomonas aeruginosa* PAO1. *Microbiol.* **148**: 923–932.

Ren, D., Sims, J.J., Wood, T.K. (2001) Inhibition of biofilm formation and swarming of *Escherichia coli* by (5*Z*)-4-bromo-5-(bromomethylene)-3-butyl-2(5*H*)- furanone. *Environ. Microbiol.* **3**: 731–736.

Ren, D., Sims, J.J., Wood, T.K. (2002) Inhibition of biofilm formation and swarming of *Bacillus subtilis* by (5*Z*)-4-bromo-5-(bromomethylene)-3-butyl-2(5*H*)-furanone. *Lett. Appl. Microbiol.* **34**: 293–299.

Schaefer, A.L., Val, D.L., Hanzelka, B.L., Cronan Jr., J.E., Greenberg, E.P. (1996) Generation of cell-to-cell signals in quorum sensing: acyl homoserine lactone synthase activity of a purified *Vibrio fischeri* LuxI protein. *Proc. Natl Acad. Sci. USA* **93**: 9505–9509.

Teplitski, M., Robinson, J.B., Bauer, W.D. (2000) Plants secrete substances that mimic bacterial *N*-acyl homoserine lactone signal activities and affect population density-dependent behaviors in associated bacteria. *Mol. Plant–Microbe Interact.* **13**: 637–648.

Watson, W.T., Minogue, T.D., Val, D.L., Beck von Bodman S., Churchill, M.E.A. (2002) Structural basis and specificity of acyl-homoserine lactone signal production in bacterial quorum sensing. *Mol. Cell* **9**: 685–694.

Welch, M., Todd, D.E., Whitehead, N.A., McGowan, S.J., Bycroft, B.W., Salmond, G.P. (2000) *N*-acyl homoserine lactone binding to the CarR receptor determines quorum-sensing specificity in *Erwinia*. *EMBO J.* **19**: 631–641.

Whitehead, N.A., Welch, M., Salmond, G.P.C. (2001a) Silencing the majority. *Nature Biotech.* **19**: 735–736.

Whitehead, N.A., Barnard, A.M.L., Slater, H., Simpson, N.J.L., Salmond, G.P.C. (2001b) Quorum-sensing in Gram-negative bacteria. *FEMS Microbiol. Rev.* **25**: 365–404.

Zhang, L.H., Murphy, P.J., Kerr, A., Tate, M.E. (1993) *Agrobacterium* conjugation and gene regulation by *N*-acyl-L-homoserine lactones. *Nature* **362**: 446–447.

Zhang, H.B., Wang, L.H., Zhang, L.H. (2002) Genetic control of quorum-sensing signal turnover in *Agrobacterium tumefaciens*. *Proc. Natl Acad. Sci. USA* **99**: 4638–4643.

Zhu, J., Winans, S.C. (1999) Autoinducer binding by the quorum-sensing regulator TraR increases affinity for target promoters in vitro and decreases TraR turnover rates in whole cells. *Proc. Natl Acad. Sci. USA* **96**: 4832–4837.

Zhu, J., Winans, S.C. (2001) The quorum-sensing transcription regulator TraR requires its cognate signaling ligand for protein folding, protease resistance, and dimerization. *Proc. Natl Acad. Sci. USA* **98**: 1507–1512.

9

Plant disease and climate change

Sukumar Chakraborty and Ireneo B. Pangga

9.1 Introduction

Over US$76 billion from a US$225 billion global harvest of rice, wheat, barley, maize, potato, soybean, cotton and coffee are lost to plant diseases (Oerke *et al.*, 1994). The costs of managing disease and growing less-profitable alternative crops are among other significant economic impacts of plant diseases. The sociopolitical repercussions of major epidemics such as the Irish potato famine (1845–1846) and the Bengal famine (Padmanabhan, 1973) and the threat to human and animal health (IARC, 1993; Payne and Brown, 1998) from mycotoxins and other fungal metabolites go far beyond simple economic impacts. In the USA alone, aflatoxin, fumonisin and deoxynivalenol cause over $1.5 billion loss (Cardwell *et al.*, 2001). Mycotoxins regularly cause suffering and loss of life in developing countries (Bhat *et al.*, 1988), where monitoring and detection are not as advanced.

Despite improved crop yields through the development of new varieties and management technologies (Amthor, 1998), losses from plant diseases have increased throughout the world since the 1940s (Oerke *et al.*, 1994) with pesticide usage increasing more than 30-fold during this period (Pimentel, 1997). With the well-documented evidence for a changed global climate spanning this period (Houghton *et al.*, 2001), it is tempting to ascribe the increasing crop loss to a changing climate. In reality, a critical shortage of relevant information does not allow meaningful analysis of climate change impacts on plant diseases. Nevertheless, some studies have demonstrated that inter-annual disease severity fluctuates according to climatic variation (Coakley, 1979; Rosenzweig *et al.*, 2001, Scherm and Yang, 1995; Yang and Scherm, 1997).

The paucity of research on climate change and plant diseases is striking given the close relationship between the host plant, weather and the pathogen forming the basic tenet of plant pathology. A virulent pathogen cannot induce disease even on the most highly susceptible host if weather conditions are not favourable. However, analysis is complicated by the large variation in pathogen and host-mediated responses to climate and the complex interactions with abiotic factors that lead to crop damage.

Plant Microbiology, Michael Gillings and Andrew Holmes

Apart from the impact of air pollution on diseases (Coakley, 1995; Darley and Middleton, 1966; Frankland *et al.*, 1996; Sandermann, 1996), interest in climate change impacts on plant diseases has been relatively recent. Since 1995, four reviews (Chakraborty *et al.*, 1998, 2000, 2000b; Coakley *et al.*, 1999; Manning and Tiedemann, 1995), many commentaries (Ando, 1994; Atkinson, 1993; Boag and Neilson, 1996; Chakraborty, 2001; Coakley and Scherm, 1996; Frankland *et al.*, 1996; Goudriaan and Zadoks, 1995; Hughes and Evans, 1996; Malmstrom and Field, 1997; Rosenzweig *et al.*, 2001) and a growing number of scientific papers highlight an emerging interest in the area. This area is now a recognised activity under the Global Change and Terrestrial Ecosystems core project of the International Geosphere-Biosphere Program (Sutherst *et al.*, 1996). This chapter builds upon past reviews and incorporates findings from recently published and unpublished research for an up-to-date and comprehensive treatise on climate change influences on plant diseases.

9.2 Climate change

Climate is a result of solar-radiation-mediated physical, biological and chemical interactions between the atmosphere, hydrosphere, biosphere and geosphere. Radiation reaching the planet is partly absorbed, causing the earth to emit thermal radiation and part of the radiation is reflected back to the atmosphere. Water vapour and radiatively active gases such as CO_2, CH_4, N_2O and O_3, partly trap the reflected radiation to warm the surface temperature to about 15°C, a natural phenomenon known as the 'greenhouse effect'. Without this greenhouse effect the surface temperature of earth would be a frigid –18°C (Rosenzweig and Hillel, 1998). The incoming solar radiation is balanced by the outgoing terrestrial radiation and factors that change the incoming radiation or its redistribution within the atmosphere, land and the oceans, influence climate (Houghton *et al.*, 2001). Climate models compute physical laws/relationships linking the interactions between atmosphere, ocean, land surface, cryosphere and biosphere for a three-dimensional grid over the globe. These Atmospheric–Ocean General Circulation Models (AOGCMs) are increasingly becoming more accurate in their predictions (Houghton *et al.*, 2001).

Palaeoclimatic records indicate that the earth's climate has always changed and the most recent striking changes have occurred in the past 18 000 years (Landsberg, 1984). A rapid melting of the continental glaciers between 15 000 and 8000 years ago has gradually made the earth warmer (Cheddadi *et al.*, 1996). What is different is human activities are increasingly modifying the global climate; burning of fossil fuel and the large-scale clearing of forests have increased the atmospheric concentration of CO_2 and other radiatively active gases. This, and the release of new halocarbons and hexafluoride (Houghton *et al.*; IPCC, 1996) have enhanced the greenhouse effect gradually warming the earth surface.

9.2.1 Change in atmospheric composition

Based on Antarctic ice core measurements, atmospheric CO_2 has ranged between 180 and 280 ppm for the past 420 000 years (Petit *et al.*, 1999), and increased from

280 to 365 ppm between 1750 and 1998. As a direct consequence of human activities since pre-industrial times, CH_4 has increased from 700 to 1745 ppb, N_2O from 270 to 314 ppb and chlorofluorocarbon-11 (CFC) from zero to 264 ppt. These radiatively active gases have different atmospheric lifetimes and contribute to different levels of warming (Houghton *et al.*, 2001).

By 2100, atmospheric CO_2 concentrations will rise to between 540 and 970 ppm, representing increases of 75–350% above the 1750 concentration depending on the emission scenario (Houghton *et al.*, 2001). The emission scenarios include future anthropogenic emissions of CO_2, CH_4, N_2O and SO_2, as modified by changes in the energy systems and community responses to rising environmental pollution.

9.2.2 Change in temperature and rainfall

The global average surface temperature has increased by 0.6 ± 0.2°C since the late 19th century and the 1990s is the warmest decade on record (Houghton *et al.*, 2001). On average, minimum temperatures are increasing at twice the rate of maximum temperatures (0.2 versus 0.1°C/decade), with an overall decrease in mountain glaciers and a rise in average sea level. Rainfall has increased in the middle and high latitudes of the northern hemisphere and has decreased over the subtropics.

Climate models project a rise of 1.4–5.8°C over the next 100 years (Houghton *et al.*, 2001) according to various scenarios of population growth, economic development, energy and land-use change. AOGCM incorporating a 1%/year rise in CO_2, projects 1.1–3.1°C rise for 2100. There will be more hot days, higher minimum temperatures and fewer cold and frost days. Average precipitation will rise by 5–15% in the same period, and become more intense over mid to high latitudes of the northern hemisphere. Regional climates will be further influenced by surface vegetation and variation in circulation such as El Niño–Southern Oscillation (ENSO) and the North Atlantic Oscillation (NAO). Rainfall in Australia is projected to change by –60% to +10% for southwest and by –35% to +35% in other parts (CSIRO, 2001b) by 2070, but there will be little change in the tropical north.

9.2.3 Change in extremes of weather

Some 'climate surprises' are truly unpredictable (Streets and Glantz, 2000); but other drivers of variability, ENSO and NAO, can be anticipated. The interannual ENSO is a self-sustained cycle, in which sea-surface temperature anomalies over the tropical Pacific Ocean cause the strengthening or weakening of trade winds to influence ocean currents and subsurface thermal structure. El Niño events cause severe damage to crops, livestock and human settlements: the prolonged drought in the Sahelian region of Africa since the late 1960s; flooding in the USA in 1993; and drought in northeast Brazil, Indonesia and northern Australia in 1997–98 are among these (Rosenzweig *et al.*, 2001). Over the past century droughts have become longer and bursts of intense precipitation more frequent (Karl *et al.*, 1995). El Niño events have become stronger and more frequent since the 1980s, potentially due to changes in global climate (Houghton *et al.*, 2001; Timmermann *et al.*, 1999). Even with little change in El Niño events for the next 100 years, continued changes in the frequency and magnitude of extreme events are projected (Fowler and Hennessy, 1994; Houghton *et al.*, 2001).

9.3 Crop plants and climate change

Change in atmospheric CO_2 has been suggested as a force in the evolutionary transition from a C_3 to C_4 photosynthetic pathway (Arens, 2001) and C_4 crops, maize, millet, sorghum and sugarcane are competitively favoured in warm and dry environments. Nevertheless, most crops are C_3 where CO_2 fixation is compensated by photorespiration at 50 ppm and increasing CO_2 concentration increases photosynthesis. C_4 plants show no major response to CO_2 enrichment since photorespiration is practically absent and photosynthesis is CO_2-saturated at ambient CO_2. Water-use efficiency of both C_3 and C_4 plants are improved at high CO_2.

Increased yield

Findings from over 1000 studies show that if other factors are non-limiting, under twice-ambient CO_2, yield of C_3 crops increases by about 30% and C_4 by about 10% (Cure and Acock, 1986; Kimball, 1983). The magnitude of the effect depends on the variety (Ziska and Bunce, 2000) and the duration of an experiment (Idso and Idso, 2001). Often yield increases are accompanied by decreased foliar nitrogen (Biswas *et al.*, 1996) and fertiliser application, irrigation and crop residue management (Rotter and van der Geijn, 1999) are necessary to realise benefits. Plant breeding strategies to capitalise on elevated CO_2 and temperature are starting to emerge (Richards, 2002).

Changed morphogenesis

Increased number of nodes, greater internode length, stimulated leaf expansion and reduced apical dominance are among influences of elevated CO_2 on plant morphology (Pritchard *et al.*, 1999; Taylor *et al.*, 1994). Changes in at least seven rules of morphogenesis make the plant canopy dense and enlarged under elevated CO_2 (Pangga, 2002).

9.3.1 Rising CO_2 is not the only driver of yield

Despite experimental evidence, a direct contribution of elevated CO_2 on crop yield is difficult to establish from historical data, where advances in nitrogen fertilisation, improved genotypes, disease resistance, and other management strategies show much clearer association with increasing yield (Amthor, 1998). Changes in temperature and other climatic factors further modify yield benefits due to elevated CO_2. Using three different climate change scenarios and three general circulation models, average yields at +2°C warming increased by 10–15% in wheat and soybeans, and by 8% in rice and maize but yields of all four crops were reduced at +4°C warming (Rosenzweig and Parry, 1994). In sorghum also, yield increases due to CO_2 are masked by temperature with overall yield reductions in drier regions of India (Rao *et al.*, 1995).

9.3.2 Regional variation in yield

There are substantial differences in regional impacts of climate change on agriculture (IPCC, 2000; Reilly and Schimmelpfennig, 1999). Summer crop yield may increase in central and eastern Europe, but decrease in western Europe.

Production for Mexico, countries of the Central American isthmus, Brazil, Chile, Argentina, and Uruguay will decrease even if moderate levels of adaptation are factored in at the farm level. Crop productivity in Australia will depend on rainfall in winter and spring (CSIRO, 2001a). In North America predictions are negative for eastern, southeastern, and corn belts but positive for northern plains and western regions. In China projected yield of rice, wheat and maize may range between –78% and +55% by 2050. Productivity will increase in northern Siberia but decrease in southwestern Siberia. Any increase in rice, wheat and sorghum yield due to CO_2 fertilisation in tropical Asia will be more than offset by reductions from temperature or moisture changes. Major weaknesses of current projections are the lack of consideration of damage from agricultural pests and the unaccounted vulnerability of agricultural areas to floods, droughts and cyclones.

9.4 Plant disease and climate change

There is mounting concern about climate-change-mediated impacts of insect pests, diseases, weeds (Ayres and Lombardero, 2000; Clifford *et al.*, 1996) and invasive/exotic species (Baker *et al.*, 2000; Pimentel *et al.*, 2000; Simberloff, 2000) on agriculture and forestry. Invasiveness of many species, including pathogens, may increase under a changing climate to alter ecosystem properties (Dukes and Mooney, 1999), but this has not been considered to any significant extent in any impact assessment (Clifford *et al.*, 1996; Houghton *et al.*, 2001; Rosenzweig and Hillel, 1998). Among agricultural pests, pathogens have received far less attention than insects (Ayres and Lombardero, 2000; Bale *et al.*, 2002; Dukes and Mooney, 1999; Patterson *et al.*, 1999, Simberloff, 2000, Sutherst *et al.*, 1995). Despite a lack of interest among plant pathologists, qualitative assessment of climate change impacts on diseases have been made for Australia (Chakraborty *et al.*, 1998), New Zealand (Prestidge and Pottinger, 1990), Finland (Carter *et al.*, 1996), Germany (von Tiedemann, 1996) and for fungal diseases of trees (Lonsdale and Gibbs, 1996). Whether a recent review (Coakley *et al.*, 1999) and several published works signal a renewed interest remains to be seen.

9.4.1 Historical links between severe epidemics and climate

Episodic weather events have aided in the explosive spread of plant disease epidemics causing famine, starvation and acute food shortages (Rosenzweig *et al.*, 2001). Floods and heavy rains caused the great Bengal famine of 1942, famine in China in the 1960s from wheat stripe rust, and record high levels of *Fusarium* mycotoxins in the USA in 1993. Tropical storms in the Gulf of Mexico rapidly spread *Helminthosporium maydis* inoculum to a genetically uniform corn crop in the Midwest and southern USA in the early 1970s, to cause more than a billion US$ loss (Campbell and Madden, 1990).

ENSO and NAO are strongly linked to serious epidemics of malaria, typhoid and cholera (Epstein, 2001). Similarly, ENSO and severe wheat scab in eastern China are linked and scab can be predicted 4 months in advance using the Southern Oscillation Index (SOI), a measure of ENSO intensity (Zhao and Yao, 1989). Association between SOI and wheat stripe rust in China shows a 2–10 year periodicity and stem rust in the USA has a 6–8 year periodicity (Scherm and Yang, 1995). Severe stripe rust in China co-oscillates with the Western Atlantic teleconnection pattern at a periodicity of 3 years (Scherm and Yang, 1998). Other systematic studies of long-term climate and plant diseases (Coakley, 1979, 1988; Jhorar *et al.*, 1997; Petit and Parry, 1996) may also prove useful to extract inter-annual trends. As with ENSO, these relationships could be useful for early warning of epidemics.

9.4.2 Plant disease under changing atmospheric CO_2

Recent reviews have summarised the influence of UV-B and O_3 on diseases (Coakley, 1995; Darley and Middleton, 1966; Frankland *et al.*, 1996; Sandermann, 1996) and this section will focus on elevated CO_2.

Host resistance

High CO_2 changes anatomy, morphology and phenology to alter host resistance. These include reduced stomatal density and conductance (Hibberd *et al.*, 1996a; Wittwer, 1995); lowered nutrient concentration (Baxter *et al.*, 1994); extra layers of epidermal cells, accumulation of carbohydrates, waxes, and increased fibre content (Owensby, 1994); production of papillae and silicon accumulation following pathogen penetration (Hibberd *et al.*, 1996b); increased production of phenolics (Hartley *et al.*, 2000; Idso and Idso, 2001); and greater number of mesophyll cells (Bowes, 1993). Accelerated ripening and senescence alter predisposition of the host and shorten exposure to pathogens (Manning and Tiedemann, 1995).

Rapid development of certain diseases under elevated CO_2 was first reported in the early 1930s (Gassner and Straib, 1930; Volk, 1931). Since then, a number of studies have recorded increased, decreased or unchanged disease severity under elevated CO_2. Of the 26 diseases studied so far, severity has increased in 13, decreased in nine and remained unchanged in four (see *Table 9.1*).

Often cultivars differ in their expression of resistance at high CO_2 (Chakraborty *et al.*, 2000a) and nutritional status strongly influences the expression of resistance (Thompson and Drake, 1994). Severity of *Erysiphe graminis* in wheat reduces with lowered plant nitrogen but increases under increased water content (Thompson *et al.*, 1993). Temperature (Rishbeth, 1991; Wilson *et al.*, 1991), UV-B (Paul, 2000; Tiedmann and Firsching, 2000), and O_3 (Karnosky *et al.*, 2002) are among other external factors that modify host resistance at elevated CO_2. A combination of high CO_2 and low light reduces symptom development by victorin, the host-selective toxin of *Cochliobolus victoriae* (Navarre and Wolpert, 1999).

Changes in resistance and its underlying mechanisms such as increased phenolics (Hartley *et al.*, 2000) and reduced nitrogen concentration (Penulas *et al.*, 1997) are often not sustained when plants are grown at elevated CO_2 over a number of generations or for a long period of time (Fetcher *et al.*, 1988; Idso and Idso, 2001; Mouseau and Saugier, 1992). In *Stylosanthes scabra*, the increased resistance to anthracnose is reversed if plants from high CO_2 are transferred to ambient CO_2 soon after inoculation (Chakraborty *et al.*, unpublished). These studies suggest that given time or a slow enough change in CO_2, plants will reach equilibrium (Newbery *et al.*, 1995). Whether these plants will differ in resistance cannot be ascertained in the absence of data from long-term studies.

Table 9.1 Disease severity at elevated levels of atmospheric carbon dioxide

Pathogen	Host	Disease severity	Reference
9 necrotrophic fungi	Various	Increase (4), decrease (1), unchanged (4)	Manning and Tiedemann (1995)
7 biotrophic fungi	Various	Increase (6) and decrease (1)	Manning and Tiedemann (1995)
Puccinia sparganoides	*Scirpus olneyi* (C3 sedge)	Decrease	Thompson and Drake (1994)
Puccinia sparganoides	*Spartina patens* (C4 grass)	Increase	Thompson and Drake (1994)
Erysiphe graminis	Wheat	Increase or decrease based on plant nitrogen content	Thompson *et al.* (1993)
Melampsora medusae fsp. *tremuloidae*	Aspen	Increase with elevated ozone	Karnosky *et al.* (2002)
Erysiphe graminis	Barley	Decrease	Hibberd *et al.* (1996a)
Colletotrichum gloeosporioides	*Stylosanthes scabra* (pasture legume)	Decrease	Chakraborty *et al.* (2000a)
Maravalia cryptostegiae	*Cryptostegia grandiflora* (rubber vine)	Decrease	Chakraborty *et al.* unpublished
Barley Yellow Dwarf	Barley	Decrease	Malmstrom and Field (1997)
Xanthomonas campestris pv. *pelargonii*	Geranium	Decrease	Jiao *et al.* (1999)

Pathogen life cycle and epidemiology

Significant changes in the onset and duration of life stages have been reported under elevated CO_2 for both biotrophic and necrotrophic pathogens. Germtube growth and appressoria formation by the necrotrophic *Colletotrichum gloeosporioides* starts within 6 h of inoculation at low CO_2 but after 8 h at high CO_2 (*Figure 9.1*). For some pathogens the latent period, time between inoculation and sporulation, does not vary at the two CO_2 levels, due to faster pathogen growth inside host tissue at high CO_2 (Chakraborty *et al.*, 2000a). Penetration of barley by the biotrophic *Erysiphe graminis* is reduced at high CO_2 but established colonies grow faster (Hibberd *et al.*, 1996a). However, in the biotrophic *Maravalia cryptostegiae* latent period is extended from 10.6 to 11.9 days at high CO_2 but there are >57% fewer pustules per leaf (Chakraborty *et al.*, unpublished).

With enhanced reproductive fitness, fecundity of *E. graminis* (Hibberd *et al.*, 1996a), *C. gloeosporioides* (Chakraborty *et al.*, 2000a) and *M. cryptostegiae* (Chakraborty *et al.*, unpublished) are increased at high CO_2. This extends to airborne fungal propagules and soil fungi on decomposing leaf litter around *Populus tremuloides* (Klironomos *et al.*, 1997). However, geranium at elevated CO_2 contain fewer *Xanthomonas campestris* pv. *pelargonii* than at ambient CO_2 (Jiao *et al.*, 1999).

Two important trends at elevated CO_2 have emerged from the limited information in the literature: resistance levels change in many plants and many pathogens produce more propagules to cause more infections in a modified

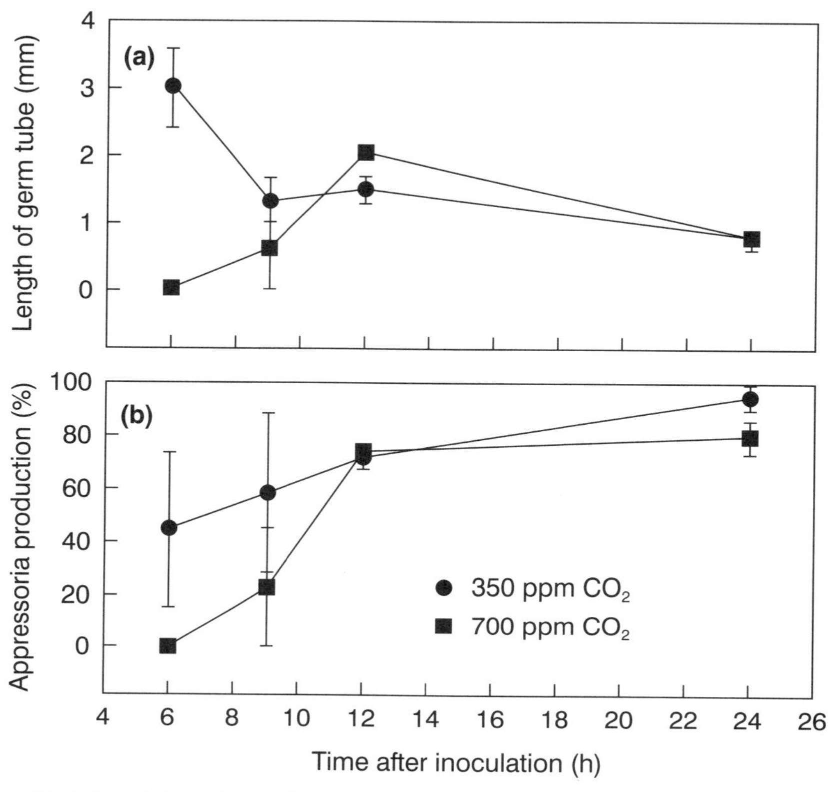

Figure 9.1 *Colletotrichum gloeosporioides* germ tube length (a) and appressoria production (b) on *Stylosanthes scabra* at 350 and 700 ppm CO_2. Reprinted from Chakraborty *et al.* (2000a) Production and dispersal of *Colletotrichum gloeosporioides* spores on *Stylosanthes scabra* under elevated CO_2. *Environ. Pollut.* **108**: 317–326. With permission from Elsevier Science.

canopy microclimate. High CO_2 *S. scabra* plants trap twice as many *C. gloeosporioides* conidia inside an enlarged canopy when exposed to natural inoculum in the field (Pangga, 2002). Although infection efficiency is reduced in high CO_2 plants due to enhanced resistance, three times as many lesions are produced on the enlarged plants (Pangga *et al.*, 2004).

Host–pathogen evolution

Enlarged plant canopy, increased fecundity and a pliant microclimate support many more generations, potentially accelerating pathogen evolution. After 25 sequential infection cycles aggressiveness of *C. gloeosporioides* increases steadily at ambient CO_2 and after an initial lag, lasting the first ten cycles, at twice ambient CO_2 (*Figure 9.2*) (Chakraborty and Datta, 2003). The initial lag represents the number of asexual pathogen generations to overcome enhanced host resistance. However, as host plants themselves will evolve, host-mediated response to pathogen aggressiveness can only be examined from long-term field studies under CO_2 enrichment (Norby *et al.*, 1997; Senft, 1995).

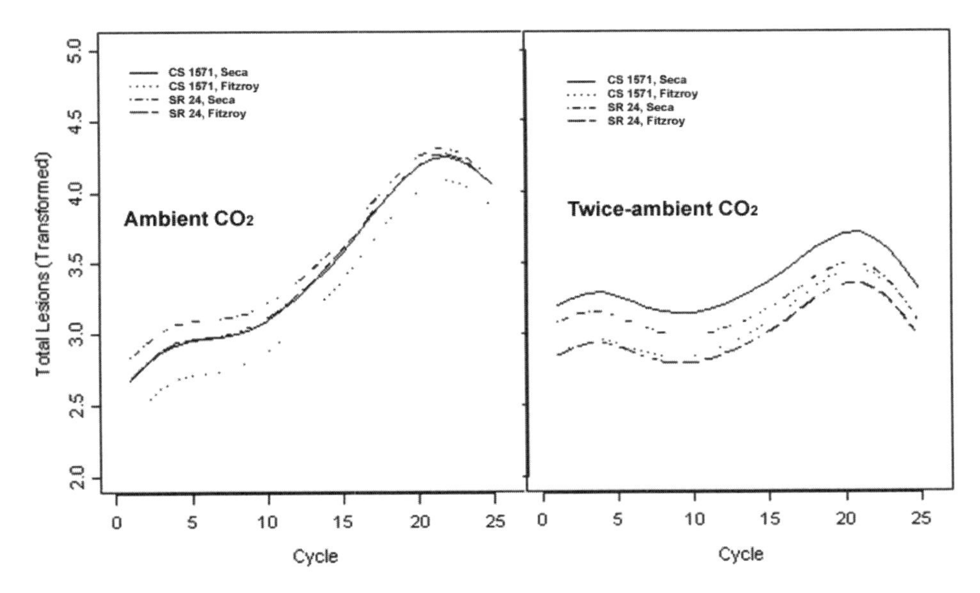

Figure 9.2 Changing aggressiveness of two *Colletotrichum gloeosporioides* isolates (sr24 and cs1571) on *S. scabra* Fitzroy and Seca with 25 sequential infection cycles at ambient and twice-ambient CO_2.

9.4.3 Plant disease in a changing climate

Both mean temperature and its variability are equally important since a modest warming can cause a significant increase in heat sums above a critical threshold to affect crop physiology and host resistance (Scherm and van Bruggen, 1994). High temperature breaks down heat-sensitive resistance genes in many plants (Bonnett *et al.*, 2002; Carver *et al.*, 1996; Dyck and Johnson, 1983; Gijzen *et al.*, 1996), increases damage from Scleroderris canker on lodgepole pine (Karlman *et al.*, 1994), but enhances resistance in some tropical species by lignification (Wilson *et al.*, 1991). Drought can stress plants to exacerbate damage from *Armillaria* sp. (Rishbeth, 1991) and *Cryptostroma corticale* (Dickenson and Wheeler, 1981).

Pathogen dispersal.

For splash-borne pathogens, long- and short-distance dispersal are controlled by rain intensity and heavy rain often reduces spread due to inoculum depletion (Fitt *et al.*, 1989; Geagea *et al.*, 2000; Huber *et al.*, 1998). Wind-dispersed pathogens rely on suitable atmospheric conditions for long-distance and intercontinental travel (Westbrook and Isard, 1999). *Mycosphaerella fijiensis* (banana black Sigatoka), *Cryphonectria parasitica* (chestnut blight), *Hemileia vastatrix* (coffee rust), *Puccinia melanocephala* (sugarcane rust) and *Puccinia striiformis* (wheat stripe rust) travel large distances to infect a crop (Brown and Hovmoller, 2002). During cool, wet, and cloudy weather, tobacco Blue Mold (Main and Spurr, 1990) and cucurbit Downy Mildew (Thomas, 1996) pathogens are transported via wind currents in the atmospheric boundary layer. In clear, dry and hot weather *Peronospora tabacina* epidemics can slow down or completely stop. Wheat rust pathogens follow a predictable pathway to match seasonal conditions to infect crops in North America and India (Nagarajan and Singh, 1990). *Blumeria graminis* fsp. *tritici* and fsp. *hordei* travels 100 km/year across Europe on prevailing westerly winds (Limpert *et al.*, 1999). Ultraviolet radiation, temperature and moisture affect survival of spores during transport and rain is important for deposition on healthy crops. Climatic factors impacting on one or more of the critical stages (Aylor, 1986) will influence long-range dispersal.

Geographical distribution

With the predicted pole-ward shift of host plant communities, pathogens will follow migrating hosts (Chakraborty *et al.*, 1998; Coakley *et al.*, 1999). Dispersal, survival, host range and population size will determine the success of migrating pathogens. There may be changes in the type, amount and prevalence of diseases; some weak pathogens may inflict serious damage following warming. Linked host–pathogen models (Teng *et al.*, 1996) and climate-matching tools (Sutherst *et al.*, 1995) have been used to predict distribution of several pathogens under climate change (see *Table 9.2*). However, some (Davis *et al.*, 1998; Lawton, 1998) have criticised the use of climate matching. Changes in temperature, and not rainfall would increase rice blast severity in cool subtropical Japan and southern China, but severity will not change for the Philippines (Luo *et al.*, 1995).

There are over 20 introduced pathogens attacking forest trees (Liebhold *et al.*, 1995) and over 60 exotic pathogens potentially threaten agriculture and forestry in the USA (Madden, 2001). New evidence suggests that many invasive species such as weeds share traits that will allow them to capitalise on climate change (Dukes and Mooney, 1999). Some species that might otherwise not have survived will

Table 9.2 Predicted changes in geographical distribution of plant pathogens predicted due to a changing climate

Pathogen	Host	Major change	Reference
Phytophthora cinnamomi	Oak, *Eucalyptus* sp.	Pole-ward shift, increased prevalence	Brasier and Scott (1994); Podger *et al.* (1990)
Xiphinema sp. *Longidorus* sp.	Various	Increased severity, migrate to northern Europe	Boag *et al.* (1991)
Pyricularia grisea	Rice	Increased risk for Japan, southern China; reduced risk for Philippines	Luo *et al.* (1988a); Luo *et al.* (1988b)
Melampsora alli-populina	Poplar	Increased risk for northern Europe	Somda and Pinon (1981)
Tilletia indica	Wheat	Increased risk of establishment over a wider geographical area	Baker *et al.* (2000)
Graemmeniella abietina	Pine	Cease to be a problem	Lonsdale and Gibbs (1996)
Fusarium foot rot	Wheat	*F. culmorum* become the dominant species in the UK	Pettit and Parry (1996)

become established (Simberloff, 2000). For instance, the dogwood anthracnose pathogen, *Discula destructiva*, is more likely to strike where acid rain is prevalent (Britton *et al.*, 1996). Climatic conditions covering much of central and southern England are suitable for the establishment of the exotic *Tilletia indica*, the Karnal bunt pathogen, and with the projected temperature increase by 2050, large areas of the UK and Europe will become suitable (Baker *et al.*, 2000).

9.5 Disease management in a changing climate

A changing climate will increase, reduce or have no effect on diseases in some regions to determine the need and appropriateness of disease management strategies. Climate will interact with disease control strategies to increase the complexity of production systems (Coakley *et al.*, 1999). The expression of resistance under a changing climate is dependent on host nutrition, water content and other factors such as O_3 (Section 9.4.2). Varieties respond differently to high CO_2, with some showing only a transient expression of augmented resistance (Chakraborty *et al.*, 2000a; Pangga *et al.*, 2004). Of particular concern is whether new virulent and aggressive strains of a pathogen may rapidly evolve to erode the usefulness of disease resistance in crop plants (Chakraborty and Datta, 2003). For host plants most at-risk, strategic pre-emptive breeding programmes, incorporating climate change related traits (Richards, 2002), will need to start early due to the long time required to develop and release cultivars. Comprehensive analysis at an appropriate spatial scale (Seem *et al.*, 2000; Strand, 2000) would be necessary to identify crops at risk.

Disease management employing chemical, physical and biological options will all be influenced by a changing climate. Disease control chemicals may be washed off the foliage reducing their effectiveness (Coakley *et al.*, 1999). Host physiology at high CO_2 will alter penetration, translocation and mode of action of chemicals (Edis *et al.*, 1996) and changes in temperature and light will influence their persistence. Similar effects may occur with biological control agents. Soilborne pathogens with broad host range, but limited spread, will damage crops as they migrate to new areas. New crops may be grown in response to changing climate, such as navy beans in the UK (Holloway and Ilberry, 1997), to disrupt disease cycle, analogous to a crop rotation. Soil solarisation will become more effective over a wider area (Strand, 2000).

Soil organic matter content will rise from increased crop residue (Tiquia *et al.*, 2002) but if inadequately stabilised, will serve as a food base for pathogens with strong saprophytic ability to increase disease; but if fully stabilised, it will suppress pathogens (Hoitink and Boehm, 1999). Increased populations of plant-growth-promoting rhizobacteria will offer protection against some insects, nematodes and diseases through induced systemic resistance (Ramamoorthy *et al.*, 2001). Changing climate will alter the composition and structure of soil communities (Tiquia *et al.*, 2002), but its effect on mycorrhizae is actively debated (Soderstrom, 2002; Staddon *et al.*, 2002).

9.6 Looking ahead

Through its influence on host, pathogen and the epidemiology and management of diseases, a changing climate will add further layers of complexity to agricultural and natural production systems. Disease severity, prevalence and distribution will be modified but accurate predictions for any region, crop or disease are not possible with the current level of knowledge. Detailed experimental studies are needed on model systems covering different host, pathogen and production systems. The limited experimental data come almost entirely from controlled environment studies. Increased production of ethylene and its precursor, aminocyclopropane-1-carboxylic acid (Grodzinski, 1992) influence plant response to elevated CO_2 in small chambers. Results from these studies are helpful to formulate hypotheses, but long-term field studies with successive generations of host plants under slowly increasing CO_2 and preferably temperature (Norby *et al.*, 1997; Senft, 1995) are desperately needed. Because of the large number of interacting factors, the impact of climate change scenarios on plant diseases is best explored using modelling approaches (Coakley and Scherm, 1996). The distribution and severity of many diseases can be modelled with the existing knowledge of weather and disease. However, modelling to extrapolate effects across spatial and temporal scales has its own drawbacks (Chakraborty *et al.*, 2000b) and examining effects at a regional level may uncover relationships not readily identified at another level. For instance, changing soil biota can influence invasiveness of plants (Klironomos, 2002). Similarly, pathogens can alter species composition and size structure of forests as well as change CO_2 flux and heat transfer to create feedbacks to climate (Ayres and Lombardero, 2000). To be more relevant, research focus must expand to ecologically relevant spatial units (Chakraborty *et al.*, 2000b); Scherm *et al.*, 2000). Since episodic weather events such as flood, drought and storm can be more catastrophic than a gradual change in atmospheric composition and climate, research has to consider both climate variability and change in developing mitigation strategies.

References

Amthor, J.S. (1998) Perspective on the relative insignificance of increasing atmospheric CO_2 concentration to crop yield. *Field Crop Res.* **58**: 109–127.

Ando, K. (1994) The influences of global environmental changes on nematodes. *Jap. J. Parasitol.* **43**: 477–482.

Arens, N.C. (2001) Climate and atmosphere interact to drive vegetation change. *Trends Eco. Evol.* **16**: 18.

Atkinson, D. (ed) (1993) *Global Climate Change, its Implication for Crop Protection*. British Crop Protection Council Monograph No 56. BCPC: Surrey, UK.

Aylor, D.E. (1986) A framework for examining inter-regional aerial transport of fungal spores. *Agric. Forest Meteorol.* **38**: 263–288.

Ayres, M.P., Lombardero, M.J. (2000) Assessing the consequences of global change for forest disturbance from herbivores and pathogens. *Sc. Total Environ.* **262**: 263–286.

Baker, R.H.A., Sansford, C.E., Jarvis, C.H., Cannon, R.J.C., MacLeod, A., Walters, K.F.A. (2000) The role of climatic mapping in predicting the potential geographical distribution of non-indigenous pests under current and future climates. *Agric. Eco. Environ.* **82**: 57–71.

Bale, J.B., Masters, G.J., Hodkinson, I.D. ***et al.*** (2002) Herbivory in global climate change research: direct effects of rising temperature on insect hervivores. *Global Change Biol.* **8**: 1–16.

Baxter, R., Ashenden, T.W., Sparks, T.H., Farrar, J.F. (1994) Effects of elevated CO_2 on three montane grass species. I. Growth and dry matter partitioning. *J. Exp. Bot.* **45**: 305–315.

Bhat, R.V., Beedu, S.R., Ramakrishna, Y., Munshi, K.L (1988). Outbreak of trichothecene mycotoxicosis associated with the consumption of mould-damaged wheat products in Kashmir Valley, India. *Lancet* **7**: 35–37.

Biswas, P.K., Hilman, D.R., Ghosh, P.P., Bhattacharya, N.C., McCrimmon, J.N. (1996) Growth and yield responses of field-grown sweetpotato to elevated carbon dioxide. *Crop Sci.* **36**: 1234–1239.

Boag, B., Neilson, R. (1996) Effects of potential climatic changes on plant–parasitic nematodes. *Aspects Appl. Biol.* **45**: 331–334.

Boag, B., Crawford, J.W., Neilson, R. (1991) The effect of potential climatic changes on the geographical distribution of the plant parasitic nematodes *Xiphinema* and *Longidorus* in Europe. *Nematologia* **37**: 312–323.

Bonnett, D.G., Park, R.F., McIntosh, R.A., Oades, J.D. (2002) The effects of temperature and light on interactions between *Puccinia coronata* f. sp. *avenae* and *Avena* spp. *Aust. Plant Pathol.* **31**: 185–193.

Bowes, G. (1993) Facing the inevitable: Plants and increasing atmospheric CO_2. *Annu. Rev. Plant Physiol. Plant Mol. Biol.* **44**: 309–332.

Brasier, C.M., Scott, J.K. (1994) European oak decline and global warming: a theoretical assessment with special reference to the activity of *Phytophthora cinnamomi*. *EPPO Bulletin* **24**: 221–232.

Britton, K.O., Berrang, P., Mavity, E. (1996) Effects of pre-treatment with simulated acid rain on the severity of dogwood anthracnose. *Plant Dis.* **80**: 646–649.

Brown, J.K.M., Hovmoller, M.S. (2002) Aerial dispersal of pathogens on the global and continental scales and its impact on plant disease. *Science* **297**: 537–541.

Campbell, C.L., Madden, L.V. (1990) *Introduction to Plant Disease Epidemiology.* John Wiley: New York.

Cardwell, K.F., Desjardins, A., Henry, S.H., Munkvold, G., Robens, J. (2001) *Mycotoxins: The Cost of Achieving Food Security and Food Quality.* American Phytopathological Society Feature, August, 2001.

Carter, T.R., Nurro, M., Torkko, S. (1996) Global climate change and agriculture in the North. *Agric. Food Sci. Finld.* 5: 223–385.

Carver, T.L.W., Zhang, L., Zeyan, R.J., Robbins, M.P. (1996) Phenolic biosynthesis inhibitors suppress adult plant resistance to *Erysiphe graminis* in oat at 20°C and 10°C *Physiol. Mol. Plant Pathol* **49**: 121–141.

Chakraborty, S. (2001) Effects of climate change. In: Waller, J.M, Lenne, J., Waller, S.J. (eds) *Plant Pathologists' Pocketbook,* 3rd edn., pp. 203–207. CABI Publishing: Wallingford, UK.

Chakraborty, S., Datta, S. (2003) How will plant pathogens adapt to host plant resistance at elevated CO_2 under a changing climate? *New Phytol.* **159**: 733–742.

Chakraborty, S., Murray, G.M., Magarey, P.A. *et al.* (1998) Potential impact of climate change on plant diseases of economic significance to Australia. *Aust. Plant Pathol.* **27**: 15–35.

Chakraborty, S., Pangga, I.B., Lupton, J., Hart, L., Room, P.M., Yates, D. (2000a) Production and dispersal of *Colletotrichum gloeosporioides* spores on *Stylosanthes scabra* under elevated CO_2. *Environ. Pollut.* **108**: 381–387.

Chakraborty, S., Tiedemann, A.V., Teng, P.S. (2000b) Climate change: potential impact on plant disease. *Environ. Pollut.* **108**: 317–326.

Cheddadi, R., Yu, G., Guiot, J., Harrison, S.P., Prentice, I.C. (1996) The climate of Europe 6000 years ago. *Climate Dynamics* **13**: 1–9.

Clifford, B.C., Davies, A., Griffith, G., Froud-Williams, R.J., Harrington, R., Hocking, T.J., Smith, H.G., Thomas, T.H. (1996) UK climate change models to predict crop disease and pests threats. *Aspects Appl. Biol.* **45**: 269–276.

Coakley, S.M. (1979) Climatic variability in the Pacific Northwest and its effect on stripe rust of winter wheat. *Climate Change* **2**: 33–51.

Coakley, S.M. (1988) Variation in climate and prediction of disease in plants. *Annu. Rev. Phytopathol.* **26**: 163–181.

Coakley, S. M. (1995) Biospheric change: will it matter in plant pathology? *Can. J. Plant Pathol.* **17**: 147–153.

Coakley, S.M., Scherm, H. (1996) Plant disease in a changing global environment. *Aspects Appl. Biol.* **45**: 227–237.

Coakley, S.M., Scherm, H., Chakraborty, S. (1999) Climate change and plant disease management. *Annu. Rev. Phytopathol.* **37**: 399–426.

CSIRO (2001a) *Climate Change Impacts for Australia*. CSIRO Sustainable Ecosystems: Aitkenvale, Queensland, Australia.

CSIRO (2001b) *Climate Change Projections for Australia*. Climate Impact Group, CSIRO Atmospheric Research: Aspendale, Victoria, Australia.

Cure, J.D., Acock, B. (1986) Crops response to carbon dioxide doubling; A literature survey. *Agric. For. Meteorol.* **38**: 127–145.

Darley, E.F., Middleton, J.T. (1966) Problems of air pollution in plant pathology. *Annu. Rev. Phytopathol.* **4**: 103–118.

Davis, A.J., Jenkins, L.S., Lawton, J.H., Shorrocks, B., Wood, S. (1998) Making mistakes when predicting shifts in species range response to global warming. *Nature* **391**: 783–786.

Dickenson, S., Wheeler, B.E.J. (1981) Effects of temperature and water stress in sycamore on growth of *Cryotostroma corticale*. *Trans. Brit. Mycol. Soc.* **76**: 181–185.

Dukes, J.S., Mooney, H.A. (1999) Does global change increase the success of biological invaders? *Tree* **14**: 135–139.

Dyck, P.L., Johnson, R. (1983) Temperature sensitivity of genes for resistance in wheat to *Puccinia recondita*. *Can. J. Plant Pathol.* 5: 229–234.

Edis, D., Hull, M.R., Cobb, A.H., Sanders-Mills, G.E. (1996) A study of herbicide action and resistance at elevated levels of carbon dioxide. *Aspects Appl. Biol.* **45**: 205–209.

Epstein, P.R. (2001) Climate change and emerging infectious diseases. *Microb. Infect.* **3**: 747–754.

Fetcher, N., Jaeger, C.H., Strain, B.R., Sionit, N. (1988) Long-term elevation of atmospheric CO_2 concentration and the carbon exchange rates of saplings of *Pinus taeda* L. and *Liquidambar styraciflua* L. *Tree Physiol.* **4**: 255–262.

Fitt, B.D.L., McCartney, H.A., Walklate, P.J. (1989) The role of rain in dispersal of pathogen inoculum. *Annu. Rev. Phytopathol.* **27**: 241–270.

Fowler, A.M., Hennessy, K.J. (1994) Potential impacts of global warming on the frequency and magnitude of heavy precipitation. *Natural Hazards* **11**: 283–303.

Frankland, J.C., Magan, N., Gadd, G.M. (1996) *Fungi and Environmental Change*. Cambridge University Press: Cambridge, UK.

Gassner, G., Straib, W. (1930) Untersuchungen uber die Abbangigkeit des Infektionsverhaltens der Getreiderostpilze vom Kohlensauregehalt der Luft. *J. Phytopathol.* **1**: 1–30.

Geagea, L., Huber, L., Sache, I., Flura, D., McCartney, H.A., Fitt, B.D.L. (2000) Influence of simulated rain on dispersal of rust spores from infected wheat seedlings. *Agric. For. Meteorol.* **101**: 53–66.

Gijzen, M., MacGregor, T. Bhattacharyya, M., Buzzell, R. (1996) Temperature induced susceptibility to *Phytophthora sojae* in soybean isolines carrying different *Rps* genes. *Physiol. Mol. Plant Pathol.* **48**: 209–215.

Goudriaan, J., Zadoks, J.C. (1995) Global climate change: modelling the potential responses of agro-ecosystems with special reference to crop protection. *Environ. Pollut.* **87**: 215–224.

Grodzinski, B. (1992) Plant nutrition and growth regulation by CO_2 enrichment. *BioScience* **42**: 517–525.

Hartley, S.E., Jones, C.J., Couper, G.C., Jones, T.H. (2000) Biosynthesis of plant phenolic compounds in elevated atmospheric CO_2. *Global Change Biol.* **6**: 497–506.

Hibberd, J.M., Whitbread, R., Farrar, J.F. (1996a) Effect of elevated concentrations of CO_2 on infection of barley by *Erysiphe graminis*. *Physiol. Mol. Plant Pathol.* **48**: 37–53.

Hibberd, J.M., Whitbread, R., Farrar, J.F. (1996b) Effect of 700 mmol per mol CO_2 and infection of powdery mildew on the growth and partitioning of barley. *New Phytol.* **1348**: 309–345.

Hoitink, H.A.J., Boehm, M.J. (1999) Biocontrol within the context of soil microbial communities: a substrate dependant phenomenon. *Annu. Rev. Phytopathol.* **37**: 427–446.

Holloway, L.E., Ilbery, B.W. (1997) Global warming and navy beans: decision making by farmers and food companies in the UK. *J. Rur. Stud.* **13**: 343–355.

Houghton, J.T., Ding, Y., Griggs, D.J., Noguer, M., van der Linden P.J., Xiaosu, D., Maskell, K., Johnson, C.A. (2001) *Climate Change 2001: The Scientific Basis.* Cambridge University Press: Cambridge, UK.

Huber, L., Madden, L.V., Fitt, B.D.L. (1998) Rain-splash and spore dispersal: a physical perspective. In: Jones, D.G. (eds) *The Epidemiology of Plant Diseases*, pp. 348–370. Kluwer Academic Publishing: Dordrecht.

Hughes, J.M., Evans, K. A. (1996) Global warming and pest risk assessment. *Aspects Appl. Biol.* **45**: 339–342.

IARC (1993) *Monograph 56: Some Naturally Occurring Substances, Some Food Items and Constituents, Heterocyclic Amines and Mycotoxins*. International Agency for Research on Cancer: Lyon, France.

Idso, S.B., Idso, K.E. (2001). Effects of atmospheric CO_2 enrichment on plant constituents related to animal and human health. *Env. Exp. Bot.* **45**: 179–199.

IPCC (1996) *Climate Change 1995: The Science of Climate Change.* Houghton, J.T., Meira Filho, L.G., Callander, B.A., Harris, N., Kattenberg, A., Maskell, K. (eds) Contribution of Working Group I to the second assessment report of the Intergovernmental Panel on Climate Change. Cambridge University Press: Cambridge, UK.

IPCC (2000) *The Regional Impacts of Climate Change, IPCC Special Reports.* Watson, R.T., Zinyowera, M.C., Moss, R.H., Dokken, D.J. (eds) IPPC secretariat: Geneva, Switzerland.

Jiao, J., Goodwin, P., Grodzinski, B. (1999) Inhibition of photosynthesis and export in geranium grown at two CO_2 levels and infected with *Xanthomonas campestris* pv. *pelargonii. Plant Cell Environ.* **22**: 15–25.

Jhorar, O.P., Mathauda, S.S., Singh, G., Butler, D.R., Mavi, H.S. (1997) Relationships between climatic variables and Ascochyta blight of chickpea in Punjab, India. *Agric For. Meteor.* **87**: 171–177.

Karl, T.R., Knight, R.W., Easterling, D.R., Quayle, R.G. (1995) Trends in U.S. climate during the twentieth century. *Consequences* **1**: 3–12.

Karlman, M., Hansson, P., Witzell, J. (1994) Scleroderris canker on lodgepole pine introduced in northern Sweden. *Can. J. For. Res.* **24**: 1948–1959.

Karnosky, D.F., Percy, K.E., Xiang, B. *et al.* (2002) Interacting elevated CO_2 and trophospheric O_3 predisposes aspen (*Populus tremuloides* Michx.) to infection by rust (*Melampsora medusae* f.sp. *tremuloidae*). *Global Change Biol.* **8**: 329–338.

Kimball, B.A. (1983) CO_2 and agricultural yield: An assemblage and analysis of 430 observations. *Agron. J.* **75**: 779–788.

Klironomos, J.N. (2002) Feedback with soil biota contributes to plant rarity and invasiveness in communities. *Nature* **417**: 67–70.

Klironomos J.N., Rillig, M.C., Allen, M.F., Zak, D.R., Kubiske, M., Pregitzer, K.S. (1997) Soil fungal-arthropod responses to *Populus tremuloides* grown under enriched atmospheric CO_2 under field conditions. *Global Change Biol.* **3**: 473–478.

Landsberg, H.E. (1984) Global climate trends. In: Simon, J.L, Kahn, H. (eds) *The Resourceful Earth, A Response to Global 2000*, pp. 272–315. Basil Blackwell: New York.

Lawton, J.H. (1998) Small earthquakes in Chile and climate change. *Oikos* **82**: 209–211.

Liebhold, A.M., McDonald, W.L., Bergdahl, D., Mastro, V.C. (1995) Invasion by exotic forest pests: a threat to forest ecosystems. *For. Sci. Monogr.* **30**: 49.

Limpert, E., Godet, F., Muller, K. (1999) Dispersal of cereal mildews across Europe. *Agric. For. Meteor.* **97**: 293–308.

Lonsdale, D., Gibbs, J.N. (1996) Effects of climate change on fungal diseases of trees. In: Frankland, J.C., Magan, N., Gadd, G.M. (eds) *Fungi and Environmental Change*, pp. 1–19, Cambridge University Press: Cambridge, UK.

Luo, Y., TeBeest, D.O., Teng, P.S., Fabellar, N.G. (1995) Simulation studies on risk analysis of rice blast epidemics associated with global climate in several Asian countries. *J. Biogeogr.* **22**: 673–678.

Luo, Y., Teng, P.S., Fabellar, N.G., TeBeest, D.O. (1998a) The effect of global temperature change on rice blast epidemics: a simulation study in three agroecological zones. *Agric. Eco. Environ.* **68**: 187–196.

Luo, Y., Teng, P.S., Fabellar, N.G., TeBeest, D.O. (1998b) Risk analysis of yield loss caused by rice blast associated with temperature changes above and below for five Asian countries. *Agric. Eco. Environ.* **68**: 197–205.

Madden, L.V. (2001) What are the non indigenous plant pathogens that threaten US crops and forests? Online APSnet Feature, American Phytopathological Society: St. Paul, MN.

Main, C.E., Spurr, H.W., Jr. (1990) Blue Mold disease of tobacco. Proceedings of the International Symposium on Blue Mold of Tobacco, Raleigh, NC, February 14–17, 1988. Delmar Publishing: Charlotte, NC.

Malmström, C.M., Field, C.B. (1997) Virus-induced differences in the response of oat plants to elevated carbon dioxide. *Plant Cell Environ.* **20**: 178–188.

Manning, W.J., Tiedemann, A.V. (1995) Climate change: potential effects of increased atmospheric carbon dioxide (CO_2), ozone (O_3), and ultraviolet-B (UVB) radiation on plant diseases. *Environ. Pollut.* **88**: 219–245.

Mouseau, M., Saugier, B. (1992) The direct effect of increased CO_2 on gas exchange and growth of forest tree species. *J. Exp. Bot.* **43**: 1121–1130.

Nagarajan, S., Singh, D.V. (1990) Long-distance dispersion of rust pathogens. *Annu. Rev. Phytopathol.* **28**: 139–153.

Navarre, D.A., Wolpert, T.J. (1999) Effects of light and CO_2 on victorin-induced symptom development in oats. *Physiol. Mol. Plant Pathol.* 55: 237–242.

Newbery, R.M., Wolfenden, J., Mansfield, T.A., Harrison, A.F. (1995) Nitrogen, phosphorus and potassium uptake and demand in *Agrostis capillaris*: the influence of elevated CO_2 and nutrient supply. *New Phytol.* **130**: 565–574.

Norby, R.J., Edwards, N.T., Riggs, J.S., Abner, C.H., Wullschleger, S.D., Gunderson, C.A. (1997) Temperature-controlled open-top chambers for global change research. *Global Change Biol.* **3**: 259–267.

Oerke, E-C., Dehne, H-W., Schonbeck, F., Weber, A. (1994) *Crop Production and Crop Protection, Estimated Losses in Major Food and Cash Crops*. Elsevier: Amsterdam, The Netherlands.

Owensby, C.E. (1994) Climate change and grasslands: ecosystem-level responses to elevated carbon dioxide. *Proceedings of XVII International Grassland Congress* 1119–1124.

Padmanabhan, S.Y. (1973). The great Bengal famine. *Annu. Rev. Phytopathol.* **11**: 11–26.

Pangga, I.B. (2002) *Effects of elevated CO_2* on plant architecture of *Stylosanthes scabra* and epidemiology of anthracnose disease. PhD thesis, University of Queensland, Australia.

Pangga, I.B., Chakraborty, S., Yates, D. (2004) Canopy size and induced resistance in *Stylosanthes scabra* determine anthracnose severity at high CO_2. *Phytopathology* **94**: (in press).

Patterson, D.T., Westerbrook, J.K., Joyce, R.J.V., Lingren, P.D., Rogasik, J. (1999) Weeds, insects and diseases. *Climatic Change* **43**: 711–727.

Paul, N.D. (2000) Stratospheric ozone depletion, UV-B radiation and crop disease. *Environ. Pollut.* **108**: 343–355.

Payne, G.A., Brown, M.P. (1998) Genetics and physiology of aflatoxin biosynthesis. *Annu. Rev. Phytopathol.* **36**: 329–362.

Penulas, J., Idso, S.B., Ribas, A., Kimball, B.A. (1997) Effects of long-term atmospheric CO_2 enrichment on the mineral content of *Citrus aurantium* leaves. *New Phytol.* **135**: 439–444.

Petit, J.R., Jouze, J., Raynaud, D. *et al.* (1999) Climate and atmospheric history of the past 420,000 years from the Vostok Ice Core, Antartica, *Nature* **399**: 429–436.

Pettit, T.R., Parry, D.W. (1996) Effects of climate change on Fusarium foot rot of winter wheat in the United Kingdom. In: Frankland, J.C., Magan, N., Gadd, G.M. (eds) *Fungi and Environmental Change*, pp. 20–31. Cambridge University Press: Cambridge, UK.

Pimentel, D. (1997) Pest management in agriculture. In: Pimentel, D. (ed) *Techniques for Reducing Pesticide Use: Environmental and Economic Benefits*, pp. 1–12. John Wiley: Chichester, UK.

Pimentel, D., Lach, L., Zuniga, R., Morrison, D. (2000) Environmental and economic costs associated with non-indigenous species in the United States. *BioScience* **50**: 53–65.

Podger, F.D., Mummery, D.C., Palzer, C.R., Brown, M.L. (1990) Bioclimatic analysis of the distribution of damage to native plants in Tasmania by *Phytophthora*. *Aust. J. Ecol.* **15**: 281–289.

Prestidge, R.A., Pottinger, R.P. (eds) (1990) *The Impact of Climate Change on Pests, Diseases, Weeds and Beneficial Organisms Present in New Zealand Agricultural and Horticultural Systems*. MAF Technology, Ruakura Agricultural Centre: Hamilton, NZ.

Pritchard, S., Rogers, H.H., Prior, S.A., Peterson, C.M. (1999) Elevated CO_2 and plant structure: a review. *Global Change Biol.* 5: 807–837.

Ramamoorthy, V., Viswanathan, R., Raguchander, T. Prakasam, V., Samiyappan, R. (2001) Induction of systemic resistance by plant growth promoting rhizobacteria in crop plants against pests and diseases. *Crop Prot.* **20**: 1–11.

Rao, G.D., Katyal, J.C., Sinha, S.K., Srinivas, K. (1995) Impacts of climate change on sorghum production in India: simulation study. In: Rosenzweig, C., Allen, L.H., Harper, L.A., Hollinger, S.E., Jones, J.W. (eds) *Climate Change and Agriculture: Analysis of Potential International Impact*, pp. 325–337. American Society of Agronomy: Madison, USA.

Reilly, J.M., Schimmelpfennig, D. (1999) Agricultural impact assessment, vulnerability and the scope for adaptation. *Climatic Change* **43**: 745–788.

Richards, R.A. (2002) Current and emerging environmental challenges in Australian agriculture – the role of plant breeding. *Aust. J. Agric. Res.* **53**: 881–892.

Rishbeth, J. (1991) *Armillaria* in an ancient broad leaved woodland. *Eur. J. For. Pathol.* **21**: 239–249.

Rosenzweig, C., Hillel, D. (1998) *Climate Change and the Global Harvest*. Oxford University Press: New York.

Rosenzweig, C., Parry, M.L. (1994) Potential impact of climate change on world food supply. *Nature* **367**: 133–138.

Rosenzweig, C., Iglesias, A., Yang, X.B., Epstein, P.R., Chivian, E. (2001) Climate change and extreme weather events implications for food production, plant diseases and pests. *Global Change Human Health* 2: 90–104.
Rotter, R., van der Geijn, S.C. (1999) Climate change effects of plant growth, crop yield and livestock. *Climatic Change* **43**: 651–681.
Sandermann, H., Jr. (1996) Ozone and plant health. *Ann. Rev. Phytopathol.* **34**: 347–366.
Scherm, H., van Bruggen, A.H.C. (1994) Global warming and nonlinear growth: how important are changes in average temperature? *Phytopathology* **84**: 1380–1384.
Scherm, H., Yang, X.B. (1995) Interannual variations in wheat rust development in China and the United States in relation to the El Niño/Southern Oscillation. *Phytopathology* **85**: 970–976.
Scherm, H., Yang, X.B. (1998) Atmospheric teleconnection patterns associated with wheat stripe rust disease in North China. *Int. J. Biometeorol.* **42**: 28–33.
Scherm, H., Newton, A., Harrington, R. (2000) Global networking for assessment of impacts of global change on plant pests. *Environ. Pollut.* **108**: 333–341.
Seem, R.C., Magarey, R.D., Zack, J.W., Russo, J.M. (2000) Estimating disease risk at the whole plant level with general circulation models. *Environ. Pollut.* **108**: 389–395.
Senft, D. (1995) FACE-ing the future. *Agric. Res.* **43**: 4–6.
Simberloff, D. (2000) Global climate change and introduced species in United States forests. *Sci. Total Environ.* **262**: 253–261.
Söderström, B. (2002) Challenges for mycorrhizal research into the new millennium. *Plant Soil* **244**: 1–7.
Somda, B., Pinon, J. (1981) Ecophysiologie du state uredien de *Melampsora larici-populina* Kleb. et de *M. alli-populina* Kleb. *Eur. J. For. Pathol.* **11**: 243–254.
Staddon, P.L., Heinemeyer, A., Fitter, A.H. (2002) Mycorrhizas and global environmental change: research at different scales. *Plant Soil* **244**: 253–261.
Strand, J.F. (2000) Some agrometeorological aspects of pest and disease management for the 21st century. *Agric. For Meteor.* **103**: 73–82.
Streets, D.G., Glantz, M.H. (2000) Exploring the concept of climate surprise. *Global Environ. Change* **10**: 97–107.
Sutherst, R.W., Maywald, G.F., Skarrate, D.B. (1995) Predicting insect distributions in a changed climate. In: Harrington, R., Stork, N.E. (eds) *Insects in a Changing Environment,* pp. 59–91. Academic Press: London, UK.
Sutherst, R.W., Ingram, J.S.I., Steffen, W.L. (1996) *Global Change Impacts on Pests, Diseases and Weeds Implementation Plan.* GCTE Core Project Office: Canberra, Australia.
Taylor, G., Ranasihghe, S., Bosac, C., Gardner, S.D.L., Ferris, R. (1994) Elevated CO_2 and plant growth: cellular mechanisms and responses of whole plants. *J. Exp. Bot.* **45**: 1761–1774.
Teng, P.S., Heong, K.L.K., Kropff, K.J., Nutter, F.W., Sutherst, R.W. (1996) Linked pest-crop models under global change. In: Walker, B., Steffen, W. (eds) *Global Change and Terrestrial Ecosystems,* pp. 291–316. Cambridge University Press: Cambridge, UK.
Thomas, C.E. (1996) Downy mildew. In: Zitter, A., Hopkins, D.L., Thomas, C.E. (eds) *Fungal Diseases of Aerial Parts. Compendium of Cucurbit Diseases,* pp. 25–27. American Phytopathological Society: St. Paul, MN.
Thompson, G.B., Brown, J.K.M., Woodward, F.I. (1993) The effects of host carbon dioxide, nitrogen and water supply on the infection of wheat by powdery mildew and aphids. *Plant Cell Environ.* **16**: 687–694.
Thompson, G.B., Drake, B.G. (1994) Insects and fungi on a C_3 sedge and a C_4 grass exposed to elevated CO_2 concentrations in open-top chambers in the field. *Plant Cell Environ.* **17**: 1161–1167.
Tiedemann, A.V., Firsching, K.H. (2000) Interactive effects of elevated ozone and carbon dioxide on growth and yield of leaf rust-infected versus non-infected wheat. *Environ. Pollut.* **108**: 357–363.
Timmermann, A., Oberhuber, J., Bacher, A., Esch, M., Latif, M., Roeckner, E. (1999) Increased El Niño frequency in a climate model forced by future greenhouse warming. *Nature* **398**: 694–696.
Tiquia, S.M., Lloyd, J., Herms, D.A., Hoitink, H.A.J., Michel, F.C., Jr. (2002) Effects of mulching and fertilization on soil nutrients, microbial activity and rhizosphere bacterial community structure determined by analysis of TRFLPs and PCR-amplified 16S rRNA genes. *Appl. Soil Ecol.* **21**: 31–48.
Volk, A. (1931) Einfluesse des Bodens, der Luft und des Lichtes auf Empfaenglichkeit der Pflanzen fuer Krankneiten. *J. Phytopathol.* **3**: 1–88.

von Tiedemann A. (1996) Globaler Wandel von Atmosphäre und Klima – welche Folgen ergeben sich für den Pflanzenschutz? *Nachrichtenbl. Deut. Pflanzenschutzd.* **48**: 73–79.
Westbrook, J.K., Isard, S.A. (1999) Atmospheric scales of biotic dispersal. *Agric. For. Meteorol.* **97**: 263–274.
Wilson, J.R., Deinum, B., Engels, F.M. (1991) Temperature effects on anatomy and digestibility of leaf and stem of tropical and temperate forage species. *Neth. J. Agric. Sci.* **39**: 31–48.
Wittwer, S.H. (1995) *Food, Climate and Carbon Dioxide – The Global Environment and World Food Production*. CRC Press: Boca Raton, FL.
Yang, X.B., Scherm, H. (1997) El Nino and infectious disease. *Science* **275**: 739.
Zhao, S., Yao, C. (1989) On the sea temperature prediction models of the prevailing level of wheat scab. *Acta Phytopathol. Sin.* **19**: 229–234.
Ziska, L.H., Bunce, J.A. (2000) Sensitivity of field-grown soybean to future atmospheric CO_2: selection for improved productivity in the 21st century. *Aust. J Plant Physiol.* **27**: 979–984.

10

Genetic diversity of bacterial plant pathogens

Mark Fegan and Chris Hayward

10.1 Introduction

The capacity to cause plant disease has evolved in a relatively small number of bacterial species which are phenotypically and genetically diverse. Below the level of the species the strains that make up these species also vary in genotype and phenotype. Traditionally phenotypic techniques such as substrate utilisation profiles and total fatty acid composition have been employed to characterise plant-pathogenic bacteria. Recently more reliable DNA-based methods have been applied which provide a more complete understanding of genetic and evolutionary relationships of bacteria.

The genetic diversity of phytopathogenic prokaryotes can be assessed by employing molecular methods which differ in the taxonomic level at which they can discriminate (*Figure 10.1*). The phylogenetic diversity of plant-pathogenic bacteria, primarily assessed by phylogenetic analysis of 16S rRNA gene sequences, is of primary importance in the description of bacterial species (Stackebrandt and Goebel, 1994). Another taxonomically important technique for the assessment of genetic diversity of bacteria is the estimation of total DNA–DNA homology. If two strains share 70% DNA–DNA homology they are considered to be related at the species level (Wayne *et al.*, 1987). However, in the absence of differential phenotypic or chemotaxonomic characteristics between strains, which exhibit less than 70% DNA–DNA homology, genomic species or genomospecies have been defined instead of a new species being described (Schloter *et al.*, 2000).

The basic premise for the assessment of the genetic diversity of any organism is to establish a taxonomic structure from which a nomenclature and classification system for the organism can be generated. The classification system thus generated can then be used to identify the organism and facilitate the prediction of the properties of new isolates, which will hopefully, in the case of plant pathogens, include plant pathogenicity. This improved taxonomy of plant-pathogenic bacteria aids in

Plant Microbiology, Michael Gillings and Andrew Holmes

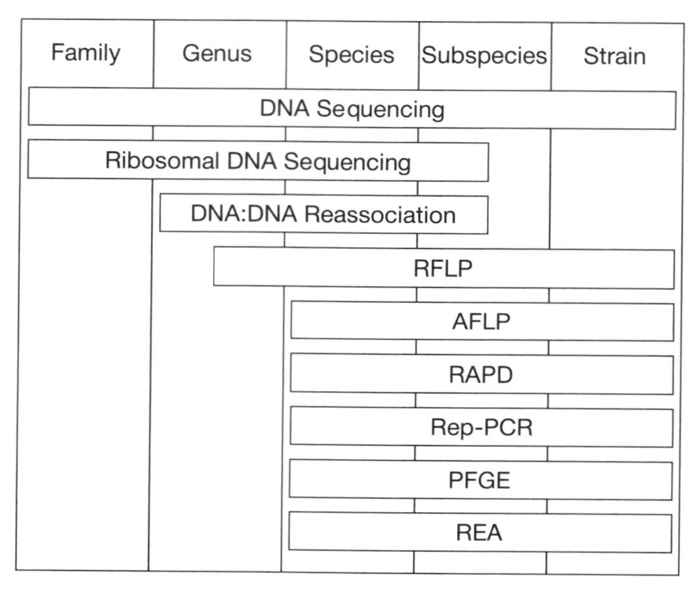

Figure 10.1. Capacity of DNA-based genetic diversity assessment methods to resolve different taxonomic levels of bacteria (adapted from Louws *et al.*, 1999 with permission, from the *Annual Review of Phythopathology*, Volume 37 © 1999 by Annual Reviews www.annualreviews.org)

the development of targeted diagnostic tests, permits the definition of subspecific groups for use in the development of quarantine regulations and is useful in the study of the epidemiology and ecology of the organisms and the study of population genetics and evolution.

Most commonly the term 'genetic diversity' is used to indicate the diversity of an organism below the species level. This infrasubspecific diversity is assessed by the use of one or more of the many high-resolution genomic fingerprinting techniques available. These techniques are based upon PCR amplification, for example arbitrarily primed PCR, or restriction digestion of either total genomic DNA or PCR-amplified genomic fragments. Sequence analysis of selected areas of the bacterial genome is also useful in the assessment of genetic diversity at the infrasubspecific level.

This chapter will assess the diversity of phytopathogenic prokaryotes beginning with a discussion of the phylogenetic diversity of plant-pathogenic bacteria followed by a description of aspects of genomic diversity and then will describe approaches and methods for the assessment of genetic diversity of plant-pathogenic bacteria at or below the level of the species.

10.2 Phylogenetic diversity of plant-pathogenic bacteria

Most bacterial plant pathogens are Gram negative and phylogenetically belong to the class *Proteobacteria* (Stackebrandt *et al.*, 1988). Within the *Proteobacteria* the majority of pathogens belong to the β and γ subdivisions (*Table 10.1*). The most

Table 10.1. List of plant pathogenic species (updated from Young (2000))

Genus	Species
Gram negative	
Alphaproteobacteria	
Acetobacter spp.	*A. aceti, A. pasteurianus*
Sphingomonas	*S. suberifaciens*
Candidatus Liberibacter	L. asiaticum, L. africana
Agrobacterium spp.[a]	*A.. rhizogenes, A.. rubi, A. tumefaciens, A. vitis*
Betaproteobacteria	
Acidovorax spp.	*A. anthurii, A. avenae, A. konjaci*
Burkholderia spp.	*B. andropogonis, B. caryophylli, B. cepacia, B. gladioli, B. glumae, B. plantarii*
Herbaspirillum	*H. rubrisubalbicans*
Ralstonia	*R. solanacearum, P. syzygii*, The Blood Disease Bacterium
Xylophilus	*X. ampelinus*
Gammaproteobacteria	
Brenneria spp.	*B. alni, B. nigrifluens, B. paradisiaca, B. quercina, B. rubrifaciens, B. salicis*
Enterobacter spp.	*E. nimipressuralis, E. cancerogenus, E. dissolvens, E. pyrinus*
Erwinia spp.	*E. amylovora, E. psidii, E. pyrifoliae, E. rhapontici, E. tracheiphila*
Pantoea spp.	*P. agglomerans, P. ananatis, P. citrea, P. dispersa, P. stewartii*
Pectobacterium spp.	*P. cacticida, P. carotovorum, P. chrysanthemi, P. cypripedii*
Pseudomonas spp.	*P. agarici, P. amygdali, P. asplenii, P. avellanae, P. beteli*[b], *P. brassicacearum, P. cannabina, P. caricapapayae, P. cichorii, P. cissicola*[b], *P. corrugata, P. ficuserectae, P. flectens*[b], *P. fluorescens, P. hibiscicola*[b], *P. marginalis, P. savastanoi, P. syringae, P. tolaasii, P. tremae, P. viridiflava*
Xanthomonas spp.	*X. albilineans, X. arboricola, X. axonopodis, X. bromi, X. campestris, X. cassavae, X. codiaei, X. cucurbitae, X. cynarae, X. fragariae, X. hortorum, X. hyacinthi, X. melonis, X. oryzae, X. pisi, X. populi, X. saccari, X. theicola, X. translucens, X. vasicola, X. vesicatoria*
Xylella	*X. fastidiosa*
Gram positive	
Actinobacteria	
Arthrobacter	*A. ilicis*
Bacillus	*B. megaterium*
Clavibacter	*C. michigensis*
Curtobacterium	*C. flaccumfaciens*
Leifsonia	*L. xyli*
Nocardia	*N. vaccinii*
Phytoplasma	
Rathayibacter spp.	*R. iranicus, R. rathayi, R. toxicus, R. tritici*
Rhodococcus	*R. fascians*
Spiroplasma spp.	*S. kunkelli, S. citri, S. phoeniceum*
Streptomyces spp.	*S. caviscabies, S. europaeiscabiei, S. acidiscabies, S. ipomoeae, S. reticuliscabei, S. scabei, S. steliiscabiei, S. turgidiscabies, S. reticuliscabiei*

[a] Reclassification of Agrobacterium species in Rhizobium has been proposed (Young *et al.* 2001), but see also van Berkum *et al.* (2003).
[b] These *Pseudomonas* spp are misclassified (Anzai *et al.*, 2000; Young, 2000); *P. beteli* and *P. hibiscicola* belong to the *Stenotrophomonas* rRNA lineage, *P. cissicola* belongs to the *Xanthomonas* rRNA lineage and *P. flectens* belongs to the *Enterobacteriaceae* rRNA lineage (Anzai *et al.*, 2000).

important genera containing plant pathogens are *Acidovorax*, *Burkholderia*, *Ralstonia*, *Agrobacterium*, *Xanthomonas*, *Pseudomonas*, *Erwinia*, *Pectobacterium* and *Pantoea*. However, there are a number of important Gram-positive plant pathogens within the class *Actinobacteria* (Stackebrandt *et al.*, 1997) (*Table 10.1*). Within the *Actinobacteria* the most important plant pathogens are the *Streptomycetes* which cause potato scab and the subspecies of *Clavibacter michiganensis* (*Table 10.1*).

The taxonomy of plant-pathogenic bacteria has been in a state of flux since 1980 when the Approved List of bacterial names was published and many accepted names of plant pathogens were discarded (Young *et al.*, 1978). Other more recent changes have led to the reclassification of the phytopathogenic erwinias into the genera *Pantoea*, *Pectobacterium* and *Brenneria*, principally on the basis of phylogenetic analysis of the 16S rRNA gene the results of which have largely been confirmed by sequencing of other areas of the genome (Brown *et al.*, 2000; Fessehaie *et al.*, 2002). Similarly the species within the genus *Xanthomonas* have been redefined primarily on the basis of DNA–DNA hybridisation (Vauterin *et al.*, 1990, 1995, 2000).

10.3 Genomic diversity of plant-pathogenic bacteria

The variation in genome size and genome organisation of plant-pathogenic bacteria has been revealed principally by the use of pulsed-field gel electrophoresis and more recently total genome sequencing (see http://www.tigr.org/~vinita/PPwebpage.html and Van Sluys *et al.* (2002) for lists of completed and ongoing sequencing projects on bacterial plant pathogens). The genome size of plant-pathogenic bacteria varies from as little as 0.53 Mb for some phytoplasmas to a huge 8.0 Mb for some strains of *Rhodococcus fascians* (*Table 10.2*). The genome of most plant-pathogenic bacteria consists of a single circular chromosome but some phytopathogens have multiple chromosomes and some even have linear chromosomes. *Agrobacterium rubi* and *A. tumefaciens* both possess two chromosomes, one of which is linear, and a varying number of plasmids (Jumas-Bilak *et al.*, 1998) (*Table 10.2*). *R. solanacearum* has two circular replicons (Salanoubat *et al.*, 2002) both of which contain a ribosomal gene cluster and tRNA genes and therefore should both be called chromosomes, although, probably for historical reasons, the second replicon is referred to as a mega-plasmid. If plasmids are present they are primarily circular but some plasmids are linear such as those found in the Gram-positive plant pathogens *R. fascians* and *C. michiganensis* subsp. *sepedonicus* (*Table 10.2*).

The genomes of strains within a species also vary in size. Strains of *X. axonopodis* pv. *phaseoli* (including the *fuscans* variant) vary in genome size by an incredible 1.5 Mb from 2.8 Mb to 4.3 Mb (Chan and Goodwin, 1999). Strains of the plant-pathogenic mollicute *Spiroplasma citri* have been found to vary in genome size from 1.65 Mb to 1.91 Mb in nine strains tested (Ye *et al.*, 1995). The genome sizes of three strains of *P. syringae* representing three pathovars (pv. *phaseolicola*, pv. *actinidae* and pv. *syringae*) vary in genome size (Sawada *et al.*, 2002), the genomes of pv. *syringae* and pv. *phaseolicola* are approximately the same size (approximately 6 Mb), while the genome of pv. *actinidiae*, is markedly smaller (4.7 Mb) (Sawada *et al.*, 2002). However, there is a question as to the taxonomic relationships of these strains at the species level with pv. *syringae* belonging to genomospecies I (Gardan *et al.*, 1999) pv. *phaseolicola* to genomospecies II (Gardan

Table 10.2. Genome sizes and genome organisation of some plant pathogenic bacteria

Species	Genome size	Genome organisation	References
A. tumefaciens	5.9 Mb	One circular and one linear chromosome	(Jumas-Bilak *et al.*, 1998)
A. rubi	5.7 Mb	One circular and one linear chromosome	(Jumas-Bilak *et al.*, 1998)
B. gladioli	6.2 Mb	Two circular chromosomes	(Wigley and Burton, 2000)
C. michiganensis	2.5–2.64 Mb	Circular chromosome circular plasmid and linear plasmid	(Brown *et al.*, 2002a)
P. syringae pathovars	4.7–6.0 Mb	Circular chromosome	(De Ita *et al.*, 1998; Sawada *et al.*, 2002)
P. tolaasii	6.7 Mb	Circular chromosome	(Rainey *et al.*, 1993)
Phytoplasmas	0.53–1.35 Mb	Circular chromosome	(Lee *et al.*, 2000)
R. fascians	5.6–8.0 Mb	Circular chromosome and linear plasmid	(Pisabarro *et al.*, 1998)
R. solanacearum	5.8 Mb	Two chromosomes	(Salanoubat *et al.*, 2002) (Ochman, 2002)
S. citri	1.6–1.9 Mb	Circular chromosome	(Ye *et al.*, 1995)
X. cucurbitae	2.6 Mb	Circular chromosome	(Chan and Goodwin, 1999)
X. axonopodis pv. *alfalfae*	4.2 Mb	Circular chromosome	(Chan and Goodwin, 1999)
X. campestris pv. *campestris*	3.0 Mb	Circular chromosome	(Chan and Goodwin, 1999)
X. campestris pv. *vesicatoria*	3.4–3.8 Mb	Circular chromosome	(Chan and Goodwin, 1999)
X. axonopodis pv. *phaseoli*	2.8–4.3 Mb	Circular chromosome	(Chan and Goodwin, 1999)
X. oryzae	4.8 Mb	Circular chromosome	(Ochiai *et al.*, 2001)

et al., 1999) and pv. *actinidiae* to genomospecies 8 (Scortichini *et al.*, 2002) (see below for a discussion of the diversity of *P. syringae*).

With the advent of genome sequencing and comparison of whole genomes, the 'holy grail' of genetic diversity assessment, has become reality. The first comparative genomics of plant pathogens was completed by da Silva *et al.* (2002) who compared the genome sequences of *X. axonopodis* pv. *citri* and *X. campestris* pv. *campestris*. Overall the genomes of those organisms show a high degree of colinearity and share approximately 80% of the total number of genes. However, several groups of strain-specific genes were identified, a large number of which were localised in an area around the putative termini of replication.

The genomes of three different strains of *X. fastidiosa* have been or are in the process of being sequenced and compared (Bhattacharyya *et al.*, 2002; Simpson *et al.*, 2000; Van Sluys *et al.*, 2003). This sequence-based approach has shown that strains share approximately 82% of open reading frames and has identified strain-specific genomic sequences. Van Sluys *et al.* (2003) have recently reported the completed genome sequence of a second strain of *X. fastidiosa* causing Pierce's disease of grapevine. Comparison of this genome sequence to that of the genome sequence of a strain causing citrus variegated chlorosis (Simpson *et al.*, 2000) has found that the genomic differences between strains are due to phage-associated chromosomal rearrangements and deletions. The areas of genomic diversity between strains tend to be clustered into genomic islands and a large proportion of the strain-specific genes are associated with mobile genetic elements (Van Sluys *et al.*, 2003).

An alternative approach for conducting comparative genome analysis, employing microarrays, is beginning to be used for the assessment of infrasubspecific genetic diversity of micro-organisms (Joyce *et al.*, 2002). The first steps in the use of microarray profiling of phytopathogenic bacteria have recently been reported for the assessment of the diversity of *X. fastidiosa* (de Oliveira *et al.*, 2002). Many problems in the use of this technology are yet to be overcome (Joyce *et al.*, 2002). One of the major stumbling blocks is the requirement for a sequenced representative to construct microarrays with the minimal amount of within-genome cross-hybridisation (Joyce *et al.*, 2002). As the number of completed genome sequences of bacterial plant pathogens is increasing rapidly this hurdle is being overcome for many important bacterial plant pathogens.

10.4 Infrasubspecific genetic diversity of plant-pathogenic bacteria

An adequate infrasubspecific taxonomy is required for ecological and epidemiological studies of plant-pathogenic bacteria to be conducted and for targeting of resistance breeding programs.

Plant pathologists have long known that bacterial plant pathogens exhibit a great deal of pathogenic diversity below the level of the species; this pathogenic diversity has led, in the cases of many *Pseudomonas* and *Xanthomonas* spp., to the establishment of pathovars (Young *et al.*, 1992). In turn this has led to the use of a trinomial nomenclature to classify pathogens at the infrasubspecific level by employing a pathovar name below the level of the species or subspecies (Dye *et al.*, 1980). Pathovars are defined as a strain or strains with the same or similar characteristics, differentiated at the infrasubspecific level from other strains of the same species or subspecies on the basis of a single characteristic, the distinctive pathogenicity to one or more plant hosts (Dye *et al.*, 1980). Not all bacterial plant pathogens can be subgrouped into pathovars, especially organisms with large and overlapping host ranges such as *Ralstonia solanacearum* and *Erwinia chrysanthemi* (Young, 2000). Problems with the pathovar system have been identified, not the least of which is the lack of extensive host range studies being completed for most pathovars of *Xanthomonas* spp. or *Pseudomonas* spp. (Vauterin *et al.*, 2000). Many studies have tried to identify molecular markers which characterise a specific pathovar with varying levels of success.

Low levels of infrasubspecific genetic variation have been equated with a recent origin of the pathogen, limited population divergence and potentially limited host range of the pathogen (Avrova *et al.*, 2002). In contrast, high levels of infrasubspecific genetic variation has been linked with an extensive geographic distribution and/or host range (Waleron *et al.*, 2002). For example the phenotypically well-defined pathogens *Erwinia amylovora* (Zhang and Geider, 1997) and *Pantoea stewartii* (Coplin *et al.*, 2002) exhibit less genetic diversity than the less well taxonomically defined *Ralstonia solanacearum* (Poussier *et al.*, 2000a) and *Erwinia chrysanthemi* (Nassar *et al.*, 1996).

Infrasubspecific genetic diversity or microdiversity, the diversity found within distinct phenotypic or genotypic groups (Schloter *et al.*, 2000), is generally assessed by high-resolution genomic fingerprinting. Genomic fingerprinting methods assess polymorphisms which accumulate relatively rapidly within the genome (Enright and Spratt, 1999) and are of great use for assessing the genetic diversity of closely related bacterial strains to identify fine-scale, short-term, epidemiological relationships between strains. The main value of genomic fingerprinting techniques lies in the assessment of the diversity of an organism at and below the level of the species. Hence, a prerequisite for using these techniques is that a reliable taxonomy at the species level is available to allow accurate and targeted use of these molecular fingerprinting methods (Schloter *et al.*, 2000). If a precise taxonomy is not available for the 'species' then two strains of different 'species' (or genomospecies) may be fingerprinted and the resulting large genetic diversity misinterpreted as indicating large infrasubspecific diversity. Conversely, if a large genetic diversity is identified between strains using genomic fingerprinting techniques this may indicate that the species is not taxonomically well-defined and other techniques with less resolution (e.g. DNA–DNA hybridisation or sequencing of conserved genes) may be more appropriate to resolve the taxonomy of the species.

10.4.1 Methods commonly employed for the assessment of diversity of plant pathogenic bacteria

Restriction fragment length polymorphism (RFLP)

Genomic DNA is digested with a restriction enzyme and the fragments are resolved by gel electrophoresis through an agarose gel. The separated fragments are transferred to a nylon or nitrocellulose membrane by Southern blotting. The membrane-bound nucleic acid is then hybridised to labelled nucleic acid probes homologous to regions of the genome of the organism studied. The probe used may either be multicopy or single/low copy. Multicopy probes commonly used include the rRNA operon in which case the procedure is called ribotyping. Ribotyping has been successfully employed to study the genetic diversity of *B. andropogonis* (Bagsic-Opulencia *et al.*, 2001), *X. campestris* (Bragard *et al.*, 1995) and *P. syringae* pathovars (Gardan *et al.*, 1999). Other multicopy probes used include insertion sequences which have been employed to assess the diversity of *X. oryzae* pv. *oryzae* (Adhikari *et al.*, 1995; Cruz *et al.*, 1996) and *R. solanacearum* (Jeong and Timmis, 2000; Lee *et al.*, 2001). Insertion sequences have also been used to develop PCR-based methods for the assessment of genetic diversity by developing outward facing primers which amplify the intervening segments of DNA between the IS elements (George *et al.*, 1997). RFLP analysis using single/low copy number probes is not as commonly used due to the increased number of probes that have to be used and therefore a higher cost. However, low copy number probes have been employed to assess the genetic diversity of plant pathogens such as *R. solanacearum* (Cook *et al.*, 1991).

PCR – Restriction fragment length polymorphism (PCR-RFLP).
Genetic loci are amplified with specific oligonucleotide primers and the amplified product subjected to RFLP analysis; differences in the molecular weight of the fragments produced are identified by gel electrophoresis and the pattern of fragments produced used to compare strains. Any PCR product can be used in this test; the most commonly used are the genes of the rRNA operon which have been used to identify genetic diversity between strains of many plant-pathogenic bacteria including 'the soft-rot erwinias' (Helias *et al.*, 1998; Seo *et al.*, 2002; Toth *et al.*, 2001) and *P. syringae* (Manceau and Horvais, 1997). The use of restriction digestion of the 16S rRNA gene is central to the taxonomy of the phytoplasmas (Lee *et al.*, 1998). PCR-RFLP analysis of pathogenicity genes has been employed for the study of the infrasubspecific diversity of *P. chrysanthemi* (Nassar *et al.*, 1996) and *R. solanacearum* (Poussier *et al.*, 1999) and the *recA* gene has been used for the genotypic characterisation of the erwinias (Waleron *et al.*, 2002).

Pulsed-field gel electrophoresis (PFGE)
PFGE is a genomic DNA fingerprinting method, which employs rare cutting restriction endonucleases to digest the genomic DNA of bacteria which is then subjected to electrophoresis using specialised conditions for the separation of large fragments of DNA. PFGE has been found to be more discriminatory than rep-PCR (Frey *et al.*, 1996) and has been termed the gold-standard of molecular-typing methods (Olive and Bean, 1999). PFGE analysis has been useful in epidemiological studies of *Erwinia amylovora* (Jock *et al.*, 2002; Zhang and Geider, 1997) and *P. stewartii* subsp. *stewartii* (Coplin *et al.*, 2002). A low level of genetic diversity was found in *C. michiganensis* subsp. *sepedonicus* by employing PFGE and the technique was able to differentiate avirulent strains from virulent strains (Brown *et al.*, 2002b).

Analysis of *X. albilineans* by PFGE (Davis *et al.*, 1997) revealed extensive diversity which in turn correlated well with previous whole-cell protein profiles and serological groupings. PFGE analysis has allowed the retrospective epidemiological identification of the source of different introductions of *X. albilineans* into the USA (Davis *et al.*, 1997). PFGE analysis of the pathogenically distinct race 3 strains of *R. solanacearum* has revealed previously unrecognised diversity (Smith *et al.*, 1995a, 1995b).

Arbitrarily primed-PCR (AP-PCR) – Random amplified polymorphic DNA analysis (RAPD)
RAPD assays are based upon the use of short random-sequence primers generally of 10 bp in length, which hybridise to genomic DNA in conditions of low stringency and initiate the amplification of random areas of the genome. The amplification products are then resolved on an agarose gel. RAPD analysis has been employed to study the diversity of *B. andropogonis* (Bagsic-Opulencia *et al.*, 2001), *E. amylovora* (Brennan *et al.*, 2002), *X. fastidiosa* (Chen *et al.*, 1995; Coletta-Filho and Machado, 2002; da Costa *et al.*, 2000; Hendson *et al.*, 2001; Lacava *et al.*, 2001), *P. syringae* (Clerc *et al.*, 1998), *Xanthomonas* sp. (Goncalves and Rosato, 2000), *X. oryzae* (Gupta *et al.*, 2001), *X. fragariae* (Pooler *et al.*, 1996), *X. campestris* (Smith *et al.*, 1994), *R. solanacearum* (Thwaites *et al.*, 1999), *Erwinia carotovora* subsp. *atroseptica* (Toth *et al.*, 1999) and *X. cynarae* (Trébaol *et al.*, 2001). Although RAPD

analysis is useful in identifying infrasubspecific genetic diversity it does suffer from a lack of reproducibility.

Repetitive element PCR (rep-PCR)

Rep-PCR is quickly becoming the most widely used method for the assessment of genetic diversity of bacteria, particularly plant-pathogenic bacteria. This genomic fingerprinting technique employs primers designed to hybridise to repetitive elements (ERIC, REP and BOX) within the genomes of bacteria and amplifies the intervening regions between these elements. These repetitive elements may play an important role in the organisation of the bacterial genome and may therefore give an indication of the structure and evolution of the bacterial genome (Lupski and Weinstock, 1992). The different primers used have been shown to reveal different levels of diversity each providing unique information (Cruz *et al.*, 1996; Louws *et al.*, 1995). For example, in the study of *X. oryzae* pv. *oryzae* the BOXA1R primer detected the least polymorphism and the REP primer pair the most (Cruz *et al.*, 1996). It has been questioned whether the rep-PCR primers are hybridising to repeat elements in the bacterial genome or are non-specifically hybridising to regions of the bacterial genome similar to AP-PCR (Gillings and Holley, 1997). However, irrespective of the regions to which the primers hybridise this technique is more reproducible than AP-PCR techniques.

Rep-PCR has been extensively applied to the assessment of diversity of *Xanthomonas* spp. (Adhikari *et al.*, 1999; Bouzar *et al.*, 1999; Cruz *et al.*, 1996; Goncalves and Rosato, 2000; Louws *et al.*, 1994; McDonald and Wong, 2001; Pooler *et al.*, 1996; Restrepo *et al.*, 2000; Sulzinski *et al.*, 1995, 1996; Vauterin *et al.*, 2000; Zhao *et al.*, 2000) and to differentiate the different pathovars of *X. populi* (McDonald and Wong, 2001). In studying the diversity of *P. syringae* this technique is able to identify the genomospecies to which a strain belongs (Marques *et al.*, 2000) and has been useful in the identification of races of *P. syringae* pv. *pisi* found in Australia (Hollaway *et al.*, 1997). However the taxonomic resolution of rep-PCR is only useful for identifying closely related strains of *P. syringae* as the differences in fingerprint patterns between more distantly related strains is too great to allow conclusions to be drawn on common ancestry (Weingart and Volksch, 1997).

Amplified fragment length polymorphism (AFLP).

The AFLP technique involves restriction of genomic DNA using two restriction endonucleases followed by ligation of double-stranded adaptors specific for each restriction endonuclease used and then amplification using the primers specific for the adaptors. The primers used for amplification include additional (to the adaptor sequence) nucleotides at the 3′ end of the primer and therefore they amplify a subset of the bacterial genome. AFLP is reported to have a greater discriminatory power than PFGE (Mougel *et al.*, 2001). Similar to other genetic fingerprinting techniques, this technique is not useful for identifying relationships between taxa that are not closely related (Avrova *et al.*, 2002; Poussier *et al.*, 2000b) and is not informative at the taxonomic level of the genus or family (Rademaker *et al.*, 2000; Savelkoul *et al.*, 1999). However, this technique is very good at discriminating closely related bacterial strains. AFLP has been employed to assess genetic diversity of *E. carotovora* and *E. chrysanthemi* (Avrova *et al.*, 2002), where it proved useful in grouping strains into species and subspecies groups and allowed the identification of unclassified strains. The technique also proved useful for the identification of diversity within *E. carotovora* subsp. *atroseptica* for epidemiological studies and for the identification of specific amplicons for development of molecular diagnostic tests (Avrova *et al.*, 2002). AFLP analysis of *Xanthomonas axonopodis* pv. *manihotis* (Gonzalez *et al.*, 2002) allowed characterisation at the pathovar and infrapathovar level. Within *P. syringae* genomospecies III intrapathovar diversity has been identified using this technique (Clerc *et al.*, 1998).

Gene sequencing.
Longer term, global epidemiological questions can be answered by the sequencing of coding regions of the genome as the variation between strains accumulates more slowly than the variation identified by genomic fingerprinting techniques (van Belkum *et al.*, 2001). A major advantage of a sequence-based approach is that it allows direct comparison between studies whereas comparisons between studies employing genomic fingerprinting approaches are generally not possible as the data are not portable or available on a global basis (Clarke, 2002).

R. solanacearum (Fegan and Prior, 2002; Fegan *et al.*, 1998a; Poussier *et al.*, 2000a) and *P. syringae* (Sawada *et al.*, 1997, 1999, 2002) are the most extensively studied phytopathogenic bacteria using analysis of gene-sequencing data. The level of genetic diversity revealed between strains depends on the genomic region sequenced. Within the rRNA operon the 16S and 23S rRNA genes reveal the least infrasubspecific diversity and the internal transcribed spacer (ITS) region between the 16S–23S rRNA genes the greatest. The ITS region is not under pressure to conserve its sequence as it is non-coding, although tRNA genes do occur in many Gram-negative organisms (Barry *et al.*, 1991; Gürtler and Stanisich, 1996). Therefore, the ITS region reveals greater variation than the rRNA genes themselves (Barry *et al.*, 1991; Gürtler and Stanisich, 1996). The ITS region has been used to assess the diversity of *Xanthomonas* sp. (Goncalves and Rosato, 2002), *R. solanacearum* (Fegan *et al.*, 1998a) and *Erwinia* sp. (Fessehaie *et al.*, 2002). Sequence analysis of protein coding genes, which tend to accumulate mutations at a faster rate than the rRNA operon, has been found to be of great use for the identification of infrasubspecific genetic diversity. In the case of *R. solanacearum* and its close relatives the 16S rRNA gene revealed two major groups of *R. solanacearum* each of which could in turn be divided into two subgroups (Poussier *et al.*, 2000b; Taghavi *et al.*, 1996). Sequence analysis of the ITS region identified the same grouping of strains but was able to resolve the two subgroups more effectively (Fegan *et al.*, 1998a). Finer resolution within these subgroups has been achieved by using the pathogenicity-related genes, the endoglucanase gene and the *hrpB* gene (Poussier *et al.*, 2000a).

10.5 Assessment of genetic diversity for clarifying infrasubspecific taxonomic relationships; the case study of *P. syringae*

In complex species encompassing a large degree of genetic diversity it is important to develop a taxonomic framework below the level of the species to allow accurate identification of strains.

P. syringae is a very diverse species being comprised of more than 50 pathovars (Rudolph, 1995). Overall the subspecific taxonomy of *P. syringae* into pathovars is complicated and makes identification of strains difficult (Rudolph, 1995; Young *et al.*, 1992). DNA–DNA hybridisation studies conducted by Gardan *et al.* (1999) identified nine DNA-hybridisation groups or genomospecies. Genomospecies 1 corresponds to *P. syringae sensu stricto*, a list of the *P. syringae* pathovars and *Pseudomonas* sp. comprising the genomovars and the proposed nomenclature is presented in *Table 10.3*. This improved taxonomy will allow accurate taxonomic classification and will present opportunities to reduce the potential for spread of pathogens internationally (Stead *et al.*, 2002).

The study of *P. syringae* by DNA–DNA hybridisation has uncovered some problems in the pathovar naming system (Gardan *et al.*, 1999). In some cases different strains of the same pathovar belong to different genomovars. Different strains of *P. syringae* pathovars *morsprunorum* and *lachrymans* belong to genomospecies 2 and 3 and pathovars *ribicola* and *primulae* are found in genomospecies 3 and 6 (Gardan *et al.*, 1999). A study employing phylogenetic analysis of sequence data from four genes (*gyrB*, *rpoD*, *hrpL* and *HrpS*) by Sawada *et al.* (1999) also found that different strains of *P. syringae* pathovars *morsprunorum* and *lachrymans* were polyphyletic and therefore some strains may be incorrectly placed in these pathovars (Gardan *et al.*, 1999; Sawada *et al.*, 1999).

Wide host range normally equates with greater genetic diversity (Louws *et al.*, 1994). Genetic diversity is greater among *P. syringae* pathovars with a wide host

Table 10.3. Genomospecies of *P. syringae* and related fluorescent *Pseudomonas* sp.

Genomospecies	Bacterial species or *P. syringae* pathovar	Proposed nomenclature
1	*P. syringae* pathovars *syringae*, *aptata*, *lapsa*, *papulans*, *pisi*, *atrofaciens*, *aceris*, *panici*, *dysoxyli* and *japonica*	*P. syringae*
2	*Pseudomonas savastanoi*, *Pseudomonas ficuserectae*, *Pseudomonas meliae*, *Pseudomonas amygdali* and *P. syringae* pathovars *phaseolicola*, *ulmi*, *mori*, *lachrymans*[a], *sesami*, *tabaci*, *morsprunorum*[a], *glycinea*, *ciccaronei*, *eriobotryae*, *mellea*, *aesculi*, *hibisci*, *myricae*, *photiniae* and *dendropanacis*	*P. amygdali*
3	*P. syringae* pathovars *tomato*, *persicae*, *antirrhini*, *maculicola*, *viburni*, *berberidis*, *apii*, *delphinii*, *passiflorae*, *morsprunorum*[a], *lachrymans*[a], *philadelphi*, *ribicola*[a] and *primulae*[a]	Unnamed
4	'*P. coronafaciens*' and *P. syringae* pathovars *porri*, *garcae*, *striafaciens*, *atropurpurea*, *oryzae* and *zizaniae*	*P. coronafaciens*
5	*P. syringae* pathovar *tremae*	*P. tremae*
6	*Pseudomonas viridiflava* and *P. syringae* pathovars *ribicola*[a] and *primulae*[a]	*P. viridiflava*
7	*P. syringae* pathovars *tagetis* and *helianthi*	Unnamed
8	*Pseudomonas avellanae* and *P. syringae* pathovars *theae* and *actinidae*	*P. avellanae*
9	*P. syringae* pathovar *cannabina*	*P. cannabina*

[a] Strains of these pathovars are found in two genomospecies.

range than those with a more restricted host range (Denny *et al.*, 1988). Strains of the pathogenically diverse *P. syringae* pv. *maculicola* exhibit greater genetic diversity than the closely related but less pathogenically diverse pathovar *P. syringae* pv. *tomato* (Zhao *et al.*, 2000). Sawada *et al.* (1999) found that various strains of the pathovar *syringae* were polyphyletic which is confirmed by other genetic diversity studies on strains of pv. *syringae* which have shown great diversity within this pathovar (Legard *et al.*, 1993; Sundin *et al.*, 1994; Weingart and Volksch, 1997). This may reflect the wide host range of *P. syringae* pv. *syringae* as it is able to cause disease in over 180 plant species in many unrelated genera of plants (Bradbury, 1986) or it may reflect that there is pathogenic specialisation of strains collectively referred to as a single pathovar (Weingart and Volksch, 1997). Indeed the strains classified as pathovar *syringae* may represent different pathovars especially as many strains of *P. syringae* have been placed in this pathovar without establishing the host range of the strains (Young, 1991). Genetic diversity studies have enabled the identification of pathogenic specialisation within this heterogeneous pathovar. *P. syringae* pv. *syringae* strains which cause disease in bean form a genetically distinct grouping (Legard *et al.*, 1993) as do strains infecting stone fruit (Little *et al.*, 1998).

Identification of *P. syringae* strains to the pathovar level has posed serious practical problems primarily due to the difficulties in carrying out the host range tests required and verification of the pathogenicity of strains on a standard set of host plants is rarely completed (Morris *et al.*, 2000). Although many researchers have attempted to identify genetic markers which will allow identification of *P. syringae* strains to the pathovar level this has only rarely been successful (Louws *et al.*, 1994; Weingart and Volksch, 1997). However, genetic diversity studies have been used as an aid to identify new pathovars of *P. syringae* (Cintas *et al.*, 2002) or to ascribe outbreaks of disease to previously identified pathovars (Koike *et al.*, 1999; Morris *et al.*, 2000). All strains of a new pathovar, *P. syringae* pv. *alisalensis* which is pathogenic for broccoli and broccoli raab were found to have the same rep-PCR profile which varied from the other *P. syringae* pathovars tested (Cintas *et al.*, 2002). However, phenotypically *P. syringae* pv. *alisalensis* belongs to genomospecies 3 but the authors failed to include other genomospecies 3 strains in the study. In identifying the cause of bacterial blight of leeks in California Koike *et al.* (1999) used rep-PCR and sequencing of the ITS region to identify the pathogen as *P. syringae* pv. *porri*.

Below the level of the pathovar there have been attempts to relate race grouping of strains to genomic fingerprints, but, in most cases this has proven to be impossible. No correlation was found between races of pv. *phaseolicola* and the genetic fingerprint produced using a ribotyping protocol save for race 2, of which only two strains were studied (González *et al.*, 2000). Using RAPD genomic fingerprinting pv. *phaseolicola* could be differentiated into two clusters of strains cluster 1 representing races 1, 5, 7 and 9 and cluster 2 representing races 2, 3, 4, 6 and 8 (Marques *et al.*, 2000). Strains of *P. syringae* pv. *pisi* from Australia representing races 2, 3 and 6 could be identified by using rep-PCR and genetic diversity within races was also identified (Hollaway *et al.*, 1997). However, strains representing races 0 and 1 of pathovar *tomato* were indistinguishable by either AFLP or RAPD techniques (Clerc *et al.*, 1998).

10.6 Genetic diversity and development of molecular diagnostics

Molecular diagnostic tests seek to identify an unknown organism by assigning it to a known taxonomic group by the use of molecular techniques. Knowledge of the diversity of a pathogen is therefore central to the development of targeted diagnostic tests to detect phytopathogenic bacteria at various taxonomic levels. For example, *R. solanacearum* race 3 which causes brown rot of potato is an important quarantine pathogen in Europe (Elphinstone *et al.*, 2000) and has been identified to belong to two closely related clonal lineages (Cook and Sequeira, 1994; Cook *et al.*, 1989). A molecular diagnostic test has been developed to identify *R. solanacearum* race 3 strains (Fegan *et al.*, 1998b).

An assessment of the diversity of species will also help in the choice of the genomic region to target for the development of a molecular diagnostic test. If an organism is genetically very diverse then a conserved area of the genome such as the 16S rRNA gene will need to be targeted. Such is the case for *R. solanacearum* where a primer pair based upon the 16S rRNA gene has proven useful in the detection of this pathogen (Seal *et al.*, 1993).

An understanding of the genetic diversity of a species has proven useful in the selection of strains for a subtractive hybridisation approach to identify genomic DNA fragments to which diagnostic oligonucleotide primer pairs can be developed (Prior and Fegan, 2002). Prior and Fegan (2002) and Woo and Fegan (unpublished) have recently used the phylogenetic relationships of strains of *R. solanacearum*, revealed by sequence analysis of the endoglucanase gene, to choose isolates for a subtractive hybridisation approach to identify specific markers for race 2 strains of *R. solanacearum*. The approach was successful in identifying specific markers for two clonal groups of strains.

Molecular markers identified by PCR-based genomic fingerprinting methods have also been used directly for the development of molecular diagnostic tests. Cloned and sequenced RAPD fragments identified as being unique to the organism of interest have been used to develop specific PCR detection methods (Catara *et al.*, 2000; Opina *et al.*, 1995; Pooler and Hartung, 1995; Pooler *et al.*, 1996; Toth *et al.*, 1998; Trébaol *et al.*, 2001) as have cloned rep-PCR fragments (Sulzinski *et al.*, 1996; Tegli *et al.*, 2002). Although these methods have proven useful for the development of diagnostic tests, the long-term reliability of the tests is unknown because the genomic DNA fragment on which the tests are based is of unknown variability (Louws *et al.*, 1999).

Insertion sequences commonly used to assess diversity in phytopathogenic bacteria have also been used to develop molecular diagnostic tests (Lee *et al.*, 1997, 2001). However, because insertion sequences are mobile genetic elements the development of PCR-based assays to detect organisms is not advisable.

10.7 Pathogen populations: deployment of resistance

In the fight against plant disease, breeding for resistance has taken centre stage. However, many resistances that have been bred into crops have broken down as the resistant varieties deployed do not provide protection against all variants of a pathogen, or in pathogens with a high genetic diversity new variants have emerged leading to a breakdown in resistant varieties (Leung *et al.*, 1993). Changes in race structure of a pathogen in a particular geographic location may be a result of genetic change within the pathogen population (mutation and recombination) or migration from other geographic areas (Leung *et al.*, 1993). Assessment of the genetic diversity of a pathogen population helps us understand population structures, from the level of the field to the global situation, and how pathogen populations evolve.

The bacterial plant pathogen most intensively studied at the level of pathogen populations is *X. oryzae* pv. *oryzae*. *X. oryzae* pv. *oryzae* causes bacterial blight of rice (Mew *et al.*, 1993). The population structure of this important pathogen has been assessed using RFLP analysis employing repetitive probes based upon insertion sequences and avirulence genes (Adhikari *et al.*, 1995), rep-PCR (Cruz *et al.*, 1996) and RAPDs (Gupta *et al.*, 2001). In comparing *X. oryzae* pv. *oryzae* strains collected from eight Asian countries Adhikari *et al.* (1995) concluded that regionally defined pathogen populations are distinct and that this probably results from slow migration or dispersal of pathogen populations or the spatial partitioning of the host genotypes with which pathogen populations are associated. Movement of rice cultivars is restricted due to political boundaries or local preference for different rice varieties. However, a cluster of strains was identified which was comprised of strains from all countries indicating that some movement of strains has occurred. Adhikari *et al.* (1999) also found that in Nepal certain haplotypes were found in different locations indicating that there may be migration of *X. oryzae* pv. *oryzae*. This was linked to the widespread cultivation of a particular variety (Mansuli) throughout Nepal.

New pathogenic variation (pathotypes or races) have been identified in *X. oryzae* pv. *oryzae* by inoculating strains representing different lineages but belonging to the same pathotype onto hosts with previously untested resistance genes (Nelson *et al.*, 1994). A similar approach has been used to identify new pathotypes of *X. axonopodis* pv. *manihotis* (Restrepo *et al.*, 2000).

Breakdown of resistance to *R. solanacearum* in tomato can be location specific (Hanson *et al.*, 1996). However, it is unknown if this is due to the genetic diversity of the pathogen population or due to differences in environmental variables in these different locations. Unlike *X. oryzae* pv. *oryzae*, *R. solanacearum* does not have a well-defined pathotype/race structure based upon the reaction of strains to differential cultivars. Hanson *et al.* (1996) reported that tomato accessions resistant to bacterial wilt in Taiwan and Malaysia are susceptible in Indonesia and the Philippines. In Indonesia strains of *R. solanacearum* phylotype IV, which are found only in Indonesia, cause bacterial wilt of tomato (Fegan, unpublished results) which may account for the breakdown in resistance.

Attempts have also been made to link the aggressiveness of isolates of *R. solanacearum* and *X. oryzae* pv. *oryzae* to genetic diversity of strains. However, no association has been found between the genetic grouping of strains and aggressiveness to a set of differential cultivars. This is not surprising as the nature of the plant–pathogen interaction is complex and the methods used to define aggressiveness of isolates are very subjective (Darrasse *et al.*, 1998; Jaunet and Wang, 1999; Mundt *et al.*, 2002).

10.8 The use of genetic fingerprinting in epidemiology

DNA fingerprinting plays a central role in the analysis of the spread and persistence of pathogenic bacteria in the environment. Genetic fingerprinting of *Xanthomonas campestris* pv. *mangiferaeindicae* by RFLP analysis identified a clone that has been widely disseminated, potentially on planting material (Gagnevin *et al.*, 1997). Similarly by using PFGE to assess the genetic diversity of *E. amylovora* the long-distance spread of the pathogen on the European continent has been traced (Jock *et al.*, 2002; Zhang and Geider, 1997).

R. solanacearum race 3/biovar 2 strains belong to two closely related clonal groups (Cook and Sequeira, 1994; Cook *et al.*, 1989). By comparison of a worldwide collection of strains of *R. solanacearum* race 3/biovar 2 using restriction endonuclease analysis of total genomic DNA Gillings and Fahy (1993, 1994) were able to show that one of these clonal lineages has been spread worldwide probably on latently infected planting material. Short-distance movement of *Xanthomonas campestris* pv. *mangiferaeindicae* has been traced by RFLP analysis employing an insertion sequence as a probe. One haplotype of *Xanthomonas campestris* pv. *mangiferaeindicae* was found to be spread from a single focus a distance of 250 m into an uninfected orchard following tropical storms.

10.9 The nature of genetic diversity

Strains of a bacterial species may diverge from each other by acquisition or loss of mobile genetic elements, by point mutation, or by insertions, deletions or inversions. All of these mechanisms contribute to the genetic diversity and genome plasticity of pathogenic bacteria (Brown *et al.*, 2001; Dobrindt and Hacker, 2001). Analysis of fully sequenced bacterial genomes indicates that horizontal gene transfer (HGT) has had a major impact on the genetic diversity of different bacterial species and strains within a species (Bhattacharyya *et al.*, 2002; Salanoubat *et al.*, 2002; Simpson *et al.*, 2000; Van Sluys *et al.*, 2003). The genomes of bacteria are thought to be comprised of a core genome and a set of strain-specific genes (Dobrindt and Hacker, 2001; Lan and Reeves, 2000). These strain-specific genes are commonly found clustered together on genomic islands associated with mobile genetic elements and are considered to be acquired via HGT (Dobrindt and Hacker, 2001; Van Sluys *et al.*, 2003). The further study of genome sequences of plant-pathogenic bacteria will help in the identification of the role of HGT in the evolution of the pathogen genome and the contribution of HGT to population structures.

Mobile genetic elements (insertion sequences, bacteriophage, transposons, etc.) play a major role in producing the genetic variability identified by genomic fingerprinting techniques (Gurtler and Mayall, 2001). Direct evidence of this has recently been identified in the human pathogen *Escherichia coli*. The polymorphisms identified by PFGE analysis of *E. coli* O157 are not due to point mutations resulting in the generation or abolition of restriction sites but are due to the presence or absence of discrete DNA segments containing the individual restriction sites (Kudva *et al.*, 2002).

The extent to which HGT and recombination occurs in bacterial populations determines if a bacterial population is clonal, weakly clonal or non-clonal (Spratt and Maiden, 1999). The level of recombination will impact on the choice of the technique used to answer short-term (e.g., tracking the spread of a pathogen during an outbreak) and long-term (e.g., tracking global spread) epidemiological questions. Genomic fingerprinting techniques will be of use for studying the short-term epidemiology irrespective of the level of clonality of a population. However, for a highly clonal population, genomic fingerprinting techniques will also be of use for studying longer-term epidemiological questions whereas gene-sequencing approaches will be of less use. In a weakly clonal population gene-sequencing approaches will be of more use in uncovering the longer-term epidemiology of the population. In non-clonal populations the longer-term epidemiological question may be impossible to identify with any technique (Spratt and Maiden, 1999).

10.10 Conclusions

Plant pathogenesis has arisen in phylogenetically and genetically diverse bacterial species. An understanding of the genetic diversity of plant-pathogenic bacteria from the level of the species to the infrasubspecific level is necessary for epidemiological and ecological studies, the development of targeted diagnostic tests, the definition of subspecific groups for use in the development of quarantine regulations and the study of population genetics and evolution.

Techniques varying in their taxonomic resolution from gene sequencing and DNA–DNA hybridisation to genomic fingerprinting methods and the whole genome approaches of genome-sequencing and microarray technologies have been employed to identify genetic diversity between strains of a pathogen. Irrespective of the methodology employed to assess the genetic diversity of an organism it is important to firstly taxonomically define the organism. Taxonomic subgrouping of strains like the genomospecies of *P. syringae* (Gardan *et al.*, 1999) allow the comparison of meaningful groups of strains by genetic fingerprinting techniques. Taxonomic subgrouping of strains also allows the more logical description of an organism and the use of this description in the identification of unknowns. For example, naming and identification of pathovars of *P. syringae* will be easier due to the genomospecies scheme. This scheme can be used to initially identify the genomospecies to which an unknown belongs followed by the comparison of the genetic diversity of the 'new' pathovar to closely related relatives by the use of genomic fingerprinting techniques. If the 'new' pathovar is different to other closely related pathovars it could be described as a 'new' pathovar after the appropriate pathogenicity tests have been conducted.

The availability of genome-scanning techniques, such as microarray analysis, will allow the identification of strain-specific loci which will in turn allow us to identify genetic diversity of pathogens. However, more importantly microarray techniques will also allow us to understand the biological significance of the genetic variation which has been observed (Joyce *et al.*, 2002). Within the next few years the use of microarray technology will undoubtedly revolutionise our understanding of the genetic diversity and evolution of plant pathogenic bacteria.

References

Adhikari, T.B., Cruz, C.M.V., Zhang, Q., Nelson, R.J., Skinner, D.Z., Mew, T.W., Leach, J.E. (1995) Genetic diversity of *Xanthomonas oryzae* pv. *oryzae* in Asia. *Appl. Environ. Microbiol.* **61**: 966–971.

Adhikari, T.B., Mew, T.W., Leach, J.E. (1999) Genotypic and pathotypic diversity in *Xanthomonas oryzae* pv. *oryzae* in Nepal. *Phytopathology* **89**: 687–694.

Anzai, Y., Kim, H., Park, J.-Y., Wakabayashi, H., Oyazu, H. (2000) Phylogenetic affiliation of the pseudomonads based on 16S rRNA sequence. *Int. J. System. Evolution. Microbiol.* **50**: 1563–1589.

Avrova, A.O., Hyman, L.J., Toth, R.L., Toth, I.K. (2002) Application of amplified fragment length polymorphism fingerprinting for taxonomy and identification of the soft rot bacteria *Erwinia carotovora* and *Erwinia chrysanthemi*. *Appl. Environ. Microbiol.* **68**: 1499–1508.

Bagsic-Opulencia, R.D., Hayward, A.C., Fegan, M. (2001) Use of ribotyping and random amplified polymorphic DNA to differentiate isolates of *Burkholderia andropogonis*. *J. Appl. Microbiol.* **91**: 686–696.

Barry, T., Colleran, G., Glennon, M., Dunican, L.K., Gannon, F. (1991) The 16S/23S ribosomal spacer region as a target for DNA probes to identify eubacteria. *PCR Meths. Appls.* **1**: 51–56.

Bhattacharyya, A., Stilwagen, S., Ivanova, N. *et al.* (2002) Whole-genome comparative analysis of three phytopathogenic *Xylella fastidiosa* strains. *Proc. Natl Acad. Sci. USA* **99**: 12403–12408.

Bouzar, H., Jones, J.B., Stall, R.E., Louws, F.J., Schneider, M., Rademaker, J.L.W., de Bruijn, F.J., Jackson, L.E. (1999) Multiphasic analysis of xanthomonads causing bacterial spot disease on tomato and pepper in the Caribbean and Central America: Evidence for common lineages within and between countries. *Phytopathology* **89**: 328–335.

Bradbury, J.F. (1986) *Pseudomonas syringae* pv. *syringae*. In: Bradbury, J.F. *Guide to Plant Pathogenic Bacteria*, pp. 175–177. CAB International Mycological Institute: Kew, England.

Bragard, C., Verdier, V., Maraite, H. (1995) Genetic diversity among *Xanthomonas campestris* strains pathogenic for small grains. *Appl. Environ. Microbiol.* **61**: 1020–1026.

Brennan, J.M., Doohan, F.M., Egan, D., Scanlan, H., Hayes, D. (2002) Characterization and differentiation of Irish *Erwinia amylovora* isolates. *J. Phytopathol.* **150**: 414–422.

Brown, E.W., Davis, R.M., Gouk, C., van der Zwet, T. (2000) Phylogenetic relationships of necrogenic *Erwinia* and *Brenneria* species as revealed by glyceraldehyde-3-phosphate dehydrogenase gene sequences. *Int. J. Syst. Evol. Microbiol.* **50**: 2057–2068.

Brown, E.W., LeClerc, J.E., Kotewicz, M.L., Cebula, T.A. (2001) Three R's of bacterial evolution: How replication, repair, and recombination frame the origin of species. *Environ. Mol. Mutag.* **38**: 248–260.

Brown, S.E., Knudson, D.L., Ishimaru, C.A. (2002a) Linear plasmid in the genome of *Clavibacter michiganensis* subsp. *sepedonicus*. *J. Bacteriol.* **184**: 2841–2844.

Brown, S.E., Reilley, A.A., Knudson, D.L., Ishimaru, C.A. (2002b) Genomic fingerprinting of virulent and avirulent strains of *Clavibacter michiganensis* subspecies *sepedonicus*. *Curr. Microbiol.* **44**: 112–119.

Catara, V., Arnold, D., Cirvilleri, G., Vivian, A. (2000) Specific oligonucleotide primers for the rapid identification and detection of the agent of tomato pith necrosis, *Pseudomonas corrugata*, by PCR amplification: evidence for two distinct genomic groups. *Eur. J. Plant Pathol.* **106**: 753–762.

Chan, J.W.Y.F., Goodwin, P.H. (1999) Differentiation of *Xanthomonas campestris* pv. *phaseoli* from *Xanthomonas campestris* pv. *phaseoli* var. *fuscans* by PFGE and RFLP. *Eur. J. Plant Pathol.* **105**: 867–878.

Chen, J., Lamikanra, O., Chang, C.J., Hopkins, D.L. (1995) Randomly amplified polymorphic DNA analysis of *Xylella fastidiosa* Pierce's disease and oak leaf scorch pathotypes. *Appl. Environ. Microbiol.* **61**: 1688–1690.

Cintas, N.A., Koike, S.T., Bull, C.T. (2002) A new pathovar, *Pseudomonas syringae* pv. *alisalensis* pv. nov., proposed for the causal agent of bacterial blight of broccoli and broccoli raab. *Plant Disease* **86**: 992–998.

Clarke, S.C. (2002) Nucleotide sequence-based typing of bacteria and the impact of automation. *Bioessays* **24**: 858–862.

Clerc, A., Manceau, C., Nesme, X. (1998) Comparison of randomly amplified polymorphic DNA with amplified fragment length polymorphism to assess genetic diversity and genetic relatedness within genospecies III of *Pseudomonas syringae*. *Appl. Environ. Microbiol.* **64**: 1180–1187.

Coletta-Filho, H.D., Machado, M.A. (2002) Evaluation of the genetic structure of *Xylella fastidiosa* populations from different *Citrus Sinensis* varieties. *Appl. Environ. Microbiol.* **68**: 3731–3736.

Cook, D., Sequeira, L. (1994) Strain differentiation of *Pseudomonas solanacearum* by molecular genetic methods. In: Hayward, A.C., Hartman, G.L. (eds) *Bacterial Wilt: The Disease and its Causative Agent*, Pseudomonas solanacearum, pp. 77–93. CAB International: Wallingford, UK.

Cook, D., Barlow, E., Sequeira, L. (1989) Genetic diversity of *Pseudomonas solanacearum*: detection of restriction fragment polymorphisms with DNA probes that specify virulence and hypersensitive response. *Mol. Plant–Microbe Interact.* **2**: 113–121.

Cook, D., Barlow, E., Sequeira, L. (1991) DNA probes as tools for the study of host-pathogen evolution: the example of *Pseudomonas solanacearum*. In: Hennecke, H., Verma, D.P.S. (eds) *Advances in Molecular Genetics of Plant–Microbe Interactions*, pp. 103–108. Kluywer Academic Publishers: Dordrecht.

Coplin, D.L., Majerczak, D.R., Zhang, Y.X., Kim, W.S., Jock, S., Geider, K. (2002) Identification of *Pantoea stewartii* subsp *stewartii* by PCR and strain differentiation by PFGE. *Plant Disease* **86**: 304–311.

Cruz, C.M.V., Ardales, E.Y., Skinner, D.Z., Talag, J., Nelson, R.J., Louws, F.J., Leung, H., Mew, T.W., Leach, J.E. (1996) Measurement of haplotypic variation in *Xanthomonas oryzae* within a single field by rep-PCR and RFLP analyses. *Phytopathology* **86**: 1352–1359.

da Costa, P.I., Franco, C.F., Miranda, V.S., Teixeira, D.C., Hartung, J.S. (2000) Strains of *Xylella fastidiosa* rapidly distinguished by arbitrarily primed-PCR. *Curr. Microbiol.* **40**: 279–282.

da Silva, A.C.R., Ferro, J.A., Reinach, F.C. *et al.* (2002) Comparison of the genomes of two *Xanthomonas* pathogens with differing host specificities. *Nature* **417**: 459–463.

Darrasse, A., Trigalet, A., Prior, P. (1998) Correlation of aggressiveness with genomic variation in *Ralstonia solanacearum* race 1. In: Prior, P., Allen, C., Elphinstone, J. (eds) *Bacterial Wilt Disease: Molecular and Ecological Aspects*, pp. 89–98. INRA Editions: Paris, France.

Davis, M.J., Rott, P., Warmuth, C.J., Chatenet, M., Baudin, P. (1997) Intraspecific genomic variation within *Xanthomonas albilineans*, the sugarcane leaf scald pathogen. *Phytopathology* **87**: 316–324.

De Ita, M.E., Marsch-Moreno, R., Guzmán, P., Alvarez-Morales, A. (1998) Physical map of the chromosome of the phytopathogenic bacterium *Pseudomonas syringae* pv. *phaseolicola*. *Microbiology* **144**: 493–501.

de Oliveira, R.C., Yanai, G.M., Muto, N.H., Leite, D.B., de Souza, A.A., Coletta Filho, H.D., Machado, M.A., Nunes, L.R. (2002) Competitive hybridization on spotted microarrays as a tool to conduct comparative genomic analyses of *Xylella fastidiosa* strains. *FEMS Microbiol. Letts.* **216**: 15–21.

Denny, T.P., Gilmour, M.N., Selander, R.K. (1988) Genetic diversity and relationships of 2 pathovars of *Pseudomonas syringae*. *J. Gen. Microbiol.* **134**: 1949–1960.

Dobrindt, U., Hacker, J. (2001). Whole genome plasticity in pathogenic bacteria. *Curr. Opin. Microbiol.* **4**: 550–557.

Dye, D.W., Bradbury, J.F., Goto, M., Hayward, A.C., Lelliott, R.A., Schroth, M.N. (1980) International standards for naming pathovars of phytopathogenic bacteria and a list of pathovar names and pathotype strains. *Rev. Plant Pathol.* **59**: 153–168.

Elphinstone, J.G., Stead, D.E., Caffier, D. *et al.* (2000) Standardization of methods for the detection of *Ralstonia solanacearum* in potato. *EPPO Bulletin* **30**: 391–395.

Enright, M.C., Spratt, B.G. (1999) Multilocus sequence typing. *Trends Microbiol.* **7**: 482–487.

Fegan, M., Prior, P. (2002) How complex is the "Ralstonia solanacearum species complex". In: *3rd International Bacterial Wilt Symposium*, pp. 88. White River, South Africa.

Fegan, M., Taghavi, M., Sly, L.I., Hayward, A.C. (1998a) Phylogeny, diversity and molecular diagnostics of *Ralstonia solanacearum*. In: Prior, P., Allen, C., Elphinstone, J. (eds) *Bacterial Wilt Disease: Molecular and Ecological Aspects*, pp. 19–33. INRA Editions: Paris, France.

Fegan, M., Hollway, G., Hayward, A.C., Timmis, J. (1998b) Development of a diagnostic test based upon the polymerase chain reaction (PCR) to identify strains of *Ralstonia solanacearum* exhibiting the biovar 2 genotype. In: Prior, P., Allen, C., Elphinstone, J. (eds) *Bacterial Wilt Disease: Molecular and Ecological Aspects*, pp. 34–43. INRA Editions: Paris, France.

Fessehaie, A., De Boer, S.H., Lévesque, C.A. (2002) Molecular characterization of DNA encoding 16S–23S rRNA intergenic spacer regions and 16S rRNA of pectolytic *Erwinia* species. *Can. J. Microbiol.* **48**: 387–398.

Frey, P., Smith, J.J., Albar, L., Prior, P., Saddler, G.S., Trigalet-Demery, D., Trigalet, A. (1996) Bacteriocin typing of *Burkholderia* (*Pseudomonas*) *solanacearum* race 1 of the French West Indies and correlation with genomic variation of the pathogen. *Appl. Environ. Microbiol.* **62**: 473–479.

Gagnevin, L., Leach, J.E., Pruvost, O. (1997) Genomic variability of the *Xanthomonas* pathovar *mangiferaeindicae*, agent of mango bacterial black spot. *Appl. Environ. Microbiol.* **63**: 246–253.

Gardan, L., Shafik, H., Belouin, S., Broch, R., Grimont, F., Grimont, P.A.D. (1999) DNA relatedness among the pathovars of *Pseudomonas syringae* and description of *Pseudomonas tremae* sp. nov. and *Pseudomonas cannabina* sp. nov. (ex Sutic and Dowson 1959). *Int. J. Syst. Bacteriol.* **49**: 469–478.

George, M.L.C., Bustamam, M., Cruz, W.T., Leach, J.E., Nelson, R.J. (1997) Movement of *Xanthomonas oryzae* pv *oryzae* in southeast Asia detected using PCR-based DNA fingerprinting. *Phytopathology* **87**: 302–309.

Gillings, M., Fahy, P. (1993) Genetic diversity of *Pseudomonas solanacearum* biovars 2 and N2 assessed using restriction endonuclease analysis of total genomic DNA. *Plant Pathology* **42**: 744–753.

Gillings, M.R., Fahy, P. (1994) Genomic fingerprinting: towards a unified view of the *Pseudomonas solanacarum* species complex. In: Hayward, A.C., Hartman, G.L. (eds) *Bacterial Wilt: The Disease and its Causative Agent,* Pseudomonas solanacearum, pp. 95–112. CAB International: Wallingford, UK.

Gillings, M., Holley, M. (1997) Repetitive element PCR fingerprinting (rep-PCR) using enterobacterial repetitive intergenic consensus (ERIC) primers is not necessarily directed at ERIC elements. *Letts. Appl. Microbiol.* **25**: 17–21.

Goncalves, E.R., Rosato, Y.B. (2000) Genotypic characterization of xanthomonad strains isolated from passion fruit plants (*Passiflora* spp.) and their relatedness to different *Xanthomonas* species. *Int. J. Syst. Evol. Microbiol.* **50**: 811–821.

Goncalves, E.R., Rosato, Y.B. (2002) Phylogenetic analysis of *Xanthomonas* species based upon 16S–23S rDNA intergenic spacer sequences. *Int. J. Syst. Evol. Microbiol.* **52**: 355–361.

González, A.J., Landeras, E., Mendoza, M.C. (2000) Pathovars of *Pseudomonas syringae* causing bacterial brown spot and halo blight in *Phaseolus vulgaris* L. are distinguishable by ribotyping. *Appl. Environ. Microbiol.* **66**: 850–854.

Gonzalez, C., Restrepo, S., Tohme, J., Verdier, V. (2002) Characterisation of pathogenic and nonpathogenic strains of *Xanthomonas axonopodis* pv. *manihotis* by PCR-based DNA fingerprinting techniques. *FEMS Microbiol. Letts.* **215**: 23–31.

Gupta, V.S., Rajebhosale, M.D., Sodhi, M., Singh, S., Gnanamanickam, S.S., Dhaliwal, H.S., Ranjekar, P.K. (2001) Assessment of genetic variability and strain identification of *Xanthomonas oryzae* pv. *oryzae* using RAPD-PCR and IS1112-based PCR. *Current Science* **80**: 1043–1049.

Gürtler, V., Mayall, B.C. (2001) Genomic approaches to typing, taxonomy and evolution of bacterial isolates. *Int. J. Syst. Evol. Microbiol.* **51**: 3–16.

Gürtler, V., Stanisich, V.A. (1996) New approaches to typing and identification of bacteria using the 16S–23S rDNA spacer region. *Microbiology* **142**: 3–16.

Hanson, P.M., Wang, J.-F., Lucardo, O., Hanudin, Mah, S.Y., Hartmen, G.L., Lin, Y.C., Chen, J.-T. (1996) Variable reactions of tomato lines to bacterial wilt evaluated at several locations in Southeast Asia. *Hortscience* **31**: 143–146.

Helias, V., Le Roux, A.C., Bertheau, Y., Andrivon, D., Gauthier, J.P., Jouan, B. (1998) Characterisation of *Erwinia carotovora* subspecies and detection of *Erwinia carotovora* subsp. *atroseptica* in potato plants, soil and water extracts with PCR-based methods. *Eur. J. Plant Pathol.* **104**: 685–699.

Hendson, M., Hildebrand, D.C., Schroth, M.N. (1992) Relatedness of *Pseudomonas syringae* pv. *tomato*, *Pseudomonas syringae* pv. *maculicola* and *Pseudomonas syringae* pv. *antirrhini*. *J. Appl. Bacteriol.* **73**: 455–464.

Hendson, M., Purcell, A.H., Chen, D.Q., Smart, C., Guilhabert, M., Kirkpatrick, B. (2001) Genetic diversity of Pierce's disease strains and other pathotypes of *Xylella fastidiosa*. *Appl. Environ. Microbiol.* **67**: 895–903.

Hollaway, G.J., Gillings, M.R., Fahy, P.C. (1997) Use of fatty acid profiles and repetitive element polymerase chain reaction (PCR) to assess genetic diversity of *Pseudomonas syringae* pv. *pisi* and *Pseudomonas syringae* pv. *syringae* isolated from field peas in Australia. *Aust. Plant Pathol.* **26**: 98–108.

Jaunet, T.X., Wang, J.F. (1999) Variation in genotype and aggressiveness of *Ralstonia solanacearum* race 1 isolated from tomato in Taiwan. *Phytopathology* **89**: 320–327.

Jeong, E.L., Timmis, J.N. (2000) Novel insertion sequence elements associated with genetic heterogeneity and phenotype conversion in *Ralstonia solanacearum*. *J. Bacteriol.* **182**: 4673–4676.

Jock, S., Donat, V., Lopez, M.M., Bazzi, C., Geider, K. (2002) Following spread of fire blight in Western, Central and Southern Europe by molecular differentiation of *Erwinia amylovora* strains with PFGE analysis. *Environ. Microbiol.* **4**: 106–114.

Joyce, E.A., Chan, K., Salama, N.R., Falkow, S. (2002) Redefining bacterial populations: A post-genomic reformation. *Nature Reviews Genetics* **3**: 462–473.

Jumas-Bilak, E., Michaux-Charachon, S., Bourg, G., Ramuz, M., Allardet-Servent, A. (1998) Unconventional genomic organization in the alpha subgroup of the *Proteobacteria*. *J. Bacteriol.* **180**: 2749–2755.

Koike, S.T., Barak, J.D., Henderson, D.M., Gilbertson, R.L. (1999) Bacterial blight of leek: A new disease in California caused by *Pseudomonas syringae*. *Plant Disease* **83**: 165–170.

Kudva, I.T., Evans, P.S., Perna, N.T., Barrett, T.J., Ausubel, F.M., Blattner, F.R., Calderwood, S.B. (2002) Strains of *Escherichia coli* O157:H7 differ primarily by insertions or deletions, not single-nucleotide polymorphisms. *J. Bacteriol.* **184**: 1873–1879.

Lacava, P.T., Araujo, W.L., Maccheroni, W., Azevedo, J.L. (2001) RAPD profile and antibiotic susceptibility of *Xylella fastidiosa*, causal agent of citrus variegated chlorosis. *Letts. Appl. Microbiol.* **33**: 302–306.

Lan, R.T., Reeves, P.R. (2000) Intraspecies variation in bacterial genomes: the need for a species genome concept. *Trends Microbiol.* **8**: 396–401.

Lee, I.M., Bartoszyk, I.M., Gundersen, D.E., Mogen, B., Davis, R.E. (1997) Nested PCR for ultrasensitive detection of the potato ring rot bacterium, *Clavibacter michiganensis* subsp. *sepedonicus*. *Appl. Environ. Microbiol.* **63**: 2625–2630.

Lee, I.-M., Gundersen-Rindal, D.E., Davis, R.E., Bartoszyk, I.M. (1998) Revised classification scheme of phytoplasmas based on RFLP analyses of 16S rRNA and ribosomal protein gene sequences. *Int. J. Syst. Bacteriol.* **48**: 1153–1169.

Lee, I.M., Davis, R.E., Gundersen-Rindal, D.E. (2000) Phytoplasma: phytopathogenic mollicutes. *Ann. Rev. Microbiol.* 54: 221–255.

Lee, Y.A., Fan, S.C., Chiu, L.Y., Hsia, K.C. (2001) Isolation of an insertion sequence from *Ralstonia solanacearum* race 1 and its potential use for strain characterization and detection. *Appl. Environ. Microbiol.* **67**: 3943–3950.

Legard, D.E., Aquadro, C.F., Hunter, J.E. (1993) DNA sequence variation and phylogenetic relationships among strains of *Pseudomonas syringae* pv. *syringae* inferred from restriction site maps and restriction fragment length polymorphism. *Appl. Environ. Microbiol.* **59**: 4180–4188.

Leung, H., Nelson, R.J., Leach, J.E. (1993) Population structure of plant pathogenic fungi and bacteria. *Adv. Plant. Pathol.* **10**: 157–250.

Little, E.L., Bostock, R.M., Kirkpatrick, B.C. (1998) Genetic characterization of *Pseudomonas syringae* pv. *syringae* strains from stone fruits in California. *Appl. Environ. Microbiol.* **64**: 3818–3823.

Louws, F.J., Fulbright, D.W., Stephens, C.T., de Bruijn, F.J. (1994) Specific genomic fingerprints of phytopathogenic *Xanthomonas* and *Pseudomonas* pathovars and strains generated with repetitive sequences and PCR. *Appl. Environ. Microbiol.* **60**: 2286–2295.

Louws, F.J., Fulbright, D.W., Stephens, C.T., de Bruijn, F.J. (1995) Differentiation of genomic structure by rep-PCR fingerprinting to rapidly classify *Xanthomonas campestris* pv. *vesicatoria*. *Phytopathology* **85**: 528–536.

Louws, F.J., Rademaker, J.L.W., de Bruijn, F.J. (1999) The three Ds of PCR-based genomic analysis of phytobacteria: Diversity, detection, and disease diagnosis. *Ann. Rev. Phytopathol.* **37**: 81–125.

Lupski, J.R., Weinstock, G.M. (1992) Short, interspersed repetitive DNA sequenses in prokaryotic genomes. *J. Bacteriol.* **174**: 4525–4529.

Manceau, C., Horvais, A. (1997) Assessment of genetic diversity among strains of *Pseudomonas syringae* by PCR-restriction fragment length polymorphism analysis of rRNA operons with special emphasis on *P. syringae* pv *tomato*. *Appl. Environ. Microbiol.* **63**: 498–505.

Marques, A.S.D., Corbière, R., Gardan, L., Tourte, C., Manceau, C., Taylor, J.D., Samson, R. (2000) Multiphasic approach for the identification of the different classification levels of *Pseudomonas savastanoi* pv. *phaseolicola*. *Eur. J. Plant Pathol.* **106**: 715–734.

McDonald, J.G., Wong, E. (2001) Use of a monoclonal antibody and genomic fingerprinting by repetitive-sequence-based polymerase chain reaction to identify *Xanthomonas populi* pathovars. *Can. J. Plant Pathol.* **23**: 47–51.

Mew, T.W., Alvarez, A.M., Leach, J.E., Swings, J. (1993) Focus on bacterial-blight of rice. *Plant Disease* **77**: 5–12.

Morris, C.E., Glaux, C., Latour, X., Gardan, L., Samson, R., Pitrat, M. (2000) The relationship of host range, physiology, and genotype to virulence on cantaloupe in *Pseudomonas syringae* from cantaloupe blight epidemics in France. *Phytopathology* **90**: 636–646.

Mougel, C., Teyssier, S., D'Angelo, C. *et al.* (2001) Experimental and theoretical evaluation of typing methods based upon random amplification of genomic restriction fragments (AFLP) for bacterial population genetics. *Genetics Selection Evolution* **33**: S319–S338.

Mundt, C.C., Nieva, L.P., Cruz, C.M.V. (2002) Variation for aggressiveness within and between lineages of *Xanthomonas oryzae* pv. *oryzae*. *Plant Pathology* **51**: 163–168.

Nassar, A., Darrasse, A., Lemattre, M., Kotoujansky, A., Dervin, C., Vedel, R., Bertheau, Y. (1996) Characterization of *Erwinia chrysanthemi* by pectinolytic isozyme polymorphism and restriction fragment length polymorphism analysis of PCR-amplified fragments of *pel* genes. *Appl. Environ. Microbiol.* **62**: 2228–2235.

Nelson, R.J., Baraoidan, M.R., Vera Cruz, C.M., Yap, I.V., Leach, J.E., Mew, T.W., Leung, H. (1994) Relationship between phylogeny and pathotype for the bacterial blight pathogen of rice. *Appl. Environ. Microbiol.* **60**: 3275–3283.

Ochiai, H., Inoue, Y., Hasebe, A., Kaku, H. (2001) Construction and characterization of a *Xanthomonas oryzae* pv. *oryzae* bacterial artificial chromosome library. *FEMS Microbiol. Letts.* **2000**: 59–65.

Ochman, H. (2002) Bacterial evolution: Chromosome arithmetic and geometry. *Current Biology* **12**: R427–R428.

Olive, D.M., Bean, P. (1999) Principles and applications of methods for DNA-based typing of microbial organisms. *J. Clin. Microbiol.* **37**: 1661–1669.

Opina, N.L., Timmis, J.N., Fegan, M., Hayward, A.C. (1995) Development of probes and primers for detection of *P. solanacearum*. *Philippine Phytopathol.* **31**: 143–144.

Pisabarro, A., Correia, A., Martín, J.F. (1998) Pulsed-field gel electrophoresis analysis of the genome of *Rhodococcus fascians*: Genome size and linear and circular replicon composition in virulent and avirulent strains. *Curr. Microbiol.* **36**: 302–308.

Pooler, M.R., Hartung, J.S. (1995) Specific PCR detection and identification of *Xylella fastidiosa* strains causing citrus variegated chlorosis. *Curr. Microbiol.* **31**: 377–381.

Pooler, M.R., Ritchie, D.F., Hartung, J.S. (1996) Genetic relationships among strains of *Xanthomonas fragariae* based on random amplified polymorphic DNA PCR, repetitive extragenic palindromic PCR, and enterobacterial repetitive intergenic consensus PCR data and generation of multiplexed PCR primers useful for the identification of this phytopathogen. *Appl. Environ. Microbiol.* **62**: 3121–3127.

Poussier, S., Vandewalle, P., Luisetti, J. (1999) Genetic diversity of African and worldwide strains of *Ralstonia solanacearum* as determined by PCR-restriction fragment length polymorphism analysis of the hrp gene region. *Appl. Environ. Microbiol.* **65**: 2184–2194.

Poussier, S., Prior, P., Luisetti, J., Hayward, C., Fegan, M. (2000a) Partial sequencing of the *hrpB* and endoglucanase genes confirms and expands the known diversity within the *Ralstonia solanacearum* species complex. *Syst. Appl. Microbiol.* **23**: 479–486.

Poussier, S., Trigalet-Demery, D., Vandewalle, P., Goffinet, B., Luisetti, J., Trigalet, A. (2000b) Genetic diversity of *Ralstonia solanacearum* as assessed by PCR- RFLP of the hrp gene region, AFLP and 16S rRNA sequence analysis, and identification of an African subdivision. *Microbiology* **146**: 1679–1692.

Prior, P., Fegan, M. (2002) Unravelling the *Musa* subgroups of *Ralstonia solanacearum* using multiplex PCR. In *3rd International Bacterial Wilt Symposium*, pp. 85. White River, South Africa.

Rademaker, J.L.W., Hoste, B., Louws, F. J., Kersters, K., Swings, J., Vauterin, L., Vauterin, P., de Bruijn, F.J. (2000) Comparison of AFLP and rep-PCR genomic fingerprinting with DNA-DNA homology studies: *Xanthomonas* as a model system. *Int. J. Syst. Evol. Microbiol.* **50**: 665–677.

Rainey, P.B., Brodey, C.L., Johnstone, K. (1993) Identification of a gene-cluster encoding three high-molecular-weight proteins, which is required for synthesis of tolaasin by the mushroom pathogen *Pseudomonas tolaasii*. *Mol. Microbiol.* **8**: 643–652.

Restrepo, S., Velez, C.M., Verdier, V. (2000) Measuring the genetic diversity of *Xanthomonas axonopodis* pv. *manihotis* within different fields in Colombia. *Phytopathology* **90**: 683–690.

Rudolph, K.W.E. (1995) *Pseudomonas syringae* pathovars. In: Singh, U.S., Singh, R.P., Kohmoto, K. (eds) *Pathogenesis and Host Specificity in Plant Diseases: Histopathological, Biochemical, Genetic and Molecular Bases; Volume 1 Prokaryotes*, pp. 47–138. Elsevier Sciences: Oxford.

Salanoubat, M., Genin, S., Artiguenave, F. *et al.* (2002) Genome sequence of the plant pathogen *Ralstonia solanacearum*. *Nature* **415**: 497–502.

Savelkoul, P.H.M., Aarts, H.J.M., de Haas, J., Dijkshoorn, L., Duim, B., Otsen, M., Rademaker, J.L.W., Schouls, L., Lenstra, J.A. (1999) Amplified-fragment length polymorphism analysis: the state of an art. *J. Clin. Microbiol.* **37**: 3083–3091.

Sawada, H., Takeuchi, T., Matsuda, I. (1997) Comparative analysis of *Pseudomonas syringae* pv. *actinidiae* and pv *phaseolicola* based on phaseolotoxin-resistant ornithine carbamoyltransferase gene (*argK*) and 16S-23S rRNA intergenic spacer sequences. *Appl. Environ. Microbiol.* **63**: 282–288.

Sawada, H., Suzuki, F., Matsuda, I., Saitou, N. (1999) Phylogenetic analysis of *Pseudomonas syringae* pathovars suggests the horizontal gene transfer of *argK* and the evolutionary stability of *hrp*, gene cluster. *J. Mol. Evol.* **49**: 627–644.

Sawada, H., Kanaya, S., Tsuda, M., Suzuki, F., Azegami, K., Saitou, N. (2002) A phylogenomic study of the OCTase genes in *Pseudomonas syringae* pathovars: The horizontal transfer of the *argK-tox* cluster and the evolutionary history of OCTase genes on their genomes. *J. Mol. Evol.* **54**: 437–457.

Schloter, M., Lebuhn, M., Heulin, T., Hartmann, A. (2000) Ecology and evolution of bacterial microdiversity. *FEMS Microbiol. Revs.* **24**: 647–660.

Scortichini, M., Marchesi, U., Di Prospero, P. (2002) Genetic relatedness among *Pseudomonas avellanae*, *P. syringae* pv. *theae* and *P.s.* pv. *actinidiae*, and their identification. *Eur. J. Plant Pathol.* **108**: 269–278.

Seal, S.E., Jackson, L.A., Young, J.P.W., Daniels, M.J. (1993) Differentiation of *Pseudomonas solanacearum*, *Pseudomonas syzygii*, *Pseudomonas pickettii* and the blood disease bacterium by partial 16S rRNA sequencing: construction of oligonucleotide primers for sensitive detection by polymerase chain reaction. *J. Gen. Microbiol.* **139**: 1587–1594.

Seo, S.T., Furuya, N., Lim, C.K., Takanami, Y., Tsuchiya, K. (2002) Phenotypic and genetic diversity of *Erwinia carotovora* ssp *carotovora* strains from Asia. *J. Phytopathol.-Phytopathologische Zeitschrift* **150**: 120–127.

Simpson, A.J.G., Reinach, F.C., Arruda, P. *et al.* (2000) The genome sequence of the plant pathogen *Xylella fastidiosa*. *Nature* **406**: 151–157.

Smith, J.J., Scott-Craig, J.S., Leadbetter, J.R., Bush, G.L., Roberts, D.L., Fulbright, D.W. (1994) Characterisation of random amplified polymorphic DNA (RAPD) products from *Xanthomonas campestris* and some comments on the use of RAPD products in phylogenetic analysis. *Mol. Phylogen. Evol.* **3**: 135–145.

Smith, J.J., Offord, L.C., Holderness, M., Saddler, G.S. (1995a) Genetic diversity of *Burkholderia solanacearum* (synonym *Pseudomonas solanacearum*) race 3 in Kenya. *Appl. Environ. Microbiol.* **61**: 4263–4268.

Smith, J.J., Offord, L.C., Holderness, M., Saddler, G.S. (1995b) Pulsed-field gel electrophoresis of *Pseudomonas solanacearum*. *EPPO Bulletin* **25**: 163–167.

Spratt, B.G., Maiden, M.C.J. (1999) Bacterial population genetics, evolution and epidemiology. *Phil. Tran. Roy. Soc. Lond. B* **354**: 701–710.

Stackebrandt, E., Goebel, B.M. (1994) Taxonomic note: a place for DNA–DNA reassociation and 16S rRNA sequence analysis in the present species definition in bacteriology. *Int. J. Syst. Bacteriol.* **44**: 846–849.

Stackebrandt, E., Murray, R.G.E., Truper, H.G. (1988) *Proteobacteria classis* nov., a name for the phylogenetic taxon that includes the 'purple bacteria and their relatives'. *Int. J. Syst. Bacteriol.* **38**: 321–325.

Stackebrandt, E., Rainey, F.A., WardRainey, N.L. (1997) Proposal for a new hierarchic classification system, *Actinobacteria classis* nov. *Int. J. Syst. Bacteriol.* **47**: 479–491.

Stead, D., Stanford, H., Aspin, A., Heeney, J. (2002) Current status of some new and some old plant pathogenic pseudomonads. In: *6th* International Conference on Pseudomonas syringae pathovars and related pathogens, pp. 42. Acquafredda di Maratea: Italy.

Sulzinski, M.A., Moorman, G.W., Schlagnhaufer, B., Romaine, C.P. (1995) Fingerprinting of *Xanthomonas campestris* pv. *pelargonii* and related pathovars using random-primed PCR. *J. Phytopathol.* **143**: 429–433.

Sulzinski, M.A., Moorman, G.W., Schlagnhaufer, B., Romaine, C.P. (1996) Characteristics of a PCR-based assay for *in planta* detection of *Xanthomonas campestris* pv. *pelargonii*. *J. Phytopathol.* **144**: 393–398.

Sundin, G.W., Demezas, D.H., Bender, C.L. (1994) Genetic and plasmid diversity within natural populations of *Pseudomonas syringae* with various exposures to copper and streptomycin Bactericides. *Appl. Environ. Microbiol.* **60**: 4421–4431.

Taghavi, M., Hayward, C., Sly, L.I., Fegan, M. (1996) Analysis of the phylogenetic relationships of strains of *Burkholderia solanacearum*, *Pseudomonas syzygii*, and the blood disease bacterium of banana based on 16S rRNA gene sequences. *Int. J. Syst. Bacteriol.* **46**: 10–15.

Tegli, S., Sereni, A., Surico, G. (2002) PCR-based assay for the detection of *Curtobacterium flaccumfaciens* pv. *flaccumfaciens* in bean seeds. *Letts. Appl. Microbiol.* **35**: 331–337.

Thwaites, R., Mansfield, J., Eden-Green, S., Seal, S. (1999) RAPD and rep PCR-based fingerprinting of vascular bacterial pathogens of *Musa* spp. *Plant Pathology* **48**: 121–128.

Toth, I.K., Hyman, L.J., Taylor, R., Birch, P.R.J. (1998) PCR-based detection of *Xanthomonas campestris* pv. *phaseoli* var. *fuscans* in plant material and its differentiation from *X. c.* pv. *phasioli*. *J. Appl. Microbiol.* **85**: 327–336.

Toth, I.K., Bertheau, Y., Hyman, L.J. ***et al.*** (1999) Evaluation of phenotypic and molecular typing techniques for determining diversity in *Erwinia carotovora* subspp. *atroseptica*. *J. Appl. Microbiol.* **87**: 770–781.

Toth, I.K., Avrova, A.O., Hyman, L.J. (2001) Rapid identification and differentiation of the soft rot erwinias by 16S–23S intergenic transcribed spacer-PCR and restriction fragment length polymorphism analyses. *Appl. Environ. Microbiol.* **67**: 4070–4076.

Trébaol, G., Manceau, C., Tirilly, Y., Boury, S. (2001) Assessment of the genetic diversity among strains of *Xanthomonas cynarae* by randomly amplified polymorphic DNA analysis and development of specific characterized amplified regions for the rapid identification of *X. cynarae*. *Appl. Environ. Microbiol.* **67**: 3379–3384.

van Belkum, A., Struelens, M., de Visser, A., Verbrugh, H., Tibayrenc, M. (2001) Role of genomic typing in taxonomy, evolutionary genetics, and microbial epidemiology. *Clin. Microbiol. Revs.* **14**: 547–560.

van Berkum, P., Terefework, Z., Paulin, L., Suomalainen, S., Lindström, K., Eardly, B.D. (2003). Discordant phylogenies within the *rrn* loci of Rhizobia. *J. Bacteriol.* **185**: 2988–2998.

Van Sluys, M.A., Monteiro-Vitorello, C.B., Camargo, L.E.A. ***et al.*** (2002) Comparative genomic analysis of plant-associated bacteria. *Ann. Rev. Phytopathol.* **40**: 169–189.

Van Sluys, M.A., de Oliveira, M.C., Monteiro-Vitorello, C.B. ***et al.*** (2003) Comparative analyses of the complete genome sequences of Pierce's disease and citrus variegated chlorosis strains of *Xylella fastidiosa*. *J. Bacteriol.* **185**: 1018–1026.

Vauterin, L., Swings, J., Kersters, K. ***et al.*** (1990) Towards an improved taxonomy of *Xanthomonas*. *Int. J. Syst. Bacteriol.* **40**: 312–316.

Vauterin, L., Hoste, B., Kersters, K., Swings, J. (1995) Reclassification of *Xanthomonas*. *Int. J. Syst. Bacteriol.* **45**: 472–489.

Vauterin, L., Rademaker, J.L.W., Swings, J. (2000) Synopsis on the taxonomy of the genus *Xanthomonas*. *Phytopathology* **90**: 677–682.

Waleron, M., Waleron, K., Podhajska, A.J., Lojkowska, E. (2002) Genotyping of bacteria belonging to the former *Erwinia* genus by PCR-RFLP analysis of a *recA* gene fragment. *Microbiology* **148**: 583–595.

Wayne, L.G., Brenner, D.J., Colwell, R.R. ***et al.*** (1987) Report of the ad hoc committee on reconciliation of approaches to bacterial systematics. *Int. J. Syst. Bacteriol.* **37**: 463–464.

Weingart, H., Volksch, B. (1997) Genetic fingerprinting of *Pseudomonas syringae* pathovars using ERIC-, REP-, and IS50-PCR. *J. Phytopathol.* **145**: 339–345.

Wigley, P., Burton, N.F. (2000) Multiple chromosomes in *Burkholderia cepacia* and *B. gladioli* and their distribution in clinical and environmental strains of *B. cepacia*. *J. Appl. Microbiol.* **88**: 914–918.

Ye, F.C., Laigret, F., Carle, P., Bove, J.M. (1995) Chromosomal heterogeneity among various strains of *Spiroplasma citri*. *Int. J. Syst. Bacteriol.* **45**: 729–734.

Young, J.M. (1991) Pathogenicity and identification of the lilac pathogen, *Pseudomonas syringae* pv. *syringae* Van Hall 1902. *Anns. Appl. Biol.* **118**: 283–298.

Young, J.M. (2000) Recent developments in systematics and their implications for plant pathogenic bacteria. In: Priest, F.G., Goodfellow, M. (eds) *Applied Microbial Systematics*, pp. 135–163. Kluwer Academic Publishers, Dordrecht.

Young, J.M., Dye, D.W., Bradbury, J.F., Panagopoulos, C.G., Robbs, C.F. (1978) A proposed nomenclature and classification for plant pathogenic bacteria. *New Zeal. J. Agric. Res.* **21**: 153–177.

Young, J.M., Takikawa, Y., Gardan, L., Stead, D.E. (1992) Changing concepts in the taxonomy of plant pathogenic bacteria. *Ann. Rev. Phytopathol.* **30**: 67–105.

Young, J.M., Kuykendall L.D., Martinez-Romero, E., Kerr, A., Sawada, H. (2001) A revision of *Rhizobium* Frank 1889, with an emended description of the genus, and the inclusion of all species of *Agrobacterium* Conn 1942 and *Allorhizobium* de Lajudie et al. 1998 as new combinations: *Rhizobium radiobacter*, *R. rhizogenes*, *R. rubi*, *R. undicolor*, and *R. vitis*. *Int. J. Syst. Bacteriol.* **51**: 89–103.

Zhang, Y., Geider, K. (1997) Differentiation of *Erwinia amylovora* strains by pulsed-field gel electrophoresis. *Appl. Environ. Microbiol.* **63**: 4421–4426.

Zhao, Y.F., Damicone, J.P., Demezas, D.H., Rangaswamy, V., Bender, C.L. (2000) Bacterial leaf spot of leafy crucifers in Oklahoma caused by *Pseudomonas syringae* pv. *maculicola*. *Plant Disease* **84**: 1015–1020.

11

Genetic diversity and population structure of plant-pathogenic species in the genus *Fusarium*

Brett A. Summerell and John F. Leslie

11.1 Introduction

The fungal genus *Fusarium* contains some of the most economically and socially important species of plant pathogens affecting agriculture and horticulture. Diseases such as head blight of wheat and *Fusarium* wilt of bananas have not only caused enormous losses to crops, such as wheat and bananas around the world, but also have had a huge impact on the communities that depend on these crops (McMullen *et al.*, 1997; Ploetz, 1990; Windels, 2000). The genus is a complex, polyphyletic grouping whose taxonomy has been controversial for at least a century, with recognized species numbers ranging from over 1000 at the beginning of the 1900s to as few as nine in the 1950s and 1960s. Current estimates are around 50 (Kirk *et al.*, 2001). While there has been considerable research on genetic diversity within many taxa in the genus, because of their economic importance, the available information is still less than for many other pathogens of similar or lesser economic import. The research that has been conducted has practical implications in terms of plant breeding and epidemiology, with effective controls now available for many important *Fusarium* diseases. Indeed, it is within this disease context that even the most basic of studies has been conducted. Studies of genetic diversity in the genus, usually in biogeographical or evolutionary biology contexts, have increased as molecular tools that detect variation with no observable impact on morphological characters have become available. To appreciate these relatively recent genetic diversity studies, an understanding of the identification, nomenclature and taxonomy of taxa within the genus *Fusarium* is needed.

Plant Microbiology, Michael Gillings and Andrew Holmes

11.2 Taxonomic history and species concepts in *Fusarium*

Studies of genetic diversity require a stable taxonomic framework. If species are poorly defined or easily confused then studies of genetic diversity will necessarily be flawed, with species defined either too broadly or too narrowly resulting in different types of significant errors. Species definitions in *Fusarium* have been problematic for most of the past two centuries. *Fusarium* was initially described and defined by Link (1809), and by the early 1900s approximately 1000 species had been defined, usually on the basis of host associations. In 1935, Wollenweber and Reinking (1935) reduced this number to 135 species and their classification system formed the basis for all subsequent taxonomic systems. Wollenweber and Reinking developed a subgeneric system based on 16 sections, many of which are still in common use, even though they probably are not monophyletic.

In the 1940s and 1950s, Snyder and Hansen (1940, 1941, 1945, 1954) radically reduced the number of species within the genus to nine. Their approach was very popular with plant disease diagnosticians, who could use them to rapidly identify a disease-causing agent to species. Unfortunately, these species were too broad to be precise and much of the work with these species definitions as a base is difficult, if not impossible, to interpret. Two of Snyder and Hansen's species, *F. oxysporum* and *F. solani*, are still in general use, but there is little doubt that both of these taxa contain more than a single species and are in need of serious taxonomic revision.

Studies by Booth (1971), Gerlach and Nirenberg (1982), and Nelson *et al.* (1983) undid most of the changes proposed by Snyder and Hansen and returned the scientific community to taxonomic systems that were based primarily on the Wollenweber and Reinking system. In these three systems more attention was paid to careful assessment of morphological features, e.g. conidiogenous cells, macro- and microconidia, and chlamydospores, on standard media while simultaneously taking into account the variation within a species that had been clearly demonstrated by Snyder and Hansen. Much has been written about the differences between these three taxonomic systems (Booth, 1971; Gerlach and Nirenberg, 1982; Nelson *et al.*, 1983); however, there are many more common features than there are differences. All three systems use many of Wollenweber and Reinking's sections and species definitions. In some cases a species may have several different names, but generally these differences were based on nomenclatural disputes rather than differences in species definitions. The majority of *Fusarium* researchers use these systems as the basis for identifying *Fusarium* species and the description of new taxa (Britz *et al.*, 2002; Klittich *et al.*, 1997; Marasas *et al.*, 2001; Nirenberg and O'Donnell 1998; Nirenberg *et al.*, 1998; Zeller *et al.*, 2004). As genetic and molecular techniques have become more sophisticated and more widely available and applied, formerly functional species concepts, definitions and relationships are being stretched. Thus, a reassessment of many *Fusarium* species and their boundaries is clearly needed, and in numerous cases already in progress. In the last 20 years, numerous species of *Fusarium* have been described, usually with sexual cross-fertility or DNA-based characters either supplementing or guiding the evaluation of morphological characters. The *Dictionary of Fungi* (Kirk *et al.*, 2001) states that 500 species of *Fusarium* are reported. This number is likely to increase in the future as new ecosystems are explored and old species are redefined.

11.3 Why is the species definition important to studies of genetic diversity?

Any study of genetic diversity within a species implicitly assumes a stable well-defined taxon. Such stability usually implies that the full range of variation within a species is being evaluated. If the taxon is poorly defined, then the extent of variation can be confused at or near the poorly defined species boundary(ies). Furthermore, measures of characters such as genetic isolation, linkage disequilibrium, and random mating can be badly flawed. Such problems do not necessarily prevent studies of genetic diversity in poorly defined taxonomic assemblages, however, and in *Fusarium* such studies have resulted in revised species descriptions. These studies often require thoughtful partitioning of the data and extra vigilance in their analysis if meaningful conclusions are to result.

As with many fungi, three species concepts – morphological, biological and phylogenetic – are currently used to define species of *Fusarium*. Traditionally, morphological species concepts have dominated in *Fusarium*, but more recently biological (Leslie, 1981) and phylogenetic (Taylor *et al.*, 2000) concepts have become much more important, and have provided different foci and new insights into the taxonomy of *Fusarium*. A more extensive review of species concepts in *Fusarium* can be found in Leslie *et al.* (2001), but it is important to briefly note the most important points of each species concept and how they relate to *Fusarium*.

11.3.1 Morphological species concepts

These species concepts are based on the hypothesis that the morphology of a 'type' (or individual) can encompass the variation present in a species. Reliable species definitions require distinct morphological characters, or combination of characters, in species, i.e., members of different species must look different (Mayr, 1940, 1963). This traditional approach has been used extensively by fungal taxonomists, is well known, and has lengthy and extensive support in the scientific literature. The Gerlach and Nirenberg (1982) and Nelson *et al.* (1983) systems are both morphological in nature, and serve as the base systems against which biological and phylogenetic species concepts currently are tested. The main problem with morphological species concepts in microfungi is that the number of readily detectable characters usually is insufficient to distinguish all of the species that warrant recognition. Despite these limitations, the current widespread utilisation of morphological criteria by many diagnosticians, and the practical need to routinely identify many *Fusarium* cultures means that these characters will remain important, if not dominant, in *Fusarium* species concepts (see Summerell *et al.*, 2003).

11.3.2 Biological species concepts

The biological species concept as articulated by Mayr (1940, 1963) considers '... species as groups of populations that actually or potentially interbreed with each other'. There are practical limitations to applying a biological species concept in *Fusarium*, as many of the species reproduce predominantly, if not exclusively, asexually. However, in some groups, most notably the *Gibberella fujikuroi* complex, application of the biological species concept has been critical to the revision of the species and targeting groups that can be analysed as populations. For those species in which this concept has been applied, standard tester strains of both mating types are available through the Fungal Genetics Stock Center (Department of Microbiology, University of Kansas Medical Center, Kansas City, Kansas) that can be used to make test crosses to determine the fertility of unknown isolates. The availability of these reference strains together with the development of PCR-based tests for mating type (Kerényi *et al.*, 1999; Steenkamp *et al.*, 2000) have resulted in the more widespread application of this species concept for identification as well as providing information on an important character in field populations that can be used to estimate the relative amount of sexual and asexual reproduction occurring in these species (Britz *et al.*, 1998; Mansuetus *et al.*, 1997).

11.3.3 Phylogenetic species concepts

The phylogenetic species concept has found relatively recent application in *Fusarium* systematics and can help resolve taxonomic difficulties or, if inappropriately applied or misinterpreted, can result in further confusion. Phylogenetic species concepts are most useful for asexual species, homothallic species, and cultures or species that lack distinctive morphological characters (Taylor *et al.*, 2000). Phylogenetic species concepts require numerous characters to be statistically powerful. Normally molecular markers, usually DNA sequence data, are utilised, so relevant characters are available regardless of the morphological status or sexual fertility of an isolate. However, the problem commonly associated with phylogenetic species is where to draw the line between 'species', i.e., 'How different must two strains be to belong to different taxa?' In practice, many fungal phylogenetic studies rely on DNA sequences from one or two loci, from one or a few representative, or well-characterized or widely distributed, isolates. This practice can lead to problems that are best avoided by ensuring that enough loci and enough individuals are studied to overcome any sampling bias that might occur.

Within *Fusarium*, molecular data have been used to help resolve groups that were later described as separate species (Geiser *et al.*, 2001; Marasas *et al.*, 2001; Nirenberg and O'Donnell, 1998; Zeller *et al.*, 2004), usually in combination with distinctive morphological characters. In some cases, e.g., the mating populations within *Gibberella fujikuroi*, the groups defined by using either a biological species concept or a phylogenetic species concept are the same (Leslie, 1995, 1999; O'Donnell *et al.*, 1998a). In contrast to this result, O'Donnell *et al.* (2000) have recently proposed that *Fusarium graminearum* be divided into at least seven (now nine) phylogenetic species (O'Donnell *et al.*, 2003). However, members of at least some of these phylogenetic species are known to be cross-fertile under laboratory conditions (Bowden and Leslie, 1999; Jurgenson *et al.*, 2002a) and putative interlineage hybrids have been found in field populations in Brazil (Bowden *et al.*, 2003), Nepal (O'Donnell *et al.*, 2000), and Korea (Jeon *et al.*, 2003). Thus, for *F. graminearum* the phylogenetic and biological species concepts do not yet yield the same result.

We think that instances in which the different species concepts appear to give different answers are very important for studies of fungal evolution and differentiation. Such groups may be intermediates in the fungal speciation process, with the evolutionary process of species separation begun, as indicated by the available molecular data, but not yet complete, as indicated by the existing cross-fertility. These cases provide an opportunity to evaluate fungal evolution and speciation through observation and analysis of their intermediates, rather than needing to infer these processes from studies of the putative starting and ending points, i.e., current well-resolved species.

11.4 The reality of current *Fusarium* taxonomy

A number of different rankings currently are used to define taxa within *Fusarium*. This anamorphic genus as a whole is polyphyletic, and several teleomorph taxa, e.g., *Gibberella*, *Haemanectria* and *Albonectria*, are associated with *Fusarium*. Traditionally the genus has been subdivided into sections, which each include one or more species with common morphological characters. It is unlikely, however, that these sections will be monophyletic when DNA sequence characters are critically analysed. At present, *Fusarium* species definitions vary significantly, with some species very well defined and others clearly species aggregates in need of further resolution. Many plant pathologists assumed that the species that were well-known pathogens also were well-defined species that contained appropriate levels of genetic diversity. This view has been challenged most seriously in *Fusarium oxysporum* and the *Gibberella fujikuroi* species complex resulting both in the definition of new species, e.g., *F. thapsinum* (Klittich *et al.*, 1997) and *F. andiyazi* (Marasas *et al.*, 2001), and a re-evaluation of the significance of plant pathogenicity as a taxonomic criterion (e.g. Baayen 2000; Baayen *et al.*, 2000; O'Donnell *et al.*, 1998a; Skovgaard *et al.*, 2001). Within *Fusarium*, genetic diversity has been most critically evaluated in *F. oxysporum*, *Gibberella zeae* (*Fusarium graminearum*), and the *Gibberella fujikuroi* species complex, which encompasses Section *Liseola* and related species not clearly associated with any of the other sections.

11.4.1 Fusarium oxysporum

The level of genetic diversity in *F. oxysporum* is of great economic importance and significant scientific interest. The species concept in *F. oxysporum* is in need of attention, as the species definition of Snyder and Hansen (1940), which combined at least 30 different taxa into a single species, has led to great confusion. Analysis of DNA sequence data (Baayen *et al.*, 2000; O'Donnell *et al.*, 1998b) identifies numerous phylogenetic lineages within this species, many of which probably are distinct biological entities. A similar problem exists in the sister species *Fusarium solani* (Suga *et al.*, 2000). Functional mating-type alleles also are known in strains of *F. oxysporum* (Yun *et al.*, 2000). Thus, although the teleomorph (sexual stage) of this fungus is unknown, its existence seems likely. We anticipate that this species will be subdivided into many species, and that effective population and classical genetic analyses will then begin. Current studies are generally limited to quantitation of genetic variation in samples from field populations and the subdivision of the variation with respect to time, geographic location and/or host.

Fusarium oxysporum encompasses a number of pathogenic strains, each of which generally has a narrow host range. For example, *F. oxysporum* f. sp.

vasinfectum infects only cotton, *F. oxysporum* f. sp. *cubense* infects only bananas, and *F. oxysporum* f. sp. *lycopersici* infects only tomatoes. Within the pathogenic strains, or *formae speciales*, various levels of genetic diversity have been detected. Some *formae speciales*, e.g. *F. oxysporum* f. sp. *albedinis*, which pathogenises date palm (Fernandez *et al.*, 1997; Tantaoui *et al.*, 1996), or *F. oxysporum* f. sp. *ciceris*, which pathogenises chickpea (Jimenez-Gasco *et al.*, 2002), have a very limited amount of variation and are effectively clonal. Others, such as *F. oxysporum* f. sp. *cubense* have significant levels of variation (Bentley *et al.*, 1998; O'Donnell *et al.*, 1998b). This variation could have two very different origins. One possibility is that pathogen diversity is the result of mutation and selection within a pathogen strain that continually overcomes new resistance sources introduced into commercial varieties. Under this hypothesis, pathogenic strains are monophyletic in origin, relatively uniform genetically, belong to a limited number of vegetative compatibility groups (VCGs), and probably spread with the host. An alternative hypothesis is that pathogenic strains arise from local non-pathogens in response to the introduction of a particular host plant, cultivar or variety. Under this hypothesis, pathogenic strains share only the capacity to cause disease on a common host. Such strains need have little genetic similarity to one another, and may not even be in the same biological or phylogenetic species. Strain populations at different locations could thus be very different from one another, even while being genetically similar, even clonal, within the local populations. *Fusarium oxysporum* f. sp. *cubense* contains examples of both types of evolution. In *F. o.* f. sp. *cubense* Race 1, a clear molecular phylogeny and distribution of the pathogen with clonally propagated planting material has been demonstrated (Moore *et al.*, 2001). Within the currently economically important *F. o.* f. sp. *cubense* Race 4, there is considerable genetic diversity with strains belonging to many VCGs, and probably of polyphyletic origin (Bentley *et al.*, 1998; Koenig *et al.*, 1997). Thus, many populations of *F. o.* f. sp. *cubense* Race 4 probably have arisen as a result of multiple independent events occurring at many locations throughout the world.

The genetic diversity in non-pathogenic saprophytic strains of *F. oxysporum* often is much greater than is the variation found in similarly collected populations of pathogen strains (Correll *et al.*, 1986; Gordon and Okamoto, 1992). If *F. oxysporum* is composed predominantly of non-pathogenic strains that co-exist with plants as colonisers of root and stem tissue without causing disease in native ecosystems, as hypothesised by Gordon and Martyn (1997), then these large, diverse populations would provide a source of strains from which pathogenicity to a newly introduced host or variety could be readily selected.

Development of such pathogenic strains within Australia appears likely. The Australian races of *F. oxysporum* f. sp. *vasinfectum*, which attacks cotton, are found nowhere else in the world, are in unique VCGs, and appear to have significant genetic differences from strains of *F. o.* f. sp. *vasinfectum* found elsewhere (Davis *et al.*, 1996). These strains could have evolved from 'non-pathogenic' strains found on native Australian species of Malvaceae (Wang *et al.*, 2003). Similar evidence is available to explain the evolution of *F. oxysporum* f. sp. *canariensis*, which pathogenises the Canary Island Date Palm, that are different from strains from outside the country (Gunn and Summerell, 2002). Although the existing genetic diversity in many *formae speciales* of *F. oxysporum* is consistent with a hypothesis of numerous, diverse, independent origins of pathogenic strains of this species, little work has been done to explicitly test this hypothesis.

11.4.2 Fusarium graminearum

Fusarium graminearum (teleomorph *Gibberella zeae*) is a geographically widely distributed fungus that is associated with *Fusarium* head blight (scab) of wheat and barley (Wiese, 1987) and stalk rot and ear rot of maize (White, 1999). In the last decade, this fungus has caused destructive epidemics on wheat and barley in the United States and Canada (Gilbert and Tekauz, 2000; McMullen *et al.*, 1997) with extraordinary cumulative losses and disruption to farming communities (Windels, 2000). Control of this pathogen has focused on breeding for disease resistance and fungicide applications, processes whose efficacy may be enhanced by knowledge of the genetic structure of the pathogen population.

Gibberella zeae is homothallic and produces perithecia under both laboratory (Bowden and Leslie, 1999; Nelson *et al.*, 1983) and field (Francis and Burgess, 1977; Nelson *et al.*, 1983) conditions. Heterozygous outcrosses with different parents can be identified easily under laboratory conditions, and occur with an unknown frequency under field conditions. Heterozygous perithecia have not been recovered from field populations, but have been inferred from studies with VCGs (Bowden and Leslie, 1992), and molecular markers (Dusabenyagasani *et al.*, 1999; Schilling *et al.*, 1997; Zeller *et al.*, 2003a, 2003b) in which the genotypic diversity in field populations was high, and evidence for linkage disequilibrium was lacking. Even relatively low rates of outcrossing can have a significant impact on population structure (Leslie and Klein, 1996; Taylor *et al.*, 2000), and on gene-flow among populations. Population structure in *G. zeae* could be particularly impacted since ascospores are produced regularly under field conditions by most *G. zeae* strains and are an important means of natural dispersal for *G. zeae* (Bai and Shaner, 1994). If sexual recombination is occurring between isolates of this homothallic fungus under field conditions, then new gene combinations for traits such as fungicide resistance or aggressiveness could be rapidly generated and dispersed in the fungal population.

Laboratory crosses with strains of *G. zeae* have been used to make genetic maps with >1000 segregating markers at >450 polymorphic loci. The most detailed map contains two chromosome rearrangements and is based on parents with ~50% similarity in amplified fragment length polymorphism (AFLP) banding patterns (Jurgenson *et al.*, 2002a). The map resulting from this cross has several areas of segregation distortion. Some of these distorted regions are attributable to the crossing protocol, in which the parents carried complementary nitrate non-utilising (*nit*) mutations and the only progeny analysed were those with wild-type alleles at both of the heterozygous *nit* loci. The number of identified linkage groups is nine, but this number is larger than the number of clearly identifiable chromosomes (4–5) based on cytological analyses (Waalwijk *et al.*, 2003). A second cross, which is a major support for the current efforts to sequence the *G. zeae* genome, is between strains that are more closely related than were the strains in the first cross (>70% AFLP banding pattern similarity), and lacks both the chromosome rearrangements and the segregation distortion described in the first cross.

Gibberella zeae has been proposed to be subdivided into a series of at least nine different phylogenetic species or lineages (O'Donnell *et al.*, 2000, 2003; Ward *et al.*, 2002) based on differences in DNA sequences of six different loci. Not all loci, e.g., several in the trichothecene gene cluster (Ward *et al.*, 2002), follow this pattern. AFLP identifies genetically separated populations in which representatives of the various lineages can be placed. In field populations analysed on the basis of AFLP data, strains that appear to be intermediate between lineages can be identified. In at least some cases, these intermediates have characteristics of more than one of the described phylogenetic lineages. Toxin production is known to vary in field populations of *G. zeae* (Ichinoe *et al.*, 1983), but there is no clear correlation between phylogenetic lineage and the type of toxin produced. Members of at least some of the different lineages are cross-fertile (Bowden and Leslie, 1999), as is expected given the existence of the intermediate strains in the field populations. Thus, *G. zeae* is best viewed as a large, but very diverse, biological species in which further speciation – as indicated by the differences in chromosome rearrangements (Jurgenson *et al.*, 2002a) and the reduced cross-fertility between members of the different subpopulations – is currently in progress. The movement of this pathogen with its hosts through agricultural spread and exchange could be sufficient to prevent the isolation needed for the final resolution of these incipient species to occur.

The risks posed by the various subpopulations to crops in regions in which all of the subpopulations are not yet known has not been adequately evaluated. For example, lineage 7 isolates dominate as the cause of *Fusarium* head blight in the United States and Australia, but lineage 6 dominates in China. Increased resistance to benzimidazole fungicides has been reported in China (Chen *et al.*, 2000), and similar resistance could develop to the triazole fungicides popular in the United States. Differences in isolate aggressiveness also have been reported in *G. zeae* (Mesterhazy *et al.*, 1999; Miedaner and Schilling 1996; Miedaner *et al.*, 2001), suggesting potential for further pathogenic adaptation and evolution, although cultivar isolate specificity has not been reported for the *G. zeae*–wheat interaction, (cf. Miedaner *et al.*, (1992) and Mesterhazy *et al.*, (1999)). Further research on these lineages/subpopulations, with strains from diverse locations and environments, is needed to determine the significance of the population subdivisions as they relate to resistance breeding programmes and other control measures.

Fusarium head blight epidemics in North America appear to be sporadic, and strongly correlated to local environmental conditions (Francl *et al.*, 1999; Paulitz, 1996; Windels, 2000). In general, *G. zeae* populations are genetically diverse, regardless of the technique used to make the assessment (Dusabenyagasani *et al.*, 1999; Gale *et al.*, 2002; McCallum *et al.*, 2001; Moon *et al.*, 1999; Schilling *et al.*, 1997; Zeller *et al.*, 2003a, 2003b). Individual wheat heads commonly are infected by multiple strains during an epidemic, with adjacent heads usually colonised by different *G. zeae* strains. Thus, wheat head infection probably is initiated by spores with distinct fungal genotypes (presumably ascospores) although some secondary infection also can occur. Note, however, that genetically identical ascospores, produced homothallically, that initiate the infection of adjacent heads, or multiple infections of a single head would not be distinguishable from secondary spread of an isolate mediated by asexually produced conidial spores.

Ten populations of *G. zeae* from the central and eastern United States collected over 8 years were examined for genetic diversity by Zeller *et al.* (2003a, 2003b) with 30 polymorphic loci whose alleles were present at a frequency between 5 and 95%. These loci also could be placed on the current genetic map of *G. zeae* (Jurgenson *et al.*, 2002a), which enables detailed analyses of linkage equilibrium. Within individual populations, 5–10% of the locus pairs were statistically in disequilibrium ($p = 0.05$). There is no clear pattern that can be used to predict which loci will be in disequilibrium. Loci on the same chromosome were generally not in disequilibrium, which suggests that these populations are randomly mating, and have been randomly mating for quite some time. Differences in genetic similarity between populations generally were small, but statistically significant. Genetic and geographic distances between populations were correlated ($r = 0.591$, $p < 0.001$). Differences within populations accounted for 97% of the observed variation, and differences between populations account for the remaining 3% of the variation. We think that these differences probably represent the time required for different alleles and genotypes to diffuse through time and across relatively large geographic distances. If genes for aggressiveness and pathogenicity are distributed in a manner similar to that observed for the AFLP loci, then host material in resistance breeding programmes grown anywhere in the central and eastern United States probably has been exposed to most of the pathogenic variation in the fungal population in the country.

11.4.3 **Gibberella fujikuroi** *species complex*

The *G. fujikuroi* species complex also is known as *Fusarium* section *Liseola* and associated species. The species concept in this group has been substantially revised in the last 20 years with all of the strains in this group assigned to a single species, *Fusarium moniliforme*, by Snyder and Hansen (1945), now distributed across a minimum of nine described biological species (Britz *et al.*, 1999; Leslie, 1999; Zeller *et al.*, 2003c), or more than 25 phylogenetic species (Nirenberg and O'Donnell, 1998; O'Donnell *et al.*, 1998a). In one case there is at least some cross-fertility between recognised species, *F. fujikuroi* and *F. proliferatum*, and field isolates cross-fertile with standard testers of both species have been identified (Zeller *et al.*, 2003d). The number of species in this group is expected to increase steadily, as more strains are associated with existing, poorly described/represented phylogenetic lineages and distinguishing characters for these species identified.

Gibberella moniliformis (*Fusarium verticillioides*) has a highly developed map (Jurgenson *et al.*, 2002b; Xu and Leslie, 1996), with >600 markers on 12 linkage groups. Markers include mating type (*MAT*), spore killer (*SK*), fumonisin production (*FUM*), auxotrophs, restriction fragment length polymorphisms (RFLPs), and AFLPs, and except for the regions linked to *SK*, marker segregation generally was not significantly different from 1:1. The genetic linkage groups have been correlated with physical chromosomes based on CHEF gel electrophoresis, and 12 chromosomes of similar size and a genome size of 40–45 Mb are known for the six mating populations within this species that have been examined (Xu *et al.*, 1995). The karyotype for all six mating populations includes a chromosome of <1 Mb. In *G. moniliformis*, this chromosome can be lost or rearranged at a rate of approximately 3% during meiosis in crosses made under laboratory conditions (Xu and Leslie, 1996), but was present in all of the field strains examined. No genes for either toxins or pathogenicity have been described from this dispensable chromosome, although it is known to carry transcribed sequences.

Members of the *G. fujikuroi* species complex have a mixed life cycle in the sense that both sexual and asexual reproduction can occur. This type of life cycle results in selection for strains that can function as only the male parent in sexual crosses, since the asexual spores also serve as spermatia in addition to being the primary means of asexual spread and reproduction. Female-fertile strains are self-sterile hermaphrodites that become male-only/female-sterile strains when they lose the ability, presumably through mutation, to form protoperithecia. Male-only/female-sterile strains increase during asexual reproduction, but contribute <50% of the gametes to succeeding sexual generations. In field populations of species in the *G. fujikuroi* species complex, the proportion of male-only strains usually is 50–90% (Leslie and Klein, 1996). Both mating-type allele frequencies and the frequency of male-only strains can be used to estimate N_e, the effective population number. N_e based on mating type decreases as the ratio of the two *MAT* alleles deviates from 1:1. The reduction of N_e in field populations due to unequal frequencies of the *MAT* alleles usually is no more than 10% of the total number of individuals counted in the population. Reduction of N_e in response to a relatively high number of male-only strains, to as little as 30% of the total number of counted individuals, usually is much more severe than is the reduction due to the deviation of the ratio of the *MAT* alleles from 1:1. The proportion of female-fertile strains in a population also can be used to estimate the relative number of asexual generations per sexual generation. This value is 35–700 asexual generations per sexual generation for *F. verticillioides*, and much higher, 60–1200 asexual generations per sexual generation for *F. thapsinum* (Mansuetus *et al.*, 1997).

Strains in the *G. fujikuroi* species complex produce a diverse spectrum of secondary metabolites (Marasas *et al.*, 1984), among the most prominent of which are gibberellic acid (Cerdá-Olmeda *et al.*, 1994; Phinney and West, 1960), fumonisins (Gelderblom *et al.*, 1988), moniliformin, fusaproliferin and beauvericin. Production of these compounds varies by species and by strains within species (Fotso *et al.*, 2002; Rheeder *et al.*, 2002; Thiel *et al.*, 1991), and additional toxins probably remain to be identified (Leslie *et al.*, 1996). Most of the species make only one, or a few, of these compounds, e.g., *F. verticillioides* produces fumonisins but none of the other compounds. *F. proliferatum* is the only species that can synthesise all of these secondary metabolites. Toxin profiles are not diagnostic for species, since mutants that do not produce toxins are known from field isolates, e.g. the *FUM1–4* mutants of *F. verticillioides* all were initially recovered from field populations (Desjardins *et al.*, 1992, 1995; Leslie *et al.*, 1992; Plattner *et al.*, 1996). The maximum amount of toxin that a strain can produce also has a significant genetic component (Desjardins *et al.*, 1996; Proctor *et al.*, 1999).

Studies of genetic diversity of field populations of species in the *G. fujikuroi* species complex have examined a number of different characters. Vegetative compatibility, mediated by a series of vegetative incompatibility (*vic*) genes, has been a commonly studied character (Leslie, 2001). In *F. verticillioides*, the number of *vic* loci segregating in a population has been estimated at 10–15 (Puhalla and Spieth, 1983), with a cross in which 8–9 loci are segregating analysed in much more detail (Zeller *et al.*, 2001). In *F. verticillioides*, the general pattern is that virtually every strain from a field population is in a different VCG, and little information can be gleaned from such analyses beyond the fact that virtually every strain is genetically unique. This diversity has been exploited to track strains within a plant and to demonstrate that individual maize plants are infected by more than one strain of *F. verticillioides* (Kedera *et al.*, 1994) and that strains that infect a planted seed colonise the plant endophytically and can be recovered from the seeds of the resulting mature plant (Kedera *et al.*, 1992). Variation for VCGs in general is much less in *F. thapsinum* than in *F. verticillioides*, with ~75% of the *F. thapsinum* strains from the United States belonging to one of ten VCGs (Klittich and Leslie, 1988). The reduction in VCG variation observed in these populations is consistent with the lack of female-fertile strains and the relatively low N_e values reported from both global (Leslie and Klein, 1996) and African (Mansuetus *et al.*, 1997) populations.

Collectively, these results have resulted in the widespread assumption that strains in the same VCG are clones. In a clonal population, e.g., those of many *formae speciales* of *Fusarium oxysporum*, this assumption may be valid, but in a sexually reproducing population, strains in the same VCG may be identical only at the *vic* loci. This constraint does not require clonality, and strains in the same VCG may be quite different at other genetic markers (Chulze *et al.*, 2000). In general, VCGs are not particularly useful for studies of populations of species in the *G. fujikuroi* species complex, and techniques that generate data on multiple discrete loci, e.g., AFLPs or RFLPs, should be used instead.

A number of other traits also are known to vary in field populations. One of these is perithecial pigmentation (Chaisrisook and Leslie, 1990) in *F. verticillioides*, a nuclearly encoded trait with female-limited expression. Isozyme variation also is known both within and between species, but generally has not been used to analyse species level variation, as each species often has only a single isozymic form for any given enzyme. Spore-killer (*SK*) variants that result in meiotic drive during meiosis are polymorphic in field populations of several of the species within this group, e.g. *F. verticillioides* (Kathariou and Spieth, 1982) and *F. subglutinans* (Sidhu, 1984). Killer alleles (SK^K) vary in the effectiveness of the killing process, with 75–95% of the progeny in a cross between strains with SK^K and spore killer sensitive (SK^S) alleles being SK^K. In *F. verticillioides* the *SK* locus has been mapped (Jurgenson *et al.*, 2002b; Xu and Leslie, 1996) and does not appear to be associated with complex chromosome rearrangements such as seen in *Neurospora crassa* (Campbell and Turner, 1987; Raju, 1994; Turner and Perkins, 1979, 1991) and *Cochliobolus heterostrophus* (Bronson *et al.*, 1990; Chang and Bronson, 1996; Raju, 1994). Curiously, intraspecific variation in pathogenicity and host line/pathogen isolate interactions is not known for any of the *Fusarium* species in the *Gibberella fujikuroi* species complex, suggesting that the classic gene-for-gene interactions that are important in many host–pathogen interactions are not of particular importance in this group of fungi.

11.5 The future of population genetic studies in *Fusarium*

The *Fusarium* species discussed above are widely dispersed, economically important, pathogens of agricultural crops (Summerell *et al.*, 2001). As a consequence, the genetic diversity displayed in such organisms is likely to be restricted as a result of being anthropogenically distributed by man and with selection pressures favouring the predominance of strains of pathogens that are adapted to the host plants that they infect. In addition, studies in which the genetic diversity of pathogenic *Fusarium* have been analysed are based on unrepresentative collections of field isolates, or, worse, on strains solely from culture collections. For these reasons it is important not to extrapolate from these findings and assume that such studies indicate the full extent of diversity in the genus. We believe that the real future of studies on population genetics and evolutionary biology in *Fusarium* will be those studies that either incorporate isolates from natural ecosystems, or that focus on isolates from wild host populations of sibling species to domesticated host species. Such populations are more likely to include the full diversity of genetic variation found within the species and to provide the insights needed to properly understand the evolution, phylogenetic relationships, and genetic diversity within this genus.

Acknowledgements

Research in JFL's laboratory is supported in part by the Sorghum and Millet Collaborative Research Support Program (INTSORMIL) AID/DAN-1254-G-00-0021-00 from the US Agency for International Development, and the Kansas Agricultural Experiment Station. Manuscript no. 03-398-B from the Kansas Agricultural Experiment Station, Manhattan.

References

Baayen, R.P. (2000) Diagnosis and detection of host-specific forms of *Fusarium oxysporum*. *Bulletin-OEPP* **30**: 489–491.

Baayen, R.P., O'Donnell, K., Bonants, P.J., Cigelnik, E., Kroon, L.P.N.M., Roebroeck, E.J.A., Waalwijk, C. (2000) Gene genealogies and AFLP analyses in the *Fusarium oxysporum* complex identify monophyletic and nonmonophyletic formae speciales causing wilt and rot disease. *Phytopathology* **90**: 891–900.

Bai, G., Shaner, G. (1994) Scab of wheat: Prospects for control. *Plant Dis.* **78**: 760–766.

Bentley, S., Pegg, K.G., Moore, N.Y., Davis, R.D., Buddenhagen, I.W. (1998) Genetic variation among vegetative compatability groups of *Fusarium oxysporum* f. sp. *cubense* analyzed by DNA fingerprinting. *Phytopathology* **88**: 1283–1293.

Booth, C. (1971) The genus *Fusarium*. Commonwealth Mycological Institute: Surrey, UK.

Bowden, R.L., Leslie, J.F. (1992) Nitrate non-utilizing mutants of *Gibberella zeae* (*Fusarium graminearum*) and their use in determining vegetative compatibility. *Exp. Mycol.* **16**: 308–315.

Bowden, R.L., Leslie, J.F. (1999) Sexual recombination in *Gibberella zeae*. *Phytopathology* **89**: 182–188.

Bowden, R.L., Zeller, K.A., Vargas, J.I., Valdovinos-Ponce, G., Leslie, J.F. (2003) Population structure of *Gibberella zeae* in North, Central and South America. Proceedings of the Ninth International Fusarium Workshop, Sydney, Australia, January 2003, p. 29.

Britz, H., Wingfield, M.J., Coutinho, T.A., Marasas, W.F.O., Leslie, J.F. (1998) Female fertility and mating type distribution in a South African population of *Fusarium subglutinans* f. sp. *pini*. *Appl. Environ. Microbiol.* **64**: 2094–2095.

Britz, H., Coutinho, T.A., Wingfield, M.J., Marasas, W.F.O., Gordon, T.R., Leslie, J.F. (1999) *Fusarium subglutinans* f. sp. *pini* represents a distinct mating population in the *Gibberella fujikuroi* species complex. *Appl. Environ. Microbiol.* **65**: 1198–1201.

Britz, H., Coutinho, T.A., Wingfield, M.J., Marasas, W.F.O. (2002) Validation of the description of *Gibberella circinata* and morphological differentiation of the anamorph *Fusarium circinatum*. *Sydowia* **54**: 9–22.

Bronson, C.R., Taga, M., Yoder O.C. (1990) Genetic control and distorted segregation of T-toxin production in field isolates of *Cochliobolus heterostrophus*. *Phytopathology* **80**: 819–823.

Campbell, J.L., Turner, B.C. (1987) Recombination block in the spore killer region of *Neurospora*. *Genome* **29**: 129–135.

Cerdá-Olmedo, E., Fernández, M.R., Avalos, J. (1994) Genetics and gibberellin production in *Gibberella fujikuroi*. *Antonie van Leeuwenhoek* **65**: 217–225.

Chaisrisook, C., Leslie, J.F. (1990) A nuclear gene controlling perithecial pigmentation in *Gibberella fujikuroi (Fusarium moniliforme)*. *J. Hered.* **81**: 189–192.

Chang, H.-R., Bronson, C.R. (1996) A reciprocal translocation and possible insertion(s) tightly associated with host-specific virulence in *Cochliobolus heterostrophus*. *Genome* **39**: 549–557.

Chen, L.-F., Bai, G.-H., Desjardins, A.E. (2000) Recent advances in wheat scab research in China. In: Raupp W.J., Ma Z., Chen P., Liu D. (eds) *Proceedings of the International Symposium on Wheat Improvement for Scab Resistance*, pp. 258–273. Nanjing Agricultural University: Jiangsu, China.

Chulze, S.N., Ramirez, M.L., Torres, A., Leslie, J.F. (2000) Genetic variation in *Fusarium* section *Liseola* from no-till maize in Argentina. *Appl. Environ. Microbiol.* **66**: 5312–5315.

Correll, J.C., Puhalla, J.E., Schneider, R.W. (1986) Vegetative compatibility groups among non-pathogenic root-colonizing strains of *Fusarium oxysporum*. *Can. J. Bot.* **64**: 2358–2361.

Davis, R.D., Moore, N.Y., Kochman, J.K. (1996) Characterization of a population of *Fusarium oxysporum* f. sp. *vasinfectum* causing wilt of cotton in Australia. *Aust. J. Agric. Res.* **47**: 1143–1156.

Desjardins, A.E., Plattner, R.D., Leslie, J.F., Nelson, P.E. (1992) Heritability of fumonisin B_1 production in *Gibberella fujikuroi* mating population "A". *Appl. Environ. Microbiol.* **58**: 2799–2805.

Desjardins, A.E., Plattner, R.D., Nelsen, T.C., Leslie, J.F. (1995) Genetic analysis of fumonisin production and virulence of *Gibberella fujikuroi* mating population A (*Fusarium moniliforme*) on maize (*Zea mays*) seedlings. *Appl. Environ. Microbiol.* **61**: 79–86.

Desjardins, A.E., Plattner, R.D., Proctor, R.H. (1996) Linkage among genes responsible for fumonisin biosynthesis in *Gibberella fujikuroi* mating population A. *Appl. Environ. Microbiol.* **62**: 2571–2576.

Dusabenyagasani, M., Dostaler, D., Hamelin, R.C. (1999) Genetic diversity among *Fusarium graminearum* strains from Ontario and Quebec. *Can. J. Plant. Pathol.* **21**: 308–314.

Fernandez, D., Ouinten, M., Tantaoui, A., Geiger, J.P. (1997) Molecular records of micro-evolution within the Algerian population of *Fusarium oxysporum* f. sp. *albedinis* during its spread to new oases. *Eur. J. Plant. Pathol.* **103**: 485–490.

Fotso, J., Leslie, J.F., Smith, J.S. (2002) Production of beauvericin, moniliformin, fusaproliferin, and fumonisins B_1, B_2 and B_3 by ex-type strains of fifteen *Fusarium* species. *Appl. Environ. Microbiol.* **68**: 5195–5197.

Francis, R.G., Burgess, L.W. (1977) Characteristics of two populations of *Fusarium roseum* 'Graminearum' in eastern Australia. *Trans. Brit. Mycol. Soc.* **68**: 421–427.

Francl, L., Shaner, G., Bergstrom, G. *et al.* (1999) Daily inoculum levels of *Gibberella zeae* on wheat spikes. *Plant Dis.* **83**: 662–666.

Gale, L.R., Chen, L.-F., Hernick, C.A., Takamura, K., Kistler, H.C. (2002) Population analysis of *Fusarium graminearum* from wheat fields in eastern China. *Phytopathology* **92**: 1315–1322.

Geiser, D.M., Juba, J.H., Wang, B., Jeffers, S.N. (2001) *Fusarium hostae* sp. nov., a relative of *F. redolans* with a *Gibberella* teleomorph. *Mycologia* **93**: 670–678.

Gelderblom, W.C.A., Jaskiewicz, K., Marasas, W.F.O., Thiel, P.G., Horak, R.M., Vleggaar, M., Kriek, N.P.J. (1988) Fumonisins – Novel mycotoxins with cancer-promoting activity produced by *Fusarium moniliforme*. *Appl. Environ. Microbiol.* **54**: 1806–1811.

Gerlach, W., Nirenberg, H. (1982) The genus *Fusarium* – a pictorial atlas. *Mitteilungen aus der Bioloischen Bundesansalt fur Land- und Forstwirschaft. Berlin – Dahlem* **209**: 1–405.

Gilbert, J., Tekauz A. (2000) Review: Recent developments in research on Fusarium head blight of wheat in Canada. *Can. J. Plant. Pathol.* **22**: 1–8.

Gordon, T.R., Martyn, R.D. (1997) The evolutionary biology of *Fusarium oxysporum*. *Annu. Rev. Phytopathol.* **35**: 111–128.

Gordon, T.R., Okamoto, D. (1992) Population structure and the relationship between pathogenic and nonpathogenic strains of *Fusarium oxysporum*. *Phytopathology* **82**: 73–77.

Gunn, L.V., Summerell, B.A. (2002) Differentiation of *Fusarium oxysporum* isolates from *Phoenix canariensis* (Canary Island Date Palm) by pathogenicity, vegetative compatibility grouping and molecular analysis. *Aust. Plant Pathol.* **31**: 351–358.

Ichinoe, M., Kurata, H., Sugiura, Y., Ueno, Y. (1983) Chemotaxonomy of *Gibberella zeae* with special reference to production of trichothecenes and zearalenone. *Appl. Environ. Microbiol.* **46**: 1364–1369.

Jeon, J.-J., Kim, H., Kim, H.-S., Zeller, K.A., Lee, T., Yun, S.-H., Bowden, R.L., Leslie, J.F., Lee, Y.-W. (2003) Genetic diversity of *Fusarium graminearum* from maize in Korea. *Fung. Genet. Newsl.* **50**(Suppl.): 142.

Jimenez-Gasco, M.M., Milgroom, M.G., Jimenez-Diaz, R.M. (2002) Gene genealogies support *Fusarium oxysporum* f. sp. *ciceris* as a monophyletic group. *Plant Pathol.* **51**: 72–77.

Jurgenson, J.E., Bowden, R.L., Zeller, K.A., Leslie, J.F., Alexander, N.J., Plattner, R.D. (2002a) A genetic map of *Gibberella zeae* (*Fusarium graminearum*). *Genetics* **160**: 1452–1460.

Jurgenson, J.E., Zeller, K.A., Leslie, J.F. (2002b) An expanded genetic map of *Gibberella moniliformis* (*Fusarium verticillioides*). *Appl. Environ. Microbiol.* **68**: 1972–1979.

Kathariou, S., Spieth, P.T. (1982) Spore killer polymorphism in *Fusarium moniliforme*. *Genetics* **102**: 19–24.

Kedera, C.J., Leslie, J.F., Claflin, L.E. (1992) Systemic infection of corn by *Fusarium moniliforme*. *Phytopathology* **82**: 1138.

Kedera, C.J., Leslie, J.F., Claflin, L.E. (1994) Genetic diversity of *Fusarium* section *Liseola* (*Gibberella fujikuroi*) in individual maize plants. *Phytopathology* **84**: 603–607.

Kerényi, Z., Zeller, K.A., Hornok, L., Leslie, J.F. (1999) Molecular standardization of mating type terminology in the *Gibberella fujikuroi* species complex. *Appl. Environ. Microbiol.* **65**: 4071–4076.

Kirk, P.M., Cannon, P.F., David, J.C., Stalpers, J.A. (2001) Dictionary of the Fungi, 9th edition. CAB International: Wallingford, UK.

Klittich, C.J.R., Leslie, J.F. (1988) Unusual isolates of *Fusarium moniliforme* from sorghum in Kansas. *Phytopathology* **78**: 1519.

Klittich, C.J.R., Leslie, J.F., Nelson, P.E., Marasas, W.F.O. (1997) *Fusarium thapsinum* (*Gibberella thapsina*): a new species in section *Liseola* from sorghum. *Mycologia* **89**: 643–652.

Koenig, R.L., Ploetz, R.C., Kistler, H.C. (1997) *Fusarium oxysporum* f. sp. *cubense* consists of a small number of divergent and globally distributed clonal lineages. *Phytopathology* **87**: 915–923.

Leslie, J.F. (1981) Inbreeding for isogeneity by backcrossing to a fixed parent in haploid and diploid eukaryotes. *Genet. Res.* **37**: 239–252.

Leslie, J.F. (1995) *Gibberella fujikuroi*: Available populations and variable traits. *Can. J. Bot.* **73**(Suppl. 1): S282–S291.

Leslie, J.F. (1999) Genetic status of the *Gibberella fujikuroi* species complex. *Plant Pathol. J.* **15**: 259–269.

Leslie, J.F. (2001) Population genetics level problems in the *Gibberella fujikuroi* species complex. In: Summerell, B.A., Leslie, J.F., Backhouse, D., Bryden, W.L., Burgess, L.W., (eds) Fusarium*: Paul E. Nelson Memorial Symposium*, pp. 113–121. APS Press: St. Paul, MN.

Leslie, J.F., Klein K.K. (1996) Female fertility and mating-type effects on effective population size and evolution in filamentous fungi. *Genetics* **144**: 557–567.

Leslie, J.F., Doe, F.J., Plattner, R.D., Shackelford, D.D., Jonz, J. (1992) Fumonisin B_1 production and vegetative compatibility of strains from *Gibberella fujikuroi* mating population "A" (*Fusarium moniliforme*). *Mycopathologia* **117**: 37–45.

Leslie, J.F., Marasas, W.F.O., Shephard, G.S., Sydenham, E.W., Stockenström, S., Thiel, P.G. (1996) Duckling toxicity and the production of fumonisin and moniliformin by isolates in the A and F mating populations of *Gibberella fujikuroi* (*Fusarium moniliforme*). *Appl. Environ. Microbiol.* **62**: 1182–1187.

Leslie, J.F., Zeller, K.A., Summerell, B.A. (2001) Icebergs and species in populations of *Fusarium*. *Physiol. Mol. Plant Pathol.* **59**: 107–117.

Link, H.F. (1809) Observationes in ordines plantarum naturals, Dissetatio I. *Mag. Ges. Naturf. Freunde, Berlin* **3**: 3–42.

Mansuetus, A.S.B., Odvody, G.N., Frederiksen, R.A., Leslie, J.F. (1997) Biological species of *Gibberella fujikuroi* (*Fusarium* section *Liseola*) recovered from sorghum in Tanzania. *Mycol. Res.* **101**: 815–820.

Marasas, W.F.O., Nelson, P.E., Toussoun, T.A. (1984) *Toxigenic* Fusarium *Species*. Pennsylvania State University Press, University Park, PA.

Marasas, W.F.O., Rheeder, J.P., Lamprecht, S.C., Zeller, K.A., Leslie, J.F. (2001) *Fusarium andiyazi* sp. nov., a new species from sorghum. *Mycologia* **93**: 1203–1210.

Mayr, E. (1940) Speciation phenomena in birds. *Am. Nat.* **74**: 249–278.

Mayr, E. (1963) *Animal Species and Evolution*. Harvard University Press: Cambridge, MA.

McCallum, B.D., Tekauz, A., Gilbert, J. (2001) Vegetative compatibility among *Fusarium graminearum* (*Gibberella zeae*) isolates from barley spikes in southern Manitoba. *Can. J. Plant. Pathol.* **23**: 83–87.

McMullen, M.P., Jones, R., Gallenberg, D. (1997) Scab of wheat and barley: A re-emerging disease of devastating impact. *Plant Dis.* **81**: 1340–1348.

Mesterhazy, A., Bartok, T., Mirocha, C.G., Komoroczy, R. (1999) Nature of wheat resistance to *Fusarium* head blight and the role of deoxynivalenol for breeding. *Plant Breeding* **118**: 97–110.

Miedaner, T., Schilling, A.G. (1996) Genetic variation of aggressiveness in individual field populations of *Fusarium graminearum* and *Fusarium culmorum* tested on young plants of winter rye. *Eur. J. Plant. Pathol.* **102**: 823–830.

Miedaner, T., Borchardt, D.C., Geiger, H.H. (1992) Genetic analysis of inbred lines and their crosses for resistance to head blight (*Fusarium culmorum, Fusarium graminearum*) in winter rye. *Euphytica* **65**: 123–133.

Miedaner, T., Schilling, A.G., Geiger, H.H. (2001) Molecular genetic diversity and variation for aggressiveness in populations of *Fusarium graminearum* and *Fusarium culmorum* sampled from wheat fields in different countries. *J. Phytopathol.* **149**: 641–648.

Moon, J.-H., Lee, Y.-H., Lee, Y.-W. (1999) Vegetative compatibility groups in *Fusarium graminearum* isolates from corn and barley in Korea. *Plant Pathol. J.* **15**: 53–56.

Moore, N.Y., Pegg, K.G., Buddenhagen, I.W., Bentley, S. (2001) Fusarium wilt of banana: A diverse clonal pathogen of a domesticated clonal host. In: Summerell, B.A., Leslie, J.F., Backhouse, D., Bryden, W.L., Burgess, L.W. (eds) Fusarium*: Paul E. Nelson Memorial Symposium*, pp. 212–224. APS Press: St. Paul, MN.

Nelson, P.E., Toussoun, T.A., Marasas, W.F.O. (1983) Fusarium *Species: An Illustrated Manual for Identification*. Pennsylvania State University Press, University Park, PA.

Nirenberg, H.I., O'Donnell, K. (1998) New *Fusarium* species and combinations within the *Gibberella fujikuroi* species complex. *Mycologia* **90**: 434–458.

Nirenberg, H.I., O'Donnell, K., Kroschel, J., Andrianaivo, A.P., Frank, J.M., Mubatanhema, W. (1998) Two new species of *Fusarium*: *Fusarium brevicatenulatum* from the noxious weed *Striga asiatica* in Madagascar and *Fusarium pseudoanthophilum* from *Zea mays* in Zimbabwe. *Mycologia* **90**: 459–464.

O'Donnell, K., Cigelnik, E., Nirenberg, H.I. (1998a) Molecular systematics and phylogeography of the *Gibberella fujikuroi* species complex. *Mycologia* **90**: 465–493.

O'Donnell, K., Kistler, H.C., Cigelnik, E., Ploetz, R.C. (1998b) Multiple evolutionary origins of the fungus causing Panama disease of banana: Concordant evidence from nuclear and mitochondrial gene genealogies. *Proc. Natl Acad. Sci. USA* **95**: 2044–2049.

O'Donnell, K., Kistler, H.C., Tacke B.K., Casper, H.H. (2000) Gene genealogies reveal global phylogeographic structure and reproductive isolation among lineages of *Fusarium graminearum*, the fungus causing wheat scab. *Proc. Natl Acad. Sci. USA* **97**: 7905–7910.

O'Donnell, K., Ward, T., Kistler, H.C. (2003) Discordant evolution trichothecene toxins and species withinthe *Fusarium graminearum* species complex: Phylogenetic evidence from multigene genealogies. Proceedings of the Ninth International Fusarium Workshop, Sydney, Australia, January 2003, p. 29.

Paulitz, T.C. (1996) Diurnal release of ascospores by *Gibberella zeae* in inoculated wheat plots. *Plant Dis.* **80**: 674–678.

Phinney, B.O., West, C.A. (1960) Gibberellins as native plant growth regulators. *Ann. Rev. Plant Physiol.* **11**: 411–436.

Plattner, R.D., Desjardins, A.E., Leslie, J.F., Nelson, P.E. (1996) Identification and characterization of strains of *Gibberella fujikuroi* mating population A (*Fusarium moniliforme*) with rare fumonisin phenotypes. *Mycologia* **87**: 416–424.

Ploetz, R.C. (ed) (1990) Fusarium *Wilt of Banana*. APS Press: St. Paul, MN.

Proctor, R.H., Desjardins, A.E., Plattner, R.D. (1999) Biosynthetic and genetic relationships of B-series fumonisins produced by *Gibberella fujikuroi* mating population A. *Natural Toxins* **7**: 251–258.

Puhalla, J.E., Speith, P.T. (1983) Heterokaryosis in *Fusarium moniliforme*. *Exp. Mycol.* **7**: 328–335.

Raju, N.B. (1994) Ascomycete spore killers: Chromosomal elements that distort genetic ratios among the products of meiosis. *Mycologia* **86**: 461–473.

Rheeder, J.P., Marasas, W.F.O., Vismer, H.F. (2002) Production of fumonisin analogs by *Fusarium* species. *Appl. Environ. Microbiol.* **68**: 2101–2105.

Schilling, A.G., Miedaner, T., Geiger, H.H. (1997) Molecular variation and genetic structure in field populations of *Fusarium* species causing head blight in wheat. *Cer. Res. Comm.* **25**: 549–554.

Sidhu, G.S. (1984) Genetics of *Gibberella fujikuroi*. V. Spore killer alleles in *G. fujikuroi*. *J. Hered.* **75**: 237–238.

Skovgaard, K., Nirenberg, H.I., O'Donnell, K., Rosendahl S. (2001) Evolution of *Fusarium oxysporum* f. sp. *vasinfectum* races inferred from multigene genealogies. *Phytopathology* **91**: 1231–1237.

Snyder, W.C., Hansen, H.N. (1940) The species concept in *Fusarium*. *Am. J. Bot.* **27**: 64–67.

Snyder, W.C., Hansen, H.N. (1941) The species concept in *Fusarium* with reference to section *Martiella*. *Am. J. Bot.* **28**: 738–742.

Snyder, W.C., Hansen, H.N. (1945) The species concept in *Fusarium* with reference to *Discolor* and other sections. *Am. J. Bot.* **32**: 657–666.

Snyder, W.C., Hansen, H.N. (1954) Variation and speciation in the genus *Fusarium*. *Proc. NY Acad. Sci* **60**: 16–23.

Steenkamp, E.T., Wingfield, B.D., Coutinho, T.A., Zeller, K.A., Wingfield, M.J., Marasas, W.F.O., Leslie, J.F. (2000) PCR-based identification of *MAT-1* and *MAT-2* in the *Gibberella fujikuroi* species complex. *Appl. Environ. Microbiol.* **66**: 4378–4382.

Suga, H., Hasegawa, T., Mitsui, H., Kageyama, K., Hyakumachi, M. (2000) Phylogenetic analysis of the phytopathogenic fungus *Fusarium solani* based on the rDNA-ITS region. *Mycol. Res.* **104**: 1175–1183.

Summerell, B.A., Leslie, J.F., Backhouse, D., Bryden, W.L., Burgess, L.W. (eds) (2001) Fusarium*: Paul E. Nelson Memorial Symposium*. APS Press: St. Paul, MN.

Summerell, B.A., Salleh, B., Leslie, J.F. (2003) A utilitarian approach to *Fusarium* identification. *Plant Dis.* **87**: 117–128.

Tantaoui, A., Ouinten, M., Geiger, J.P., Fernandez, D. (1996) Characterization of a single clonal lineage of *Fusarium oxysporum* f. sp. *albedinis* causing Bayoud disease of date palm in Morocco. *Phytopathology* **86**: 787–792.

Taylor, J.W., Jacobson, D.J., Kroken, S., Kasuga, T., Geiser, D.M., Hibbett, D.S., Fisher M.C. (2000) Phylogenetic species recognition and species concepts in fungi. *Fung. Genet. Biol.* **31**: 21–32.

Thiel, P.G., Marasas, W.F.O., Sydenham, E.W., Shephard, G.S., Gelderblom, W.C.A., Nieuwenhuis, J.J. (1991) Survey of fumonisin production by *Fusarium* species. *Appl. Environ. Microbiol.* **57**: 1089–1093.

Turner, B.C., Perkins, D.D. (1979) Spore killer, a chromosomal factor in *Neurospora* that kills meiotic products not containing it. *Genetics* **93**: 587–606.

Turner, B.C., Perkins, D.D. (1991) Meiotic drive in *Neurospora* and other fungi. *Am. Nat.* **137**: 416–429.

Waalwijk, C., Taga, M., Sato, T., Kema, G.H.J. (2003) Cytological karyotyping of *Fusarium* spp. by the germ tube burst method. Proceedings of the Ninth International Fusarium Workshop, Sydney, Australia, January 2003, p. 57.

Wang, B., Brubaker, C.L., Burdon, J.J. (2003) Incidence of Fusarium wilt pathogens in the rhizosphere of Australian native cottons. Proceedings of the Eighth International Congress of Plant Pathol. (CD-ROM), Christchurch, New Zealand, February 2003.

Ward, T.J., Bielawski, J.P., Kistler, H.C., Sullivan, E., O'Donnell, K. (2002) Ancestral polymorphism and adaptive evolution in the trichothecene mycotoxin gene cluster of phytopathogenic *Fusarium*. *Proc. Natl Acad. Sci. USA* **99**: 9278–9283.

White, D.G. (ed) (1999) *Compendium of Corn Diseases*, 3rd edn. APS Press: St. Paul, MN.

Wiese, M.V. (ed) (1987) *Compendium of Wheat Diseases*, 2nd edn. APS Press: St. Paul, MN.

Windels, C.E. (2000) Economic and social impacts of Fusarium head blight: Changing farms and rural communities in the Northern Great Plains. *Phytopathology* **90**: 17–21.

Wollenweber, H.W., Reinking, O.A. (1935) *Die Fusarien, ihre Beschreibung, Schadwirkung und Bekampfung*. Verlag Paul Parey: Berlin, Germany.

Xu, J.-R., Leslie, J.F. (1996) A genetic map of *Fusarium moniliforme* (*Gibberella fujikuroi* mating population A). *Genetics* **143**: 175–189.

Xu, J.-R., Yan, K., Dickman, M.B., Leslie, J.F. (1995) Electrophoretic karyotypes distinguish the biological species of *Gibberella fujikuroi* (*Fusarium* section *Liseola*). *Mol. Plant–Microbe Interact.* **8**: 74–84.

Yun, S.-H., Tsutomu, A., Kaneko, I., Yoder, O.C., Turgeon, B.G. (2000) Molecular organization of mating type loci in heterothallic, homothallic, and asexual *Gibberella/Fusarium* species. *Fung. Genet. Biol.* **31**: 7–20.

Zeller, K.A., Jurgenson, J.E., Leslie J.F. (2001) Simultaneous mapping of multiple *vic* loci in *Gibberella fujikuroi* Mating Population A (*Fusarium verticillioides*). *Phytopathology* **91**: s99.

Zeller, K.A., Bowden, R.L., Leslie, J.F. (2003a) Diversity of epidemic populations of *Gibberella zeae* from small quadrats in Kansas and North Dakota. *Phytopathology* **93**: 874–880.

Zeller, K.A., Bowden, R.L., Leslie, J.F. (2004) Population differentiation and recombination in wheat scab populations of *Gibberella zeae* from the United States. *Molecular Ecology*, (in press).

Zeller, K.A., Summerell, B.A., Bullock, S., Leslie, J.F. (2003c) *Gibberella konza* (*Fusarium konzum*) sp. nov. from prairie grasses, a new species in the *Gibberella fujikuroi* species complex. *Mycologia* **95**: 943–954.

Zeller, K.A., Wohler, M.A., Gunn, L.V., Bullock, S., Summerell, B.A., Leslie, J.F. (2003d) Interfertility and marker segregation in hybrid crosses of *Gibberella fujikuroi* and *Gibberella intermedia*. *Fung. Genet. Newsl.* **50**(Suppl): 144.

12

Genome sequence analysis of prokaryotic plant pathogens

Derek W. Wood, Eugene W. Nester and Joao C. Setubal

12.1 Introduction

Genome sequence data provide valuable insights into many aspects of biological science. The explosion of genome sequencing activity and the need to categorise and extract information from these large data sets has led to the formation of the rapidly expanding field of bioinformatics. Bioinformatics tools allow us to define genes, infer metabolic pathways, and compare organisms, all of which provide insights into biological processes. The need to streamline and improve existing tools as well as to develop new tools that can extract useful information from these vast repositories of information becomes more pressing as additional genomes are sequenced. As we progress into the genomics era, the availability of such data and its interpretation will become both more complex and commonplace. This chapter seeks to (i) summarize common bioinformatic approaches to the analysis and interpretation of primary genome sequence data and (ii) provide examples of data generated using these approaches derived from published genome analyses of plant-pathogenic bacteria. Such data serve as a starting point from which a reasonable subset of candidate genes can be defined and targeted for more precise genetic and biochemical analyses. It is hoped that this review will provide insight into the methodologies of genome analysis that will facilitate genome analysis for new researchers and provide a framework upon which others can interpret the conclusions derived from these analyses.

12.2 Background

Three hundred and nineteen prokaryotic genome sequencing projects have been completed (73) or are in progress (246) as of the writing of this chapter (http://wit.integratedgenomics.com/GOLD/, http://www.tigr.org/tdb/mdb/mdbcomplete.html,

Plant Microbiology, Michael Gillings and Andrew Holmes

http://www.ncbi.nlm.nih.gov/PMGifs/Genomes/bact.html). More than 800 viral genomes have been sequenced with greater than 40% (346) being plant pathogens (http://www.ncbi.nlm.nih.gov/PMGifs/Genomes/vis.html). In contrast, only five prokaryotic phytopathogens have been completely sequenced with 15 additional projects in progress (see *Table 12.1* and http://www.tigr.org/~vinita/PPwebpage.html). Efforts are underway to sequence additional phytopathogens including eukaryotic oomycetes, fungi (Soanes *et al.*, 2002) and nematodes, although the larger size of the latter genomes necessitates longer time frames and unique analytical approaches (see *http://www.tigr.org/~vinita/PPwebpage.html* for a complete list).

Many pathogenic strategies are shared between phytopathogenic bacteria that provide targets for identification and further characterisation by genome researchers. Among these targets are genes that confer the ability to survive and compete in their particular habitat. Genes encoding such functions are indirect pathogenicity factors that allow the pathogen to survive until it has access to the host. Targets in this category encompass a broad range and include those involved in the uptake and utilisation of nutrients, degradation of toxic compounds, the uptake of iron, survival under various environmental stresses and antagonistic factors such as bacteriocins or antibiotics. Once established in a particular habitat, the bacterium must find ways to access its host directly or through an association with insects or other vectors. Factors influencing this stage include chemotaxis systems and, for insect-borne pathogens, gene products that allow association with the host. Once the host is located, the development of an intimate association is often a prerequisite to pathogenesis. Products involved in this facet include fimbrial and afimbrial adhesins, pili and other structures that mediate attachment to the host (Soto and Hultgren, 1999). Entry into the host occurs by a variety of mechanisms that may be indirect, such as damage to the host caused by insects or natural or artificial wounding, or direct, such as those mediated by the pathogen, including chemotaxis to natural openings or the production of degradative enzymes. Once inside the host the bacterium must evade host defences and disseminate. Products involved at this stage include enzymes that detoxify plant defence compounds or manage oxidative stress as well as those that degrade host tissues to facilitate dissemination of the pathogen. Interaction with the host throughout the disease process requires effective translocation of products from the bacterium to the environment or host. Five protein secretion systems have been identified in bacteria that serve this function (Harper and Silhavy, 2001). Among these, the type II, III and IV systems are commonly associated with pathogenicity and virulence. Type II systems encode the general secretion pathway and are key to the export of many degradative enzymes. Type III systems likely mediate transfer of bacterial products to the host in a contact-dependent manner. These include products that may alter virulence and so-called 'avirulence' proteins that limit host range when recognised by corresponding host proteins that activate plant defence pathways (Kjemtrup *et al.*, 2000). Collectively, type III secreted proteins are called effectors to reflect their putative interaction with host systems. These and other factors that may mediate host range are key targets in comparative genomic analyses. Type IV systems transit either protein or protein and DNA to the host. Although such systems are required for virulence in many bacterial mammalian pathogens, the best-characterised example is the T-DNA transport system of *A. tumefaciens* (Baron *et al.*, 2002). Type IV systems are ancestrally related to conjugal transfer systems.

Table 12.1 Genome sequencing of phytopathogenic bacteria[a]

Organism	Disease	Status	Website
***Agrobacterium tumefaciens* C58**	Crown gall	Complete (Goodner *et al.*, 2001; Wood *et al.*, 2001)	http://www.agrobacterium.org
***Burkholderia cepacia* J2315**	Sour skin	In progress	http://www.sanger.ac.uk/Projects/B_cepacia/
Clavibacter michiganensis* subsp. *michiganensis	Bacterial wilt and canker	In progress	http://www.genetik.uni-bielefeld.de/GenoMik/partner/bi_eichen.html
Clavibacter michiganensis* subsp. *sepedonicus	Ring rot	In progress	http://www.sanger.ac.uk/Projects/C_ michiganensis
Leifsonia xyli* subsp. *Xyli	Ratoon stunting disease	In progress	http://aeg.lbi.ic.unicamp.br
Pectobacterium (Erwinia) carotovora* subsp. *atroseptica	Soft rot and blackrot	In progress (Bell *et al.*, 2002)	http://www.sanger.ac.uk/Projects/E_carotovora/
***Pectobacterium (Erwinia) chrysanthemi* 3937**	Soft rot	In progress	http://www.ahabs.wisc.edu/~pernalab/erwinia/index.html
***Pseudomonas syringae* pv. *tomato* DC3000**	Bacterial speck	Complete (Buell *et al.*, 2003)	http://www.tigr.org
***Pseudomonas syringae* B728a**	Bacterial speck	In progress	http://www.jgi.doe.gov/JGI_microbial/html/pseudomonas_syr/pseudo_syr_homepage.html
***Ralstonia solanacearum* GMI1000**	Bacterial wilt	Complete (Salanoubat *et al.*, 2002)	http://sequence.toulouse.inra.fr/R.solanacearum
***Spiroplasma kunkelii* CR2-3X**	Corn stunt	In progress	http://www.genome.ou.edu/spiro.html
Xanthomonas axonopodis pv. aurantifolii B	Cancrosis B	In progress	http://www.lbm.fcav.unesp.br/
Xanthomonas axonopodis pv. aurantifolii C	Cancrosis C	In progress	http://www.lbm.fcav.unesp.br/
Xanthomonas axonopodis* pv. *citri	Asiatic canker	Complete (da Silva *et al.*, 2002)	http://cancer.lbi.ic.unicamp.br/xanthomonas/
Xanthomonas campestris* pv. *campestris	Black rot	Complete (da Silva *et al.*, 2002)	http://cancer.lbi.ic.unicamp.br/xanthomonas/
Xylella fastidiosa 9a5C	Citrus variegated chlorosis	Complete (Simpson *et al.*, 2000)	http://aeg.lbi.ic.unicamp.br/xf/
***Xylella fastidiosa pv. almond* Dixon**	Almond leaf scorch	In progress (Bhattacharyya *et al.*, 2002)	http://www.jgi.doe.gov/JGI_microbial/html/xylella_almond/xyle_almnd_homepage.html
***Xylella fastidiosa pv. oleander* Ann1**	Oleander leaf scorch	In progress (Bhattacharyya *et al.*, 2002)	http://www.jgi.doe.gov/JGI_microbial/html/xylella_oleander/xyle_olean_homepage.html
***Xylella fastidiosa* Temecula 1**	Pierce's disease	Complete (Van Sluys *et al.*, 2002a)	http://aeg.lbi.ic.unicamp.br/xf-grape/

[a]Modified from Van Sluys *et al.* (2002b). See http://www.tigr.org/~vinita/PPwebpage.html for an updated list of plant pathogen sequencing projects.

The recent discovery of a new type IV system in *A. tumefaciens* required for conjugal transfer of the pAtC58 plasmid, but not virulence, emphasises the need to experimentally characterise such putative pathogenicity factors (Chen *et al.*, 2002). Genes that encode pathogenicity and virulence factors directly responsible for disease symptoms are also key targets and include toxins and exopolysaccharides. Finally, identifying regulatory proteins that influence production of pathogenicity factors is key since both the timing and level of the expression of these genes is likely essential to effective competition and pathogenesis.

12.3 Pathogen background and disease mechanism

12.3.1 Agrobacterium tumefaciens

Agrobacterium is a diverse genus within the *Rhizobiaceae* whose members belong to the alpha subgroup of proteobacteria. Infection by these soil bacteria results in the production of galls arising at the site of infection (Tzfira and Citovsky, 2002). As these galls often occur at the stem–root interface, the disease is referred to as crown gall. *Agrobacterium* has an extremely broad host range and affects a wide range of agriculturally important plants including stone fruit and nut trees, grapevines and ornamentals. During disease initiation, agrobacteria living in the soil may chemotax towards plant wound sites and attach to host tissues. Conditions present at the plant wound site, including low pH, sugars and plant phenolic compounds, stimulate expression of the virulence regulon (*vir* regulon). The products of the *vir* regulon mediate the excision and transfer of a specific segment of DNA present on the Ti plasmid (T-DNA) to the plant host. This transfer requires a type IV secretion system encoded by the *virB* operon of this regulon. The T-DNA transits to the plant cell nucleus where it is integrated into the genome and expressed. The production of plant growth regulators encoded on the T-DNA leads to tumour formation. Nitrogenous compounds, called opines, are produced by the incorporated T-DNA and used by the surrounding agrobacteria as a nutrient source. These opines are specific to the infecting *Agrobacterium* strain and fall into a number of classes, including octopine and nopaline. Control of crown gall is achieved primarily through quarantine and cultural practices, although an effective biological control can be achieved for some strains using *A. rhizogenes* K84 that produces bacteriocins (McClure *et al.*, 1998). The sequenced strain, *A. tumefaciens* C58 (Goodner *et al.*, 2001; Wood *et al.*, 2001), contains a nopaline type Ti plasmid and is unusual in that it contains both a circular and linear chromosome (Jumas-Bilak *et al.*, 1998).

12.3.2 Ralstonia solanacearum

Ralstonia solanacearum is a soilborne pathogen belonging to the beta subgroup of proteobacteria. The species is diverse and contains three races and six biovars based on host range, molecular and phenotypic analyses (Hayward, 2000). *R. solanacearum*, the causative agent of bacterial wilt, has an extremely wide host range that includes over 200 plant families and numerous economically important species including tomato, potato and banana. The pathogen causes latent infections in weeds and other hosts making its eradication difficult and necessitating more stringent quarantine inspections. *R. solanacearum* colonises roots and enters its host via wound sites (Schell, 2000). Extracellular degradative enzymes likely facilitate the pathogens entry into the vascular system where it accumulates in the xylem. During its rapid multiplication in this tissue, the organism produces an extracellular acidic polysaccharide (EPS I) which blocks nutrient and water flow in the host resulting in wilting and death. *R. solanacearum* controls expression of these virulence factors in response to the specific environment it inhabits (i.e. host vs soil). This response is controlled by the LysR homologue, PhcA. The PhcB quorum-sensing system activates PhcA in response to high localised cell densities achieved in the host. Activation of PhcA leads to production of EPS I and extracellular degradative enzymes. In its inactive state, PhcA allows the expression of siderophores and other products that may enhance competition in soil or rhizosphere environments. A second regulatory system activated by host contact (Brito *et al.*, 2002) controls expression of the type III secretion system required for pathogenicity, survival in the host and recognition by plant defence systems (hypersensitive response). Control of bacterial wilt is achieved primarily via quarantine and the use of appropriate cultural practices, although efforts to breed effectively resistant hosts continue. The genome of strain GMI1000 has been sequenced (Salanoubat *et al.*, 2002).

12.3.3 Xanthomonas

The genus *Xanthomonas* is composed primarily of plant pathogens (Mew and Swings, 2000). Members of this genus belong to the gamma subgroup of proteobacteria and infect a wide range of plant species including the economically important crop species rice, wheat, corn, tomato, rapeseed and citrus. Twenty species of *Xanthomonas* have been classified whose host range varies from highly specific to broad. Xanthomonads are generally poor soil competitors and exist primarily in pathogenic or epiphytic association with their plant hosts and on seeds. The genomes of *X. axonopodis* pv. *citri* and *X. campestris* pv. *campestris* have been sequenced (da Silva *et al.*, 2002). *X. axonopodis* pv. *citri* is the causative agent of citrus canker which is manifested by the appearance of canker lesions that lead to the loss of both fruit and leaves. *X. campestris* pv. *campestris* causes black rot of crucifers resulting in leaf chlorosis and vascular infection leading to wilting and necrosis. Entry into the plant host commonly occurs via natural openings (hydathodes) or wound sites from which bacteria enter the host and multiply. Virulence is mediated in part by extracellular polysaccharides in conjunction with proteases, cellulases and pectinases that degrade plant tissues. Type II, or sec-dependant, secretion mediates the export of the extracellular degradative enzymes and is therefore a critical virulence determinant. The expression of genes that encode these products is mediated by the *rpf* gene cluster (Dow and Daniels, 2000). This system is not well characterised but appears to encode a complex heirarchical sensory mechanism in which both a quorum-sensing system (RpfF, RpfB) and two-component regulators (RpfC) are present. Type III secretion systems have been well characterised in *Xanthomonas* that are also required for pathogenicity (Bonas and Van den Ackerveken, 1999; Lahaye and Bonas, 2001). Antibiotic treatments are commonly used for disease control although resistant strains are becoming more common (McManus *et al.*, 2002).

12.3.4 Xylella fastidiosa

Xylella fastidiosa is a Gram-negative bacterium belonging to the gamma subgroup of proteobacteria. Individual pathovars are responsible for diseases of economically important crops including citrus variegated chlorosis of orange, Pierces disease of grape, phony peach disease and leaf scorch in a wide variety of tree species (Purcell and Hopkins, 1996). *X. fastidiosa* also infects a wide range of other plants in which symptoms are not apparent, making effective control difficult. The pathogen is not known to survive in soils and is transmitted by xylem-feeding sharpshooter leafhoppers. The bacteria colonise the cibarial pump and oesophageal lining of these insect vectors. Little else is known about this essential phase in the life cycle of the pathogen. Once inside the plant host, the bacteria move through and colonise the vascular system where they multiply and produce fibrous aggregates. The xylem of infected plants is blocked causing symptoms of water stress, although it is unclear if a host response or bacterial products are responsible. Control methods include careful removal of diseased tissues and the use of insecticides to reduce vector transmission. The genome of *X. fastidiosa* 9a5c, the pathovar responsible for citrus variegated chlorosis, has been sequenced (Simpson *et al.*, 2000).

12.4 Genome sequence analyses of phytopathogenic bacteria

Successful infection by any pathogen is mediated through complex interactions between pathogen, host and environment. Each facet of this disease triangle must be addressed if effective control measures are to be developed. Genomics promises to provide new information to aid in the study of these interactions. Genome sequence is available for a number of plant pathogens (*Table 12.1*) as well as the model plant host *Arabidopsis thaliana* (The Arabidopsis Genome Initiative, 2000). Genetic systems that allow efficient molecular characterisation are available for each of the sequenced pathogens with the exception of *Xylella*, although efforts are underway to develop these tools in *X. fastidiosa* (da Silva Neto *et al.*, 2002). The combination of genome data for both host and pathogens coupled with the power of genetic analysis should allow researchers to quickly address key questions related to pathogenesis and host interactions.

In this section we will discuss four analyses commonly performed on genome data: the identification of *features*, *anomalous regions* and *systems* and *comparisons* between organisms. For each analysis we will describe basic *informatics* approaches used to generate these data and give examples relating to the *biology* of the organism suggested by published genome analyses of phytopathogenic bacteria.

12.4.1 Features

Genome analysis requires the accurate identification of information modules embedded within primary sequence. Such modules are referred to as features of the genome and include protein-coding genes, RNA species and mobile elements. The identification, annotation and categorisation of these genome features produces a tremendous amount of data. To use this information to its best advantage, most genome teams assemble a website that allows easy access to these data by the scientific community (*Table 12.1*). These websites provide access to the annotated information for each feature, sequence data for each gene and protein, maps, and search methods. Such tools are invaluable to the bench researcher as they allow the identification of genes of interest and facilitate molecular analyses. The National Center for Biotechnological Information sponsored by the National Institutes of Health (NCBI, http://www.ncbi.nlm.nih.gov/Genomes/index.html) also provides an increasing number of tools to facilitate genome analysis.

Informatics

The initial analysis of primary sequence has three stages: identification, annotation and categorisation of features. Informatics tools have been developed to facilitate each of these stages (see *Table 12.2* and Mount, 2001).

Identification of features

Protein-coding genes

A start codon that defines the first amino acid of the protein, and a stop codon that truncates protein synthesis delimit an open reading frame (ORF). In the vast

Table 12.2. Examples of informatics programs available for genome analysis

Program	URL
Metabolic pathways BioCyc Kyoto Encyclopaedia of Genes and Genomes (Kanehisa *et al.*, 2002)	http://BioCyc.org/ http://www.genome.ad.jp/kegg/metabolism.html
Open reading frame identification Glimmer (Delcher *et al.*, 1999a) GeneMark (Lukashin and Borodovsky, 1998)	http://www.tigr.org/software/glimmer/ http://opal.biology.gatech.edu/GeneMark/
Phylogeny Cluster of Orthologous Groups (Tatusov *et al.*, 2001) PHYLIP	http://www.ncbi.nlm.nih.gov/COG/ http://evolution.genetics.washington.edu/phylip.html
Profile alignment HMMer (Durbin *et al.*, 1997)	http://hmmer.wustl.edu/
Protein localization and substructure Psort (localization) (Nakai and Kanehisa, 1991) SignalP (signal peptides) (Nielsen *et al.*, 1997) TMHMM (transmembrane domains) (Krogh *et al.*, 2001)	http://psort.nibb.ac.jp/ http://www.cbs.dtu.dk/services/SignalP/ http://www.cbs.dtu.dk/services/TMHMM/
Protein motifs Pfam (Bateman *et al.*, 2002)	http://pfam.wustl.edu/
tRNA identification tRNAscan (Lowe and Eddy, 1997)	http://www.genetics.wustl.edu/eddy/tRNAscan-SE/
Whole genome comparisons MUMmer (Delcher *et al.*, 2002)	http://www.tigr.org/software/mummer/
Other annotation tools INTERPRO (Mulder *et al.*, 2002) Gene Ontology Consortium (TheGeneOntologyConsortium, 2001)	http://www.ebi.ac.uk/interpro/ http://geneontology.org/

majority of cases, genes in newly sequenced genomes are described solely in terms of their ORFs, rather than the larger region that includes upstream regulatory sequences. Protein-coding genes can be identified in two ways: by similarity to existing sequences and by intrinsic methods.

Similarity-based methods compare the predicted amino acid sequence of an ORF to previously identified sequences in a database. Significant matches suggest that the ORF encodes a protein, especially if the match is to a protein with an experimentally defined function. The most widely used similarity-detection tool is the Basic Local Alignment Search Tool (BLAST) program (Altschul *et al.*, 1997). BLAST compares the input or 'query' sequence (BLASTN-nucleotide; BLASTP-amino acid) to sequences in a database and identifies matches or 'subject sequences' that are similar to the given sequence. The BLAST algorithm is designed to find significant local alignments. A local alignment may include only a fraction of one or both sequences being compared, for example, when they only share a conserved domain. For this reason it is important to consider the 'coverage' of the sequences, i.e. the portion of each sequence that participates in the alignment. The degree of similarity (the score) is reported together with a measure of the statistical significance of the alignment (the expect or e-value). The score reflects the similarity of aligned pairs of nucleotides or amino acids and the presence of gaps in the alignment. The e-value is based on the similarity score and database size with smaller e-values representing more statistically significant alignments. In a database the size of Genbank expect values greater than 10^{-5} are generally not considered significant.

Using sequence similarity to identify genes has the obvious drawback that it will fail if the feature in hand does not resemble anything in the database. Among sequenced genomes a large percentage of genes (20–40%) fall into this category and are called hypothetical genes, a subset of which may consist of rapidly evolving genes for which similarity-based tools fail. Intrinsic methods are uniquely valuable for identifying hypothetical genes, although they can identify all gene classes. Such approaches attempt to capture statistical patterns that are representative of genes within an organism (e.g. codon usage). A program based on such a method needs a *training set:* a set of sequences known to encode proteins in the organism being studied. In the absence of experimental data, one way to define a training set is to scan the genome for ORFs longer than some predetermined threshold. The statistical properties of the codons in these ORFs are then tabulated and used to search for additional ORFs with similar properties. Such programs assign a probability to each newly identified ORF that can be used to select those likely to represent real protein-coding genes. As described, the programs may fail to find laterally transferred genes since these will tend to have different statistical properties. These limitations are mostly overcome with the use of other tools during the genome annotation process (see 'Anomalous regions' below). Compared to eukaryotic gene finding, the success rate of these programs is fairly high; they likely identify more than 90% of the genes in a prokaryotic genome, while having a low rate (less than 5%) of falsely predicted ORFs.

Other genes

Non-protein-coding genes are found in a variety of ways. Ribosomal RNA genes and other RNA species are usually identified by their similarity to examples in the databases (using, for example, the BLASTN program). A precise determination of their borders typically requires the use of a secondary structure prediction program. Transfer RNA genes are found by intrinsic methods implemented by such programs as tRNAscan-SE (Lowe and Eddy, 1997).

Annotation

For the purposes of this review we define annotation to mean the assignment of *putative* function to genes, a task that relies extensively on sequence similarity. Function is assigned to new genes based on their similarity to sequences available in databases such as Genbank. If the new gene is found by BLAST comparison to be similar to sequences with assigned function, and if the e-value is significant (typically $<e = -10$), the same function is assigned to the new gene. This occurs fairly often and allows putative functions to be assigned for most of the genome. Many genes, however, will be similar to hypothetical genes or genes of unknown function, and are collectively called 'conserved' hypothetical genes. Complications arise that must be resolved by the judgement of a curator when significant matches occur to multiple proteins with distinct functions, when the e-value is marginally significant (roughly $-5 > e > -10$), or when only segments of the new gene match database sequences. The accuracy of annotation relies predominantly on the quality of previous annotations. Many curators now qualify their functional assignment by assigning a confidence level that reflects the quality of the match used to assign function (e.g. 'function experimentally determined' or 'function based on domain similarity').

Categorisation

Categorising features provides insight into the biological potential of an organism. The various features are usually assigned to categories (e.g. regulation, amino acid biosynthesis) based on those originally defined by Monica Riley for *Escherichia coli* gene classification (Riley, 1993). Neither the categories nor the criteria for assigning genes to them are standardised, however, and rely on decisions made by individual curators, a process that results in discrepancies between genomes. In order to compensate for this, a number of more uniform approaches are being developed. Standards being defined by The Gene Ontology Consortium (The Gene Ontology Consortium, 2001) provide a common and controlled vocabulary to name molecular functions, biological processes, and cellular components. This standard is being used in a number of major eukaryote genome projects, but so far it has not been widely adopted for prokaryotic genomes.

Mobile elements

Insertion sequences (IS) are transposable elements typically composed of a transposase flanked by inverted repeat sequences (Mahillon and Chandler, 1998). These repeats are usually 15–25 bp in length and may not be exact. Short direct repeats (as small as 2 bp) usually flank the inverted repeats. Compound transposons are formed by two IS elements flanking a group of genes that move together as a unit and are thought to facilitate transmission of gene blocks both within and between species (lateral transfer). IS elements are initially identified by BLAST comparison to conserved transposase sequences. Complete characterisation and classification of IS elements, however, requires identification of the direct and inverted repeats. The IS database is a valuable reference for identifying and classifying IS elements (http://www-IS.biotoul.fr/is.html).

Biology

As previously noted, the bulk of biological information garnered from genome sequencing is provided by feature identification. In addition to virulence factors found in other plant pathogens (discussed below), a wide range of genes similar to those that encode pathogenicity determinants of mammalian pathogens were identified in the sequenced phytopathogens. This is not unexpected since previous work has shown that plant and animal pathogens share a number of common pathogenicity and virulence factors (Cao *et al.*, 2001). Examples among the sequenced phytopathogens include numerous proteins similar to adhesins of mammalian pathogens. These include type IV fimbriae that are required for adherence, twitching or gliding motility and virulence (Liu *et al.*, 2001; Mattick, 2002). These fimbriae in *X. fastidiosa* may play a role in attachment to both the host and the hindgut of the sharpshooter vector (Simpson *et al.*, 2000). Consistent with this, fimbriae are observed in both host and vector. Bacterial adhesion in the sharpshooter vector is ordered, consistent with the polar attachment mediated by type IV pili. Since no flagellar systems were identified in *X. fastidiosa* (Bhattacharyya *et al.*, 2002), spread within the plant host may be mediated by bacterial growth, plant fluid mechanics within the xylem or by retracting type IV pili. Five type IV fimbrial operons were identified in *R. solanacearum*, at least one of which plays a role in virulence (Liu *et al.*, 2001). A large number of filamentous haemagglutinins (14 total) were also identified in *R. solanacearum*, although representatives were found in all sequenced phytopathogens. Filamentous haemagglutinins are pathogenicity determinants required for attachment in *Bordetella pertussis* (Alonso *et al.*, 2002). These proteins have recently been shown to play a similar role in the plant pathogen *Erwinia chrysanthemi* suggesting that their presence and function is conserved among plant and animal pathogens (Rojas *et al.*, 2002). Other orthologues of animal virulence genes include haemolysins (*At*, *Rs*, *Xac*, *Xcc*, *Xf*, discussed below), IalA and IalB invasion-related proteins found in *Brucella melitensis* (*At*) and an LpxO orthologue that mediates lipid A modification and virulence in *Salmonella* (*Xf*, see Bhattacharyya *et al.* (2002)).

12.4.2 Anomalous regions

Having the complete sequence of a genome makes it possible to look for regions that are distinctive in some way with respect to the genome as whole. Usually this is done to identify genomic islands: regions containing genes with a related function that may have been laterally (horizontally) acquired. When such islands contain genes related to pathogenesis, they are called pathogenicity islands (Hacker and Carniel, 2001).

Informatics

Due to the degeneracy of the genetic code, multiple codon choices are available to denote most amino acids. The frequency at which bacteria use particular codons, the distribution of dinucleotide pairs and GC content within a genome are each characteristic for a given organism (Karlin *et al.*, 1998). Genome-wide analysis of such factors can therefore be used to predict subsets of genes that may have recently arrived from different species (i.e. laterally transferred genes) (Karlin, 2001). Characteristic features such as tRNA genes, direct repeats or mobile genetic elements often flank genomic islands and are used to define them.

Biology

In addition to evolutionary pressures that reassort and modify genes at the nucleotide level, acquisition of new traits via lateral transfer from other organisms is common among bacteria. Such transfer is mediated by a number of mechanisms including phage transduction, transformation and conjugation. Examples of such transfer can be inferred from genome analyses. In *X. fastidiosa* 7% of the genome consists of phage remnants (Simpson *et al.*, 2000). The *vapA* gene similar to that found in the sheep pathogen *Dichelobacter nodosus* was found associated with phage genes suggesting that it entered the *Xylella* genome via transduction. Natural competence for DNA uptake has been observed for *R. solanacearum* and may account for the large number of genes predicted to have entered the genome via lateral transfer (Salanoubat *et al.*, 2002). An analysis of regions with differential GC content and codon usage identified 93 regions likely to have entered the genome via lateral transfer, 43 of which were associated with mobile elements such as insertion sequences (Salanoubat *et al.*, 2002). In addition, the Tra and Trb systems associated with conjugal transfer were identified as part of a conjugative transposon. Acquisition of novel traits from other organisms has been proposed as a primary mechanism in the evolution of bacteria into pathogenic and symbiotic lifestyles (Ochman and Moran, 2001). Evidence for this was seen in the genomes of *A. tumefaciens* and the closely related nitrogen-fixing legume symbiont *Sinorhizobium meliloti*. Although the genomes of these organisms were similar suggesting that they shared a recent common ancestor, they differ in their complements of genes involved in pathogenic (*vir* genes and T-DNA) and symbiotic (*nod* genes) associations with plants. These genes had unusual GC content and codon usage consistent with their recent acquisition via lateral transfer. Differential acquisition of such traits by the ancestral progenitor of these organisms likely led to their divergence into pathogenic and symbiotic lifestyles.

12.4.3 Systems

Availability of a well-annotated genome sequence allows the automated identification of entire systems including those involved in metabolism, transport, secretion and regulation.

Informatics

Metabolism

Given a set of reference metabolic pathways it is possible to determine if similar pathways are present in newly sequenced genomes given accurate annotation data. A key aspect of this process is the assignment of enzyme commission (EC) numbers that reflect the enzymatic function of the predicted proteins. It should be noted that steps in these enzymatic pathways predicted to be absent by automated analyses may in fact be performed by novel genes. A number of reference metabolic pathways are available that provide tools to facilitate automated pathway analyses including the Kyoto Encyclopaedia of Genes and Genomes (http://www.genome.ad.jp/kegg/metabolism.html).

Transport
Classification of transport systems has been pioneered by M. Saier and collaborators who have devised a transport classification (TC) scheme similar to the EC system that exists for enzymes (http://tcdb.ucsd.edu/tcdb/background.php (Saier, 1999)). Transporters are identified by sequence similarity to known transporter gene sequences in specialised databases that include TC classifications.

Regulation
Two methods are commonly used to identify regulatory proteins: (i) sequence similarity with regulatory proteins in databases or (ii) identification of regulatory motifs or domains within the candidate protein by comparison to specialised databases (e.g., http://pfam.wustl.edu). Regulatory roles can be predicted for new proteins given a statistically significant match against one of the regulatory domains. Genome data also allow the identification of regulatory sequence motifs, such as binding sites targeted by transcriptional regulators. Analysing the promoters of co-regulated gene sets for related sequence motifs can identify candidate binding or regulatory sites (Fouts *et al.*, 2002). The presence of such a motif in the promoter of a gene suggests that the transcriptional regulator that interacts with that site *directly* mediates its expression. Once the motif is identified, and its function experimentally defined, additional occurrences in the genome can be located. The occurrence of such a motif in the promoter of new genes suggests that they may be co-regulated with the initial set.

Biology

Metabolism and transport
Analyses of predicted metabolic and transport systems in phytobacteria reveal many that could promote survival in the rhizosphere and in association with plant hosts. *A. tumefaciens* and *S. meliloti* harbour extensive transport and metabolic capabilities to utilise sugars, amino acids and peptides commonly found in the rhizosphere. *A. tumefaciens* also contains the largest number of ATP-Binding Cassette (ABC) transporters found among sequenced bacteria to date. These high-affinity transporters make up more than 60% of the transport complement of this species and may enhance the ability of *Agrobacterium* to compete for nutrients in rhizosphere and soil environments. Iron acquisition systems present in each of the sequenced phytopathogens likely confer competitive advantages to organisms in iron-sequestered or limited environments. Although no secreted iron-binding proteins (siderophores) were identified in *X. fastidiosa*, ferrous transport systems are present and ferric forms may be imported as complexes with citrate or malate that are found naturally in the xylem of plants (supplemental data (Bhattacharyya *et al.*, 2002)). It has been suggested that the large number of iron transporters in *X. fastidiosa* may deplete iron stores in the plant leading to the variegation seen in infected leaves (Simpson *et al.*, 2000).

Secretion
Type I secretion systems, commonly associated with the secretion of haemolysins in pathogenic bacteria (Ludwig, 1996), are present in each of the sequenced

phytopathogens. Haemolysins belonging to the Repeat-In-Toxin (RTX) family, virulence factors in mammalian pathogens, were identified in the sequenced phytopathogens. The role of such proteins in plant pathogens is unclear, however, deletion of the single RTX family member in *A. tumefaciens* had no effect on either virulence or haemolytic activity (Peterson and Wood, unpublished results).

As noted above, type III secretion systems and the effectors they translocate are key targets of genome researchers because of their critical role in pathogenicity. Genome analyses revealed 17 putative effector proteins in the xanthomonads. Many of these were species-specific suggesting that they may mediate host range or differences in virulence. Forty effector candidates were identified in *R. solanacearum*, of which 14 are similar to previously identified Avr proteins in other phytopathogenic bacteria. The authors noted that these findings were surprising since host range restrictions mediated by single genes had not previously been identified in this pathogen. Surprisingly, type III secretion systems were not identified in *A. tumefaciens* or *X. fastidiosa* indicating that such systems are not ubiquitous in phytopathogenic bacteria. It has been speculated that flagellar synthesis machinery present in *A. tumefaciens* could translocate putative virulence proteins as has been described in *Yersinia enterocolitica* (Goodner *et al.*, 2001). In contrast, no flagellar biosynthesis components are present in the three *X. fastidiosa* pathovars analysed to date (Bhattacharyya *et al.*, 2002). Complete or partial type IV systems are present in each of the sequenced phytopathogens (Van Sluys *et al.*, 2002b), however, with the exception of those in *A. tumefaciens* (Chen *et al.*, 2002), their role in pathogenesis remains to be determined.

Regulation

Tight regulation of gene expression is likely important for survival and competition in complex environments such as the rhizosphere. It has been noted that the regulatory complement of an organism increases in relation to the complexity of the environments that it inhabits (Stover *et al.*, 2000). This same trend is seen among the sequenced phytopathogens. Many genes of plant-pathogenic bacteria are induced in response to host signals suggesting that virulence systems are also tightly regulated. One reason for this may be to prevent detection by host defence systems.

Identification of new components of key virulence regulons has been facilitated by genome data. Plant-Inducible-Promoter boxes (PIP boxes) are found upstream of genes regulated by HrpX in *Xanthomonas* that encode type III structural and effector proteins (Fenselau and Bonas, 1995). Genome analysis has identified PIP boxes in the promoters of 17 genes in *X. campestris* pv. *campestris* and 20 genes in *X. axonopodis* pv. *citri* (da Silva *et al.*, 2002). Products encoded by these genes include putative proteases and cell-wall-degrading enzymes. In *R. solanacearum*, six PIP boxes were identified in the promoters of genes whose products are similar to Avr candidates suggesting that they may be exported to host cells (Salanoubat *et al.*, 2002). A similar approach that included genetic and biochemical validation of secreted candidates was used to identify effector proteins secreted by the Hrp system of *Pseudomonas syringae* pv. *tomato* (Fouts *et al.*, 2002; Petnicki-Ocwieja *et al.*, 2002).

12.4.4 Comparisons

Comparative genomics is a powerful tool that compares the total information content of two or more genomes. Such comparisons allow detailed phylogenetic analyses that complement 16s rRNA analyses. A number of efforts are underway to use comparative genomics to identify unique systems expected to define host range, mechanisms for survival in specific habitats and common virulence mechanisms. Although currently limited by the paucity of genome sequences available, comparative genomics promises to yield significant data as more organisms are sequenced.

Informatics

Phylogenetic relationships between organisms can be defined using a number of approaches, the most common of which is 16S rRNA comparisons (see *Figure 12.1*). Additional approaches include analyses of orthologous proteins and identification of conserved gene order between genomes.

An important tool for the identification and analysis of orthologous proteins (i.e. those conserved between species) is the Clusters of Orthologous Groups (COG) database available at NCBI (Tatusov *et al.*, 2001). A COG is defined by at least three orthologous proteins from three organisms for which a certain phylogenetic distance is seen between any two. NCBI researchers have categorised and assigned a putative function to each COG with the goal of predicting function based on COG membership. The COGnitor program provided by NCBI allows genome researchers to obtain COG classifications for most proteins. A number of proteins will not be assigned COGs as they are not sufficiently similar to those from other organisms. These may be among the most interesting, as they may be responsible for the unique properties exhibited by the organism. Once COGs have been assigned, the new genome can be immediately compared to all other genomes available in the COG database. This provides a useful overview of the genome in relation to the others in the database and makes COG classification a useful complement to more standard phylogenetic analyses.

A whole genome alignment between closely related organisms provides insight into the extent of nucleotide and gene order conservation and allows the detection of genomic rearrangements (inversions or translocations). Alignments of large genomes (>3 Mb) require specially designed tools (e.g. MUMer (Delcher *et al.*,

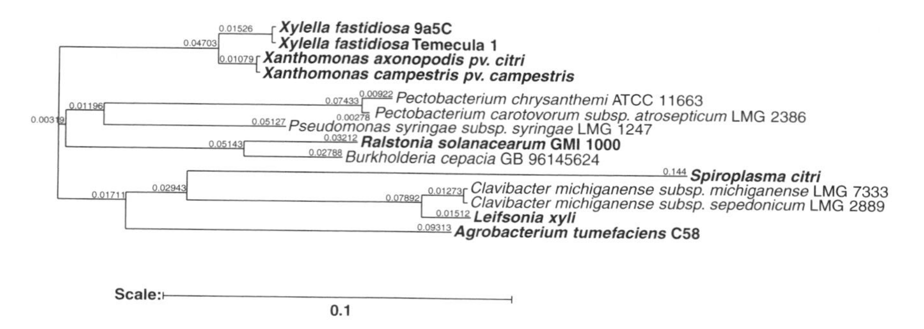

Figure 12.1. Phylogenetic analysis of phytopathogenic bacteria targeted for genome sequencing. Neighbour joining tree constructed using available 16s rRNA sequences from phytopathogenic bacteria for which genome sequencing is complete (shown in bold) or in progress. Tree analysis was performed with standard parameters using the online PHYLIP package at http://rdp.cme.msu.edu/html/.

1999b)) since typical similarity programs such as BLAST cannot cope with large sequences. Comparison at the protein level can identify regions of conserved gene order that reflect both the evolutionary relatedness of two organisms as well as a likely conservation of function between the two gene sets. Such comparisons can also be used to identify organism-specific genes that may be responsible for any unique phenotypes.

Biology

Phylogeny

Genome sequence data can be used to define evolutionary relationships between organisms that extend and complement traditional 16S rRNA comparisons. The relationship between *A. tumefaciens* and the nitrogen-fixing plant symbiont *Sinorhizobium meliloti* serves as an example of the power of such tools. Analyses of their genomes confirm their close evolutionary relationship (Wood *et al.*, 2001). Significant similarities were identified within the predicted proteomes of these organisms. This similarity extends to the nucleotide level and includes extensive conservation of gene order between the circular chromosomes. Although the predicted proteins encoded by the other replicons are quite well conserved, gene order conservation is not evident. The latter finding suggests that the other replicons (two megaplasmids in *S. meliloti* and two plasmids and a linear chromosome in *A. tumefaciens*) were subject to pressures that resulted in the rapid assortment of their gene complements. One might speculate that this was due to the conjugative properties of the replicons that would allow the transfer and acquisition of new genes. These findings support a recent common ancestor for these plant-associated bacteria from which the lineages rapidly diverged into pathogenic and symbiotic lifestyles.

Other studies suggest that the identification of markers in broadly conserved proteins within genomes will facilitate studies of evolutionary divergence (Gupta and Griffiths, 2002). While evolutionary relationships are difficult to define, the availability of complete genomes is certain to provide additional insights into the relationships between organisms.

Comparative genomics

A recent comparison of the genomes of available plant-associated bacteria highlights both the value and difficulties of comparative genomics (Van Sluys *et al.*, 2002b). The authors compared the complete genome sequences of the phytopathogens *A. tumefaciens*, *R. solanacearum*, *X. axonopodis* pv. *citri*, *X. campestris* pv. *campestris*, *X. fastidiosa* and the nitrogen-fixing symbionts *Sinorhizobium meliloti* and *Mesorhizobium loti*. These Gram-negative bacteria represent three phylogenetic branches within the proteobacteria and exhibit distinct plant interactions. Those organisms with the most diverse life cycles were found to be more metabolically complex and had extensive regulatory systems to manage this complexity. The smallest and least complex genome, that of *X. fastidiosa*, was speculated to have evolved in response to the limited environments which it inhabits.

Consistent with the massive destruction of plant tissue seen with black rot, *X. campestris* pv. *campestris* has the potential to produce the widest spectrum of degradative enzymes. The export of these and other extracellular proteins depends on type II secretion systems. In agreement with the importance of such products to their mode of virulence, *X. axonopodis* pv. *citri*, *X. campestris* pv. *campestris* and *R. solanacearum* each contain two type II secretion systems as compared to the single system found in the other plant-associated bacteria. In contrast, the authors noted that *X. fastidiosa* contained only a single polygalaturonase gene that was likely to be non-functional. As previous work has linked the presence of such proteins to vascular spread, the authors speculated that the loss of this function could be responsible for the long incubation period of CVC. Consistent with this, recent work has shown that the more aggressive *X. fastidiosa* pathovar responsible for Pierces disease of grape has an intact copy of this gene (Van Sluys *et al.*, 2002a). With the exception of a syringomycin synthase found in *R. solanacearum*, no known phytotoxins were identified in this group.

Genes involved in resistance to oxidative stress were found in all pathogens, but were limited in *X. fastidiosa*. This xylem-limited pathogen contains only a single copy of the antioxidant glutathione-S-transferase (as compared with 17 copies in *M. loti* and *S. meliloti*) and lacks both the OxyR and SoxRS systems that mediate the expression of products which protect the bacterium against oxidative stress. In addition, *X. fastidiosa* did not contain DNA polymerase IV (DinP), a member of the SOS regulon mediating DNA repair, which occurs in multiple copies in *A. tumefaciens*, *S. meliloti* and *M. loti*. The authors speculate that these differences may be due to the increased exposure to DNA-damaging agents that the latter organisms are expected to encounter in their diverse habitats.

The authors also attempted to identify genes unique to plant-associated bacteria. Nineteen genes were identified in these organisms that were not found in the non-plant-associated reference group composed of *Escherichia coli*, *N. meningitidis* and *Caulobacter crescentus*. Many of these genes appeared to be localised to the membrane, suggestive of proteins involved in initial host interactions.

As noted by the authors, an obvious drawback to these analyses was the limited number of, and extensive diversity among, the organisms under study. Examining genome sequences of closely related organisms will facilitate similar studies in the future. The more similar the genomes and lifestyles being compared, the more likely we are to find meaningful distinctions responsible for specific phenotypic variations in host range or disease. Two examples highlight the value of this approach.

The first is an analysis of the closely related pathogens *X. axonopodis* pv. *citri* and *X. campestris* pv. *campestris* (da Silva *et al.*, 2002). Each organism harbours distinct complements of type III effectors/avirulence genes, a finding that may reflect in part the host range distinctions of these pathogens. *X. axonopodis* pv. *citri* contains fewer genes involved in plant cell wall degradation consistent with the limited tissue maceration evidenced by this strain. Further, *X. axonopodis* pv. *citri* was found to lack the regulatory components, *rpfH* and *rpfI*, which mediate expression of extracellular degradative enzymes in *X. campestris* pv. *campestris*. Such insights are possible due to the extensive similarity of the two organisms and reflect the benefit of sequencing closely related organisms.

The second example is provided by Bhattacharyya *et al.* (2002) who examined commonalities and differences between three *X. fastidiosa* pathovars. The analysis compared partial sequences of *X. fastidiosa* pv. *almond* and *X. fastidiosa* pv. *oleander* to the previously published genome of *X. fastidiosa* pv. *citri*. The authors identified a set of 130 genes common to all three pathovars. Gene sets unique to each pathovar were also identified to the extent possible. These sets, although large (Xfa-132, Xfo-180, Xfc-375), provide candidates for investigators studying the mechanism of host range and symptomatic variation. The authors concede that genes missing in either of the two partially sequenced genomes, or unique to the finished genome, cannot be accurately predicted at this point highlighting the need for completely sequenced genomes.

12.5 Conclusions

The identification of putative pathogenicity factors by the genome researcher is the first step in a long process. The functional role of these factors must be defined using genetic and biochemical analyses. Genome data complement such classical approaches by identifying candidates for further investigation, including those not easily found using genetic screens (Giaever *et al.*, 2002). Given the value of such information, priority should be placed on completing additional genome sequences of key plant-pathogenic bacteria. These projects should include finished sequence data for closely related pathogens that allow effective comparative analyses. An intimate understanding of the factors that influence disease development provided by these methods will provide the basis for effective control of many economically devastating diseases of plants.

References

Alonso, S., Reveneau, N., Pethe, K., Locht, C. (2002) Eighty-kilodalton N-terminal moiety of *Bordetella pertussis* filamentous hemagglutinin: adherence, immunogenicity, and protective role. *Infect. Immun.* **70**: 4142–4147.

Altschul, S.F., Madden, T.L., Schaffer, A.A., Zhang, J., Zhang, Z., Miller, W., Lipman, D.J. (1997) Gapped BLAST and PSI-BLAST: a new generation of protein database search programs. *Nucleic Acids Res.* **25**: 3389–3402.

Baron, C., O'Callaghan, D., Lanka, E. (2002) Bacterial secrets of secretion: EuroConference on the biology of type IV secretion processes. *Mol. Microbiol.* **43**: 1359–1365.

Bateman, A., Birney, E., Cerruti, L. *et al.* (2002) The Pfam protein families database. *Nucleic Acids Res.* **30**: 276–280.

Bell, K.S., Avrova, A.O., Holeva, M.C. *et al.* (2002) Sample sequencing of a selected region of the genome of *Erwinia carotovora subsp. atroseptica* reveals candidate phytopathogenicity genes and allows comparison with *Escherichia coli*. *Microbiology* **148**: 1367–1378.

Bhattacharyya, A., Stilwagen, S., Ivanova, N. *et al.* (2002) Whole-genome comparative analysis of three phytopathogenic *Xylella fastidiosa* strains. *Proc. Natl Acad. Sci. USA* **99**: 12403–12408.

Bonas, U., Van den Ackerveken, G. (1999) Gene-for-gene interactions: bacterial avirulence proteins specify plant disease resistance. *Curr. Opin. Microbiol.* **2**: 94–98.

Brito, B., Aldon, D., Barberis, P., Boucher, C., Genin, S. (2002) A signal transfer system through three compartments transduces the plant cell contact-dependent signal controlling *Ralstonia solanacearum hrp* genes. *Mol. Plant–Microbe Interact.* **15**: 109–119.

Buell, C.R., Joardar, V., Lindeberg, M. *et al.* (2003). The complete genome sequence of the *Arabidopsis* and tomatoe pathogen *Pseudomonas syringae* pv. *tomato* DC3000. *Proc. Natl Acad. Sci. USA* **100**: 10181–10186.

Cao, H., Baldini, R.L., Rahme, L.G. (2001) Common mechanisms for pathogens of plants and animals. *Ann. Rev. Phytopathol.* **39**: 259–284.

Chen, L., Chen, Y., Wood, D.W., Nester, E.W. (2002) A new type IV secretion system promotes conjugal transfer in *Agrobacterium tumefaciens*. *J. Bacteriol.* **184**: 4838–4845.

da Silva, A.C., Ferro, J.A., Reinach, F.C. *et al.* (2002) Comparison of the genomes of two *Xanthomonas* pathogens with differing host specificities. *Nature* **417**: 459–463.

da Silva Neto, J.F., Koide, T., Gomes, S.L., Marques, M.V. (2002) Site-directed gene disruption in *Xylella fastidiosa*. *FEMS Microbiol. Lett.* **210**: 105–110.

Delcher, A.L., Harmon, D., Kasif, S., White, O., Salzberg, S.L. (1999a) Improved microbial gene identification with GLIMMER. *Nucleic Acids Res.* **27**: 4636–4641.

Delcher, A.L., Kasif, S., Fleischmann, R.D., Peterson, J., White, O., Salzberg, S.L. (1999b) Alignment of whole genomes. *Nucleic Acids Res.* **27**: 2369–2376.

Delcher, A.L., Phillippy, A., Carlton, J., Salzberg, S.L. (2002) Fast algorithms for large-scale genome alignment and comparison. *Nucleic Acids Res.* **30**: 2478–2483.

Dow, J.M., Daniels, M.J. (2000) *Xylella* genomics and bacterial pathogenicity to plants. *Yeast* **17**: 263–271.

Durbin, R., Eddy, S.R., Krogh, A., Mitchison, G. (1997) *Biological Sequence Analysis: Probabilistic Models of Proteins and Nucleic Acids.* Cambridge University Press: Cambridge, UK.

Fenselau, S., Bonas, U. (1995) Sequence and expression analysis of the *hrpB* pathogenicity operon of *Xanthomonas campestris pv. vesicatoria* which encodes eight proteins with similarity to components of the Hrp, Ysc, Spa, and Fli secretion systems. *Mol. Plant–Microbe Interact.* **8**: 845–854.

Fouts, D.E., Abramovitch, R.B., Alfano, J.R. *et al.* (2002) Genomewide identification of *Pseudomonas syringae pv. tomato* DC3000 promoters controlled by the HrpL alternative sigma factor. *Proc. Natl Acad. Sci. USA* **99**: 2275–2280.

Giaever, G., Chu, A.M., Ni, L. *et al.* (2002) Functional profiling of the *Saccharomyces cerevisiae* genome. *Nature* **418**: 387–391.

Goodner, B., Hinkle, G., Gattung, S. *et al.* (2001) Genome sequence of the plant pathogen and biotechnology agent *Agrobacterium tumefaciens* C58. *Science* **294**: 2323–2328.

Gupta, R.S., Griffiths, E. (2002) Critical issues in bacterial phylogeny. *Theor. Popul. Biol.* **61**: 423–434.

Hacker, J., Carniel, E. (2001) Ecological fitness, genomic islands and bacterial pathogenicity. A Darwinian view of the evolution of microbes. *EMBO Rep.* **2**: 376–381.

Harper, J.R., Silhavy, T.J. (2001) Germ warfare: The mechanisms of virulence factor delivery. In: Groisman, E.A. (ed) *Principles of Bacterial Pathogenesis*, pp. 43–74. Academic Press, San Diego, CA.

Hayward, A.C. (2000) *Ralstonia solanacearum*. In: Lederberg, J. (ed) *Encyclopedia of Microbiology*. Vol. 4, pp. 32–42. Academic Press: New York.

Jumas-Bilak, E., Michaux-Charachon, S., Bourg, G., Ramuz, M., Allardet-Servent, A. (1998) Unconventional genomic organization in the alpha subgroup of the *Proteobacteria*. *J. Bacteriol.* **180**: 2749–2755.

Kanehisa, M., Goto, S., Kawashima, S., Nakaya, A. (2002) The KEGG databases at GenomeNet. *Nucleic Acids Res.* **30**: 42–46.

Karlin, S. (2001) Detecting anomalous gene clusters and pathogenicity islands in diverse bacterial genomes. *Trends Microbiol.* **9**: 335–343.

Karlin, S., Campbell, A.M., Mrazek, J. (1998) Comparative DNA analysis across diverse genomes. *Annu. Rev. Genet.* **32**: 185–225.

Kjemtrup, S., Nimchuk, Z., Dangl, J.L. (2000) Effector proteins of phytopathogenic bacteria: bifunctional signals in virulence and host recognition. *Curr. Opin. Microbiol.* **3**: 73–78.

Krogh, A., Larsson, B., von Heijne, G., Sonnhammer, E.L.L. (2001) Predicting transmembrane protein topology with a hidden Markov model: Application to complete genomes. *J. Mol. Biol.* **305**: 567–580.

Lahaye, T., Bonas, U. (2001) Molecular secrets of bacterial type III effector proteins. *Trends Plant Sci.* **6**: 479–485.

Liu, H., Kang, Y., Genin, S., Schell, M.A., Denny, T.P. (2001) Twitching motility of *Ralstonia solanacearum* requires a type IV pilus system. *Microbiology* **147**: 3215–3229.

Lowe, T.M., Eddy, S.R. (1997) tRNAscan-SE: a program for improved detection of transfer RNA genes in genomic sequence. *Nucleic Acids Res.* **25**: 955–964.

Ludwig, A. (1996) Cytolytic toxins from gram-negative bacteria. *Microbiologia* **12**: 281–296.

Lukashin, A.V., Borodovsky, M. (1998) GeneMark.hmm: new solutions for gene finding. *Nucleic Acids Res.* **26**: 1107–1115.

Mahillon, J., Chandler, M. (1998) Insertion sequences. *Microbiol. Mol. Biol. Rev.* **62**: 725–774.

Mattick, J.S. (2002) Type IV pili and twitching motility. *Annu. Rev. Microbiol.* **56**: 289–314.

McClure, N.C., Ahmadi, A.R., Clare, B.G. (1998) Construction of a range of derivatives of the biological control strain *Agrobacterium rhizogenes* K84: a study of factors involved in biological control of crown gall disease. *Appl. Environ. Microbiol.* **64**: 3977–3982.

McManus, P.S., Stockwell, V.O., Sundin, G.W., Jones, A.L. (2002) Antibiotic use in plant agriculture. *Annu. Rev. Phytopathol.* **40**: 443–465.
Mew, T.W., Swings, J. (2000) *Xanthomonas*. In: Lederberg, J. (ed) *Encyclopedia of Microbiology*. Vol. 4, pp. 921–929. Academic Press: New York.
Mount, D.W. (2001) *Bioinformatics, Sequence and Genome Analysis*. Cold Spring Harbor Laboratory Press: Cold Spring Harbor, New York.
Mulder, N.J., Apweiler, R., Attwood, T.K. *et al.* (2002) InterPro: an integrated documentation resource for protein families, domains and functional sites. *Brief Bioinform.* **3**: 225–235.
Nakai, K., Kanehisa, M. (1991) Expert system for predicting protein localization sites in gram-negative bacteria. *Proteins* **11**: 95–110.
Nielsen, H., Engelbrecht, J., Brunak, S., von Heijne, G. (1997) Identification of prokaryotic and eukaryotic signal peptides and prediction of their cleavage sites. *Protein Eng* **10**: 1–6.
Ochman, H., Moran, N.A. (2001) Genes lost and genes found: evolution of bacterial pathogenesis and symbiosis. *Science* **292**: 1096–1099.
Petnicki-Ocwieja, T., Schneider, D.J. *et al.* (2002) Genomewide identification of proteins secreted by the Hrp type III protein secretion system of *Pseudomonas syringae pv. tomato* DC3000. *Proc. Natl Acad. Sci. USA* **99**: 7652–7657.
Purcell, A.H., Hopkins, D.L. (1996) Fastidious xylem-limited bacterial plant pathogens. *Annu. Rev. Phytopathol.* **34**: 131–151.
Riley, M. (1993) Functions of the gene products of *Escherichia coli*. *Microbiol. Rev.* **57**: 862–952.
Rojas, C.M., Ham, J.H., Deng, W.L., Doyle, J.J., Collmer, A. (2002) HecA, a member of a class of adhesins produced by diverse pathogenic bacteria, contributes to the attachment, aggregation, epidermal cell killing, and virulence phenotypes of *Erwinia chrysanthemi* EC16 on *Nicotiana clevelandii* seedlings. *Proc. Natl Acad. Sci. USA* **99**: 13142–13147.
Saier, M.H., Jr. (1999) A functional-phylogenetic system for the classification of transport proteins. *J. Cell Biochem.* **Suppl**: 84–94.
Salanoubat, M., Genin, S., Artiguenave, F. *et al.* (2002) Genome sequence of the plant pathogen *Ralstonia solanacearum*. *Nature* **415**: 497–502.
Schell, M.A. (2000) Control of virulence and pathogenicity genes of *Ralstonia solanacearum* by an elaborate sensory network. *Annu. Rev. Phytopathol.* **38**: 263–292.
Simpson, A.J., Reinach, F.C., Arruda, P. *et al.* (2000) The genome sequence of the plant pathogen *Xylella fastidiosa*. The *Xylella fastidiosa* Consortium of the Organization for Nucleotide Sequencing and Analysis. *Nature* **406**: 151–157.
Soanes, D.M., Skinner, W., Keon, J., Hargreaves, J., Talbot, N.J. (2002) Genomics of phytopathogenic fungi and the development of bioinformatic resources. *Mol. Plant–Microbe Interact.* **15**: 421–427.
Soto, G.E., Hultgren, S.J. (1999) Bacterial adhesins: common themes and variations in architecture and assembly. *J. Bacteriol.* **181**: 1059–1071.
Stover, C.K., Pham, X.Q., Erwin, A.L. *et al.* (2000) Complete genome sequence of *Pseudomonas aeruginosa* PA01, an opportunistic pathogen. *Nature* **406**: 959–964.
Tatusov, R.L., Natale, D.A., Garkavtsev, I.V. *et al.* (2001) The COG database: new developments in phylogenetic classification of proteins from complete genomes. *Nucleic Acids Res.* **29**: 22–28.
The Arabidopsis Genome Initiative (2000) Analysis of the genome sequence of the flowering plant *Arabidopsis thaliana*. *Nature* **408**: 796–815.
The Gene Ontology Consortium (2001) Creating the gene ontology resource: Design and implementation. *Genome Res.* **11**: 1425–1433.
Tzfira, T., Citovsky, V. (2002) Partners-in-infection: host proteins involved in the transformation of plant cells by *Agrobacterium*. *Trends Cell Biol.* **12**: 121–129.
Van Sluys, M.A., de Oliveira, M.C., Monteiro-Vitorello, C.B. *et al.* (2003) Comparative analyses of the complete genome sequences of Pierce's disease and citrus variegated chlorosis strains of *Xylella fastidiosa*. *J. Bacteriol.* **185**: 1018–1026.
Van Sluys, M.A., Monteiro-Vitorello, C.B., Camargo, L.E. *et al.* (2002b) Comparative genomic analysis of plant-associated bacteria. *Annu. Rev. Phytopathol.* **40**: 169–189.
Wood, D.W., Setubal, J.C., Kaul, R. *et al.* (2001) The genome of the natural genetic engineer *Agrobacterium tumefaciens* C58. *Science* **294**: 2317–2323.

13

Analysis of microbial communities in the plant environment

Andrew J. Holmes

13.1 Introduction

Plants are surrounded by complex microbial communities (Tiedje *et al.*, 1999; Torsvik *et al.*, 1996). Direct interactions between plants and microbiota are well-known in the form of symbioses and pathogenesis and are discussed in other chapters of this volume. Of interest here is that microbial communities also exert indirect effects on plants. Indirect effects include such phenomena as soil formation, nutrient cycling (especially phosphorous and nitrogen mobilisation), acidification, disease suppression, detoxification and many more (Zhou *et al.*, 2002). These phenomena are emergent properties of microbial activity in the plant environment rather than any specific plant–microbe interaction (Kent and Triplett, 2002; Robertson *et al.*, 1997). Whether direct or indirect interactions are involved there are obviously numerous opportunities for feedback interactions between plant and microbial communities (Bever, 2003).

A series of simple observations underpin questions regarding the importance of the soil microbial community to plant biology. These may be summarised as follows:

1. Plant environments include abiotic, microbiotic and plant components.
2. Change in one of these components will influence the others. This results in natural variation (Baer *et al.*, 2003; Girvan *et al.*, 2003; Zhou *et al.*, 2002), but also provides the potential for engineering change.
3. There is spatial variation in plant productivity and in soil biochemical activity, at least some of which is apparently independent of soil geology (Broughton and Gross, 2000; Cavigelli and Robertson, 2000, 2001; Robertson *et al.*, 1997).
4. This variation is therefore likely to reflect differences in the activity and structure of soil microbial communities. Understanding of soil microbial

Plant Microbiology, Michael Gillings and Andrew Holmes

ecology should provide additional options for engineering the plant environment (Chellemi and Porter, 2001; Ettema and Wardle, 2002; Johnson *et al.*, 2003; Klironomos, 2002; Robertson *et al.*, 1997; Sen, 2003).

The objectives of microbial ecology could be summarised as description, explanation and management (although it is easy to gain the impression that the objective of soil microbial community analysis is to collect lots of different rRNA sequences!). *Description* refers to the task of categorising the component members of an ecosystem, their relevant properties and their variation in space and time. In the case of plant microbiology this includes plants, soil microbiota, soil biochemistry and soil geology. *Explanation* refers to characterisation of the ecological role of these components; the members of the microbial community, their interaction with the soil matrix, their interaction with other microbiota, and their interaction with plants. It enables prediction of ecosystem responses to environmental change. *Management* refers to the exploitation of this knowledge to monitor change in the ecosystem and predict the outcome of manipulations targeted at engineering specific aspects of the plant environment. The vision for soil microbial community analysis includes both its role as a research tool in understanding the basis of plant responses to environmental change and as a management tool to provide new means of monitoring 'soil health' and new opportunities for engineering ecosystem productivity (Broughton and Gross, 2000; Filip, 2002; Hill *et al.*, 2000; Kowalchuk *et al.*, 2003; Kuske *et al.*, 2002; Marschner *et al.*, 2003; McCaig *et al.*, 2001; van Bruggen and Semenov, 2000).

In this review I focus on the key challenges that remain to be solved for microbial ecology to become an established part of plant management. The focus is on bacteria, but most issues are equally relevant to other soil microbiota, including Archaea and fungi.

13.2 The challenges

In a nutshell the principal challenge for soil microbial community analysis is the useful integration of single-microbial-cell properties with community-scale observations. That is, to go from characterising organisms to characterising communities. There are four distinct aspects to this challenge (*Figure 13.1*).

13.2.1 Representation

The comparative approach is a powerful means of identifying factors that differ between systems, but is reliant on having the statistical power to identify significant variation (Ettema and Wardle, 2002). Plant–soil-microbe interactions occur within a highly spatially structured system (Bever, 2003; Treves *et al.*, 2003). As a consequence, collecting a representative sample of soil microbiota across a plant ecosystem is not a trivial task. The sampling effort required is immense and there is little understanding of the appropriate spatial scale for sample collection (Franklin and Mills, 2003; Molofsky *et al.*, 2002). The large sample sizes required also create the issue of sample-processing speed.

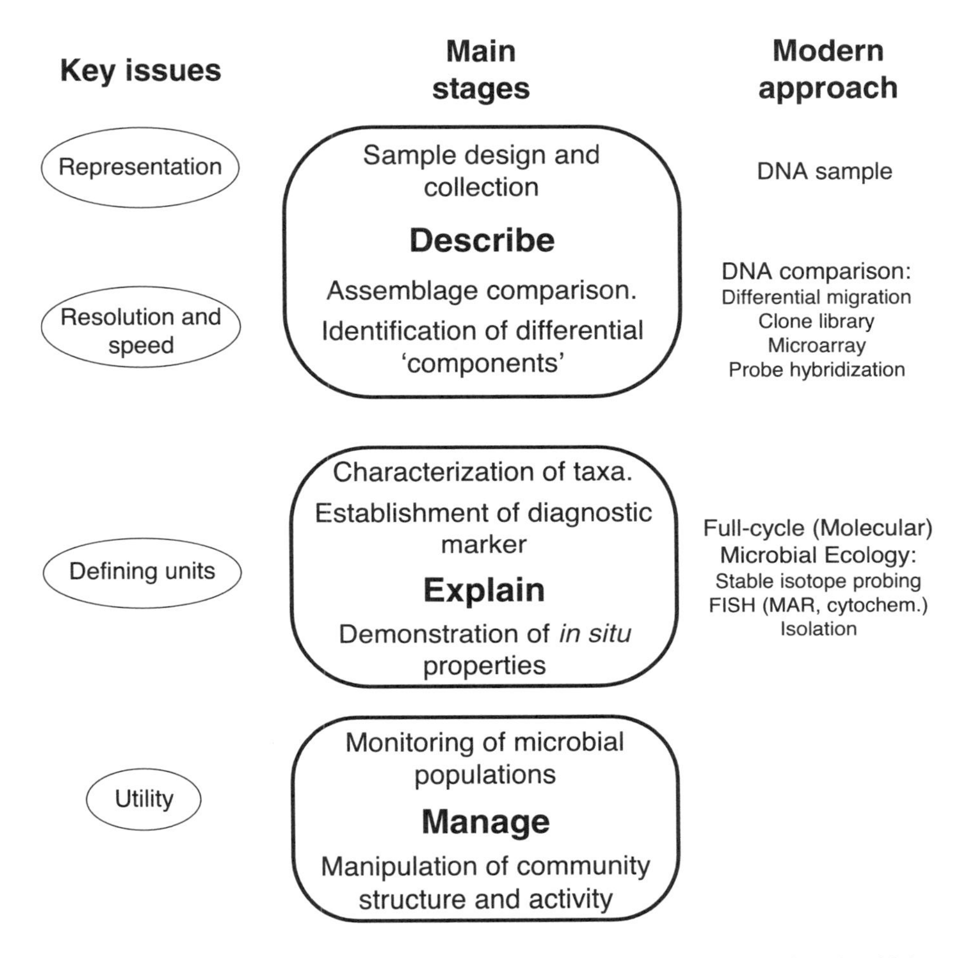

Figure 13.1. Contributions of different data types to polyphasic taxonomy in microbial systematics

13.2.2 Speed

The high diversity of soil microbiota means that a major rate-limiting step is identification, consequently one of the simplest means of improving this is to 'decrease diversity' by employing simple operational classification schemes based on a limited set of easily measured characters. For example, specimens that require a single identification test, or samples that only require sorting into ten broad groups instead of 10 000 species, will obviously be processed faster. This leads to the issue of resolution.

13.2.3 Resolution

Whilst a large soil bacterial sample may be rapidly processed if the bacteria are classified into a small number of operational taxonomic units (OTUs), there is a significant trade-off with respect to the ability to identify biological differences (resolving power). This gets us to the ultimate problem, how do we define microbiological units that are useful for community comparisons? It is useful to consider these issues of speed and resolution in community analysis by analogy to study of invertebrate community structure (Oliver and Beattie, 1996; Pik *et al.*, 1997).

13.2.4 Defining microbiological units

Characterisation of microbiota and their classification represents an important interface between the problem of making measurements on single cells and getting data on a macroscale. Arbitrary (operational) taxonomic units are perfectly adequate for the purposes of comparing biological assemblages, so long as they are reproducible (Bohannan and Hughes, 2003; Hughes *et al.*, 2001). However, to adequately explain the contribution of community members to an emergent property of an ecosystem, ecologically meaningful units must eventually be identified via detailed characterisation of micro-organisms.

The challenge for microbial community comparison could be thought of as: to define OTUs that are amenable to rapid comparison of large biological assemblages *and* compatible with the practice of microbial systematics.

13.3 Characterising micro-organisms

The two major goals of microbial systematics are: (i) to define ecologically meaningful units – populations where the members have equivalent ecological roles; and, (ii) to construct an internally consistent taxonomic hierarchy. A taxonomic hierarchy is considered internally consistent if organisms can only belong to one taxonomic lineage within the hierarchy. Where this is so, one may use higher taxa to improve processing efficiency in assemblage comparisons with minimal information loss and to improve the efficiency of data-mining.

The modern polyphasic approach to microbial systematics recognises that these goals cannot be adequately met by a single aspect of micro-organisms' life history (Gillis *et al.*, 2001). As shown in *Figure 13.2* the different aspects of polyphasic taxonomy could be thought of as: (i) phylogeny – characterising the evolutionary history of the cell, (ii) morphology – the structure and composition of the cell, and (iii) physiology – the activity of the cell. Although all three aspects may be considered equally important for microbial systematics, however for the objectives of community structure analysis molecular sequence data are of particular importance and will be the focus of this discussion. The reasons for this are briefly discussed below.

The traditional approach to assessing microbial diversity was to isolate cells for characterisation of eco-physiological properties in pure culture. To this day it remains the most effective means of comparing ecologically relevant differences

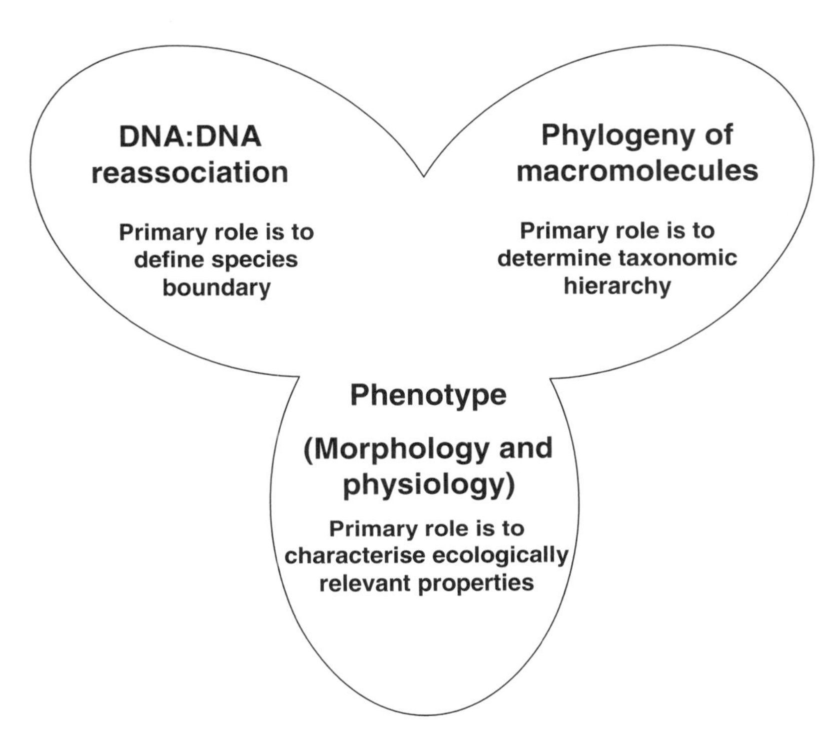

Figure 13.2. Schematic representation of the objectives and challenges of microbial community structure analysis.

between two strains. It is, however, ineffective for exploration of community structure. There are two main reasons underpinning this problem, sampling inefficiency and processing inefficiency.

It is frequently argued that many soil bacteria are 'unculturable' and that this is the major stumbling block to culture-based approaches to community analysis. This is not the case, and recent advances in sampling strategies have given rise to dramatic increases in 'culturability' (Janssen *et al.*, 2002; Leadbetter, 2003; Zengler *et al.*, 2002). The principal reason for the inadequacy of culture-based approaches for community comparison is processing inefficiency; these methods are inherently unsuitable for appropriate sampling strategies for the very large numbers of bacteria present in soil. Each colony requires a reasonable area on a solid growth medium, a relatively long period of time for growth, and multiple independent tests must be performed to classify the colonies once they are obtained. The time, laboratory space and testing regime for even a single soil sample with one billion cells of several thousand species are obviously problematic. Thus even with 'perfect' cultivability, comparison of culture collections would remain as an ineffective basis for comparative community structure analysis.

In conclusion, although characterisation in pure culture remains an integral part of microbial systematics, there is a clear need for reliance on culture-independent means of microbial identification for community comparisons. As is hinted at by

Figure 13.2, of the three broad classes of data used in microbial systematics, only macromolecule sequence data offer the possibility of a culture-independent route for rapid microbial identification that is broadly compatible with the taxonomic hierarchy.

13.4 Molecular surrogates for characterising bacteria

13.4.1 Which macromolecules?

The requirements of a sequence for use as a molecular marker include the following of clock-like behaviour, information content, and vertical transfer. In simple terms these requirements are that the molecule has maintained the same function over evolutionary time, consequently experiencing constant selective pressures (clock-like behaviour). The molecule is sufficiently long and shows sufficient conservation that it retains a large number of sites that are not expected to have undergone multiple changes (useful information content). The molecule is not subject to lateral transfer between lineages. An excellent review of this area may be found in Ludwig and Kenk (2001).

Most discussion of the advantages of the 'molecular era' focuses on the capacity to sample previously unsampled micro-organisms by culture-independent means. This has undoubtedly revolutionised microbial ecology (Kent and Triplett, 2002; O'Donnell and Gorres, 1999; Ogram, 2000; Prosser, 2002; Tiedje *et al.*, 1999; Torsvik and Ovreas, 2002). Nevertheless, in the opinion of the present author the primary advantages of using molecular surrogates for microbial characterisation revolve around their capacity to facilitate rapid identification techniques with high taxonomic discrimination. It is this property that makes them (uniquely) suitable to characterising complex communities but, ironically, it has yet to be fulfilled.

In practice, the culture-independent use of molecular sequences in community structure simply presented new technical challenges with respect to obtaining an effective sample of microbial assemblages. The major difference was that the sampled units were macromolecules rather than cells. These technical challenges are reflected by the proliferation of different strategies for isolation of nucleic acids and for cloning of target sequences throughout the 1990s (see the reviews cited above for details). Major concerns that emerged from this period were the efficiency of nucleic acid recovery, possibility of biased nucleic acid recovery, possibility of biased cloning, and the possibility of introduction of artifacts in the cloning procedure. Although a comprehensive survey remains to be done, direct DNA extraction using a combined physical and detergent lysis method has proven more effective in most tested cases (for examples of method comparisons see Frostegard *et al.*, 1999; Kuske *et al.*, 1998; Zhou *et al.*, 1996). There is no doubt that PCR represents the simplest means by which to recover a specific nucleic acid sequence. There is however considerable debate regarding the relative merits of different primers for recovery of genes (Schmalenberger *et al.*, 2001), the possibilities of amplification bias (Farrelly *et al.*, 1995; Reysenbach *et al.*, 1992; Suzuki and Giovannoni 1996; Suzuki *et al.*, 1998) and the introduction of sequence artifacts (Liesack *et al.*, 1991; Wang and Wang, 1996).

With careful experimentation none of these factors presents a serious impediment to the use of molecular sequence markers in comparative investigations of microbial community structure (Bohannan and Hughes, 2003). They do however emphasise the need for verification of preliminary findings based on molecular sequence data. This is sometimes termed 'full-cycle molecular ecology' (Kowalchuk and Stephen, 2001; Murrell and Radajewski, 2000; Wellington *et al.*, 2003).

To restate the vision for microbial community structure analysis in plant and soil microbiology in terms of molecular ecology: where multiple sites show variation in emergent properties, relevant to plants, that are not able to be accounted for by geographical features, the objective is to be able to test the hypothesis that microbial activity can explain the phenomenon. This involves:

1. Testing for variation in the pattern of an appropriate macromolecular sequence marker across sites.
2. Identifying those aspects of the molecular view of the microbial assemblage that are correlated with the environmental property of interest.
3. Linkage of the molecular marker (operational taxon) to a set of organisms.
4. Characterisation of the properties (phylogenetic, physiological and morphological) of members of this group, definition of ecologically relevant units (species) within the group, and definition of a diagnostic marker(s) for these species.
5. Demonstration that the *in situ* activity of the species contributes to the environmental property of interest.
6. Exploitation of the marker(s) within a management plan.

13.5 Measuring community richness

13.5.1 Macromolecule sequences

The molecular markers used in microbial ecology are frequently classified into two broad groups, phylogenetic genes and functional genes. These terms are perhaps unfortunate since no gene is phylogenetic and virtually all genes are functional. They do however reflect different applications; in that molecular sequences tend to be used either as markers to identify organisms within a 'universal' *phylogenetic framework* or as markers for a particular *metabolic function*. Here I will refer to universal markers or metabolic markers.

Universal markers

The recent explosion in genome sequences has allowed a comprehensive survey for potential universal molecular markers. It has been estimated that there are less than 100 candidates (Ludwig and Kenk, 2001). There are numerous, more practical requirements for the use of these genes in bacterial classification meaning that, to date, relatively few have been used. These include the three ribosomal RNAs, RecA (Fiore *et al.*, 2000), catalytic subunit of ATPase (Ludwig *et al.*, 1998), RpoB (Dahllof *et al.*, 2000), and elongation factor Tu/1a (Jenkins and Fuerst, 2001). Of these, the SSU rRNA was the first used, is the only one with a comprehensive database, and is by far the most widely used in the soil environment.

Metabolic markers

In almost all cases the organisms contributing to the key biogeochemical process do not constitute a physiologically coherent, monophyletic group in phylogenetic analyses with universal markers. It is thus difficult to target molecular ecology studies to the key biogeochemical groups using universal markers such as 16S rRNA. This limitation provided the impetus to develop alternative macromolecular markers diagnostic for metabolic functions. The metabolic pathways for a number of key biogeochemical activities have been characterised in detail. In most cases at least one enzyme has been found to be diagnostic, enabling genes encoding it to be exploited for investigation of communities participating in specific biogeochemical processes. Processes and their marker genes that have been reported in the literature are summarised in *Table 13.1*.

Interpreting taxa defined by sequences

Inevitably, a sequence cannot reflect the full properties of the organism. If we consider the 'real' biological units to be populations of cells that are ecologically equivalent, then any taxonomic unit based exclusively on sequence data from a single marker should *always* be thought of as an operational taxonomic unit (OTU) at an unspecified rank. Arguably one of the greatest problems in using sequence data to estimate microbial diversity is the clear and consistent application of OTU definitions (Bohannan and Hughes, 2003). In phylogenetic trees a monophyletic group reflects the data set selected to build the tree (subjectively) not an objective view of the total population. In practical terms, if sequences are classified by phylogenetic analysis, then statistically significant (monophyletic) groups in one data set do not necessarily represent ecologically significant groups in the total microbiota.

Table 13.1. Metabolic marker macromolecules that have been used in microbial community structure studies

Process	Gene or enzyme	Selected references
Nitrogen fixation	*NifH*	Poly *et al.* (2001); Zehr *et al.* (2003)
Methanotrophy	*pmoA*, *mmoX*	Holmes *et al.* (1999); Heyer *et al.* (2002); Horz *et al.* (2001)
Ammonia oxidation	*AmoA*	Horz *et al.* (2000); Kowalchuk *et al.* (2000); Rotthauwe *et al.* (1997)
Methanogenesis	*McrA*	Luton *et al.* (2002)
Autotrophy	*RbcL*	Alfreider *et al.* (2003)
Sulfate reduction	*DsrA*	Castro *et al.* (2002); Dhillon *et al.* (2003)
Aromatic hydrocarbon degradation	*BphA*	Futumata *et al.* (2001); Taylor *et al.* (2002); Yeates *et al.* (2000)
Methylotrophy	*MxaF*	McDonald *et al.* (1997); Morris *et al.* (2003)
Acetogenesis	FTHFS	Leaphart and Lovell (2001); Leaphart *et al.* (2003)
Chitin degradation	*Chi*	Metcalfe *et al.* (2002)
Denitrification	*nirK*, *nirS*, *narG*, *nosZ*	Avrahami *et al.* (2002); Braker *et al.* (2000); Philippot *et al.* (2002); Scala and Kerkhof (1999); Taroncher-Oldenburg *et al.* (2003)

It is not just phylogenetic trees that need to be interpreted carefully. There are limits to the resolution of the sequence data itself. In the case of bacterial 16S rRNA various workers have chosen 1, 2 or 3% sequence divergence as arbitrary limits to define OTUs for the purposes of sequence comparison (Bornemann and Triplett, 1997; Hughes *et al.*, 2001; McCaig *et al.*, 1999). This variation reflects a natural limitation to the taxonomic informativeness of 16S rRNA in bacteria: both organisms with greater than 5% variation between multiple copies of the 16S rRNA in the same cell and ecologically distinct organisms with identical 16S rRNA sequences are known. The bottom line is that *the only unambiguously definable sequence-based OTUs are identical sequences*, but this is neither biologically realistic nor practical to use for community comparisons.

There are additional problems with the metabolic markers. Where novel sequences showing homology to the marker are recovered this is no guarantee that they will represent physiologically related organisms. A good example of this is supplied by the case of the membrane-associated mono-oxygenases. Members of this enzyme family may participate in either methane oxidation pathways or ammonia oxidation (Holmes *et al.*, 1995), making it impossible to assign a physiological role to novel forms in the absence of additional information (Holmes *et al.*, 1999).

In conclusion, there are two main problems associated with the use of sequences for community comparisons. The first is the difficulty in obtaining clear and consistent OTU definitions. The second is the time and expense in obtaining significant samples for comparative purposes (Dunbar *et al.*, 1999; McCaig *et al.*, 1999). A paper by Hughes *et al.* (2001) goes into this issue in more detail, concluding that advances in sequencing technology mean that for some systems sequencing clone libraries may be a useful means of comparing community structure. Such applications are likely to be restricted to simple communities where relatively few samples are to be analysed. The clone library/sequence approach is unlikely to ever be a useful strategy where multiple samples must be processed in routine fashion. The most extensive study to date compared 9000 clones from 29 soil samples using RFLP (Zhou *et al.*, 2002). In comparison, a major invertebrate diversity study collected 1536 samples to process over 150 000 specimens (Oliver *et al.*, 2003). There is a clear need for alternatives methods to process microbial community data if they are to be comparable.

13.5.2 Other macromolecule-based OTUs

Macromolecular sequences are valuable tools in microbial ecology and have revolutionised microbial systematics. Nevertheless, it is presently impractical to attempt comparative analysis of microbial communities by sequence analysis. For this reason measurement of community turnover must be by methods that permit rapid extraction of the essential information from molecular markers. Ideally such methods will retain a large proportion of the information content of sequences and produce data that can be integrated with existing databases.

Differential migration methods

These methods are all based on the resolution of bands within an electrophoresis gel and have become extremely popular over the last few years. They have the advantage that they are extremely simple and rapid. Each OTU is simply a unique band position in a gel, making assemblage comparison easy (Fromin *et al.*, 2002).

This simplistic basis for classification also gives them a common theoretical resolution limit determined by the number of resolvable positions on a gel; this is unlikely to exceed 1000 in presently available gel formats.

Similarly there is a practical sensitivity limit for all gel-based methods. This is effectively determined by the loading capacity of the gel in combination with the detection method. Thus if the sensitivity of detection is 5 ng of DNA, and the loading capacity of a gel is 1 mg, then organisms whose sequences have a relative abundance of less than 0.5% are below the limit of detection. Note, that this example would effectively impose a maximum resolution limit of 200 taxa for any one sample. Although this figure will differ for different gel and detection formats, the example illustrates that the theoretical resolution of 1000 taxa is unlikely to ever be reached by gel-based methods. Given these common features, all differential migration methods are most appropriate for analysis of simple communities, or of the relatively abundant components of diverse communities.

The methods are based on different principles and consequently differ in their suitability for use with different molecular sequence markers, the way in which they may be integrated with sequence databases, and the extent to which OTUs will be consistent with 'standard' classification.

Sequence-conformation-based methods; DGGE, TGGE, and SSCP.

The thermodynamic stability of the DNA double helix is highly sequence dependent. Denaturing and thermal gradient gel electrophoresis (hereafter referred to as DGGE for simplicity) exploit this by discriminating between sequences with different melting behaviour on the basis of differential migration in polyacrylamide gels with denaturing gradients (Muyzer and Smalla, 1998). It has been shown that single nucleotide polymorphisms can be detected under optimised conditions and this has been frequently cited as evidence for the sensitivity of DGGE in community studies. It is important to note that this does not mean DGGE preserves the information content of the sequence. A DGGE with a resolving distance of 150 mm, band thicknesses of 1 mm, and accuracy of +/– 0.5 mm can separate a maximum of 75 OTUs – this is nowhere near the thousands of bacterial species estimated to occur in soil communities (Torsvik *et al.*, 1996).

DGGE has been successfully applied to comparative analysis of soil bacterial communities using both universal and metabolic markers (Dahllof *et al.*, 2000; Garbeva *et al.*, 2003; Kowalchuk *et al.*, 2003; Nakatsu *et al.*, 2000; Nicol *et al.*, 2003; Ovreas and Torsvik, 1998; Peixoto *et al.*, 2002; Wieland *et al.*, 2001). There are two common sources of error with DGGE: heterogeneity and heteroduplexes. As discussed above, heterogeneous copies of the 16S rRNA within a genome are relatively common. The high capacity of DGGE to distinguish similar sequences can create a problem in this circumstance since it may lead to the presence of multiple bands (OTUs) from a single cell (Dahllof *et al.*, 2000). Heteroduplexes can be formed in the later rounds of PCR where imperfectly complementary single DNA strands come together. Heteroduplexes present a particular problem for analysis of communities by DGGE since they almost inevitably exhibit different melting behaviour to either homoduplex parent (Lowell and Klein, 2000; Ward *et al.*, 1998). It is likely that the pattern of sequence conservation in the non-protein-coding rRNA genes increases the propensity for heteroduplex formation relative to protein-coding genes, although this has not been experimentally tested.

The electrophoretic migration of a DNA fragment in DGGE can not be predicted from its sequence alone. Consequently fragments are classified into OTUs according to their relative migration rate against internal standards, rather than on any absolute property. Since relative migration rate changes with different electrophoretic conditions, this places extreme technical limitations on the reproducibility of the denaturing gradient if reliable gel-to-gel comparisons are to be made. Furthermore relative migration bears no relationship to phylogenetic affiliation, so OTUs defined from DGGE data are not consistent with phylogenetically defined taxa. Consequently, DGGE does not lead to the generation of a cumulative database that is compatible with existing taxonomic databases. Comparison to databases usually occurs via the additional step of excising bands of interest from the gel and sequencing them for identification purposes (Felske *et al.*, 1998; Smalla *et al.*, 2001; Ward *et al.*, 1998). It is worth noting that more than one DNA 'species' may have contributed to a DGGE band and in most cases workers seldom report proof that the sequence determined for a recovered band was the sole (or even major) contributing one to the band. DGGE is very successful for studies that involve the comparison of a relatively small number of samples.

Single-strand conformation polymorphism (SSCP) operates on different principles to DGGE (Lowell and Klein, 2001; Schwieger and Tebbe, 1998, 2000; Stach *et al.*, 2001). DNA fragments are resolved on the basis of sequence-dependent conformational differences in single strands that lead to changes in electrophoretic mobility. With the exception of the heteroduplex issue most of the above comments are equally applicable to SSCP.

Sequence-length-based methods; RISA, and LH-PCR.

Few sequences obey the requirements of molecular markers for taxonomy and show sufficient length variation to be taxonomically informative. One of the exceptions is sequences within the ribosomal RNA operons, including the SSU rRNA, LSU rRNA and their intergenic spacers. Ribosomal intergenic spacer analysis (RISA) involves amplification of either IGS1, IGS2 or both by PCR and resolution of the fragments by polyacrylamide gel electrophoresis (Guertler and Stanisich, 1996; Ranjard *et al.*, 2001). LH-PCR typically refers to amplification of a length-variable segment of the 16S rRNA (Ritchie *et al.*, 2000). In both cases, fragments are classified according to 'absolute' length rather than relative migration facilitating both gel-to-gel comparisons and the accumulation of a database that can be shared by workers in other laboratories.

The theoretical resolving power depends on the electrophoresis format and amplification primers being used, but is unlikely to exceed 1000 taxa. This limit is based on achieving 1 bp resolution over the expected size range of IGS amplicons, *ca* 200–1200 bp (Ranjard *et al.*, 2001). As with DGGE, the heterogeneity of ribosomal RNA operons can also lead to the presence of multiple bands from a single genome and sequence length does not correlate to phylogenetic relationships. Therefore, while OTUs may be absolutely defined by RISA they are not consistent with 'standard' taxa. Correlation to a taxonomic database can be made by performing the additional step of recovery of bands from a gel to enable sequencing.

Although its greater resolving power means RISA offers significant advantages for large-scale comparative studies this approach has not been as popular as DGGE with bacteriologists. This almost certainly reflects the fact that the IGS spacer of bacteria is seldom sequenced and thus the capacity to link RISA data to the major bacterial databases is presently limited. In contrast, it is seen as the method of choice by mycologists and ITS sequences are increasingly being used in fungal taxonomy leading to a useful database (Borneman and Hartin, 2000; Viaud *et al.*, 2000).

It is also worth mentioning amplified fragment length polymorphism (AFLP) here, since this technique could also be considered a length-based method. AFLP has been applied to investigation of spatial patterns in microbial community structure in soil (Franklin *et al.*, 2003). However AFLP represents a random sampling of the total metagenome, rather than sampling a specific marker from each organism. As such it is limited to general questions of spatial patterns and not effective within the broad objectives of soil community structure analysis outlined earlier.

Restriction site methods; T-RFLP

Restriction digests (like AFLP) are unsuitable for completing the objectives of community analysis because they do not yield a single comparable character for each organism within the community. Terminal restriction fragment length polymorphism (T-RFLP) overcomes this limitation by detection of only the terminal fragment from a restriction digest (Marsh, 1999). T-RFLP has been widely used with both universal and metabolic markers in microbial community structure (Blackwood *et al.*, 2003; Girvan *et al.*, 2003; Horz *et al.*, 2000, 2001; Liu *et al.*, 1997; Rousseaux *et al.*, 2003).

There are some notable differences between T-RFLP and the other differential migration methods in that sequence length is not the only piece of information. In T-RFLP the information points include the presence of a specific tetranucleotide, the distance of this site from a defined terminus, and the absence of this tetranucleotide from the preceding positions. Like RISA, T-RFLP lends itself to generation of a cumulative database that is easily shared by multiple laboratories, but it also shows a greater level of internal consistency with taxonomic hierarchies derived from phylogenetic analysis of 16S rRNA (Marsh *et al.*, 2000). Nevertheless this correlation is not perfect and is offset by the limitation that it is not practical to recover T-RFLP fragments from the electrophoresis gel for subsequent sequencing. Thus, of the three methods, T-RFLP offers the greatest potential for inference of a taxon directly from the electrophoresis gel with no further experimentation, facilitating rapid sample processing; but it is also the most limited for subsequent unambiguous linkage to a taxonomic database.

Differential migration methods – conclusions

Where multiple differential migration methods have been employed in the same study they have yielded similar results (e.g. Girvan *et al.*, 2003). This probably reflects that they share one advantage and two limitations in common. The common advantage is that they allow simultaneous classification of multiple DNA fragments in a single analytical test, an electropherogram. The first of the common disadvantages is that they are limited to resolving 100–1000 OTUs (practical limitations mean the lower end of this range is typical). This does provide a very useful means of comparing microbial assemblages, but falls well short of the hundreds of thousands of naturally occurring microbial species that are expected to occur in soils (Curtis *et al.*, 2002; Hughes *et al.*, 2001). The second common disadvantage is that the resulting OTUs are either poorly, or not at all, consistent with microbial taxonomic hierarchies. These methods will always give cases where organisms that are effectively unrelated are placed in the same OTU and organisms that are quite closely related will be placed in different OTUs. This creates an undesirable complication for subsequent analyses. The first of these limitations can be alleviated by prefractionation of the microbial sample to reduce its complexity.

13.5.3 Comparison after prefractionation

The commonly applied rapid techniques for diversity assessment simply can not resolve the diversity adequately for detailed comparative purposes. One approach to bypass this problem is to reduce the complexity of communities via selective sampling. Very often this increases the ecological insight. There are several broad routes to prefractionation, which could broadly be considered as either targeted to specific ecological questions or non-targeted.

G+C gradient

DNA with different average %G+C contents can be separated in caesium chloride density gradients. This has been exploited to obtain DNA samples that represent reproducible fractions of the total community (Nusslein and Tiedje, 1998, 1999). G+C fractionation is essentially a non-targeted means of reducing the complexity of environmental communities. As a consequence its major advantage is in facilitating meaningful comparisons between samples.

Stable isotope probing

For some applications, stable isotope probing (Radajewski *et al.*, 2000, 2003) may be considered a significant advance on the G+C fractionation approach. In this method a community is supplied with an isotopically labelled growth substrate. Organisms that assimilate the substrate (typically a carbon source) incorporate the heavy isotope into their DNA. The heavy DNA is separated in a density gradient facilitating the recovery of DNA that specifically targets the members of a selected physiological group. SIP has been used to examine the diversity of methylotrophs in soils by supplying ^{13}C-labelled methane or methanol (Morris *et al.*, 2002; Radajewski *et al.*, 2002), or autotrophic ammonia oxidisers by supplying $^{13}CO_2$ (Whitby *et al.*, 2001), and phenol degraders by supplying $^{13}C_6$-phenol (Manefield *et al.*, 2002). This approach could be easily adapted to identify those members of the soil community that are most directly dependent on plant-derived carbon by supplying isotopically labelled plant exudates, either artificially or via feeding labelled CO_2 to plants.

Group-specific PCR

By far the most generally applicable means of fractionating environmental community samples is by group-specific PCR. Where the organisms of interest to the question are predictable, the sequences to be collected may be targeted by use of either phylogenetic group-specific (PGS) primers or the use of metabolic group-specific (MGS) primers.

By far the majority of group-specific applications have targeted specific processes rather than phylogenetic groups. Examples include the autotrophic ammonia oxidising bacteria (Bruns *et al.*, 1999; Horz *et al.*, 2000; Kowalchuk *et al.*, 2000), methane oxidising bacteria (Holmes *et al.*, 1999; Horz *et al.*, 2001), sulphate-reducing bacteria (Dhillon *et al.*, 2003), methanogens (Luton *et al.*, 2002) and nitrogen-fixing bacteria (Zehr *et al.*, 2003). In most cases the specificity has been sufficient to reduce the level of diversity such that even a clone library strategy is a reasonable means of comparing samples.

A less-utilised path, but one that is likely to be fruitful, is to target phylogenetic groups for comparative analysis of soil communities. There is a relatively small set of higher taxa that are characteristically strongly represented in samples of soil communities (Buckley and Schmidt, 2003; Valinsky *et al.*, 2002) (*Figure 13.3*). The use of PGS-PCR targeting groups known to be diverse and abundant in soil is likely to result in sample sets with much greater discriminatory powers. This approach has already been taken with *Pseudomonas* (Stach *et al.*, 2001), *Bacillus* (Garbeva *et al.*, 2003), *Nitrosomonadaceae* (Bruns *et al.*, 1999; Webster *et al.*, 2002) and Archaea (Nicol *et al.*, 2003) and found to significantly improve the ability of the differential migration techniques to identify differences between soil communities. It could easily be expanded to those higher taxa for which probe or sequence data have already indicated environmental variation in abundance and diversity. These include the Acidobacteria (Barns *et al.*, 1999), Rubrobacteridae (Holmes *et al.*, 2000) and Verrucomicrobia (Buckley and Schmidt 2001).

13.6 Microarrays – the final solution?

The challenges for microbial community structure analysis are to perform rapid assemblage comparison via unambiguously defined OTUs that are easily correlated to existing taxonomic databases. The principal requirement for rapid comparison is that multiple specimens can be simultaneously classified into OTUs using a single analytical test. The differential migration methods largely solve the problem of rapid comparison, but are not based on OTUs that are consistent with taxonomic databases. Sequence analysis does give OTUs that are compatible with databases, but is not suitable for rapid processing of thousands of specimens. Microarray technology is widely anticipated to overcome these limitations.

Microarrays theoretically offer the possibility of a single solution to the major challenges of microbial community structure analysis: (i) the array format permits simultaneous assay of thousands of different molecules; (ii) probes are theoretically capable of unambiguous identification of taxa that correlate with standard taxonomic databases. In environmental microbiology the development of microarrays is still in its infancy and microarrays have been broadly classified into three distinct types; phylogenetic oligonucleotide arrays (POA), functional gene arrays (FGA), and community genome arrays (CGA) (Zhou and Thompson, 2002).

FGA microarrays are targeted towards study of particular biogeochemical processes, rather than attempting to address the broad challenge of community

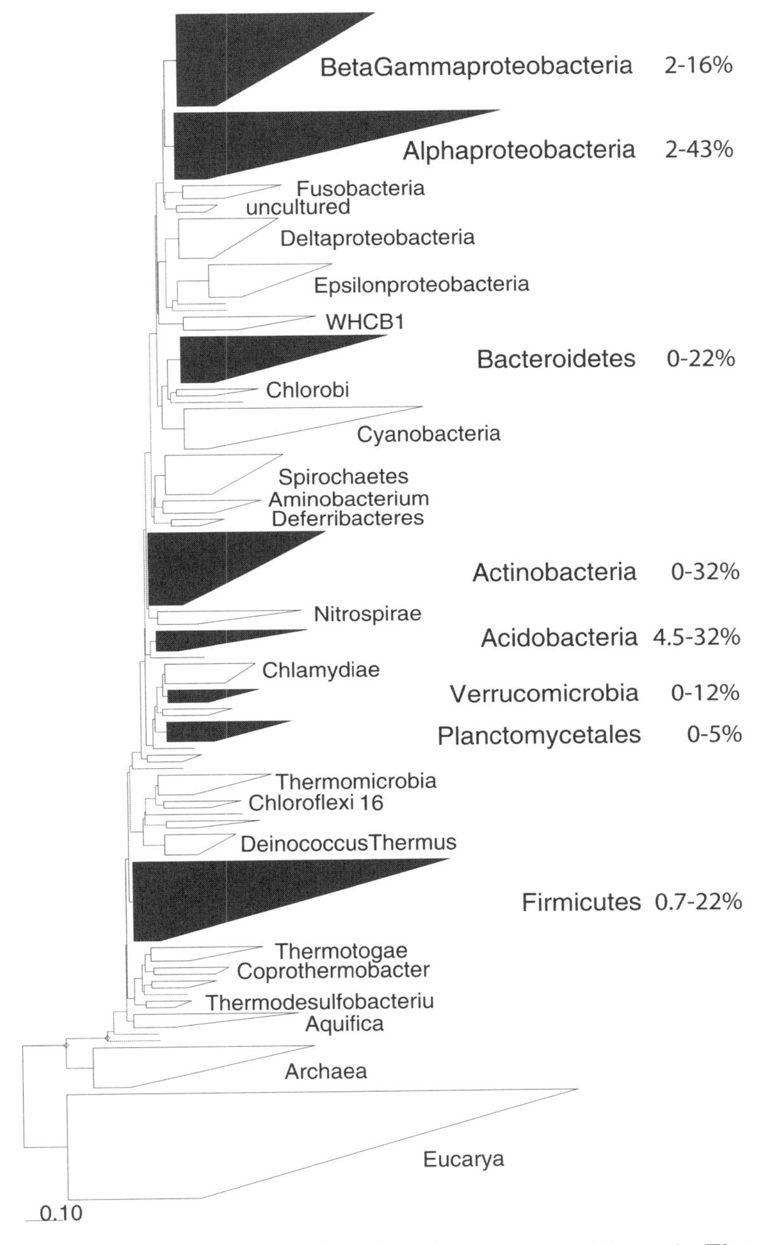

Figure 13.3. Phylogenetic tree illustrating the major groups of soil bacteria. The tree is modified from the tree distributed with the ARB software package (Ludwig and Kenk, 2001) based on available near-complete SSU rRNA sequences. The bacterial phylogenetic groups that are consistently highly represented in soil rDNA clone libraries, or which consistently give strong hybridisation signals against total soil rDNA are shown in black. Note that these groups are phylogenetically diverse (indicated by depth of the triangle) and not all members of each group are typical soil inhabitants. The numbers at the right of the group names indicate the range of relative abundance for the group reported from soils (see also Buckley and Schmidt, 2003).

analysis. Zhou and Thompson (2002) used the term FGA to refer to arrays where the bound probe is DNA (polynucleotide) encoding a metabolic marker gene rather than oligonucleotides (Wu *et al.*, 2001). Recently, microarrays that target a specific functional group, but use oligonucleotides targeting either universal markers (Loy *et al.*, 2002) or metabolic markers (Bodrossy *et al.*, 2003; Taroncher-Oldenburg *et al.*, 2003) have been described.

Regardless of whether microarrays are based on oligonucleotides or longer DNA probes, targeting universal or metabolic markers, the potential advantages they offer are the same: simultaneous testing of all members of a nucleic acid sample with fine-scale taxonomic discrimination. In practice, these advantages of microarrays are difficult to realise and they present a new suite of technical challenges that are largely unique to microarrays. These technical challenges are extensively discussed in a number of recent publications and are only briefly summarised here (see Bodrossy *et al.*, 2003; Cook and Sayler 2003; Smalla *et al.*, 2001; Urakawa *et al.*, 2003; Wilson *et al.*, 2002; Zhou and Thompson, 2002, for more detail).

The principal limitation to the application of microarrays to community structure investigations is the capacity to simultaneously apply different probes under conditions of universally high stringency and sensitivity. Oligonucleotide probes are typically 18–24 bases long. There are over one billion possible combinations for an 18-base sequence. Although evolutionary constraints on sequences mean that this is a gross overestimation of the number of possible probe-based OTUs, it is obvious that oligonucleotide probes have a high potential to discriminate biological differences. The difficulty is that this potential for specific discrimination is only realised under stringent hybridisation conditions and conditions for stringent hybridisation vary considerably between different probe-target pairs. Among the many avenues being explored to address this issue are the use of internal reference mismatch controls (Wilson *et al.*, 2002), selection of probes with similar thermal melting behaviour (Bodrossy *et al.*, 2003), and incorporation of greater analysis of probe hybridisation behaviour in data interpretation (Urakawa *et al.*, 2003).

13.7 Patterns of microbial diversity in soil

Despite the technical challenges of sampling microbial diversity, the last 5 years have seen the emergence of the capacity to process large numbers of samples and are beginning to revolutionise our view of soils (Hill *et al.*, 2000). As little as 10 years ago the prevailing view was that the microbiota were uniformly distributed and could be treated as 'background noise' within the (plant) environment. There is now considerable evidence that this is not always so. Soil type-specific communities (Gelsomino *et al.*, 1999; Girvan *et al.*, 2003) and plant species-specific communities (Johnson *et al.*, 2003; Smalla *et al.*, 2001), have been reported. At 'field scale' most studies show remarkable levels of homogeneity in soil microbiota (Felske and Akkemans, 1998; Gelsomino *et al.*, 1999; Lukow *et al.*, 2000). However, given that spatial isolation does influence microbial community structure in soil (Fierer *et al.*, 2003; Treves *et al.*, 2003), that 'within-field' variability of microbial activity, biomass and soil physical properties occurs (Lopez-Granados *et al.*, 2002;

Robertson *et al.*, 1997), and that microbial communities are known to be highly diverse (Borneman and Triplett, 1997; McCaig *et al.*, 1999), these observations of homogeneity must be considered surprising. This highlights the difficulty in demonstrating covariance of microbial community parameters with either plant productivity or plant diversity, which is so far largely restricted to defined experimental systems (Horner-Devine *et al.*, 2003; van der Heijden *et al.*, 1998). Clearly soil microbiota do have a spatially explicit structure, but our capacity to observe this and relate it to plant properties, is dependent on the technique used and the scale of observation (see Ettema and Wardle, 2003 for a review).

Clearly there is now a need for new emphasis on the challenge of how to sample effectively. This includes both the size of sample to be collected (Ellingsoe and Johnsen, 2002) and the spatial scale of sample collection (Franklin *et al.*, 2002). The statistical procedures used to compare variables such as microbial community structure and plant properties assume independence of observations. Violations of sample independence can lead to incorrect conclusions. Physical and chemical properties likely to influence microbiota are known to be spatially dependent (Ettema and Wardle, 2002; Robertson *et al.*, 1997; Stoyan *et al.*, 2000). There is now convincing evidence that microbial communities show spatial dependence at a number of different scales (Franklin *et al.*, 2003; Nunan *et al.*, 2002, 2003; Saetre and Bååth, 2000). If we accept that microbial communities do show spatial dependence at a number of scales, then pseudo-correlation of samples is almost impossible to avoid in attempts at comparative analysis. Given that very few studies have employed a spatially explicit sampling design this is highly likely to have contributed to the difficulty in demonstrating a clear relationship between microbiota and plant properties.

A less-well-recognised, but equally important issue, is taxonomic independence. The differential migration techniques for resolving microbial community structure rely on operational taxonomic units that are not capable of resolving all taxa present in a sample. A consequence of this is that as the richness (number of taxa) of a community increases the independence of classification decreases and the capacity to distinguish the two samples declines. Put simply, if the electrophoretic technique can only resolve 500 positions on a gel, then communities of >500 species can never be separated. Although it seems obvious, this aspect of autocorrelation appears to have been ignored, since it is not a significant issue for the macrobiota where essentially all theoretical development of ecological sampling has occurred.

13.8 Concluding remarks

Of all the variables that impact upon plant growth soil microbial activity is arguably the one least taken into account for agricultural (or conservation) management. The importance of the microbiota to biogeochemistry has long been appreciated (Conrad, 1996). Interactions between plants and microbes have long been known and we are increasingly aware of inter-kingdom communication signals across a broader range of ecological interactions than simple two-species mutualisms. Few would argue the point that the microbiota are an intimate part of the plant ecosystem and that understanding their roles will lead to new management opportunities. Through describing patterns of variation in soil microbiota, and explaining the basis of their ecological interactions with plants, soil microbial ecologists aim to develop new management tools for plant systems.

There are still many challenges to achieving this goal. Arguably the three biggest gaps at present are: (i) a comprehensive database of soil microbial diversity where diagnostic characters are linked to eco-physiological properties for each species; (ii) a sound theoretical basis for the comparison of samples to identify properties that may co-vary at multiple spatial scales; and, (iii) the capacity to rapidly and economically process very large sample sets. Progress in the first of these challenges is being steadily made by the application of the full range of microbial ecology techniques and is strongly complemented by community structure analyses utilising the present generation of differential migration strategies. The second of these challenges has arguably only just begun to be appreciated by microbiologists. There is a rich theory of spatial pattern analysis, developed by ecologists working with macro-organisms, that should enable rapid progress. The final challenge is largely unmet at present. It is however widely anticipated that the processing power of microarray technology will largely solve this within the next decade.

One notable distinction between macro-organism and micro-organism data sets is that the morphological criteria used to recognise operational (or formal) taxonomic units of macro-organisms gives rise to an infinite number of OTUs. In contrast, the present generation of differential migration techniques for sorting microbial specimens has a finite number of OTUs. This problem of 'taxonomic autocorrelation' is peculiar to microbiology and not accounted for by existing theory. An additional advantage of the successful use of PGS probes in very large microarrays is that they would largely circumvent this issue. We are already seeing the development of microbial community structure as a tool for management of 'purely' microbial systems such as sewage treatment plants (Daims *et al.*, 2001). It seems inevitable that we will see similar developments in plant/soil microbiology in the near future.

Acknowledgements

I would like to acknowledge the following co-workers who have collaborated with me in the study of soil microbial diversity over the past 6 years at Macquarie University and The University of Sydney; Andrew Beattie, Jocelyn Bowyer, David Briscoe, Mark Dangerfield, Michael Gillings, Jessica Green, Marita Holley, Ian Oliver, Madeline Raison, and Christine Yeates. Our work on soil microbial diversity has been supported by the Australian Research Council, Key Centre for Biodiversity and Bioresources and the Resource Conservation and Assessment Council of NSW.

References

Alfreider, A., Vogt, C., Hoffmann, D., Babel, W. (2003) Diversity of ribulose-1,5-bisphosphate carboxylase/oxygenase large-subunit genes from groundwater and aquifer micro-organisms. *Microb. Ecol.* **45**: 317–328.

Baer, S.G., Blair, J.M., Collins, S.L., Knapp, A.K. (2003) Soil resources regulate productivity and diversity in newly established tallgrass prairie. *Ecology* **84**: 724–735.

Barns, S.M., Takala, S.L., Kuske, C.R. (1999) Wide distribution and diversity of members of the bacterial kingdom Acidobacterium in the environment. *Appl. Environ. Microbiol.* **65**: 1731–1737.

Bever, J.D. (2003) Soil community feedback and the coexistence of competitors: conceptual frameworks and empirical tests. *New Phytol.* **157**: 465–473.

Blackwood, C.B., Marsh, T., Kim, S.H., Paul, E.A. (2003) Terminal restriction fragment length polymorphism data analysis for quantitative comparison of microbial communities. *Appl. Environ. Microbiol.* **69**: 926–932.

Bodrossy, L., Stralis-Pavese, N., Murrell, J.C., Radajewski, S., Weilharter, A., Sessitsch, A. (2003) Development and validation of a diagnostic microbial microarray for methanotrophs. *Environ. Microbiol.* 5: 566–582.

Bohannan, B.J.M., Hughes, J. (2003) New approaches to analyzing microbial biodiversity data. *Curr. Opin. Microbiol.* **6**: 282–287.

Borneman, J., Hartin, R.J. (2000) PCR primers that amplify fungal rRNA genes from environmental samples. *Appl. Environ.Microbiol.* **66**: 4356–4360.

Borneman, J., Triplett, E.W. (1997) Molecular microbial diversity in soils from eastern Amazonia: Evidence for unusual micro-organisms and microbial population shifts associated with deforestation. *Appl. Environ. Microbiol.* **63**: 2647–2653.

Braker, G., Zhou, J.Z., Wu, L.Y., Devol, A.H., Tiedje, J.M. (2000) Nitrite reductase genes (nirK and nirS) as functional markers to investigate diversity of denitrifying bacteria in Pacific northwest marine sediment communities. *Appl. Environ. Microbiol.* **66**: 2096–2104.

Broughton, L.C., Gross, K.L. (2000) Patterns of diversity in plant and soil microbial communities along a productivity gradient in a Michigan old-field. *Oecologia* **125**: 420–427.

Bruns, M.A., Stephen, J.R., Kowalchuk, G.A., Prosser, J.I., Paul, E.A. (1999) Comparative diversity of ammonia oxidizer 16S rRNA gene sequences in native, tilled, and successional soils. *Appl. Environ. Microbiol.* **65**: 2994–3000.

Buckley, D.H., Schmidt, T.M. (2001) Environmental factors influencing the distribution of rRNA from Verrucomicrobia in soil. *FEMS Microbiol. Ecol.* **35**: 105–112.

Buckley, D.H., Schmidt, T.M. (2003) Diversity and dynamics of microbial communities in soils from agro-ecosystems. *Environ. Microbiol.* 5: 441–452.

Castro, H., Reddy, K.R., Ogram, A. (2002) Composition and function of sulfate-reducing prokaryotes in eutrophic and pristine areas of the Florida Everglades. *Appl. Environ. Microbiol.* **68**: 6129–6137.

Cavigelli, M.A., Robertson, G.P. (2000) The functional significance of denitrifier community composition in a terrestrial ecosystem. *Ecology* **81**: 1402–1414.

Cavigelli, M.A., Robertson, G.P. (2001) Role of denitrifier diversity in rates of nitrous oxide consumption in a terrestrial ecosystem. *Soil Biol. Biochem.* **33**: 297–310.

Chellemi, D.O., Porter, I.J. (2001) The role of plant pathology in understanding soil health and its application to production agriculture. *Austral. Plant Pathol.* **30**: 103–109.

Conrad, R. (1996) Soil micro-organisms as controllers of atmospheric trace gases (H-2, CO, CH4, OCS, N2O, and NO). *Microbiol. Rev.* **60**: 609–643.

Curtis, T.P., Sloan, W.T., Scannell, J.W. (2002) Estimating prokaryotic diversity and its limits. *Proc. Natl Acad. Sci. USA* **99**: 10494–10499.

Dahllof, I., Baillie, H., Kjelleberg, S. (2000) *rpoB*-based microbial community analysis avoids limitations inherent in 16S rRNA gene intraspecies heterogeneity. *Appl. Environ. Microbiol.* **66**: 3376–3380.

Daims, H., Purkhold, U., Bjerrum, L., Arnold, E., Wilderer, P.A., Wagner, M. (2001) Nitrification in sequencing biofilm batch reactors: Lessons from molecular approaches. *Water Sci. Technol.* **43**: 9–18.

Dhillon, A., Teske, A., Dillon, J., Stahl, D.A., Sogin, M.L. (2003) Molecular characterization of sulfate-reducing bacteria in the Guaymas Basin. *Appl. Environ. Microbiol.* **69**: 2765–2772.

Dunbar, J., Takala, S., Barns, S.M., Davis, J.A., Kuske, C.R. (1999) Levels of bacterial community diversity in four arid soils compared by cultivation and 16S rRNA gene cloning. *Appl. Environ. Microbiol.* **65**: 1662–1669.

Ellingsoe, P., Johnsen, K. (2002) Influence of soil sample sizes on the assessment of bacterial community structure. *Soil Biol. Biochem.* **34**: 1701–1707.

Ettema, C.H., Wardle, D.A. (2002) Spatial soil ecology. *Trends Ecol. Evol.* **17**: 177–183.

Farrelly, V., Rainey, F.A., Stackebrandt, E. (1995) Effect of genome size and rrn gene copy number on PCR amplification of 16S ribosomal-RNA genes from a mixture of bacterial species. *Appl. Environ. Microbiol.* **61**: 2798–2801.

Felske, A., Akkermans, A.D.L. (1998) Spatial homogeneity of abundant bacterial 16S rRNA molecules in grassland soils. *Microb. Ecol.* **36**: 31–36.

Felske, A., Wolterink, A., Van Lis, R., Akkermans, A.D.L. (1998) Phylogeny of the main bacterial 16S rRNA sequences in Drentse A grassland soils (The Netherlands). *Appl. Environ. Microbiol.* **64**: 871–879.

Fierer, N., Schimel, J.P., Holden, P.A. (2003) Variations in microbial community composition through two soil depth profiles. *Soil Biol. Biochem.* **35**: 167–176.

Filip, Z. (2002) International approach to assessing soil quality by ecologically-related biological parameters. *Agric. Ecosyst. Environ.* **88**: 169–174.

Fiore, A., Laevens, S., Bevivino, A., Dalmastri, C., Tabacchioni, S., Vandamme, P., Chiarini, L. (2001) *Burkholderia cepacia* complex: distribution of genomovars among isolates from the maize rhizosphere in Italy. *Environ. Microbiol.* **3**: 137–143.

Franklin, R.B., Blum, L.K., McComb, A.C., Mills, A.L. (2002) A geostatistical analysis of small-scale spatial variability in bacterial abundance and community structure in salt marsh creek bank sediments. *FEMS Microbiol. Ecol.* **42**: 71–80.

Franklin, R.B., Mills, A.L. (2003) Multi-scale variation in spatial heterogeneity for microbial community structure in an eastern Virginia agricultural field. *FEMS Microbiol. Ecol.* **44**: 335–346.

Fromin, N., Hamelin, J., Tarnawski, S. *et al.* (2002) Statistical analysis of denaturing gel electrophoresis (DGE) fingerprinting patterns. *Environ. Microbiol.* **4**: 634–643.

Frostegard, A., Courtois, S., Ramisse, V., Clerc, S., Bernillon, D., Le Gall, F., Jeannin, P., Nesme, X., Simonet, P. (1999) Quantification of bias related to the extraction of DNA directly from soils. *Appl. Environ. Microbiol.* **65**: 5409–5420.

Futamata, H., Harayama, S., Watanabe, K. (2001) Group-specific monitoring of phenol hydroxylase genes for a functional assessment of phenol-stimulated trichloroethylene bioremediation. *Appl. Environ. Microbiol.* **67**: 4671–4677.

Garbeva, P., van Veen, J.A., van Elsas, J.D. (2003) Predominant *Bacillus* spp. in agricultural soil under different management regimes detected via PCR-DGGE. *Microb. Ecol.* **45**: 302–316.

Gelsomino, A., Keijjzer-Wolters, A.C., Cacco, G., van Elsas, J.D. (1999) Assessment of bacterial community structure in soil by polymerase chain reaction and denaturing gradient gel electrophoresis. *J. Microbiol. Meths.* **38**: 1–15.

Gillis, M., Vandamme, P., De Vos, P., Swings, J., Kersters, K. (2001) Polyphasic taxonomy. In: Boone, D.R., Castenholz, R.W. (eds) *Bergey's Manual of Systematic Bacteriology, 2nd* Ed, Volume One, The Archaea and the deeply branching and phototrophic Bacteria. 43–48.

Girvan, M.S., Bullimore, J., Pretty, J.N., Osborn, A.M., Ball, A.S. (2003) Soil type is the primary determinant of the composition of the total and active bacterial communities in arable soils. *Appl. Environ. Microbiol.* **69**: 1800–1809.

Guertler, V., Stanisich, V.A. (1996) New approaches to typing and identification of bacteria using the 16S-23S rDNA spacer region. *Microbiology* **142**: 3–16.

Heyer, J., Galchenko, V.F., Dunfield, P.F. (2002) Molecular phylogeny of type II methane-oxidizing bacteria isolated from various environments. *Microbiology* **148**: 2831–2846.

Hill, G.T., Mitkowski, N.A., Aldrich-Wolfe, L., Emele, L.R., Jurkonie, D.D., Ficke, A., Maldonado-Ramirez, S., Lynch, S.T., Nelson, E.B. (2000) Methods for assessing the composition and diversity of soil microbial communities. *Appl. Soil Ecol.* **15**: 25–36.

Holmes, A.J., Costello, A., Lidstrom, M.E., Murrell, J.C. (1995) Evidence that particulate methane monooxygenase and ammonia monooxygenase may be evolutionarily related. *FEMS Microbiol. Lett.* **132**: 203–208.

Holmes, A.J., Roslev, P., McDonald, I.R., Iversen, N., Henriksen, K., Murrell, J.C. (1999) Characterization of methanotrophic bacterial populations in soils showing atmospheric methane uptake. *Appl. Environ. Microbiol.* **65**: 3312–3318.

Holmes, A.J.,. Bowyer, J., Holley, M.P., O'Donoghue, M., Montgomery, M., Gillings, M.R. (2000) Diverse, yet-to-be-cultured members of the Rubrobacter subdivision of the Actinobacteria are widespread in Australian arid soils. *FEMS Microbiol. Ecol.* **33**: 111–120.

Horner-Devine, M.C., Leibold, M.A., Smith, V.H., Bohannan, B.J.M. (2003) Bacterial diversity patterns along a gradient of primary productivity. *Ecol. Lett.* **6**: 613–622.

Horz, H.P., Rotthauwe, J.H., Lukow, T., Liesack, W. (2000) Identification of major subgroups of ammonia-oxidizing bacteria in environmental samples by T-RFLP analysis of *amoA* PCR products. *J. Microbiol. Methods* **39**: 197–204.

Horz, H.P., Yimga, M.T., Liesack, W. (2001) Detection of methanotroph diversity on roots of submerged rice plants by molecular retrieval of pmoA, mmoX, mxaF, and 16S rRNA and ribosomal DNA, including pmoA-based terminal restriction fragment length polymorphism profiling. *Appl. Environ. Microbiol.* **67**: 4177–4185.

Hughes, J.B., Hellmann, J.J., Ricketts, T.H., Bohannan, B.J.M. (2001) Counting the uncountable: Statistical approaches to estimating microbial diversity. *Appl. Environ. Microbiol.* **67**: 4399–4406.

Janssen, P.H., Yates, P.S., Grinton, B.E., Taylor, P.M., Sait, M. (2002) Improved culturability of soil bacteria and isolation in pure culture of novel members of the divisions Acidobacteria, Actinobacteria, Proteobacteria, and Verrucomicrobia. *Appl. Environ. Microbiol.* **68**: 2391–2396.

Jenkins, C., Fuerst, J.A. (2001) Phylogenetic analysis of evolutionary relationships of the planctomycete division of the domain bacteria based on amino acid sequences of elongation factor Tu. *J. Mol. Evol.* **52**: 405–418.

Johnson, M.J., Lee, K.Y., Scow, K.M. (2003) DNA fingerprinting reveals links among agricultural crops, soil properties, and the composition of soil microbial communities. *Geoderma* **114**: 279–303.

Kent, A.D., Triplett, E.W. (2002) Microbial communities and their interactions in soil and rhizosphere ecosystems. *Annu. Rev. Microbiol.* **56**: 211–236.

Klironomos, J.N. (2002) Feedback with soil biota contributes to plant rarity and invasiveness in communities. *Nature* **417**: 67–70.

Kowalchuk, G.A., Stephen, J.R. (2001) Ammonia-oxidizing bacteria: A model for molecular microbial ecology. *Annu. Rev. Microbiol.* **55**: 485–529.

Kowalchuk, G.A., Stienstra, A.W., Heilig, G.H.J., Stephen, J.R., Woldendorp, J.W. (2000) Changes in the community structure of ammonia-oxidizing bacteria during secondary succession of calcareous grasslands. *Environ. Microbiol.* **2**: 99–110.

Kowalchuk, G.A., Os, G.J., Aartrijk, J., Veen, J.A. (2003) Microbial community responses to disease management soil treatments used in flower bulb cultivation. *Biol. Fertil. Soils* **37**: 55–63.

Kuske, C.R., Banton, K.L., Adorada, D.L., Stark, P.C., Hill, K.K., Jackson, P.J. (1998) Small-scale DNA sample preparation method for field PCR detection of microbial cells and spores in soil. *Appl. Environ. Microbiol.* **64**: 2463–2472.

Kuske, C.R., Ticknor, L.O., Miller, M.E., Dunbar, J.M., Davis, J.A., Barns, S.M., Belnap, J. (2002) Comparison of soil bacterial communities in rhizospheres of three plant species and the interspaces in an arid grassland. *Appl. Environ. Microbiol.* **68**: 1854–1863.

Leadbetter, J.R. (2003) Cultivation of recalcitrant microbes: cells are alive, well and revealing their secrets in the 21st century laboratory. *Curr. Opin. Microbiol.* **6**: 274–281.

Leaphart, A.B., Lovell, C.R. (2001) Recovery and analysis of formyltetrahydrofolate synthetase gene sequences from natural populations of acetogenic bacteria. *Appl. Environ. Microbiol.* **67**: 1392–1395.

Leaphart, A.B., Friez, M.J., Lovell, C.R. (2003) Formyltetrahydrofolate synthetase sequences from salt marsh plant roots reveal a diversity of acetogenic bacteria and other bacterial functional groups. *Appl. Environ. Microbiol.* **69**: 693–696.

Liesack, W., Weyland, H., Stackebrandt, E. (1991) Potential risks of gene amplification by PCR as determined by 16S rDNA analysis of a mixed-culture of strict barophilic bacteria. *Microb. Ecol.* **21**: 191–198.

Liu, W.T., Marsh, T.L., Cheng, H., Forney, L.J. (1997) Characterization of microbial diversity by determining terminal restriction fragment length polymorphisms of genes encoding 16S rRNA. *Appl. Environ. Microbiol.* **63**: 4516–4522.

Lopez-Granados, F., Jurado-Exposito, M., Atenciano, S., Garcia-Ferrer, A., de la Orden, M.S., Garcia-Torres, L. (2002) Spatial variability of agricultural soil parameters in southern Spain. *Plant Soil* **246**: 97–105.

Lowell, J.L., Klein, D.A. (2000) Heteroduplex resolution using T7 endonuclease I in microbial community analyses. *Biotechniques* **28**: 676–682.

Lowell, J.L., Klein, D.A. (2001) Comparative single-strand conformation polymorphism (SSCP) and microscopy-based analysis of nitrogen cultivation interactive effects on the fungal community of a semiarid steppe soil. *FEMS Microbiol. Ecol.* **36**: 85–92.

Loy, A., Lehner, A., Lee, N., Adamczyk, J., Meier, H., Ernst, J., Schleifer, K.-H., Wagner, M. (2002) Oligonucleotide microarray for 16S rRNA gene-based detection of all recognized lineages of sulfate-reducing prokaryotes in the environment. *Appl. Environ. Microbiol.* **68**: 5064–5081.

Ludwig, W., Kenk, H.P. (2001) Overview: A phylogenetic backbone and taxonomic framework for prokaryotic systematics. In: Boone, D.R., Castenholz, R.W. (eds) *Bergey's Manual of Systematic Bacteriology, 2nd* Ed, Volume One, The Archaea and the deeply branching and phototrophic Bacteria. 49–65.

Ludwig, W., Strunk, O., Klugbauer, S. Klugbauer, N., Weizenegger, M., Neumaier, J., Bachleitner, M., Schleifer, K.H. (1998) Bacterial phylogeny based on comparative sequence analysis. *Electrophoresis* **19**: 554–568.

Lukow, T., Dunfield, P.F., Liesack, W. (2000) Use of the T-RFLP technique to assess spatial and temporal changes in the bacterial community structure within an agricultural soil planted with transgenic and non-transgenic potato plants. *FEMS Microbiol. Ecol.* **32**: 241–247.

Luton, P.E., Wayne, J.M., Sharp, R.J., Riley, P.W. (2002) The mcrA gene as an alternative to 16S rRNA in the phylogenetic analysis of methanogen populations in landfill. *Microbiology* **148**: 3521–3530.

Manefield, M., Whiteley, A.S., Griffiths, R.I., Bailey, M.J. (2002) RNA stable isotope probing, a novel means of linking microbial community function to Phylogeny. *Appl. Environ. Microbiol.* **68**: 5367–5373.

Marschner, P., Kandeler, E., Marschner, B. (2003) Structure and function of the soil microbial community in a long-term fertilizer experiment. *Soil Biol. Biochem.* **35**: 453–461.

Marsh, T.L. (1999) Terminal restriction fragment length polymorphism (T-RFLP): an emerging method for characterizing diversity among homologous populations of amplification products. *Curr. Opin. Microbiol.* **2**: 323–327.

Marsh, T.L., Saxman, P., Cole, J., Tiedje, J. (2000) Terminal restriction fragment length polymorphism analysis program, a web-based research tool for microbial community analysis. *Appl. Environ. Microbiol.* **66**: 3616–3620.

McCaig, A.E., Glover, L.A., Prosser, J.I. (1999) Molecular analysis of bacterial community structure and diversity in unimproved and improved upland grass pastures. *Appl. Environ. Microbiol.* **65**: 1721–1730.

McCaig, A.E., Grayston, S.J., Prosser, J.I., Glover, L.A. (2001) Impact of cultivation on characterisation of species composition of soil bacterial communities. *FEMS Microbiol. Ecol.* **35**: 37–48.

McDonald, I.R., Murrell, J.C. (1997) The methanol dehydrogenase structural gene mxaF and its use as a functional gene probe for methanotrophs and methylotrophs. *Appl. Environ. Microbiol.* **63**: 3218–3224.

Metcalfe, A.C., Krsek, M., Gooday, G.W., Prosser, J.I., Wellington, E.M.H. (2002) Molecular analysis of a bacterial chitinolytic community in an upland pasture. *Appl. Environ. Microbiol.* **68**: 5042–5050.

Morris, S.A., Radajewski, S., Willison, T.W., Murrell, J.C. (2002) Identification of the functionally active methanotroph population in a peat soil microcosm by stable-isotope probing. *Appl. Environ. Microbiol.* **68**: 1446–1453.

Murrell, J.C., Radajewski, S. (2000) Cultivation-independent techniques for studying methanotroph ecology. *Res. Microbiol.* **151**: 807–814.

Muyzer, G., Smalla, K. (1998) Application of denaturing gradient gel electrophoresis (DGGE) and temperature gradient gel electrophoresis (TGGE) in microbial ecology. *Antonie Van Leeuwenhoek* **73**: 127–141.

Nakatsu, C.H., Torsvik, V., Ovreas, L. (2000) Soil community analysis using DGGE of 16S rDNA polymerase chain reaction products. *Soil Sci. Soc. Am. J.* **64**: 1382–1388.

Nicol, G.W., Glover, L.A., Prosser, J.I. (2003) The impact of grassland management on archaeal community structure in upland pasture rhizosphere soil. *Environ. Microbiol.* 5: 152–162.

Nunan, N., Wu, K., Young, I.M., Crawford, J.W., Ritz, K. (2002) In situ spatial patterns of soil bacterial populations, mapped at multiple scales, in an arable soil. *Microb. Ecol.* **44**: 296–305.

Nunan, N., Wu, K.J., Young, I.M., Crawford, J.W., Ritz, K. (2003) Spatial distribution of bacterial communities and their relationships with the micro-architecture of soil. *FEMS Microbiol. Ecol.* **44**: 203–215.

Nusslein, K., Tiedje, J.M. (1998) Characterization of the dominant and rare members of a young Hawaiian soil bacterial community with small-subunit ribosomal DNA amplified from DNA fractionated on the basis of its guanine and cytosine composition. *Appl. Environ. Microbiol.* **64**: 1283–1289.

Nusslein, K., Tiedje, J.M. (1999) Soil bacterial community shift correlated with change from forest to pasture vegetation in a tropical soil. *Appl. Environ. Microbiol.* **65**: 3622–3626.

O'Donnell, A.G., Gorres, H.E. (1999) 16S rDNA methods in soil microbiology. *Curr. Opin. Biotechnol.* **10**: 225–229.

Ogram, A. (2000) Soil molecular microbial ecology at age 20: methodological challenges for the future. *Soil Biol. Biochem.* **32**: 1499–1504.

Oliver, I., Beattie, A.J. (1996) Designing a cost-effective invertebrate survey: A test of methods for rapid assessment of biodiversity. *Ecol. Appl.* **6**: 594–607.

Ovreas, L., Torsvik, V. (1998) Microbial diversity and community structure in two different agricultural soil communities. *Microb. Ecol.* **36**: 303–315.

Peixoto, R.S., Coutinho, H.L.D., Rumjanek, N.G., Macrae, A., Rosado, A.S. (2002) Use of *rpoB* and 16S rRNA genes to analyse bacterial diversity of a tropical soil using PCR and DGGE. *Lett. Appl. Microbiol.* **35**: 316–320.

Philippot, L., Piutti, S., Martin-Laurent, F., Hallet, S., Germon, J.C. (2002) Molecular analysis of the nitrate-reducing community from unplanted and maize-planted soils. *Appl. Environ. Microbiol.* **68**: 6121–6128.

Pik, A.J., Oliver, I., Beattie, A.J. (1997) Taxonomic sufficiency in ecological studies of terrestrial invertebrates. *Aust. J. Ecol.* **24**: 555–562.

Poly, F., Ranjard, L., Nazaret, S., Gourbiere, F., Monrozier, L.J. (2001) Comparison of nifH gene pools in soils and soil microenvironments with contrasting properties. *Appl. Environ. Microbiol.* **67**: 2255–2262.

Prosser, J.I. (2002) Molecular and functional diversity in soil micro-organisms. *Plant Soil* **244**: 9–17.

Radajewski, S., Webster, G., Reay, D.S., Morris, S.A., Ineson, P., Nedwell, D.B., Prosser, J.I., Murrell, J.C. (2002) Identification of active methylotroph populations in an acidic forest soil by stable isotope probing. *Microbiology* **148**: 2331–2342.

Ranjard, L., Poly, F., Lata, J.C. Mougel, C., Thioulouse, J., Nazaret, S. (2001) Characterization of bacterial and fungal soil communities by automated ribosomal intergenic spacer analysis fingerprints: Biological and methodological variability. *Appl. Environ. Microbiol.* **67**: 4479–4487.

Reysenbach, AL., Giver, L.J., Wickham, G.S., Pace, N.R. (1992) Differential amplification of ribosomal-RNA genes by Polymerase Chain-Reaction. *Appl. Environ. Microbiol.* **58**: 3417–3418.

Ritchie, N.J., Schutter, M.E., Dick, R.P., Myrold, D.D. (2000) Use of length heterogeneity PCR and fatty acid methyl ester profiles to characterize microbial communities in soil. *Appl. Environ. Microbiol.* **66**: 1668–1675.

Robertson, G.P., Klingensmith, K.M., Klug, M.J., Paul, E.A., Crum, J.R., Ellis, B.G. (1997) Soil resources, microbial activity, and primary production across an agricultural ecosystem. *Ecol. Appl.* **7**: 158–170.

Rotthauwe, J.H., Witzel, K.P., Liesack, W. (1997) The ammonia monooxygenase structural gene amoA as a functional marker: Molecular fine-scale analysis of natural ammonia-oxidizing populations. *Appl. Environ. Microbiol.* **63**: 4704–4712.

Rousseaux, S, Hartmann, A., Rouard, N., Soulas, G. (2003) A simplified procedure for terminal restriction fragment length polymorphism analysis of the soil bacterial community to study the effects of pesticides on the soil microflora using 4,6-dinitroorthocresol as a test case. *Biol. Fertil. Soils* **37**: 250–254.

Saetre, P., Bååth, E. (2000) Spatial variation and patterns of soil microbial community structure in a mixed spruce-birch stand. *Soil. Biol. Biochem.* **32**: 909–917.

Scala, D.J., Kerkhof, L.J. (1999) Diversity of nitrous oxide reductase (nosZ) genes in continental shelf sediments. *Appl. Environ. Microbiol.* **65**: 1681–1687.

Schmalenberger, A., Schwieger, F., Tebbe, C.C. (2001) Effect of primers hybridizing to different evolutionarily conserved regions of the small-subunit rRNA gene in PCR-based microbial community analyses and genetic profiling. *Appl. Environ. Microbiol.* **67**: 3557–3563.

Schwieger, F., Tebbe, C.C. (1998) A new approach to utilize PCR-single-strand-conformation polymorphism for 16s rRNA gene-based microbial community analysis. *Appl. Environ. Microbiol.* **64**: 4870–4876.

Schwieger, F., Tebbe, C.C. (2000) Effect of field inoculation with *Sinorhizobium meliloti* L33 on the composition of bacterial communities in rhizospheres of a target plant (*Medicago sativa*) and a non-target plant (*Chenopodium album*) – Linking of 16S rRNA gene-based single-strand conformation polymorphism community profiles to the diversity of cultivated bacteria. *Appl. Environ. Microbiol.* **66**: 3556–3565.

Smalla, K., Wieland, G., Buchner, A., Zock, A., Parzy, J., Kaiser, S., Roskot, N., Heuer, H., Berg, G. (2001) Bulk and rhizosphere soil bacterial communities studied by denaturing gradient gel electrophoresis: Plant-dependent enrichment and seasonal shifts revealed. *Appl. Environ. Microbiol.* **67**: 4742–4751.

Stach, J.E.M., Bathe, S., Clapp, J.P., Burns, R.G. (2001) PCR-SSCP comparison of 16S rDNA sequence diversity in soil DNA obtained using different isolation and purification methods. *FEMS Microbiol. Ecol.* **36**: 139–151.

Stoyan, H., De-Polli, H., Bohm, S., Robertson, G.P., Paul, E.A. (2000) Spatial heterogeneity of soil respiration and related properties at the plant scale. *Plant Soil* **222**: 203–214.

Suzuki, M.T., Giovannoni, S.J. (1996) Bias caused by template annealing in the amplification of mixtures of 16S rRNA genes by PCR. *Appl. Environ. Microbiol.* **62**: 625–630.

Suzuki, M., Rappe, M.S., Giovannoni, S.J. (1998) Kinetic bias in estimates of coastal picoplankton community structure obtained by measurements of small-subunit rRNA gene PCR amplicon length heterogeneity. *Appl. Environ. Microbiol.* **64**: 4522–4529.

Taroncher-Oldenburg, G., Griner, E.M., Francis, C.A., Ward, B.B. (2003) Oligonucleotide microarray for the study of functional gene diversity in the nitrogen cycle in the environment. *Appl. Environ. Microbiol.* **69**: 1159–1171.

Taylor, P.M., Medd, J.M., Schoenborn, L., Hodgson, B., Janssen, P.H. (2002) Detection of known and novel genes encoding aromatic ring-hydroxylating dioxygenases in soils and in aromatic hydrocarbon-degrading bacteria. *FEMS Microbiol. Lett.* **216**: 61–66.

Tiedje, J.M., Asuming-Brempong, S., Nusslein, K., Marsh, T.L., Flynn, S.J. (1999) Opening the black box of soil microbial diversity. *Appl. Soil Ecol.* **13**: 109–122.

Torsvik, V., Ovreas, L. (2002) Microbial diversity and function in soil: from genes to ecosystems. *Curr. Opin. Microbiol.* 5: 240–245.

Torsvik, V., Sorheim, R., Goksoyr, J. (1996) Total bacterial diversity in soil and sediment communities – A review. *J. Indust. Microbiol.* **17**: 170–178.

Treves, D.S., Xia, B., Zhou, J., Tiedje, J.M. (2003) A two-species test of the hypothesis that spatial isolation influences microbial diversity in soil. *Microb. Ecol.* **45**: 20–28.

Urakawa, H., El Fantroussi, S., Smidt, H., Smoot, J.C., Tribou, E.H., Kelly, J.J., Noble, P.A., Stahl D.A. (2003) Optimization of single-base-pair mismatch discrimination in oligonucleotide microarrays. *Appl. Environ. Microbiol.* **69**: 2848–2856.

Valinsky, L., Della Vedova, G., Scupham, A.J. *et al.* (2002) Analysis of bacterial community composition by oligonucleotide fingerprinting of rRNA genes. *Appl. Environ. Microbiol.* **68**: 3243–3250.

van Bruggen, A.H.C., Semenov, A.M. (2000) In search of biological indicators for soil health and disease suppression. *Appl. Soil Ecol.* **15**: 13–24.

Viaud, M., Pasquier, A., Brygoo, Y. (2000) Diversity of soil fungi studied by PCR-RFLP of ITS. *Mycol. Res.* **104**: 1027–1032.

Wang, G.C.Y., Wang, Y. (1996) The frequency of chimeric molecules as a consequence of PCR co-amplification of 16S rRNA genes from different bacterial species. *Microbiology* **142**: 1107–1114.

Ward, D.M., Ferris, M.J., Nold, S.C., Bateson, M.M. (1998) A natural view of microbial biodiversity within hot spring cyanobacterial mat communities. *Microbiol. Mol. Biol. Rev.* **62**: 1353–1372.

Webster, G., Embley, T.M., Prosser, J.I. (2002) Grassland management regimens reduce small-scale heterogeneity and species diversity of β-Proteobacterial ammonia oxidizer populations. *Appl. Environ. Microbiol.* **68**: 20–30.

Wellington, E.M.H., Berry, A., Krsek, M. (2003) Resolving functional diversity in relation to microbial community structure in soil: exploiting genomics and stable isotope probing. *Curr. Opin. Microbiol.* **6**: 295–301.

Whitby, C.B., Hall, G., Pickup, R. Saunders, J.R., Ineson, P. Parekh, N.R., McCarthy, A. (2001) C-13 incorporation into DNA as a means of identifying the active components of ammonia-oxidizer populations. *Lett. Appl. Microbiol.* **32**: 398–401.

Wieland, G., Neumann, R., Backhaus, H. (2001) Variation of microbial communities in soil, rhizosphere, and rhizoplane in response to crop species, soil type, and crop development. *Appl. Environ. Microbiol.* **67**: 5849–5854.

Wilson, K.H., Wilson, W.J., Radosevich, J.L., DeSantis, T.Z., Viswanathan, V.S., Kuczmarski, T.A., Andersen, G.L. (2002) High-density microarray of small-subunit ribosomal DNA probes. *Appl. Environ. Microbiol.* **68**: 2535–2541.

Wu, L.Y., Thompson, D.K., Li, G.S., Hurt, R.A., Tiedje, J.M., Zhou, J.Z. (2001) Development and evaluation of functional gene arrays for detection of selected genes in the environment. *Appl. Environ. Microbiol.* **67**: 5780–5790.

Yeates, C., Holmes, A.J., Gillings, M.R. (2000) Novel forms of ring-hydroxylating dioxygenases are widespread in pristine and contaminated soils. *Environ. Microbiol.* **2**: 644–653.

Zehr, J.P., Jenkins, B.D., Short, S.M., Steward, G.F. (2003) Nitrogenase gene diversity and microbial community structure: a cross-system comparison. *Environ. Microbiol.* **5**: 539–554.

Zengler, K., Toledo, G., Rappe, M., Elkins, J., Mathur, E.J., Short, J.M., Keller, M. (2002) Cultivating the uncultured. *Proc. Natl Acad. Sci. USA* **99**: 15681–15686.

Zhou, J.Z., Thompson, D.K. (2002) Challenges in applying microarrays to environmental studies. *Curr. Opin. Biotechnol.* **13**: 204–207.

Zhou, J.Z., Bruns, M.A., Tiedje, J.M. (1996) DNA recovery from soils of diverse composition. *Appl. Environ. Microbiol.* **62**: 316–322.

Zhou, J.Z., Xia, B.C., Treves, D.S., Wu, L.Y., Marsh, T.L., O'Neill, R.V., Palumbo, A.V., Tiedje, J.M. (2002) Spatial and resource factors influencing high microbial diversity in soil. *Appl. Environ. Microbiol.* **68**: 326–334.

14

The importance of microbial culture collections to plant microbiology

Eric Cother

14.1 Introduction

Micro-organisms are important to the continued existence of life on planet Earth, through their role in recycling of organic matter and maintenance of the geochemical cycles of nitrogen, sulphur, carbon and metals. Their importance in animal, plant and human welfare and their multitude of industrial uses from beer production to biosyntheses is legend. Micro-organisms are the source of many forms of natural crop control, are increasingly being used in bioremediation, and are the source of processes that have driven the biotechnology industry. Their association with plants as symbionts, pathogens, endophytes, epiphytes or antagonists, or in litter decomposition is such that it is fair to say that the total value of microbial life forms to the human race is immeasurable (Artuso, 1998). Despite this, the need for continued maintenance of characteristic strains and isolates of microbes in virtual perpetuity is far from obvious to many, and resources devoted to microbial culture collections compared with their importance are trivial. Their mere size relegates them, in most cases, to 'out of sight, out of mind'. Even in legislative tomes aimed at conserving biodiversity, micro-organisms rarely, if ever, get a mention (Davison *et al.*, 1999; Kirsop, 1996; Sands, 1996).

14.2 Why microbial culture collections?

Microbial culture collections are a living library of reference strains and a continuing reference source (Anon, 1981; Keyser, 1987; Malik and Claus, 1987) of well-characterised cultures that are potentially continually available to

Plant Microbiology, Michael Gillings and Andrew Holmes

researchers. They are germ plasm banks contributing invaluable resource pools for biotechnological research (Sands 1996), they have been described as the seed banks and biological gardens of the microbiologist (Hawksworth and Mound, 1991) and they are an archive with an open-ended quality, containing both recognised and as-yet-unrecognised data (Scott, 1991). Moreover, culture collections represent a microcosm of the world's biodiversity (Blaine, 1998).

Microbial culture collections are taken for the purpose of this review to include all microscopic organisms from bacteria and fungi to micro-algae, viruses and protozoa. Such collections are a source of authenticated strains for research, inoculants, pathogenicity testing, media performance and quality control. The criteria for maintaining diverse, stable and identifiable genotypes of the more familiar bacterial, fungal and virus collections are also the basis for the less common collections of protozoa (Daggett, 1980) and microalgae (Blackburn *et al.*, 1998). These collections of microscopic life may in future extend to such forms as nanobes (Pyper, 2001) and prions.

Microbial culture collections are also, in part, museums 'without the dust' (American Committee on Arthropod-borne Viruses, 2001). Museums are the archives of biodiversity (Brain, 1994) on a macro-organism scale. Museums are conservative by the very nature of their mission, which is to collect and preserve things historical or endangered (Davison, 1994). They are cultural institutions in that they are products of sociocultural forces, operating within society and serving social and cultural ends. Likewise, culture collections are dominated by organisms that are integral to the economic life of mankind be they as pathogens, beneficials (e.g. inoculants) or curiosities with potential industrial benefits (e.g., thermophiles or biological control agents). Although culture collections do not exhibit their holdings to an admiring public through display and interpretation, they exist as essential reference libraries and are the museums of microbiology. Deposited cultures ensure the uninterrupted passage of knowledge to successive generations (Lamanna, 1976). Culture collections have a large array of genome diversity among their accessions and this is the only means by which this germ plasm can be preserved with any degree of certainty for future generations.

It is worth looking briefly at the history of libraries and museums as their initiation and development have much in common with culture collections. The need for government to have a wide reference source was one of the driving forces for the establishment of the Library of Congress in the USA. The Library of Congress, one of the great libraries of the world, was '... envisioned by learned men as beacons to the new nation' (Conaway, 2000). This institution now houses over 115 million items and is a collection of the history of human activity that shaped America's development and that of the world. It is a reference collection, historical and irreplaceable.

Just as the Library of Congress got a huge impetus from the collections (6487 volumes) of Jefferson in 1815, the British Museum came into existence in 1753 with the acquisition of the collection (79 575 objects) of Sir Hans Sloan and was the first to open to the general public. Prior to this, libraries were for the benefit of specialist groups, much in the same manner that the early culture collections belonged to individuals (Porter, 1976).

The similarity with libraries has its limits, however. Ainsworth (1961) wrote 'Biology differs from most other branches of science in that an important part of its information is stored not as literature but as specimens'. Walker (1975) continued this theme noting that in a book, only a description of a disease or a fungus can be seen, but in a plant pathology specimen, the disease itself and the pathogen are there to examine.

14.3 Why deposit cultures?

Why is there a need to deposit cultures in a collection? 'In earlier periods of scientific curiosity and endeavour, collection of natural history specimens was a *cause celebre*' (Main, 1990). A valid publication of a new microbial species requires that the type strain (or in the case of some non-culturable species, the genomic sequence) should be deposited in one or more permanent culture collections. For adequate documentation of newly isolated strains, there is an expectation that editors of scientific journals require deposition of new isolates before publication (Allsopp, 1985; Malik and Claus, 1987; Smith and Waller, 1992). While this is mandatory for new species descriptions, few journals require it for new records of hosts or previously undescribed diseases. Instructions to authors for relevant journals published by Blackwell Scientific (e.g., *Journal of Applied Microbiology*, *Letters in Applied Microbiology*, *Molecular Microbiology*, *Molecular Ecology*, *Cellular Microbiology*) and a number of other journals require that nucleotide sequence data should be deposited in the EMBL/GenBank/DDBJ Nucleotide Sequence Data Libraries and the accession number referenced in the manuscript. However, a DNA sequence not linked to a specimen or culture has limited value. If the sequence data of particular loci are used for identification and subsequent research reveals the organism to be a cryptic species, a specimen is an essential basis for further study (Crous and Cother, 2003). There is, however, no mention in most author guidelines of depositing a living culture of the microbe in a recognised culture collection. All authors are expected to adhere to the Vancouver guidelines outlining best practice (www.nejm.org/general/text/requirements/1.htm). While dealing mainly with editorial issues, the inclusion of a mandatory requirement in these guidelines for deposition of new or distinctive cultures/isolates would be a progressive step in ensuring the survival of these organisms for future researchers (Ward *et al.*, 2001). Surprisingly few journals require deposition of isolates that are the subject of the published paper.

Reports of new diseases, new hosts, or most commonly, first reports of a disease in a state, province or country appear regularly in the plant pathology literature, but there is little indication that the great majority of new records are deposited in a herbarium or microbial culture collection. Of the 716 new disease records reported in *Plant Disease* from January 2000 to November 2003, only 32 reported deposition of herbarium specimens and only 31 reported deposition of cultures, with 31 of these reports giving the accession numbers. A similar situation exists for new disease reports in *Plant Pathology*. Several virology reports gave GenBank accession numbers but none make reference to deposition of voucher specimens. An occasional reference is made to identity of the causal organism being confirmed by CABI or CBS but there is no guarantee that those institutions accessioned these isolates. Hence, for all intents and purposes, they are lost to science or only available from the authors in personal collections (assuming they were kept) for a limited time. Cultures may well have been deposited, but in the absence of this information and a readily available worldwide electronic database, access to these cultures years down the track will be severely impeded. The future lies in electronic linking of cultures, sequences, authors and papers (Crous and Cother, 2003). Of 13 international plant pathology journals scrutinised, only four encourage authors to deposit specimens and cultures and only two make this mandatory prior to publication. As the instructions to authors for *Mycological Research* so eloquently state, '... your work may become little more than waste paper if it is impossible to verify what fungi were actually studied.' Reproducibility in mycology and other branches of microbiology is irreversibly and inextricably connected to the unequivocal citation of voucher specimens and cultures (Agerer *et al.*, 2000, Padmanaban, 2002).

14.4 Culture collections for patent deposits

Culture collections provide the biological annex of the patent office and are the repository for patent material (Fritze, 1994) under the Budapest Treaty on the International Recognition of the Deposit of Micro-organisms for the Purposes of Patent Procedure. Where an invention involves biological material, words alone cannot adequately describe how to make and use the invention. Much biotechnological innovation that is based on novel microbes requires patent protection and the need for such protection is increasing as the possible applications to industry of the microbial gene pool become better understood (Kirsop, 1987).

Worldwide, there are many collections designated as an International Depository Authority under the Budapest Treaty where patented cultures are securely stored for 30 years. The corner stone of patent protection is the obligation of disclosure of the invention by the applicant. The main reason for the deposit is to render it available to entitled parties for trials and examination, thus allowing reworkablility (Fritze and Weihs, 2001). For the depositor, the micro-organism is stored in a neutral place, access to it is regulated and the depositor does not have the responsibility of maintaining the biological material for 30 years (Fritze, 1998). While some people believe that DNA sequences might be exact enough to substitute for the complete biological material, a comprehensive deposition means that the invention is completely disclosed and rendered workable. It is thus less likely to be successfully challenged.

14.5 Culture collections, biotechnology and biodiversity

The rapid progress in biotechnology has dramatically increased the demand for better preservation methods and rapid availability of reliable information on microbial properties (Aguilar, 1991). Until recently, accessions in culture collections have been living and reproducible but collections could, in future, become repositories of whole cell native DNA. This may be preferable to the sole deposition of nucleotide sequences in GenBank. However, access to the whole genome rather than fragments may be essential in understanding the breadth of diversity. In this case, the whole is greater than the sum of the parts (Rainey, 2000). There is little doubt that culture collections will ultimately develop into gene libraries (Kirsop, 1987) and many have already progressed from mere assemblages of living cultures to become Biological Resource Centres. The Resource Centres 'contain collections of culturable organisms, replicable parts of these (e.g. genomes, plasmids, viruses, cDNAs) viable but not culturable organisms, cells and tissues, as well as databases containing molecular, physiological and structural information relevant to these collections...' (Anon, 2002).

The emphasis on the genomic alphabet should not replace the awareness of the need for an holistic concept of the organism and the usefulness of studying pure cultures (Stackebrandt and Tindall, 2000). The requirement by the journal *Mycological Research* for fungi that 'reference material from which sequences deposited in GenBank or other molecular databases were obtained must be preserved if they are to be published' should be mandatory for all scientific journals.

Knowledge of the extent to which pathogens vary is essential for the selection of stable crop resistance. The use of collections of widely diverse and virulent pathotypes is essential in revealing genetic diversity in potential breeding material for the development of new cultivars. An understanding of pathogen population structure and variability, and the relationship between pathogenic races and host resistance can help identify cultivars with broad-based disease resistance (e.g., Raymundo *et al.*, 1990). Microbial culture collections can supply baseline reference pathogens against which changes can be monitored and the durability of resistance genes tested. Trends essential to rational crop protection can be predicted (Smith and Waller, 1992). Moreover, culture collections provide the raw material for much basic scientific research on life cycles, genetics, host/pathogen relationships, etc., of an organism (Ingram, 1999).

The extent of microbial biodiversity is yet to be realised (Bull *et al.*, 1992) and the base line of microbial diversity remains undescribed (Aguilar, 1991). Thus one can never hope to capture the full breadth of the gene pool in culture collections, especially considering that probably only about 1% of extant organisms have been cultured (Stackebrandt, 1994). Nevertheless, existing microbial resources are unique and replacement by reisolation can never be guaranteed (Kirsop, 1987).

14.6 Culture collections vs *in-situ* conservation

The only way to ensure that a specific novel trait, especially in a microbe from an endangered ecosystem or plant, is readily available for study and development is to safely conserve microbial strains possessing such traits in a microbial culture collection. Conservation of germ plasm is a basic responsibility of collections. Hosts and their associated micro-organisms are in a state of co-evolution in natural ecosystems. *In-situ* conservation is the ideal situation but the activities of man are such that many ecosystems, large and small, are disappearing or being drastically modified. The only way to guarantee long-term access to the culturable microbial components of diversity (Stackebrandt, 1996b) is to conserve cultures in an appropriate collection. Using again the library analogy, microbial culture collections provide, at least, a refuge for the 'book' when the unique 'printing press' is destroyed. Similar to natural history collections (Main, 1990), the cumulative value of culture collections as a reference source is increasing inversely with the diminution of natural landscapes and habitats. 'We do not know what will be required in future and therefore we have a responsibility to future generations to preserve as much as we reasonably can' (Arnold and Scott, 1991).

Cultures not only obviate the need to collect new isolates from natural habitats and hosts (if they still exist), they also provide a historical perspective that is impossible to replace (Grgurinovic and Walker, 1993). Such isolation is expensive in time and resources and may not be possible due to human disturbance and modification of the landscape. Returning to the similarities between libraries and culture collections, Thomas Jefferson said of his library that he passed on to the Library of Congress '...such a collection was made as probably can never again be effected, because it is hardly probable that the same opportunities, the same time, industry, perseverance, and expense ... would again happen to be in concurrence' (quoted by Conaway, 2000). He could well have been speaking of many micro-organisms in a culture collection.

The arguments of Grgurinovic and Walker (1993) for the maintenance of herbaria and a diverse range of authenticated specimens are even more valid for culture collections as the organisms (other than obligate biotrophs) can be grown and studied in ways not possible with dried voucher specimens. Moreover, correct preservation of infected host tissue may provide obligate pathogens for future infection studies (Smith and Waller, 1992). Examples are cited by Grgurinovic and Walker (1993) and Samson and Staplers (1991) of the importance of herbarium specimens and deposited cultures for accurate distinction of species to overcome confusion in identification. If living cultures of these organisms are also available, even more information on comparative biology can be obtained. Culture collections 'allow taxa established many years ago to be re-examined in the light of characters then overlooked or unknown ... while there may be inaccuracies in published descriptions, the specimens reveal the facts to later workers' (Grgurinovic and Walker, 1993). Voucher specimens and accessioned cultures are essential if workers are to be sure that they are studying the same species from one generation to the next. Even if an organism is wrongly named, or its name changes, the work retains its value as the subject's identity can be verified (Daggett, 1980; Green, 1992; Walker, 1975). 'In the absence of voucher specimens and cultures, there is no certainty that concurrent study elsewhere or later studies with the organism involve the same species' (Agerer *et al.*, 2000; Grgurinovic and Walker, 1993). Agerer *et al.* (2000) wrote, 'It is a fundamental principle of science that research work must be reproducible. Reproducibility requires that studies can be made using the same material or cultures as the original study used'.

Specialised collections may be limited in the variety of species they hold but comprehensive in the number of strains of these species held, acquiring strains from a wide variety of hosts, geographic locations and substrates over time (Anon, 1981). These are indispensable in the study of variability and genetic diversity and useful as reference strains in the comparative taxonomy of unknown cultures (Anon, 1981). On the other hand, the larger, more formal service collections usually contain a more diverse range of taxa but fewer strains of each. Many culture collections have a vast array of genetic diversity among their accessions and are often the only places where such germ plasm is preserved for future generations.

Data that accompany cultures are as important as the culture itself: date of isolation, host or substrate of origin, pathogenicity, geographical origin, culture requirements, novel properties, all provide valuable information. Within reason, irrespective of the amount of information accompanying a strain, they are prime conservation targets for archival studies.

14.7 The economic value of microbial culture collections

In the era of economic rationalism, even culture collections can be the target of scrutiny. The economic value of a particular culture is difficult to assess. While economic value through reductionist thinking may be nil, the scientific value may still be high (Stackebrandt, 1998). However, even economists are coming to realise that genetic diversity in plant germ plasm banks has value in terms of as-yet-unidentified demand ('option value') and the sheer value of its very existence as opposed to extinction ('existence value') (Pardey *et al.*, 2001). The observation that crop breeders infrequently use gene banks does not in itself imply that marginal accessions have low value (Gollin *et al.* 2000). The same would apply to microbial culture collections and those isolates that are infrequently requested (Stackebrandt, 1998). The majority of microbial culture collections were not established for commercial reasons and, as their main function is taxonomic excellence and provision of microbial diversity, the degree of economic value is difficult to assess (Stackebrandt, 1998). The value of a culture collection includes the potential commercial value of strains of no immediate interest. In addition, culture collections provide a safe depository. The substantial time and resources invested by industry in development of strains with particular properties are in part protected by the knowledge that production strains are expertly and confidentially reserved in a culture collection (Kirsop, 1987).

Despite their importance, microflora receive scant attention by governments because of the lack of ability to give economic values to micro-organisms. The value of micro-organisms only becomes measurable when market values are applied to commercial products (Kirsop, 1996) or a specific role in an ecosystem can be defined. For example, the economic value of the fungal arbuscular mycorrhizal (AM) symbionts is estimated at US$549 billion for the phosphorus input that would be needed to substitute for native AM fungi that have co-evolved with more than 80% of all terrestrial plants from almost any habitat which are dependent to varying degrees on the fungus. The running costs of the international culture collection of AM fungi (INVAM) are approximately 0.0002% of that value, yet it recovers, maintains and researches all the appropriate germ plasm. (Morton, 1988; Pérez and Schenck, 1989).

14.8 Importance of collections to taxonomy and biosystematics

'Taxonomy has a special significance in understanding biodiversity by inferring ordered relationships from a mass of unordered detail' (Scott, 1991). May (1990) wrote that 'without taxonomy to give shape to the bricks, and systematics to tell us how to put them together, the house of biological science is a meaningless jumble'. The name of an organism provides the point of access for interpreting information about biodiversity (Scott, 1991) and the name is the crucial key to communication on any aspect of its properties (Hawksworth, 1985). Verification of the scientific names of isolates with which biologists and biotechnologists work is crucial before results are published. Culture collections are dependent on competent taxonomy for accurate classification and systematics (Heywood, 1995). The taxonomic biological literature is extremely long lived (Ainsworth, 1987) but to be useful, it is dependent, in the case of microbes, on living cultures or dried specimens for analytical and comparative purposes. A broad consistently competent level of expertise based on well-maintained reference collections must be the basis for any informed statements on biodiversity (Arnold and Scott, 1991).

14.9 Will molecular biology replace the need for culture collections?

While it is not realistic to believe that representatives of all living micro-organisms will be preserved *ex situ*, culture collection databases can be used as templates on which to measure the diversity of environmental isolates or nucleic acid samples. Genomics can help to identify unique genes and functional pathways present in the preserved strains. Moreover, knowledge of existing strains offers opportunities for the design of media for selective isolation and detection of novel, rare or economically important organisms (Bull, 1991).

'One prerequisite for the assessment of microbial diversity in natural environments is the availability of an extensive molecular database of cultured organisms which can serve as a reference for comparison of sequences from both isolates and uncultured strains' (Stackebrandt, 1996a). Cultured organisms also act as positive controls in the evaluation and optimisation of molecular methods (Blackall, 2002). There is now greater awareness that much of the microbial diversity exists in a viable but non-culturable state and this has motivated the use of various techniques to target these populations in various ecological niches (Harayama, 2000; Hugenholtz *et al.*, 1998). The identification of novel organisms *in situ* has facilitated the isolation of these strains as pure cultures. New isolation strategies, not necessarily in axenic culture, will add to the storehouse of biodiversity in culture collections (Harayama, 2000). In addition to individual species in isolation, collections may need to deal with microbial communities as 'complex multicellular organisms' for which understanding of individual cellular activities is not necessary in order to develop an understanding of a total community in a particular ecosystem (Molin *et al.*, 2000).

Molecular techniques have seen a tripling of the identifiable bacterial divisions from those based on cultivated organisms. An RNA sequence study has identified many apparently important bacterial divisions that have few cultivated representatives and about which little is known (Hugenholtz *et al.*, 1998). Biotechnology has become dominated by the spectacular development of molecular biology and it is seen by some that genetic diversity can be stored artificially as nucleic acid sequence data (Bull, 1991). However, others argue that molecular biology is reductionism in the extreme and needs to be integrated in cross-disciplinary studies

(Rainey, 2000). Stackebrandt and Tindall (2000) argue that the characterisation of isolates in complex environments requires more than just a fleeting glance at the molecular data, and classical methods of characterisation still have an important part to play. The unusually strong weight given to phylogenetic 16S rRNA sequence analysis compared with other genetic and phenotypic characters is the most obvious reason why the publication of new names at genus and species level is often based on characterisation of a single isolate (Christensen *et al.*, 2001). The creation of new taxa, however, should depend on the existence of more than one strain. Characterisation of a single strain is insufficient to represent variability without *a priori* knowledge of population structure (Christensen *et al.*, 2001). Access to many strains of an organism allows the breadth of characteristics associated with those taxa to be better documented (Bull *et al.*, 1992; Sneath, 1977).

An illustration of the reductionism of molecular biology can be seen in the heavy reliance on molecular techniques for identification of arboviruses. As many viruses are detected by molecular means only, few new viruses are being registered in catalogues, although hundreds of genomic sequences are lodged in databases. The sequence data provide little if any phenotypic information (American Committee on Arthropod-borne Viruses, 2001). The genomic sequence provides the foundation for phenotypic expression but it is not yet possible to deduce completely the phenotype of a virus or any organism solely from its genomic sequence. Detection of viral nucleic acid is not equivalent to isolating a virus. To accurately phenotype a newly discovered virus, infectious virus must be available (American Committee on Arthropod-borne Viruses, 2001). For studies on ecology, pathogenesis and disease potential, living material must be deposited in reference collections. Genetic diversity can be stored artificially as nucleic acid sequence data but these will never replace the living organism until the true nature and functions of genes are known. Biodiversity at the molecular level cannot provide information on morphological and physiological activities or how these genes work in the environment (Stackebrandt, 1996a). Functional biodiversity must also be studied (Yeates, 1996) and Davies (2002) laments that the study of microbial physiology is often replaced in many university departments by more trendy molecular biology courses.

Commoner (2002) argues that experiments on a series of protein-based processes have contradicted the hypothesis that a DNA gene exclusively governs the molecular processes that give rise to a particular inherited trait. The DNA gene clearly exerts an important influence on inheritance, but it is not unique in that respect and acts only in collaboration with a multitude of protein-based processes. The net outcome is that no single DNA gene is the sole source of a given protein's genetic information and therefore of the inherited trait. In other words, a gene's effect on inheritance cannot be predicted simply from its nucleotide sequence. Commoner (2002) further argues that most molecular biologists operate under the assumption that DNA is the secret of life, whereas the careful observation of the hierarchy of living processes strongly suggests that it is the other way around: DNA did not create life; life created DNA. We must avoid the mistake of reducing life to a master molecule in order to satisfy our emotional need for unambiguous simplicity (Commoner, 2002).

Any biotechnology based upon natural organisms, especially if they are deployed in the environment (e.g. for rhizosphere inoculants or biocontrol agents), requires a supply of authenticated living cultures. Microbial culture collections are the 'bank' from which these cultures can be reliably drawn. For many biotechnological products and processes, the molecular biology option is either unnecessary or inappropriate. A preoccupation with cells and molecules, rather than whole organisms and populations, has resulted in part in a decline of our knowledge of plant diseases among wild populations be they in the tropics (Smith and Waller, 1992) or even in the oceans where algae constitute the wild population. Without a holistic, integrated approach to microbiology, we risk tunnel vision and a flawed concept of microbes and their function (Rainey, 2000).

Molecular biology has, however, become an essential tool to be used in culture collections. It is now accepted that bacterial classification is only determined at the molecular level because the genetic diversity responsible for adaptation to particular habitats is present at the molecular level. A range of DNA typing methods and sequence analyses has facilitated determination of intra- and inter-species relatedness (Stackebrandt *et al.*, 2002). Although species definition now relies on molecular techniques, it is recommended that genomic methods be validated with collections of strains for which extensive DNA–DNA similarity data are available, and strain collections representative of the phylogenetic lineage(s) of the species should be studied (Stackebrandt and Ludwig, 1994; Stackebrandt *et al.*, 2002). Culture collections can provide this basic and unchanging material.

14.10 Culture collections as archives

It has been argued that culture collections halt evolutionary processes (Bull *et al.*, 1992) but they may also prevent, if cultures are properly maintained, loss of, or damage to, vital characteristics such as pathogenicity (Smith and Waller, 1992). However, even with the greatest care, living cultures can change over time. It is thus essential, in all living culture collections, that dried dead cultures of the fresh isolates be prepared and stored for future reference. Such dried cultures, if properly prepared, can be used in future as a check on cultural and morphological characteristics of the living culture to check agreement with the isolate as it was originally. Moreover, the dried cultures can serve as a DNA source of the isolate as it was when fresh, for comparison years later with the DNA of the living culture as it is now. The same is true of the original specimen from which the culture was isolated. Thus it can be seen that the close association of culture collections and herbaria adds greatly to the value and reference accuracy of each. Additionally, the archival nature of collections provides snapshots in time. They provide unchanging static evidence of what was studied 50, 100 or 200 years ago and allow work to be reassessed in the light of later knowledge (Walker, 1975). Microbiologists investigating antibiotic resistance conferred by plasmids have studied cultures accessioned prior to World War II and the widespread use of antibiotics. Plasmids existed before such widespread use and the findings cast a different light on the argument that antibiotics gave rise to plasmids. Such 'before' and 'after' snapshots of organisms help us understand the genetic events surrounding the evolution of traits. The archival nature of herbaria was utilised when Fraile *et al.* (1997) tracked the evolution of two tobamoviruses infecting *Nicotiana glauca* by isolating viruses from herbarium specimens accessioned between 1899 and 1972. Sequence analyses showed no increase in the genetic diversity among isolates of *Tobacco mild green mosaic tobamovirus* during that time.

Similarly, Penrose (1993) and Penrose and Senn (1995) used fungal cultures deposited in a collection before the release of specific fungicides to set baseline sensitivities of 'wild' biotypes of fruit tree pathogens. This permitted the accurate detection of population shifts towards fungicide resistance, should there be selection pressure under normal orchard practice. Resistance management strategies can then be implemented to extend the useful life of registered plant protection chemicals.

The usefulness of traditional collections is being continually enhanced by new technology. Microbial culture collections are a rich source of archival material that can be profoundly important in mapping genetic drift in a particular species. Even a non-viable isolate in a herbarium specimen or freeze-dried ampoule can yield valuable DNA for analysing molecular phylogenetic relationships. Small amounts of DNA from a herbarium specimen can be cloned into a library or used for PCR amplification without any damage to the voucher specimen (e.g. Fraile *et al.*, 1997). This is no less true for microbes as it is for other branches of science (Baker, 1994; Cooper *et al.*, 1998; Rivers and Ardren, 1998). 'The fact that DNA sequences can be amplified from minute samples ... means that such specimens have now acquired a research potential undreamed of at the time of their collection' (Brain, 1994). The safe repository of archival isolates is important for later examination by, as yet, undeveloped technologies (Rivers and Ardren, 1998). The molecular biology revolution that commenced in the 1980s could not have been foreshadowed 50 years ago. Likewise, techniques, unforeseen when cultures were deposited (e.g. Fourier transform infrared spectroscopy (Tindall *et al.*, 2000); integrons (Stokes *et al.*, 2001)) are now becoming available for differentiating large numbers of isolates such as those associated with environmental sampling (Tindall *et al.*, 2000). These techniques allow microbial culture collections, as with museums, to be perceived as conduits of information (Drinkrow *et al.*, 1994), and not merely as terminal repositories of collections.

Microbial culture collections are the dynamic Dead Sea Scrolls of microbial form, function and biodiversity. The Dead Sea Scrolls have provided a continuing source of scholarship for Christianity and new approaches to their interpretation have in turn provided new evidence from the period of history captured in their writings (Thiering, 1992). So too, the archived microbes in a culture collection allow future researchers to return again and again for renewed study (Lamanna, 1976) not only on the written word (the data accompanying the isolate), but also *on the living cell* (von Arx and Schipper, 1978). This information far exceeds the most comprehensive descriptions of the organism (Scott, 1991).

14.11 Culture collections, quarantine and trade

Culture collections also act, in the case of plant pathogens, as a record of disease occurrence in a local, regional or national context. For export purposes, freedom from a known or specific disease is vital to many primary industries, especially the seed trade. Accurate records of diseases and the causal organism are the basis for provision of Phytosanitary Certificates.

Under the GATT (now the World Trade Organisation) Uruguay Round negotiations, quarantine was recognised as an area that could be abused as a technical barrier to trade. Quarantine barriers can now only be erected if they are technically justified (Iken, 1997). Knowledge of existing pests is based on sound records in collections and herbaria (Grgurinovic and Walker, 1993) and ability to identify exotic diseases of quarantine significance is dependent on rapid diagnostic capabilities, which in turn are generally associated with reference collections. Culture collections provide positive controls for many diagnostic tests such as ELISA and PCR and a collection of exotic pathogens allows rapid response in diagnosing suspected exotic disease outbreaks. For example, viral antisera collections at the Department of Primary Industries, Queensland have been particularly useful for diagnosing recent outbreaks in the field or in quarantine interceptions (Davis *et al.*, 2000; Persley *et al.*, 2001; Thomas and Dodman, 1993)

In Australia, the Australian Quarantine and Inspection Service perceives the maintenance of collections as vital to all their activities (Iken, 1997). Miller and Moran (1997) and Sly (1998) recently detailed the importance of reference collections in Australia. The linking of collections through the Australian Plant Pest Database is well under way but there has been no move to arrest the state of decline identified in many reference collections. The need for national guidelines and mechanisms to rescue the essential elements of these collections has been highlighted (Anon, 1981; Sly, 1998) but it is usually an *ad hoc*, last minute and uncoordinated emergency response (Sly, 1998). These problems, detailed as far back as 1984, are common to culture collections around the world but they have not diminished in the intervening years (Batra and Iijima, 1984).

The long-term survival and availability of cultures in many small collections are threatened by the retirement of the researcher or by project completion (Anon, 1981; Ingram, 1999; Simmons *et al.*, 1984). Even during its active growth and maintenance, the funds on which the collection depended for its on-going maintenance may have been siphoned off from ancillary projects. Thus they are extremely vulnerable to loss (Ingram, 1999). The World Federation of Culture Collections has an endangered collections committee to assist collections that may be abandoned (http://www.wfcc.info/committee/endangered/home.html).

14.12 Conclusion

The genetic resource, be it on a macro or micro scale, is the heritage of future generations. Those who laid the foundations to establish the first national parks, the great botanic gardens, libraries and museums showed a foresight that gave future generations access to resources and facilities that may not otherwise have existed. In today's world of immediacy and instant gratification, the present-day microbiologist should pause to reflect on the current extinction rates (however speculative) and make a conscious effort to archive their work so that those who follow in this discipline can access their microbial past as well as their present biota.

Acknowledgements

Eleanor Fairbairn-Wilson, Vicki Glover, Jane Thompson and Wendy Robinson ably assisted in sourcing references. Virology examples from Andrew Geering and constructive comments on the draft manuscript by John Walker and Pedro Crous are gratefully acknowledged.

References

Agerer, R., Ammirati, J., Blanz P. *et al.* (2000) Always deposit vouchers. *Mycol. Res.* **104**: 642–644.

Aguilar, A. (1991) New scientific challenges for microbial culture collections. *World J. Microbiol. Biotechnol.* **7**: 289–291.

Ainsworth, G.C. (1961) Storage and retrieval of biological information. *Nature* **191**: 12–14.

Allsopp, D. (1985) Fungal culture collections – for the biotechnology industry. *Indust. Biotechnol.* 5: 2.

American Committee on Arthropod-borne Viruses, Subcommittee on Interrelationships among Catalogued Arboviruses (2001) Identification of arboviruses and certain rodent-borne viruses: Reevaluation of the paradigm. *Emerg. Infect. Dis.* **7**: 756–758.

Anonymous (1981) *National Work Conference on Microbial Collections of Major Importance to Agriculture.* American Phytopathological Society: St Paul, MN.

Anonymous (2002) Evolving from culture collections to Biological Resources Centres. *BCCM News* **11**.

Arnold M.H., Scott, P.R. (1991) General discussion. In: Hawksworth, D.L. (ed) *The Biodiversity of Micro-organisms and Invertebrates: Its Role in Sustainable Agriculture*, pp. 63–70. CAB International: Wallingford, UK.

Artuso, A. (1998) Economic valuation of microbial diversity – is it worth it? In: *World Federation for Culture Collections Workshop The Economic Value Of Microbial Genetic Resources".* Eighth International Symposium on Microbial Ecology, Halifax, Canada, August, 1998.

Baker, R.J. (1994) Some thoughts on conservation, biodiversity, museums, molecular characters, systematics, and basic research. *J. Mammol.* **75**: 277–287.

Batra, L R., Iijima, T. (1984) *Critical Problems of Culture Collections*. Institute for Fermentation: Osaka, Japan.

Blackall, L.L. (2002) New technologies in the characterisation of microbial populations. *Microbiol. Australia* **23**: 27–30.

Blackburn, S.I., Bolch, C.J.S., Brown M.R., LeRoi J-M., Volkman J.K. (1998) The CSIRO collection of living microalgae: the importance of culture collections and microalgal biodiversity to aquaculture and biotechnology. *Actes de colloques (Brest)* **21**: 150–159.

Blaine, L. (1998) The economic value of microbial diversity information. In: *World Federation for Culture Collections Workshop The Economic Value Of Microbial Genetic Resources".* Eighth International Symposium on Microbial Ecology, Halifax, Canada, August, 1998.

Brain, C.K. (1994) Research collections in South African natural history museums: their scope, relevance and coordination. *South Afr. J. Sci.* **90**: 439–442.

Bram, L. (2001) Bioprospecting – glorified and abjectified. *Todays Life Sci.* **13**: 14.

Bull, A.T. (1991) Biotechnology and biodiversity. In: Hawksworth, D.L. (ed) *The Biodiversity of Micro-organisms and Invertebrates: Its Role in Sustainable Agriculture*, pp. 203–219. CAB International: Wallingford, UK.

Bull, A.T., Goodfellow, M., Slater, J.H. (1992). Biodiversity as a source of innovation in biotechnology. *Ann. Rev. Microbiol.* **46**: 219–252.

Christensen, H., Bisgaard, M., Frederiksen, W., Mutters, R., Kuhnert, P., Olsen, J.E. (2001) Is characterization of a single isolate sufficient for valid publication of a new genus or species? Proposal to modify Recommendation 30b of the *Bacteriological Code* (1990 Revision). *Int. J. System. Evol. Microbiol.* **51**: 2221–2225.

Commoner, B. (2002) Unravelling the DNA myth: The spurious foundation of genetic engineering. *Harper's Magazine* **304**: 39–47.

Conaway, J. (2000) *America's Library. The story of the Library of Congress, 1800–2000*. Yale University Press: New Haven, CT.

Cooper, J.E., Dutton, C.J., Allchurch, A.F. (1998) Reference collections: their importance and relevance to modern zoo management and conservation biology. *J. Wildlife Pres. Trusts* **34**: 159–166.

Crous, P.W., Cother, E.J. (2003) The reality of a virtual laboratory: reference to personal herbaria and culture collections is bad science. *Phytopathol. News* **37**: 138.

Daggett, P-M. (1980) Why a reference collection of living protozoa. *J. Protozool.* **27**: 507–508.

Davies, J. (2002) Re-birth of microbial physiology. *Environ. Microbiol.* **4**: 6.

Davis, R.I., Parry, J.N., Geering, A.D.W., Thomas, J.E., Rahamma, S. (2000) Confirmation of the presence of *Rice tungro bacilliform virus* in Papua (formerly Irian Jaya), Indonesia. *Aust. Plant Pathol.* **29**: 223.

Davison, P. (1994) Museum collections as cultural resources. *South Afr. J. Sci.* **90**: 435–436.

Davison, A.D., Yeates, C., Gillings, M.R., de Brabandere, J. (1999) Micro-organisms, Australia and the Convention on Biological Diversity. *Biodiversity Conserv.* **8**: 1399–1415.

Drinkrow, D.R., Cherry, M.I., Siegfried, W.R. (1994) The role of natural history museums in preserving biodiversity in South Africa. *South Afr. J. Sci.* **90**: 470–479.

Fraile, A., Escriu, F., Aranda, M.A., Malpica, J.M. Gibbs, A.J., Garcia-Arenal, F. (1997) A century of tobamovirus evolution in an Australian population of *Nicotiana glauca*. *J. Virol.* **71**: 8316–8320.

Fritze, D. (1994) Patent aspects of the convention at the microbial level. In: Kirsop, B., Hawksworth, D.L (eds) *The Biodiversity of Micro-organisms and the Role of Microbial Resource Centres*, pp. 37–43. World Federation of Culture Collections: Braunschweig, Germany.

Fritze, D. (1998) The view of IDAs (International Depositary Authorities) under the Budapest Treaty for the International Recognition of Micro-organisms for the Purpose of Patent Protection. In: *World Federation for Culture Collections Workshop The Economic Value Of Microbial Genetic Resources"*. Eighth International Symposium on Microbial Ecology, Halifax, Canada, August, 1998.

Fritze, D., Weihs, V. (2001) Deposition of biological material for patent protection in biotechnology. *Appl. Microbiol. Biotechnol.* **57**: 443–450.

Gollin, D., Smale, M., Skovmand, B. (2000) Searching an *ex-situ* collection of wheat genetic resources. *Am. J. Agri. Econ.* **82**: 812–827.

Green, D.C. (1992) Ecology and conservation: the role of biological collections. *Aust. Biol.* **5**: 48–56.

Grgurinovic, C., Walker, J. (1993) Herbaria and their place in science: a mycological and plant pathological perspective. *Aust. Plant Pathol.* **22**: 14–18.

Harayama, S. (2000) Acquisition of larger microbial genetic resources. *Environ. Microbiol.* **2**: 4.

Hawksworth, D.L. (1985) Fungus culture collections as a biotechnological resource. *Biotechnol. Genet. Engineer. Rev.* **3**: 417–453.

Hawksworth, D.L., Mound, L.A. (1991) Biodiversity databases: the crucial significance of collections. In: Hawksworth, D.L. (ed) *The Biodiversity of Micro-organisms and Invertebrates: Its Role in Sustainable Agriculture*, pp 17–29. CAB International: Wallingford, UK.

Heywood, V.H. (1995) *Global Biodiversity Assessment*. Cambridge University Press: Cambridge, UK.

Hugenholtz, P., Goebel, B.M., Pace, N.R. (1998) Impact of culture-independent studies on the emerging phylogenetic view of bacterial diversity. *J. Bacteriol.* **180**: 4765–4774.

Iken, R. (1997) The diagnostic and taxonomic needs of the pest risk analysis process as a result of the WTO sanitary and phytosanitary agreement. In: Millar, J. and Moran, J. (eds) *An Evaluation of the Disease Diagnostic Capabilities of Australian Plant Industries*, pp. 10–11. Final Report for RIRDC Project No. DAV 107A. Rural Industries Research and Development Corporation, Canberra.

Ingram, D.S. (1999) Biodiversity, plant pathogens and conservation. *Plant Pathol.* **48**: 433–442.

Keyser, H.H. (1987) The role of culture collections in biological nitrogen fixation. In: Elkan, G.H. (ed) *Symbiotic Nitrogen Fixation Technology*, pp. 413–428. Marcel Dekker: New York.

Kirsop, B.E. (1987) Culture collections: repositories for microbial germ plasm. *Nature Res.* **23**: 3–9.

Kirsop, B.E. (1996) The Convention on Biological Diversity: some implications for microbiology and microbial culture collections. *J. Indust. Microbiol.* **17**: 505–511.

Lamanna, A. (1976) Role of culture collections in the era of molecular biology. In: Colwell, R.R. (ed) *The Role of Culture Collections in the Era of Molecular Biology*, p. 3. American Society for Microbiology: Washington, DC.

Main, B.Y. (1990) Restoration of biological scenarios: the role of museum collections. *Proc. Ecol. Soc. Aust.* **16**: 397–409.

Malik, K.A., Claus, D. (1987) Bacterial culture collections: their importance to biotechnology and microbiology. *Biotechnol. Gen. Engineer. Rev.* **5**: 137–167.

Mathur, E. (1998) Biodiversity, bioprospecting and combinatorial biomolecule discovery. In: *World Federation for Culture Collections Workshop The Economic Value Of Microbial Genetic Resources".* Eighth International Symposium on Microbial Ecology, Halifax, Canada, August, 1998.

May, R.M. (1990) Taxonomy as diversity. *Nature* **347**: 129–130.

Millar, J., Moran, J. (1997) *An Evaluation of the Disease Diagnostic Capabilities of Australian Plant Industries*. Final Report for RIRDC Project No. DAV 107A. Rural Industries Research and Development Corporation, Canberra.

Molin, S., Nielsen, A.T., Heydorn, A., Tolker-Nielsen, T., Sternberg, S. (2000) Environmental microbiology at the end of the second millennium. *Environ. Microbiol.* **2**: 6.

Morton, J.B. (1988) *Global Economic Value of the Arbuscular Mycorrhizal Symbiosis and a Living International Culture Collection (INVAM). World Federation for Culture Collections Workshop The Economic Value Of Microbial Genetic Resources"* Eighth International Symposium on Microbial Ecology, Halifax, Canada, August, 1998.

Padmanaban, D.A. (2002) Non-deposit of voucher cultures. *Mycol. Res.* **106**: 644.

Pardey, P.G., Koo, B., Wright, B.D., van Dusen, E., Skovmand, B., Taba, S. (2001) Costing the conservation of genetic resources: CIMMYT's ex situ maize and wheat collection. *Crop Science* **41**: 1286–1299.

Penrose, L.J. (1993) Baseline sensitivities of preserved isolates of *Penicillium* spp. from pome fruit to imazalil and iprodione. *Aust. Plant Pathol.* **22**: 37–39.

Penrose, L.J., Senn, A.A. (1995) Baseline sensitivities of preserved isolates of *Sclerotinia fructicola* from various host species to propiconazole and iprodione. *Aust. Plant Pathol.* **24**: 9–14.

Pérez, Y., Schenck, N.C. (1989) The international culture collection of VA mycorrhizal fungi (INVAM). In: Vancura, V., Kunc, F. (eds) *Interrelationships between micro-organisms and plants in soil*, pp. 171–175. Proceedings of an International Symposium, Liblice, Czechoslovakia, June 1987. Elsevier Science: Amsterdam, The Netherlands.

Persley, D.M., McMichael, L., Spence, D. (2001) Detection of *Peanut stripe virus* in post-entry quarantine in Queensland. *Aust. Plant Pathol.* **30**: 337.

Porter, J.R. (1976) The world view of culture collections. In: Colwell, R.R. (ed) *The Role of Culture Collections in the Era of Molecular Biology*, pp. 62–72. American Society for Microbiology: Washington, DC.

Pyper, W. (2001) Life, the universe and nanobes. *Today's Life Sci.* **13**: 12–13.

Rainey, P. (2000) An organism is more than its genotype. *Environ. Microbiol.* **2**: 8.

Raymundo A.K., Nelson, R.J., Ardales, E.Y., Baraoidan, M.R., Mew, T.W. (1990) A simple method for detecting genetic variation in *Xanthomonas campestris* pv *oryzae* by restriction fragment length polymorphism. *Int. Rice Res. News.* **15**: 8–9.

Rivers, P.J., Ardren, W.R. (1998) The value of archives. *Fisheries* **23**: 6–9.

Samson, R.A., Staplers, J.A. (1991) Culture collections: their role and importance. In: Champ, B.R., Highley, E., Hocking A.D., Pitt, J.I. (eds) *Fungi and Toxins in Stored Products*, pp. 73–77. Proceedings of an international conference, Bangkok 1991. ACIAR Proceedings No 36. Australian Centre for International Agricultural Research, Canberra.

Sands, P. (1996) Microbial diversity and the 1992 Biodiversity Convention. *Biodiver. Conserv.* **5**: 473–491.

Scott, P.R. (1991) The universal issue: information transfer. In: Hawksworth, D.L. (ed) *The Biodiversity of Micro-organisms and Invertebrates: Its Role in Sustainable Agriculture*, pp. 245–266. CAB International: Wallingford, UK.

Simmons, E.G., Schipper, M.A.A., Jong, S.C., Onions, A.H.S., Smith, D., Kurtzman, C., Iijima, T., Stevenson, R.E. (1984) Critical problems of culture collections: issued in commemoration of the 40th anniversary of the Institute for Fermentation, Osaka. Batra, L.R., Iijima, T. (eds). Institute for Fermentation, Osaka, Japan.

Sly, L.I. (1998) Australian microbial diversity. *Aust. Mycol. News.* **17**: 3–15.

Smith, D., Waller, J.M. (1992) Culture collections of micro-organisms: their importance in tropical plant pathology. *Fitopatologia Brasileira* **17**: 5–12.

Stackebrandt, E. (1994) The uncertainties of microbial diversity. In: Kirsop, B., Hawksworth, D.L. (eds) *The Biodiversity of Micro-organisms and the Role of Microbial Resource Centres*, pp. 59–64. World Federation of Culture Collections: Braunschweig, Germany.

Stackebrandt, E. (1996a) From strains to domains: measuring the degree of phylogenetic diversity among prokaryote species and as yet uncultured strains. In: Colwell, R.R., Simidu, U., Ohwada, K. (eds) *Microbial Diversity in Time and Space*, pp. 19–24 Plenum Press: New York.

Stackebrandt, E. (1996b) Are culture collections prepared to meet the microbe-specific demands of the articles of the Convention on Biological Diversity? In: Sampson, R.A., Stalpers, J.A., van der Mei, D, Stouthamer, A.H. (eds) *Culture Collections to Improve the Quality of Life*, pp. 74–78. Ponsen & Looyen: Wageningen, The Netherlands.

Stackebrandt, E. (1998) The value of ex-situ microbial resource centres. In: *World Federation for Culture Collections Workshop The Economic Value Of Microbial Genetic Resources"*. Eighth International Symposium on Microbial Ecology, Halifax, Canada, August, 1998.

Stackebrandt, E., Ludwig, W. (1994) The importance of using outgroup reference organisms in phylogenetic studies: the atopbium case. *System. Appl. Microbiol.* **17**: 39–43.

Stackebrandt, E., Tindall, B.J. (2000) Appreciating microbial diversity: rediscovering the importance of isolation and characterisation of micro-organisms. *Environ. Microbiol.* **2**: 9–10.

Stackebrandt, E., Frederiksen, W., Garrity, G.M. *et al.* (2002) Report of the ad hoc committee for the re-evaluation of the species definition in bacteriology. *Int. J. System. Evol. Microbiol.* **52**: 1043–1047.

Stirton, C.H., Boulos, L., MacFarlane, T.D., Singh, N.P., Nicholas, A. (1990) In defence of taxonomy. *Nature* **347**: 223–224.

Sneath, P.H.A. (1977) The maintenance of large numbers of strains of micro-organisms and the implications for culture collections. *FEMS Microbiol. Lett.* **1**: 333–334.

Stokes, H.W., Holmes, A.J., Nield, B.S., Holley, M.P., Nevalainen, K.M.H., Mabbutt, B.C., Gillings, M.R. (2001) Gene cassette PCR: sequence-independent recovery of entire genes from environmental DNA. *Appl. Environ. Microbiol.* **67**: 5240–5246.

Thiering, B. (1992) *Jesus the Man: a New Interpretation of the Dead Sea Scrolls*. DoubleDay, Sydney.

Thomas, J.E., Dodman, R.L. (1993) The first record of papaya ringspot virus-type P from Australia. *Aust. Plant Pathol.* **22**: 2–7.

Tindall, B.J, Brambilla, E., Steffen, M., Neumann, R., Pukall, R., Kroppenstedt, R.M., Stackebrandt, E. (2000) Cultivatable microbial biodiversity: gnawing at the Gordian knot. *Environ. Microbiol.* **2**: 310–318.

von Arx, J.A., Schipper, M.A.A. (1978) The CBS fungus collection. *Adv. Appl. Microbiol.* **24**: 215–236.

Walker, J. (1975) Mutual responsibilities of taxonomic mycology and plant pathology. *Ann. Rev. Phytopathol.* **13**: 335–355.

Ward, N., Eisen, J., Fraser, C., Stackebrandt, E. (2001) Sequence strains must be saved from extinction. *Nature* **414**: 148.

Yeates, C. (1996). Microbial biodiversity. *Microbiol. Aust.* **17**: 29–32.

Index

ties are smaller both as a proportion of the overall population and in numerical terms. As (and if) Coquitlam's Chinese community grows, we may yet see a reaction from long-standing residents, especially if developers build malls that long-standing residents perceive as catering to Asians exclusively. Coquitlam may yet experience community debates similar to those that developed in Richmond and Markham in the 1990s. Changes in the built environment that run counter to the established cultural norms of a community appear to increase cultural insecurities and to politicize tensions.

INFORMAL URBAN GOVERNANCE RELATIONSHIPS AND MULTICULTURALISM GOALS

In all the biracial municipalities, new governance arrangements have emerged to create a joint public–private capacity to accommodate and manage changes in ethnocultural demographics.

In Vancouver, the governance arrangements that have emerged to manage immigration and multiculturalism policy are stronger than those in the suburban biracial municipalities. First, the city has many more settlement and multiculturalism resources. Second, Vancouver's innovations in multiculturalism policy are linked to a proactive economic-development agenda that focuses on the city's ties with Pacific Rim countries (Hutton 1998, 97). Vancouver's powerful business community participates in developing the capacity to manage social change, because that change supports its economic growth objectives. Kris Olds observes that in Vancouver, the Pacific Rim–specific institutions (such as the Asia-Pacific Foundation, the International Finance Centre Vancouver, and the Hong Kong Business Association) and the mainstream institutions (such as the Greater Vancouver Real Estate Board and the Vancouver Board of Trade) are interlinked and 'command considerable public and private resources that are used to *structure the nature of policies and processes* which influence Vancouver's future' (Olds 2001, 92; emphasis added). The city's institutionalized commitment to supporting multiculturalism, coupled with a strong, proactive group of private-sector leaders, has contributed to its 'power to' manage ethnocultural relations.

In Markham and Richmond, lasting public–private governance relationships have developed around the goal of fostering positive relations between the largely Chinese immigrant community and the long-standing (and largely White) community. These informal institutions developed out of race relations crises and intercultural misunderstandings – that is, out of changing forms of political pluralism. In

Markham and Richmond, governance relationships are anchored in advisory committees but are also supported by strong informal channels of communication, as well as by resource pooling between the city and leaders in civil society. As in Vancouver, the city and the business community work together in managing race relations.

In Richmond and Markham, the emergence of governance arrangements to manage social change has been facilitated by the development of strong community-based institutions representing the local Chinese community. The strength of that community in Richmond (and in other municipalities in Greater Vancouver) is evident in the strength of S.U.C.C.E.S.S. Similarly, in Markham, the Federation of Chinese Canadians commands extensive resources. The leader of that organization (Dr Ken Ng) was one of the co-chairs of the ad hoc committee and its successor, the Coalition of Concerned Canadians, which organized in response to Carole Bell's controversial remarks.

Furthermore, in Richmond and Markham the business community supports a municipal role in managing race relations. In both places, many prominent developers and other business owners are members of the largely Chinese immigrant community. Work remains to be done in integrating various sectors of the local bifurcated community (including the business community) – recall the president and CEO of the Markham Board of Trade's comment that it is 'very hard to cross-pollinate' with the Chinese community in Markham and that there are times when 'one might get the feeling that Chinese business is not interested in serving the clientele from the existing community' (Bray 2005, interview). However, it is clear that the business communities in these locales recognize that poor ethnocultural relations are simply bad for business. And in both places, a great deal of the new business activity and development has been spurred by immigration (on the demand *and* supply sides).

In Surrey, public–private relationships have emerged at the departmental level. However, these relationships are more limited than those in Richmond and Markham, as well as more tenuous. Cleavages within Surrey's largely South Asian immigrant community make it hard for civil servants and 'street-level' bureaucrats (Lipsky 1976) to identify the community's leaders. The lack of cohesion among Surrey's large South Asian immigrant population seems to explain why stronger governance arrangements have not emerged. In addition, because of these divisions, the community has not been able to translate its numbers into collective action to pressure the city to respond more comprehensively to its concerns. Furthermore, whereas backlash against immigration is clearly a problem – a community leader commented that some long-standing residents 'proudly resent multiculturalism' – in the absence of intragroup solidarity, the South Asian community has been unable to fight its racialization in the community.

Though many developers and other business people in Surrey are of South Asian origin, they do not seem to be pressuring the municipality to adapt its services and governance structures to Surrey's new demographics. As one community leader in Greater Vancouver observed, developers in Surrey 'interact with [the city] on an individual basis' rather than as a cultural community (Sanghera 2004, interview).

Governance arrangements were limited in Coquitlam as well. However, this was not owing to divisions within the immigrant community. Rather, civil society remains underdeveloped, and the city is largely a bedroom community. Nevertheless, the strength of the Chinese community elsewhere in the Greater Vancouver area has spilled over into Coquitlam to a certain extent. S.U.C.C.E.S.S. has opened a branch there. A representative of S.U.C.C.E.S.S. had a seat on Coquitlam's Multiculturalism Advisory Committee (1999–2003). Furthermore, the willingness of the Taiwanese and other Chinese communities to donate books to the library facilitated the city's efforts to adapt library services.

The 'social diversity interpretation' leads one to expect a correlation between the ethnic configuration of a municipality and the way in which institutional goals are oriented (Hero 1998, 20). Like multiculturalism policies in other locales and at other levels of government, the goal of biracial municipalities' multiculturalism initiatives is integration. However, given the concentration of a single immigrant community in the city, fears of ethnic segregation and a perceived unwillingness among the immigrant community to integrate have emerged. Thus, biracial municipalities' initiatives address these goals. Multiculturalism policies must respond to majority concerns as well. For instance, Richmond's multiculturalism policy mandate is just as much a response to the concerns of long-standing residents as it is to the concerns of immigrants and ethnocultural minorities. For instance, the issue of non-English signage is one of its key communications issues (RIAC 2004, 2). Markham's Task Force on Race Relations identified similar community goals following the race relations crisis that was triggered by Carole Bell's comments in 1995 (Mayor's Advisory Committee 1996). Members of Richmond Intercultural Advisory Committee spent time debating the philosophy that should guide ethnic relations in Richmond, opting to reject *multiculturalism*, a term that in their view had become synonymous with ethnic segregation, in favour of *interculturalism* to reflect the need for bridges between communities and, most important, integration (Schroeder 2004, interview). Similarly, the report of Surrey's Parks and Recreation Department uses the word 'interculturalism' in its title. In Coquitlam the city's main concern is to create a bridge between its newcomer community and the municipality to encourage access and equity with respect to municipal services. In all cases, governance arrangements help broker intercultural groups and stress integration into a common local community. Influenced by neo-liberalism, multiculturalism policy discourse has shifted to the private sector concept of 'managing diversity' (Abu-Laban and Gabriel 2002). In biracial contexts, a great deal of what municipalities do involves 'managing' social conflict and cultural insecurities.

Multiracial Municipalities

MUNICIPAL RESPONSIVENESS

Multiracial municipalities are at one end or the other of the 'responsiveness' spectrum. The City of Toronto has been responsive to immigrants and ethnocultural minorities, whereas the cities of Mississauga and Brampton have both been unresponsive.

The City of Toronto has developed a comprehensive range of multiculturalism policies and institutionalized supports for those policies at the apex of power in the municipal civil service – the City Manager's Office. In Toronto, the Diversity Management and Community Engagement Unit (DMU) in the City Manager's Office supports and monitors the implementation of the city's multitude of formal (written) multiculturalism policies; the DMU is also a flexible unit that launches action when unanticipated needs arise.

In sharp contrast, the primary responses to social change in the suburban multiracial municipalities of Mississauga and Brampton are community festivals and annual 'multicultural' (Mississauga) and 'multifaith' (Brampton) community breakfasts with the local mayor. Responses in these suburban municipalities are mainly symbolic.

COMMUNITY DYNAMICS AND TYPES OF POLITICAL PLURALISM

It seems that in Mississauga and Brampton, neither immigration nor multiculturalism has altered community dynamics in an obvious way. In general, the community and local officials seem indifferent to the dramatic changes that have occurred in their populations. There has been no general backlash against immigration. Hazel McCallion, the mayor of Mississauga, has observed that Mississauga simply does not have the 'racial confrontation' that exists in Markham (McCallion 2004, interview).

One can observe some competition in the immigrant settlement sector, owing largely to the scarcity of settlement funding (Seepersaud 2004, interview). A *Toronto Star* article described a 'turf war' between Brampton's two most prominent ethnocultural organizations over which organization will do what in the settlement field (White 1992b). However, this competition for resources does not reflect a more general debate about the impact of immigration and multiculturalism on the community.

Also, in Mississauga and Brampton as in Surrey, there seem to be significant religion-based cleavages within the South Asian communities. McCallion mentioned a conflict between two Sikh factions in Mississauga's South Asian community when illustrating her approach to race relations (McCallion 2004, interview). A community leader suggested that if an Indian candidate were to run in a municipal election, Pakistanis would mobilize against that person (Seepersaud 2003, interview). Local municipal officials in Brampton cited the tendency of the South Asian community to run many candidates in each ward in municipal elections as a reason for the community's lack of electoral success (Moore 2003, interview). Another community leader remarked on the number of newspapers in Brampton that serve the Sikh community (Jeffrey 2004, interview). Unfortunately, the statistical category 'South Asian' is a too vague to allow the tracking of changes in the ethnoracial composition of Canada. Municipalities with large concentrations of 'South Asians' may not have cohesive immigrant communities.

In Brampton, former mayor Peter Robertson (1988–2000) was unable to sustain community interest in the Brampton Race Relations Action Committee because of the city's diversity. There was more interest in the committee within Brampton's Hindu, Sikh, and Black communities than there was within the French, German, Croatian, Greek, and other communities (Biggs, in White 1992a). Note that the Hindu and Sikh communities – two religions that are common in South Asia (though they constitute a single group in Statistics Canada's South Asian category) – are mentioned separately. Brampton's diversity was a barrier to developing a local agenda for multiculturalism policy. In addition, unlike the biracial municipalities, Brampton's 'White' community did not seem to see the value in a multiculturalism policy agenda.

Thus, at the time of the interviews, pluralism in Mississauga and

Brampton appeared to be competitive rather than cooperative, as well as highly limited.

In contrast, Toronto is characterized by a dynamic form of pluralism. Its political pluralism is conducive to both competition and cooperation, as evidenced in the strong governance arrangements that have developed there as well as in the city's broad-based urban autonomy coalition. The political strategies of local leaders in Toronto are more radical than those of leaders in other municipalities in the sample. For instance, unlike organizations with ethnocultural equity mandates in other locales, which prefer to influence the political process through informal communication channels and quiet negotiation, in Toronto these organizations meet both 'in the boardroom [and] on the street' (Douglas 2003, interview). As a community leader in Greater Vancouver noted, the Ontario Council of Agencies Serving Immigrants (OCASI), the umbrella organization for settlement organizations in Ontario, stood up to the province during the conservative era and lost a significant portion of its funding (Welsh 2004, interview). Toronto city councillor Shelley Carroll described the more general tendency of organizations – even small-c conservative organizations like the Toronto Board of Trade – to engage in such strategies, especially in their efforts to secure a 'New Deal' for the city from upper levels of government. She gave the impression that there is a great deal of cooperation and solidarity among a multisectoral group of local leaders and organizations in Toronto (Carroll 2004, interview). Toronto's diversity does not appear to be a barrier to cooperation.

In addition, though there are surely cleavages within immigrant communities in Toronto, ethnic conflicts have not become citywide issues. Furthermore, in Toronto there is strong evidence of ethnospecific social capital. Strong organizations represent the city's major visible-minority and immigrant communities. There are so many such organizations that they are impossible to list. Furthermore, the leaders of these organizations know one another and cooperate to advocate collectively on behalf of immigrants.

INFORMAL URBAN GOVERNANCE RELATIONSHIPS AND MUNICIPAL INSTITUTIONAL PURPOSES

Multiracial municipalities diverge in yet another respect. Urban governance arrangements to respond to multiculturalism did not emerge in Mississauga and Brampton, whereas in Toronto, local leaders built strong and inclusive governance arrangements – an urban regime – to create the capacity to respond.

Toronto has strong immigrant settlement organizations and other multiculturalism-related community organizations. Furthermore, its business community views the successful settlement of immigrants as central to the city's global economic competitiveness. As in Vancouver, Toronto's multiculturalism policies are tied to a powerful economic-development agenda. The consensus in Toronto – as its motto, 'Diversity Our Strength,' implies – is that multiculturalism is a competitive advantage in the global marketplace. Also, attracting and retaining highly skilled immigrants is considered central to Toronto's growth agenda. Its urban regime includes city officials, prominent leaders in the business community, and representatives of labour, the social service sector, immigrant settlement organizations, and other groups. Under the umbrella of the Toronto City Summit Alliance (TCSA), high-powered leaders have cooperated to address a number of challenges, including 'becoming a centre of excellence in the integration of immigrants' (TCSA 2003). The TCSA established the Toronto Region Immigrant Employment Council (TRIEC) to address barriers to immigrant integration into the labour force. The council is co-chaired by the president and vice-president of Manulife Financial, one of Canada's most powerful financial institutions.

As in Vancouver, Toronto's business leaders and city officials have tied immigrant settlement goals to the city's economic-development objectives. Local community and municipal resources are pooled within its public–private governance arrangements to address policy goals. However, coalitions in the city are not just concerned with developing cooperative responses to diversity at the local level. Rather, Toronto's urban regime has expanded to encompass the regional and cross-provincial scales. Regime partners are fighting for greater levels of autonomy for the municipality and for a new status for the City of Toronto within Canadian federalism – a New Deal for the city (Good 2007). The lobbying efforts of this dynamic alliance of city leaders did a great deal to push the 'New Deal for Cities' onto the national agenda (Broadbent 2003). Immigrants' leaders, including their policy preferences, are included in the process of setting the local regime's agenda, including its urban autonomy goals.

In Brampton and Mississauga, immigrant settlement organizations operate in a resource-scarce environment that inhibits cooperation. Also, diversity seems to prevent immigrants from forming a common front to pressure those municipalities to respond. In Mississauga, immigrant integration and multiculturalism did not seem to be on the business community's agenda. In Brampton, the business community seemed interested in helping integrate immigrants into local governance. The city may be poised to become more active again in multiculturalism policy making.

Summarizing the Lessons

To what extent does the social diversity interpretation contribute to our understanding of the politics of immigration and multiculturalism at the local level? We can observe some clear patterns (albeit imperfect) in the biracial municipalities. However, in some ways the multiracial municipalities seem to fall into two distinct groups. Table 6.4 summarizes the findings.

What lessons can one draw from the findings discussed in this chapter and throughout this book? First, there is evidence that the collective action problem within immigrant populations is more easily overcome in biracial municipalities. Concentration appears to facilitate the development of community. For instance, in municipalities with a concentration of Chinese residents in Greater Vancouver we see strong community settlement capacity in the form of S.U.C.C.E.S.S. In Greater Vancouver, this capacity was so strong that it 'spilled over' into newer immigrant-receiving communities in the region. The organization began in Vancouver and then opened satellite locations in Richmond and (more recently) in Coquitlam, which did not yet have home-grown capacity in immigrant settlement. Thus, ethnic social capital seems to be an intervening variable that explains how ethnic configurations influence policy responsiveness. The importance of ethnospecific social capital to a group's capacity to influence local politics was one of Orr's (1999) main points in his work on the Black community in Baltimore.

Second, one can observe the emergence of new community dynamics – a backlash on the part of long-standing residents – in most biracial municipalities. In three of the biracial municipalities with a concentration of Chinese immigrants, a power struggle between the long-standing and ethnoracial-minority communities developed that centred on changes in the built environment that became symbols of the cultural evolution of the municipality. In Surrey, too, there was backlash against immigration, but a power struggle was absent, as the South Asian community had not mobilized against its racialization to the same degree. Thus, a dynamic of community backlash and – where the immigrant community possesses sufficient social capital to mobilize – a counter-reaction from the immigrant community, appears more likely to de-

Table 6.4
Summary of patterns of influence of ethnic configurations in multiculturalism policy making

Types of ethnic configurations	**Multiracial**		**Biracial**
Visible minorities (as proportion of population)	High		High
Composition of visible minority population	Diverse		Homogeneous
Backlash	No		Yes
Types of political pluralism	Competitive		Limited
Active in multiculturalism	No	Yes	Yes
Urban governance relationships to support multiculturalism initiatives	No	Yes	Yes
Dominant multiculturalism purposes	Celebrating multiculturalism	Celebrating multiculturalism Equitable access to services and inclusion in governance	Bridging, manage backlash, avoid segregation/social isolation Equitable access to services and inclusion in governance (to varying degrees)
Municipalities	Mississauga and Brampton (2001)	Toronto	Vancouver, Richmond, Surrey, Coquitlam, and Markham

velop in municipalities where there is a large concentration of a single ethnic group. These new community dynamics in biracial municipalities increase the likelihood that the community will agree that there is an ethnic relations problem, which pushes the issue onto the municipal agenda. Rather unexpectedly, backlash leads to greater levels of responsiveness in biracial municipalities. As Orr (1999) also found, ethnospecific social capital solves only one element of the local collective-action problem. In order to achieve education reform that benefited Baltimore's Black community, local leaders had to build bridges among interethnic pools of 'social capital.' According to his study, 'bonding' (ethnospecific) social capital is insufficient. In his view, processes of urban regime building are one way in which interethnic bridging occurs. The new community dynamics in many biracial municipalities create incentives for local political leaders to intervene. Furthermore, the business community tends to support initiatives that will lead to resolving any conflict that might undermine the local community's economic well-being. Poor ethnic relations are simply bad for business.

Third, when a single group settles in a municipality, there is less of an immediate need for it to integrate, since it is more likely that the group will develop an extensive array of ethnospecific institutions. This has positive and negative implications. On the one hand, it seems to contribute to backlash on the part of long-standing residents, who contend that the immigrant group is not integrating. On the other, it provides the immigrant community with resources with which to organize to pressure the municipality to respond to its concerns and to counter backlash if it occurs. However, concentration can also result in exclusion and marginalization, which seems to be the case in Surrey, where, according to one informant, many in the South Asian community are in 'survival mode' owing to their social isolation (Woodman 2004, interview).

Finally, in biracial municipalities it is easier and potentially also less costly for local leaders to respond to a single immigrant group or ethnoracial minority than to a multiplicity of groups. For instance, translations have to be made in only one language, interpretation is in a single language, and employment equity efforts have to address barriers for only one group. Bridging through regime development is also easier in biracial contexts than in multicultural ones.

Nevertheless, though the municipality's ethnic configuration matters, the clear lesson to emerge is that it is not the sole factor. The *distribution of resources within the municipality and among ethnic groups* is important as well (Good 2005). At the local level, given municipalities' institutional limits, lack of resources, and perceived need to compete with other municipalities for residents and business investment (Peterson 1981), cities require the support of the business community to manage ethnoracial change. Clarence Stone (1989) has observed that 'as a practical matter, given the important resources and activities controlled by business organizations, business interests are almost certain to be one of the elements [of an urban regime], which is why regimes are best understood as operating within a political economy context' (1989, 179). Thus, municipalities in which the business community supports the development of multiculturalism policies are more responsive.

The importance of this factor is constant, regardless of the municipality's ethnic configuration. However, its importance to multiculturalism policy is structured by the community's ethnic configuration. A biracial municipality in which there is a prominent and cohesive immigrant business community provides an especially powerful combination – one that brings the weight of the business community *and* the weight of numbers to bear on municipal responsiveness to immigrants. This suggests why, in the present sample, biracial municipalities in which the ethnocultural 'minority' is predominantly Chinese have been more responsive than those in which the ethnocultural minority is South Asian. Specifically, in municipalities where Chinese immigrants predominate, there is a powerful and cohesive Chinese business community (one that includes developers). In addition, the Chinese business community is willing to facilitate municipal responsiveness to the Chinese community as a whole by donating or pooling resources. In Vancouver, multiculturalism policies are a by-product of a dominant, even 'hegemonic' (Olds 2001) economic-development model. In Toronto, too, we see how powerful the business community's support of multiculturalism can be and how this has influenced the city's capacity to respond to ethnocultural change. Tying immigrant settlement outcomes to the objectives of economic development has strongly propelled collective action around multiculturalism policies in that city, which has had to mobilize a highly diverse community, including a business community. In the Toronto case we also see the impact of tying issues to growth coalitions.

The *ethnic distribution of resources in civil society* also affects how the long-standing community perceives large-scale immigration. In all biracial municipalities with a concentration of Chinese immigrants, the ability of the Chinese business community to alter the cultural face of the municipality (for instance, through Asian malls) has led to backlash on the part of some members of the long-standing community.

At the same time, in all of the *biracial* municipalities with large Chinese immigrant populations, the immigrant community has developed an extensive network of Chinese-specific institutions. The Chinese community seems to possess a great deal of social capital. Yet as Orr (1999) observes, social capital is not a substitute for economic capital (1999, 185). In Vancouver, Richmond, and Markham the Chinese community possesses high levels of both forms of capital. In Coquitlam, civil society is poorly developed, as immigrants began arriving in large numbers only after Westwood Plateau was developed. Given the strength of social networks in Greater Vancouver, S.U.C.C.E.S.S. has recently stepped in to fill this void in Coquitlam.

In contrast, Surrey is the only biracial municipality in the sample with a predominantly South Asian immigrant community. That community seems to be more divided than Chinese immigrant communities generally are and has not developed ethnospecific South Asian social capital. Furthermore, Surrey's business community does not seem to be pressuring the municipality to adapt its services, even though many of that city's developers are South Asian. At the local level, immigrant inclusion in local governance constitutes a two-stage problem in collective action: first, the immigrant group must be able to mobilize for collective action at the level of civil society; and second, bridges must be created by developing urban regimes. The South Asian community in Surrey has failed to overcome the first-level collective action problem. Even the racialization of the community has not spurred it to mobilize.

Immigrant communities that fail to overcome the first-order collective action problem (i.e., to organize as a community) will find themselves (a) unable to pressure municipalities to respond to their concerns, and (b) unable to participate in policy-productive 'urban regimes.' Furthermore, it is clear that Statistics Canada's category 'South Asian' is too imprecise – specifically, it does not reflect how most South Asian people 'self-identify,' nor does it reflect how they construct their ethnoracial identity. Further research would shed light on the dynamics within this disparate group.

Similarly, the increased complexity of the first-order collective action problem in multiracial municipalities means it is less likely that immigrants will be included in municipal governance in those locales. One can observe the effects of this factor in Mississauga and Brampton, the two suburban multiracial municipalities in the present sample.

One might be tempted to conclude that, owing to the cleavages in its South Asian population, Surrey should be considered a 'multiracial' municipality to reflect this diversity. After all, that city shares the divisions within the immigrant community that are inherent to a multiracial context. However, the way in which the *overall* community *perceives* immigration differs in Surrey from the way in which it is perceived in multiracial municipalities. Backlash against immigration seems to be less likely in heterogeneous, multiracial municipalities – that is, those in which the perception is missing that a single immigrant group is redefining the municipality's cultural norms. It makes a difference how majorities perceive minorities; it also makes a difference how minorities perceive themselves. A municipality with a reactive policy style needs something to react *to* before it can begin developing multiculturalism policies.

There are cleavages within Surrey's immigrant community; at the same time, long-standing residents of that city have reacted to what they see as a large and monolithic group by racializing the community. In other words, social context influences relationships between the long-standing community and the immigrant population even while it influences relations *within* the immigrant population. Hero (1998) contends that 'politics and policies are products of the cooperation, competition and/or conflict between and among dominant and subordinate minority groups, not only of the dominant group within a state' (1998, 10). Therefore we must understand how social context is perceived by *all* members of political communities.

The lack of visible and widespread backlash in the multiracial suburban municipalities does not mean that exclusion from local governance does not matter to immigrant populations. Former Brampton councillor Garnett Manning (2003–6), a member of Brampton's Black community, described Brampton's race relations climate as at a 'boiling point,' noting that members of his community felt excluded from the city's power structures (Manning 2004, interview). Nevertheless, in the absence of political mobilization and community pressure, municipalities – suburban municipalities in particular – tend to resist involving themselves in new policy areas. As former Brampton councillor (and current provincial MPP) Linda Jeffrey put it: 'Councils are paralysed by the thought that they're going to set precedents that they have to continue later on'; put another way, if they do something for one group they will 'have to do that for everybody' (Jeffrey 2004, interview). In sum, multiracial suburban municipalities have not had to manage community backlash, have not been pressured to respond, and contend that the diversity of their populations and the limits on their resources make responding very difficult.

When one removes the influence of uneven community resources from the analysis by removing the two resource-rich urban-core municipalities from the sample, the pattern is straightforward, as evidenced in Table 6.5.

Toronto may be an 'exceptional case' in the sense that it has an exceptional concentration of the country's settlement resources (as well as resources more generally) with which to overcome the difficult collective-action problems that its diversity entails. These resources upset the straightforward patterns that social scientists like to identify. The Toronto case provides the clearest evidence for the need for a complex, resource-based analysis of urban politics and power – factors that urban regime theory highlights. Furthermore, in Toronto we see evidence of the central importance of *leadership* through community and municipal efforts to co-build capacity.

If one removes only Toronto from the sample, the pattern is quite clear. All *biracial* municipalities are either 'responsive' or 'somewhat responsive' and share common forms of political pluralism and institutional development as well as common debates about multiculturalism. All *multiracial* municipalities are 'unresponsive,' share a limited and sometimes competitive form of political pluralism, and lack the initiative to develop governance arrangements.

Why is Toronto an exceptional case? Is it possible that the city has overcome some of the features of ethnic configurations that would normally structure their effects? Differences between Toronto and the other multiracial municipalities, and similarities between Toronto and Vancouver, support an affirmative answer to this question.

Toronto differs from its multiracial counterparts and has more in common with Vancouver in that it is the central city in its city-region. This has many implications. One is that, as an older city, it has benefited from the investment of resources (including settlement resources) from upper levels of government for a longer period of time. In addition, Toronto's ethnoracial minorities have had more *time* to organize and to create bridges than have Mississauga's or Brampton's.

Myer Siemiatycki and his colleagues (2003) have documented the development of interethnic bridges in Toronto through time. The influence of time can also be seen in Vancouver, where large-scale immigration from China resulted in political dynamics similar to those

Table 6.5
Summary of patterns of influence of ethnic configurations in multiculturalism policy making in suburbs

Types of ethnic configurations	**Multiracial**	**Biracial**
Visible minorities (as proportion of population)	High	High
Composition of visible minority population	Diverse	Homogeneous
Backlash	No	Yes
Types of political pluralism	Competitive	Limited
Municipal multiculturalism policies	Highly Limited	Limited
Policy styles	Inactive	Reactive
Urban governance relationships to support multiculturalism initiatives	No	Yes
Dominant multiculturalism purposes	Celebrating multiculturalism	Community bridging, manage backlash, avoid segregation/social isolation, and access to services
Municipalities	Mississauga and Brampton (2001)	Markham, Richmond, Surrey, and Coquitlam

in suburban biracial municipalities such as Richmond and Markham. In Vancouver, however, the municipality and civil-society leaders were able to build capacity to manage demographic change over time, which contributed to a greater level of responsiveness than in Richmond or Markham. We also see the influence of time in Coquitlam, which has not had time to develop immigrant settlement and advocacy capacity, because immigration to the city has occurred relatively recently.

Toronto is also exceptional in relation to its multiracial suburban counterparts because its population is so much larger – about 2.5 million people – which means that its three largest visible-minority groups as well as its immigrant population as a whole are very large (see Table 6.1). Rodney Hero's (1998) categories of ethnic configurations measure the proportion of ethnic groups ('Whites,' 'White ethnics,' and ethnoracial 'minorities') in relation to the overall state population. However, he also acknowledges that 'it may be the case that a group's size might also be considered in terms of raw numbers' (1998, 152). Toronto has very large Chinese, South Asian, and Black communities. Each is larger than the entire population of either Richmond or Coquitlam, and the Chinese and South Asian communities are each larger than that population of Markham (Statistics Canada 2007b). Toronto's immigrant population is over one million – that is, its immigrant population is larger than the total population of most Canadian cities.

Finally, the City of Toronto has experienced an exceptional level of institutional upheaval over the past decade (Sancton 2000), including a major amalgamation. This has led to new patterns of political mobilization in the community. For instance, leaders in the anti-amalgamation movement – Citizens for Local Democracy (C4LD) – created bridges with Toronto's visible-minority communities (Siemiatycki et al. 2003). Toronto also suffered disproportionately from budgetary decisions at the federal and provincial levels that led to 'downloading' in the mid-1990s (MacMillan 2006; Good 2007). The social impacts of decisions at the federal and provincial levels led to the development of a strong urban-autonomy movement that is exceptional in Canada. Municipal leaders in many other major cities in Canada support municipal autonomy goals; in Toronto, the movement also has the support of a broad-based group of elites in civil society. Immigrant leaders are included in this alliance and have the support of some of Canada's most powerful business leaders. The impact of neoliberal ideology and the 'politics of restraint' has been more pronounced in Canada's largest city. The important role of intergovernmental decisions – arguably in response to global forces of rescaling – in Toronto's evolution is discussed further in chapter 7.

Toronto is the 'face' of the city-region nationally and globally. Toronto and Vancouver are the central cities of two of Canada's largest city-regions, and their local leaders perceive their municipalities as qualitatively different – as 'global' or 'world' cities. Both cities have linked the attraction, retention, and settlement of immigrants to their capacity to compete for investment in a global economy – a capacity that is supported by stronger levels of civil-society leadership than are found in their surrounding suburbs. The 'selling diversity' paradigm that they have adopted informs a multitude of multiculturalism-related policies (policies that Yasmeen Abu-Laban and Christina Gabriel [2002] have described at the national level). Together, these factors contribute to the 'proactive' approach that both cities have taken to immigrant settlement and multiculturalism policy.

These factors provide a strong case that Toronto has 'overcome' its ethnic configuration; but they also suggest that as a case, Toronto cannot be used to generalize the effects of an ethnic configuration on a municipal society and government. Similarly, the fact that Toronto does not fit the overall pattern should not be seen as a 'falsification' of the social diversity interpretation presented above.

In addition, the divergence between the central city and suburban multiracial municipalities, and the significance of this divergence for the overall value of the social diversity perspective, must be considered in light of strong evidence that in *biracial* municipalities, ethnic configuration has causal effects on local communities and their municipal governments. In Richmond and Markham – two biracial suburban municipalities in two different provinces – the ways in which residents and leaders reacted to immigration, the community debates that arose therefrom, and the ways in which immigrants perceived community reactions, are so strikingly similar that if we were to take one of these municipalities and transplant it into another province or even country, similar community dynamics and responses to multiculturalism might well occur.

As the social diversity interpretation implies, 'the context within which individuals and/or groups are situated is as, if not more, important than the values or ideas that people "bring with them" or "have within"' (Hero 1998, 10). The context is 'transsubjective' or 'transindividual' (ibid.). The strong similarities among biracial municipalities in the sample imply that one could transplant a long-standing resident of a Canadian city into a biracial social context and his or her opinion on immigration and multiculturalism would likely be more similar to those of others in the biracial context than to those of an individual who was situated in another social context (multiracial or homogeneous, for instance). Similarly, one could expect individual immigrants to react correspondingly to the type of racialization that occurs in these locales. The biracial cases in the sample provide especially strong evidence that social context has an impact on political pluralism, on institutional development, and, ultimately, on policy outputs.

In the future, the effects of social context could also be explored at the individual level of analysis, as Hero (1998) does in his work by drawing from existing secondary literature on the influence of race on White attitudes and support for particular issues and public policies. Unfortunately, individual-level data on attitudes towards social change in the eight *municipalities* discussed here are not available. However, Daniel Hiebert (2003) conducted a study of attitudes towards multiculturalism, immigration, and discrimination in five *residential areas* in Greater Vancouver.[4] These residential areas were located in the cities in the Greater Vancouver sample. Hiebert notes a 'curious' finding with respect to the data on discrimination – a finding that supports those discussed here: though immigrants report suffering discrimination more often than do the Canadian born, the pattern is reversed for discrimination in shopping, banking, and dining out (ibid., 33). He does not present his findings by neighbourhood. He does, though, note that feelings of discrimination were more notable in areas such as Richmond, East Vancouver, and Surrey-Delta, where there are many ethnic businesses catering to immigrant and minority clienteles (ibid.). Hiebert's data were collected as part of a larger collaborative project. When the data he and his collaborators collected have been published, it will be interesting to see whether they cast further light (a) on the role played by social context in shaping individual attitudes towards multiculturalism, and (b) from a theoretical perspective, on the value of a social diversity interpretation of local politics. In light of the analysis offered in this book, a comparison of the same data in biracial and multiracial municipalities in Greater Vancouver and the GTA would be very interesting as well. Based on this chapter's theoretical ideas, one might predict that the sense of discrimination among Canadian-born residents is heightened when there is a perceived threat of cultural 'takeover' by a concentrated immigrant minority. One might predict that the issue of concern to the Canadian-born resident would be that many minority businesses in these neighbourhoods cater to a single 'racial' group – an 'Asian' clientele.

A Research Note on the Brampton Case

The 2006 Census indicates that Brampton is now a biracial municipality with a dominant South Asian community. Based on the interpretation offered above, what might one expect of the future politics of multiculturalism in Brampton? The findings discussed in this book suggest that as the South Asian community becomes increasingly concentrated in Brampton, feelings of cultural threat may increase among Canadian-born residents. Therefore, one might expect to see evidence of an emerging community backlash. This could create an incentive for the municipality to intervene by establishing a race relations committee to bridge the immigrant and Canadian-born communities. In other words, we might see the revival of the Brampton Race Relations Action Committee. However, to the extent that Surrey's experience is instructive, Brampton's South Asian community might become isolated and have a more difficult time mobilizing for collective action owing to internal, religion-based cleavages. This could create an incentive for municipal policies that would increase immigrants' access to services and governance. Together these factors, coupled with the fact that the business community has taken an interest in diversity issues, foreshadow a more responsive municipality in the future.

Evidence beyond Canada: American 'Ethnoburbs'

A growing literature in geography provides further evidence to support the theoretical value of a social diversity interpretation of local politics, and of a cross-national research agenda that examines the relationship between immigrant settlement patterns and political processes. The literature on 'ethnoburbs' shows that at least some 'bifurcated' localities in the United States are characterized by similar political dynamics as the 'biracial' Canadian municipalities with concentrations of Chinese residents. The political dynamics that have resulted from demographic change in Richmond and Markham are similar to those characterizing American 'ethnoburbs' – a concept that American geographer Wei Li (1999) has developed to characterize new suburban ethnic clusters of Chinese immigrants.[5]

Li distinguishes the 'ethnoburb' from traditional ethnic ghettos or urban enclaves such as Chinatowns, insofar as actors with economic power deliberately create ethnoburbs, whereas in ethnic ghettos and urban enclaves, 'ethnic people do not have economic power' (Li 1997, 4). The first American ethnoburb emerged in Monterey Park, a suburb of Los Angeles, as a result of large-scale Chinese immigration. The political dynamics that developed in Monterey Park are similar to the dynamics that emerged in Richmond and Markham. For instance, an 'English-only' movement developed, and Chinese immigrant business owners were accused of using Chinese signage to deliberately exclude longstanding residents (ibid.). Changes in the built environment, including the construction of a Buddhist temple, became 'racialized' in Monterey Park (ibid.), just as Asian malls were controversial in Richmond and Markham. In response, the City of Monterey Park initiated a number of multicultural events, including festivals, as well as community round tables that brought together community leaders to share their opinions on issues facing the city (ibid., 19). These community round tables appear to have served the same function as 'ethnic advisory committees' in biracial Canadian suburbs. Furthermore, these responses were similar even though the United States does not have an official policy of multiculturalism.

More recent scholarship has documented the emergence of ethnoburbs in other Pacific Rim countries (Ip 2005; Li 2007). According to Li, the transformation of suburbs in these countries results in 'similar kinds of resistance from longtime local residents,' produces 'racialized incidents,' and often leads to 'similar solutions.' For example, the 'monster homes' controversy is 'a well-known and well-publicized issue in Los Angeles, Silicon Valley, Vancouver, and Auckland' (Li 2007, 14).

In essence, this literature suggests that the social diversity interpretation of local politics can serve as a comparative framework for studying urban governance in high-immigration, multicultural environments. Internationally, what geographers call the ethnoburb appears to be a common type of immigrant settlement, one that results in similar forms of political pluralism that are of interest to political scientists.

The literature on ethnoburbs strengthens rather than undermines the social diversity interpretation. According to Rodney Hero's framework, 'English-only' movements should be more common in ethnoburbs and racialization of the minority group should occur in these locales. However, the question arises: Would these dynamics occur owing to 'bifurcation' (in Hero's typology) or to concentration of a single immigrant community? And to what extent does it matter that many immigrants in ethnoburbs are powerful and deliberately create these communities?

These questions speak to the differences between the way in which Hero theorizes ethnic configurations in the United States and the way in which they are theorized in this study of Canadian communities. Both the ethnoburb literature and the findings discussed in this book suggest that it matters how powerful immigrant communities are. Power influences their ability to contribute to 'social production' and to change the face of local communities in ways that generate a particular reaction from long-standing residents. Together, the findings in this chapter and in the ethnoburb literature point to the potential theoretical fruitfulness of a cross-national social-diversity interpretation of the impact of immigration on the politics of cities.

Concluding Thoughts

To what extent does the social diversity interpretation cast light on the politics of multiculturalism policy development at the local level in Canada? There is convincing evidence that the ethnic configuration of political societies matters. Building on Rodney Hero's work in the United States, this chapter has developed a social diversity framework through which to examine how immigrant settlement patterns have contributed to urban regime building and the politics of multiculturalism at the local level. It has offered two new categories of ethnic configurations – biracial and multiracial – and it has presented convincing evidence that many aspects of the social contexts inherent in these analytical constructs matter to urban regime building and ultimately to multiculturalism policy development.

The ethnic configuration of a community affects the nature of political pluralism. In the biracial cases, we saw that large numbers of a single group appeared to make the immigrant group more threatening. However, this on its own did not result in publicly visible crises in ethnic relations. It was during debates over changes to the built environment that we saw intense intercultural conflict.

Ethnic configurations must be viewed as contextual factors shaping the distribution of resources in local communities. They affect the ways in which civil society develops, and they have an impact on collective-action problems, the resources available in the community to contribute to the management of multiculturalism, and the incentives for the immigrant community to integrate. They also affect how the business community views immigration, as well as its willingness to pool capacity to manage multiculturalism and help municipalities adapt their services.

However, we must also look at intracommunity dynamics to understand the local politics of multiculturalism. In Surrey, for instance, we must look at diversity *within* communities in order to assess whether a group's ability to act collectively is enhanced by geographic concentration. The ethnic configuration of a municipality can also affect the development of a cohesive immigrant business community. We have found that in biracial municipalities with a concentration of Chinese immigrants, the combination of a powerful and well-organized business elite and large numbers of a single group of immigrants can create the conditions for highly visible backlash on the part of long-standing residents.

Furthermore, the ethnic configuration of a municipality affects how difficult it will be for a municipality to respond. In multiracial municipalities a greater investment of time and resources is required to implement multiculturalism policies effectively. Thus, in the absence of backlash and community pressure, in many ways it is no wonder that multiracial suburbs such as Mississauga and Brampton have not responded.

With this in mind, one can appreciate Toronto's exceptionality. Despite the difficulties of building policy capacity in a highly diverse context and the need to settle and integrate more than one million immigrants, Toronto has been relatively responsive and wide-scale immigration has happened relatively smoothly. A broad range of local leaders, including prominent business leaders, have contributed to Toronto's ability to settle immigrants and achieve other long-term multiculturalism objectives. This outcome is both puzzling and interesting for the urban regime scholar. How and why have so many community leaders worked together to manage multiculturalism? What explains the exceptional leadership effort at the local level in Toronto? Parts of the answers to these questions can be found by looking at the external contexts of local governance – that is, at the intergovernmental and (to some extent) global contexts.

7 Multiculturalism and Multilevel Governance: The Role of Structural Factors in Managing Urban Diversity

In their review of the urban regime literature, Karen Mossberger and Gerry Stoker (2001) identify several factors that seem to cause regimes to form and change, including demographic shifts, economic restructuring, federal grant policies, and political mobilization (2001, 811). In chapter 6 we saw that demographic change shapes the urban governance of multiculturalism from 'below,' because the ethnoracial configuration of municipal societies affects the nature of political pluralism as well as patterns of political mobilization at the local level. It also structures the distribution of resources. This chapter explores what Jeffrey Sellers (2002a, 2002b, 2005) calls a 'national infrastructure' – that is, the bundle of ways in which national institutions, policies and cultures structure local politics, policy making, governance, and citizen preferences. In particular, it explores some of the ways in which elements of the intergovernmental system have influenced the local governance of multiculturalism. The notion of 'national infrastructure' brings to light the dynamic relationship between changes in the intergovernmental context and political mobilization at the local level on the one hand and the way in which local developments affect federal and provincial policy decisions on the other. In other words, it stresses the importance of structural factors in urban analysis without negating the influence of local agency.

As this book has pointed out, structural factors are not completely decisive. Furthermore, there are a multitude of structural factors, which result from multiple levels of governments and policies and which intersect in myriad ways that have uneven spatial effects. The concept of 'national infrastructure' captures the multilevel dimensions of decision making and also the idea that local leaders can use structures at multiple levels as supports to achieve particular ends. In this sense, these infrastructures are 'resources' to be used (or not) in local agenda building and implementation.

With respect to crafting multiculturalism policy, though municipalities are a provincial responsibility, federal multiculturalism policies and the normative standard that the federal government has established are important parts of the national infrastructure that one might expect to influence local multiculturalism policy. Yet owing to municipalities' constitutional relationship with provinces, provincial multiculturalism policies matter in a more direct way. Federal and provincial policies furnish concrete policies, which are then emulated at other levels. They also provide resources in local communities through, for example, the funding of immigrant settlement organizations and organizations that serve other multiculturalism-related purposes. One might speak of a 'multiculturalism infrastructure'[1] in Canada that includes the symbolic and material resources furnished by legislation as well as the many government departments and community organizations involved in creating the capacity to manage diversity. For instance, Statistics Canada is an important part of this infrastructure since it collects ethnospecific data as well as data on immigrants. The national 'multiculturalism infrastructure' can also influence the preferences of citizens on the ground. For instance, the consistency of immigrant settlement leaders' preferences for a multicultural model of municipal citizenship can reflect the norms established by the national infrastructure.[2] However, as this book establishes, municipal multiculturalism policy making varies significantly across municipal jurisdictions. Thus, this chapter explores how elements of Canada's national infrastructure influence local governance in a more nuanced way. It does so by examining how a variety of changes in the intergovernmental system have altered the local governance of multiculturalism.

Fiscal federalism is an important gauge of power and an important source of change and flexibility in Canada's federation. By determining the distribution of public financial resources, it structures policy making significantly. The mid-1990s represented a turning point in the evolution of the Canadian federation, for during those years, fiscal relationships among governments were radically restructured. Furthermore, differences in provincial policy decisions in the 1990s in Ontario and British Columbia contributed to different local governance dynamics in those two provinces. Specifically, in Ontario, turbulent provincial–municipal relationships from 1995 onwards injected greater dynamism into urban governance arrangements in municipalities in the GTA relative to those in Greater Vancouver. Decisions by Ontario's Conservative governments (1995–2003)[3] – including a 'local services realignment' or 'disentanglement' exercise and amalgamation – fomented an urban autonomy movement in Toronto. Among the cases in the sample, Toronto experienced the greatest level of municipal stress in the mid to late 1990s. During this turbulent period, B.C.'s multiculturalism infrastructure was more stable and supportive of multiculturalism initiatives and supports to immigrants than Ontario's.

This chapter demonstrates that adjustments to several aspects of the national infrastructure affected Toronto disproportionately, changing the political dynamics of Toronto's regime. More specifically, in response to amalgamation and downloading, local leaders in Toronto increased their efforts to build capacity in several policy areas, including immigrant settlement and integration. Furthermore, immigrant settlement issues made it onto the agenda of the federal and Ontario governments partly because these concerns were a priority of Toronto's powerful urban regime, which began lobbying upper levels of government for a New Deal for Cities. Toronto's urban regime coalition spilled over into the national and regional scales, thus becoming what Julie-Anne Boudreau (2006) calls 'polyscalar.' Local leaders in Toronto became aware that pooling local resources was insufficient to achieve policy goals. They therefore developed national and regional coalitions to pressure upper levels of government to institutionalize new forms of intergovernmental sharing of resources.

The election of Dalton McGuinty's Liberal government in Ontario in 2003 represented an important shift in the provincial–municipal relationship. It followed a dramatic period of municipal reorganization of Toronto's formal governing institutions in the mid to late 1990s that saw 'disentanglement' as well as downloading from upper levels of government. Then in 2003 a 'policy window' in the intergovernmental system opened for Toronto's regime, as political leadership changed simultaneously at the municipal, provincial, and federal levels. The opening of this policy window led to new legislative and political status in Canadian federalism for Canada's largest city. It has also led to a more supportive provincial environment for managing immigration in Toronto.

This analysis must be placed in a context of global economic restructuring. As happened in countries around the world, in the 1980s and 1990s Canadian governments at all levels were influenced by neoliberal ideas and attempted to reduced spending to eliminate budgetary deficits and repay debts. As urbanization progresses internationally, cities have emerged as the central engines of global capitalism. In this environment many urban scholars have noted a rise in the competitiveness of cities. More broadly, scholars argue that there has been a 'rescaling' of the state, with the global economy becoming reterritorialized at the urban scale (Kipfer and Keil 2002; Brenner 2004). Others, including Thomas Courchene (2005a), have described these economic processes as a 'glocalization' of the economy, with economic space becoming simultaneously global and local (urban). These economic processes have in turn contributed to greater competitiveness in cities as well as to new forms of political mobilization.

In Canada, as in many other countries (Uitermark, Rossi, and Houtum 2005), there has also been a rescaling of multiculturalism. The world's migrants have been flowing into a handful of Canadian cities; Toronto and Vancouver are the top two immigrant-receiving city-regions in English-speaking Canada. *How* migrants are incorporated into cities matters greatly to globalization processes (Sassen 1998; Keil 2002, 230). Susan Clarke and Gary Gaile (1998) describe proactive economic-development strategies as part of the new 'work of cities' in a highly competitive global environment. In this new economic context, another important job of municipal governments is to manage multiculturalism. As this book has demonstrated, some local leaders tie this role of cities to economic-development objectives.

This chapter describes broad trends in Canadian fiscal federalism. It also discusses the general nature of provincial–municipal relationships and of provincial multiculturalism infrastructures in the 1990s in Ontario and British Columbia. Both of these factors have had important implications for the local governance of multiculturalism, and their impact has been most apparent in the City of Toronto. The chapter then describes new patterns of political mobilization that developed in Toronto in response to the local implications of the intersection of federal and provincial policy decisions in the mid-1990s. These new forms of mobilization had two interrelated goals: to achieve greater municipal empowerment in the intergovernmental system, and to increase local capacity to address the city's policy challenges, including the challenges of integrating immigrants through local multiculturalism initiatives. This provides part of the explanation of how and why immigrant settlement concerns were included in the goals of Toronto's wide-ranging capacity-building regime. Changes in intergovernmental relationships also contribute to our understanding of why the settlement sector in Toronto supports municipal autonomy initiatives – in other words, why that sector pursues territorial *and* sectoral strategies when advocating on behalf of immigrants.

Like previous ones, this chapter uses urban regime theory as an analytical lens for characterizing and understanding the new patterns of mobilization in Toronto, which have extended beyond the municipal level to include multiple scales. It also describes some recent developments in Canada's intergovernmental system that have grown out of the lobbying efforts of participants in Toronto's regime. This section is meant to illustrate how urban governance arrangements are affected by changes in the intergovernmental system and how local leaders and governance arrangements can bring about changes to the intergovernmental system.

The Changing National Infrastructure: Rescaling the Canadian Federation?

'Rescaling' the state refers to two processes: the territorial reorganization of political authority, and the actions that have been taken by civil-society leaders as part of neoliberal state restructuring since the 1980s. Changes in the spatial dimensions of state authority have been driven by changes in global capitalism that have concentrated capitalism's global infrastructure in specific places. This infrastructure is located in some of the world's largest cities. Globalization has created economic winners and losers so that a hierarchy of cities has emerged. In Canada, high-immigration cities such as Toronto and Vancouver are among the 'winners'; indeed, their status as immigrant-magnet city-regions both reflects and is driven by their national global competitiveness.

The literature on rescaling would have us assume that as cities become the key nodes in the global economy, we will see greater levels of investment in cities by upper levels of government, as well as the emergence of highly competitive urban-governance coalitions. For instance, in *New State Spaces*, Neil Brenner (2004) describes a shift in Europe towards economic-development policies that privilege cities and the rise of new forms of political organization at the metropolitan scale; he refers to these new configurations of state authority as a 'Rescaled Competition State Regime (RCSR).' Rescaling processes have multiple dimensions and implications. The previous chapters described powerful urban regimes in Toronto and Vancouver wherein local leaders tied multiculturalism objectives to growth agendas. One might expect upper levels of government to support these local economic-development agendas. One might also expect strong support for multiculturalism initiatives, which are necessary in order to manage and sustain growth. Indeed, the federal government's own policy discourse links economic growth with diversity (Abu-Laban and Gabriel 2002). Yet there is considerable variation in the extent to which upper levels of government have supported cities and municipalities in their efforts to compete and to manage the integration of large-scale influxes of immigrants.

Thus the literature on rescaling provides a causal explanation for the emergence of *multilevel governance arrangements* – a term used by an interdisciplinary group of scholars to conceptualize these complex relationships and processes. Gary Marks (1993) defines multilevel governance as 'a system of continuous negotiation among nested governments at several territorial tiers – supranational, national, regional, and local – as a broad process of institutional creation and decision reallocation that has pulled some previously centralized functions of the state up to the supranational level and some down to the local/regional level' (Marks 1993, 392, in Young and Leuprecht 2006, 12). According to Julie-Anne Boudreau (2006), rescaling occurs both intentionally and unintentionally (2006, 162); in other words, it is a result of both structural changes and political agency. In particular, Boudreau highlights civil society's important role in the emergence of multilevel, 'rescaled' configurations of political authority. In her work on social mobilization in response to amalgamations in Toronto and Montreal, she uses the term 'polyscalar' to describe new types of coalitions and mobilization strategies that both citizens and leaders in civil society developed at multiple scales in these cities. She observes: 'In both Montreal and Toronto, opponents to mergers and coalitions claiming a general reform of intergovernmental relations in Canada, have developed a series of mobilizing strategies at multiple scales, striking alliances with various levels of government and pitting them against one another.' In her view, 'while rarely directly discussed in intergovernmental relations studies, these types of polyscalar outlooks exploited by civil society actors have an important impact on the kinds of institutional and territorial reorganizations undertaken by state actors, particularly in a context where decision-making processes have been opened by a variety of non-state actors' (ibid.). What is different in her view is that civil-society leaders are not using sector-specific strategies to achieve their ends; rather, they are using territorial strategies and acting at multiple territorial scales at once (ibid., 164). As she explains, 'sectoral strategies of political claims channel efforts into specific policy sectors (housing, language, health, education, etc.) [whereas] jurisdictional and territorial strategies of political claims are attempts by civil society to use one level of government against another or to create a new level of government altogether by asking for a remapping of political and administrative boundaries' (ibid., 166). These strategies are part of a global state of 'territorial flux' created by a 'restructuring-generated crisis' in capitalism (Soja 2000, in ibid., 162). However, municipal and civil-society responses have been strongly shaped by how upper levels of government have responded to global conditions, just as they have shaped the intergovernmental reaction.

Changes in fiscal federalism are a good measure of the intergovernmental system's overall response to global restructuring. Processes of decentralization and fiscal 'downloading' began in the late 1970s in Canada. These processes reached a critical juncture in the 1990s, when major reforms to fiscal transfers were introduced. The most important event in the 1990s with respect to fiscal federalism was Ottawa's decision to combine two major transfers to provinces for social-policy responsibilities – the Canada Assistance Plan (CAP)[4] and Established Programs Financing (EPF)[5] – into a single block grant called the Canada Health and Social Transfer (CHST). With the introduction of the CHST, the federal government eased the conditions on grants and transferred tax points to the provinces; it also dramatically reduced the grant's monetary value.[6] This decision was widely seen as the root cause of a resulting 'fiscal imbalance' in the federation – a situation in which one level of government now had too few revenues to meet its constitutional responsibilities while the other level had more than it needed (Advisory Panel on the Fiscal Imbalance 2006, 12–14). More specifically, the federal decision was seen as a means of 'downloading' fiscal burdens onto provinces. Downloading also had important tri-level implications.

Downloading refers to several manoeuvres on the part of provincial and federal governments. First, upper levels of government can give municipal governments new responsibilities without transferring commensurate resources – what in the United States is commonly referred to as 'unfunded mandates.' A second form of downloading occurs when upper levels of government withdraw unilaterally from a policy field and local governments are left to decide whether to fill the gap in services. A third form occurs when upper levels of government fail to consider the place-specific consequences of their public-policy decisions. Municipalities' de facto role in multiculturalism and immigrant settlement policy in Canada's most important immigrant-receiving urban and suburban municipalities is an important example of this form of downloading.

In federal systems, downloading to municipalities has multiple dimensions. The federal government can decide to download to provinces; the provinces can in turn decide to download onto municipalities. In addition, because of their place-specific consequences, decisions in a variety of policy areas – including multiculturalism and immigrant settlement policy – made at the provincial *and* federal levels may result in unfunded de facto municipal mandates.

In the mid-1990s, federal reductions in transfers to provinces were so dramatic that they led economist Thomas Courchene to argue that the Canadian federation was moving towards a model of 'hourglass federalism' (Courchene 2005a). After introducing the CHST in 1995, Ottawa replaced part of its provincial transfers with direct transfers to citizens[7] and to cities.[8] Of particular interest here, the CHST's implementation in 1995 coincided with Ottawa's decision to create the national Metropolis Project, a research initiative the goal of which was to investigate immigration's national and international impact on cities. As part of this initiative, research centres were established in Toronto and Vancouver (see earlier chapters). The activities of these centres have become important nodes in local-governance arrangements and have contributed to capacity building in immigrant settlement and ethnocultural relations. Given the reductions in federal transfers to provinces for social spending and 'medicare's voracious appetite,' as Ottawa began to fund city initiatives, the provinces 'were forced to starve virtually every other provincial policy area' (ibid., 14). As a consequence, the provinces became irrelevant in many policy areas of vital importance to cities. This in turn created incentives for local leaders to bypass the provinces and to seek direct relations with the federal government.

Courchene contends that the federal government's decisions in this respect were strategic (i.e., intentional) manoeuvres in response to the increased importance of global city-regions in a global knowledge-based economy.[9] This interpretation of the motives underpinning federal decision making in fiscal policy is consistent with what Neil Brenner's (2004) analysis of 'rescaling' in Europe would lead one to expect. Specifically, Brenner describes a shift away from 'spatial Keynesianism' whereby the state redistributes economic activity across space towards a focus on investments in the economic winners – which tend to be a select group of cities. Canada's immigrant-magnet city-regions are clearly within the top economic tier of cities in Canada. As this chapter will show, whatever the federal government's intentions, its actions have contributed to new patterns of political mobilization at different scales and to the use of territorial strategies by local leaders to achieve their desired ends.

However, the provinces (and Ontario in particular) were not simply passive observers that were 'squeezed out' of relevance in the new fiscal arrangements. Provincial decisions in response to these changes varied in Ontario and B.C. This variation has had important implications for municipal governance and for the place of municipalities on the national agenda and within federalism.

The Influence of the Provincial–Municipal Relationship on Urban Multiculturalism Policy Making

Ontario in the 1990s: Municipal Turbulence and Multiculturalism Policy Retrenchment in the Conservative Era

In Ontario, the transition from the New Democratic Party (1990–95) to the Conservative Party (1995–2003) was a time of dramatic change in municipal governance in that province. In 1995, the election of the Conservatives under Mike Harris coincided with the decision by the Liberal government under Jean Chrétien to implement the CHST. The Harris government dramatically reorganized the division of powers between the province and municipalities; it also amalgamated many municipalities in Ontario (including Metro Toronto), in part to prepare them for their new responsibilities; and it reduced services within the province in line with the neoliberal ideology that underpinned the Conservatives' 'Common Sense Revolution.'[10]

Myer Siemiatycki (1998) estimates that as a result of the reorganization of provincial and municipal responsibilities, the new City of Toronto assumed close to $400 million of additional fiscal responsibilities per fiscal year (1998, 5). The City of Toronto approved its operating budget of $7.6 billion for 2006 in March of that year. Of this budget, 36 per cent was devoted to provincially mandated services, yet only 25 per cent of the city's revenues came from provincial grants and subsidies (City of Toronto 2006, 2–3). In his comparative analysis of municipal finance, Melville MacMillan found that downloading to municipalities in the 1990s was primarily an Ontario phenomenon (MacMillan 2006).

The Harris government had a dramatic and more direct effect on the politics of immigrant integration and accommodation in Ontario. Through policy withdrawal, it downloaded responsibility for immigrant integration. The NDP provincial government that preceded the Conservatives had supported proactive policies to include immigrants and ethnoracial minorities in public, social, and economic institutions. During its single term in office the NDP had founded the Ontario Anti-Racism Secretariat and passed the Employment Equity Act (1993). Both were terminated shortly after the Harris government was elected. Frisken and Wallace (2000) summarize the policy change in the multiculturalism/immigrant settlement sector under the Harris government:

> Among the more significant casualties of government cutbacks were Ontario Welcome Houses, of which there were three in Toronto, which provided such settlement services as translation, interpretation, and the publication of information brochures in various languages; a Multilingual Access to Social Assistance Program (MASAP); the Ontario Anti-Racism Secretariat; a Community and Neighbourhood Support Services Program (CNSSP); and the Employment Equity Act. The government also stopped funding Newcomer Language Orientation Classes (NLOC). (Frisken and Wallace 2000, 226)

In addition, the Ontario government replaced the Ontario Settlement and Integration Program (OSIP) with the Newcomer Settlement Program (NSP), reducing its funding dramatically – by almost 50 per cent in 1995 (Simich 2000, 7). It shifted its funding mechanisms from core to project funding and terminated many immigrant settlement programs (ibid.). By withdrawing from this policy field, it indirectly downloaded responsibility for multiculturalism policy and immigrant settlement to the local level. The province left local leaders to choose whether to fill the gap in multiculturalism policy and settlement services; moreover, they were to make this choice in a context of increasing fiscal pressure on already strapped municipal revenues.

Provincial downloading of immigrant settlement and multiculturalism policy has had important consequences at all three levels of government. During the 1990s all of the provinces except Ontario negotiated federal–provincial agreements on immigration. There was thus a significant fiscal gap in Ontario's share of national settlement funding until November 2005, when the province signed an immigration agreement with Ottawa. As the TCSA's manifesto 'Enough Talk' highlighted in 2003, Ontario was receiving 53.2 per cent of Canada's immigrants but only 38 per cent of federal funding for immigrant settlement. In contrast, Quebec received only 15 per cent of Canada's immigrants but 33 per cent of federal funding for immigrant settlement (TCSA 2003, 20). Ontario receives more immigrants than any other province, yet it has been allotted the lowest per capita share for settlement in the country (OCASI 2005, 1). Quebec received $3,806 per immigrant from the federal government for settlement costs, whereas Ontario received only $819 per immigrant (Courchene 2005b, 4). The lack of intergovernmental parity in fiscal transfers in the settlement sector has strongly affected local governance of multiculturalism in the GTA. At the time of the interviews, leaders in the settlement sector on the ground in Toronto often pointed to the disparity between Ontario's share of settlement funds and that of Quebec (Douglas 2003, interview).

The City of Toronto's financial difficulties have been exacerbated by the fact that Toronto receives a disproportionate share of the country's immigrants. In other words, provincial downloading through the redistribution of responsibilities between provincial and local levels has intersected with the place-specific consequences of Canada's national decisions in immigration policy. When upper levels of government cut back on immigrant settlement and other social programs in general, the City of Toronto had to fill the gap. As Peter Clutterbuck and Rob Howarth (2002) put it, the city takes on 'residual responsibilities' in settlement when upper levels of government fail to meet the requirements of successful immigrant settlement (2002, 53). For instance, they note that the city spent approximately '$24 million to provide social assistance to refugees and immigrants whose sponsorships had broken down, and an additional $1.9 million to house refugees in emergency shelfter' (ibid). Furthermore, cutbacks in social services and immigrant settlement programs occurred in a context in which poverty had been racialized. Michael Ornstein's study of both the 1996 and 2001 censuses documents a clear gap between the socio-economic condition of Europeans and non-Europeans in Toronto (Ornstein 2001, 2006). According to Michael Feldman, who was deputy mayor when he was interviewed, the city has a points system for social housing that gives priority to those residents in greatest need. Immigrants tend to have the greatest need in Toronto and are therefore placed at the top of the waiting list. According to him, this creates resentment among long-standing residents, who argue that they have paid taxes all their lives and therefore are more entitled to social benefits than immigrants. In Feldman's view, that is how the city has contributed to a 'those people' syndrome in Toronto (Feldman 2005, interview). Provincial downloading has combined with the place-specific implications of federal decisions in immigration policy to place severe pressure on Toronto's municipal services and budget.

The Harris government also massively restructured Toronto's institutions by amalgamating the former Metro Toronto with its six constituent municipalities – the largest municipal merger in Canadian history.

In sum, the Harris years in the mid to late 1990s were characterized by extraordinary change that strained the municipal system. Downloading to municipalities had been a widespread trend in Canada, one that had placed a great deal of pressure on municipal systems. But as Andrew Sancton (2000) observes: '*Nowhere has the stress been greater than in Ontario,* where the Progressive Conservative government led by Premier Mike Harris has simultaneously forced the creation of Canada's most populous municipality [the new City of Toronto] and drastically reorganized the functional and financial framework for all of the province's municipalities and school boards' (2000, 425; emphasis added). Clearly, between 1995 and 2003, municipalities in Ontario were operating in an extremely volatile provincial environment.

Under amalgamation, the City of Toronto experienced far more stress that other municipal governments in the sample. According to one explanation, the Harris government amalgamated Toronto in order to shift the city's balance of power to the right by flooding council with conservative representatives from the suburbs (Sancton 2003, 13). In other words, the city was targeted for political reasons.

Furthermore, the Harris years left unresolved the issue of regional governance in the GTA. In 1998, around the time of Toronto's amalgamation, the province established a special-purpose body, the Greater Toronto Services Board, to coordinate social, infrastructure, and transportation (including GO Transit, the commuter rail service) policy in the GTA (Siegel 2005, 131). However, the province abolished this body in 2002. Local leaders in Toronto and its outer suburbs were on their own in devising ways to act collectively (ibid.). The lack of a multipurpose regional coordinating body in the GTA constitutes a major institutional gap in the region. In response, forms of 'new regionalism' have developed in the GTA to address gaps in the region's formal governance institutions and in local policy-capacity deficits. These new patterns of political mobilization are part of an ongoing process of rescaling state space in urban Canada.

B.C. in the 1990s: Support for Municipalities and Multiculturalism

There were some similarities in political and policy change in Ontario and B.C. in the 1990s. Generally, though, provincial–municipal relationships in B.C. were comparatively stable and cooperative. Furthermore, the province was consistent in its support of immigrant settlement and other multiculturalism initiatives. B.C. governments did not pursue disentanglement or forced municipal amalgamations. Also, in the late 1990s the trend in municipal legislation in that province was towards greater municipal empowerment.

In B.C., the Municipal Act was amended in 2003 with the passing of the Community Charter. This municipal legislation, touted as the most empowering of its kind in Canada (Tindal and Tindal 2004, 202), applies to three of the four municipalities in the Greater Vancouver sample: Richmond, Surrey, and Coquitlam. It recognizes municipalities as an 'order of government' with concurrent authority with the province in thirteen spheres; it also requires municipal consent to amalgamations (ibid.).

The cooperative nature of provincial–municipal relationships in B.C. builds on a long history of municipal empowerment in that province (Sancton 2002, 272). A separate municipal charter – the Vancouver Charter – which has been in place since 1886, provides the foundation for Vancouver's legislative and fiscal authority. According to Heather Murray (2006), Vancouver 'has had a long history of local government empowerment dating back to its earliest days [that] seems to defy the constitutional limits thesis' (2006, 13, 15). John Lorinc notes B.C.'s 'enduring and far-sighted tradition of provincial forbearance in the affairs of its largest city' (Lorinc 2006, 202).

Community leaders in B.C. agree that downloading of immigrant integration challenges to municipalities has occurred to a certain extent. However, several differences emerged in interviews with leaders in the immigrant settlement sector in Greater Vancouver and the GTA.

In Greater Vancouver, local leaders differ in their perceptions of the shift from the NDP (which governed B.C. from 1996 to 2001) to the Liberals (2001–present) with regard to immigrant settlement. According to Patricia Woroch, Executive Director of the Immigrant Services Society (ISS), both governments made decisions that negatively affected local capacity building in that field (Woroch 2004, interview). Another leader in Vancouver's immigrant settlement sector suggested that the Campbell government (Liberal) has been motivated by an 'ideologically loaded' neoconservative agenda to cut back on services to vulnerable groups, including immigrants (community leader 2004, interview). But the same informant tempered this view by noting that 'to the province's credit,' under Campbell the province continued to provide services to refugees that the federal government does not 'allow.' In other words, the Campbell government has been more flexible with respect to the services it permits settlement agencies to offer refugees than the federal government was when it was the settlement agencies' primary patron.

Settlement leaders seemed to consider the sector to be better off in B.C. than in Ontario. Woroch suggested that the impact of downloading was much less dire in Greater Vancouver than in Ontario during the Conservative years. She noted that settling immigrants in Toronto is more of a 'life-and-death situation' because of the sheer numbers of immigrants arriving annually (Woroch 2004, interview). In other words, whereas the *proportion* of immigrants in relation to the municipal population is similar in the two cities, because Toronto's population is so much larger, so too is the *absolute* number of immigrants.

Another central difference, according to leaders in the settlement sector, is that municipal governments do not play a significant role in B.C.'s immigration and settlement policy. According to them, B.C.'s municipal governments are responding to 'diversity, not immigration' (Welsh 2004, interview). It seems that a more stable relationship with the province has led settlement leaders in B.C. to accept the jurisdictional status quo in immigration and settlement policy to a greater degree than leaders in Toronto.

Academic research also suggests that in B.C. in the 1990s, the provincial environment for immigrant settlement and multiculturalism policy was much more stable and even proactive in addressing the concerns of immigrants and ethnocultural minorities. For instance, David Edgington, Bronwyn Hanna, and colleagues (2001) write that even while the Ontario government was retrenching provincial multiculturalism policies, the B.C. government's policy orientation was 'more committed to multicultural policies and programmes since the early 1990s' (2001, 10). They note that B.C. passed its own Multiculturalism Act in 1993 and began to provide a number of new services to immigrants – for example, it offered settlement councillors and English-language instructors at various language institutions, and it supported community-based programs and heritage-language instructors (ibid.). In B.C., provincial leadership in multiculturalism policy lessened the burden of immigrant integration on municipal governments in Greater Vancouver's principal immigrant-receiving municipalities.

B.C.'s more proactive approach to immigration and settlement is evident in the fact that it negotiated an immigration and settlement agreement with the federal government in 1998, under which it took over Ottawa's traditional role in managing short-term settlement programs. The agreement was renegotiated in 2004. In contrast to Ontario, B.C.'s share of national settlement resources in 2003 exceeded its share of immigrants. B.C. receives about 17 per cent of Canada's immigrants and collects about 20 per cent of Canada's national settlement resources (TCSA 2003, 20).

Also, regional governance institutions have been more stable in B.C. than in Ontario. The district model of regional municipal governance was 'gently imposed' in B.C. in 1965 (Tindal and Tindal 2004, 87). The Greater Vancouver Regional District (GVRD) – the regional district in which Vancouver, Richmond, Surrey, and Coquitlam participate – is a regional coordination body consisting of twenty-one partner municipalities and one electoral district. It delivers essential utility services such as drinking water, sewage treatment, recycling, and garbage disposal. Its mandate includes planning growth and development and protecting air quality and green spaces. Today, regional districts in B.C. (including the GVRD) exist in a 'striking contrast to [forced] municipal consolidations or mergers ... so prevalent in other parts of the country,' including Ontario in the 1990s (Tindal and Tindal 2004, 91).

In sum, the story of the impact of Canada's changing national infrastructure on municipalities in B.C. is one of relative stability compared to municipalities in Ontario. Furthermore, as we saw above, as a result of amalgamation the City of Toronto has experienced more volatility that other municipalities in Ontario.

The Effect of Changing National Infrastructure on the Local Governance of Multiculturalism

Differences in provincial decisions in municipal and multiculturalism policy in the 1990s led to very different governance dynamics in the two cities. In Ontario, the combined effects of downloading in the GTA, municipal restructuring in Toronto, and the Ontario government's failure to provide leadership on several policy issues led to the development of an urban regime in Toronto that is advocating for greater urban empowerment through new fiscal, legislative, and political relationships between the City of Toronto and upper levels of government.

This movement has been popularly referred to as a 'New Deal' for the GTA's largest city. The coalition has mobilized to seek new fiscal, political, and legislative relationships between the city and upper levels of government. Furthermore, multiculturalism goals have become important to the movement's agenda. An important group within this coalition – the Toronto City Summit Alliance (TCSA) – has identified excellence in immigrant settlement as one of Toronto's top five policy priorities (TCSA 2001).

A 'New Deal for Cities' as a 'Broad Purpose' of Toronto's Regime

The term 'New Deal for Cities' has only recently entered Canadian political discourse. It appeared in a page-one editorial by the former publisher of the *Toronto Star* – John Honderich – published on 12 January 2002. Honderich listed a variety of policy challenges facing the City of Toronto and then argued that 'nowhere is the [urban] malaise more evident [in Canada] than in Greater Toronto.' In his view, the central problem was governance – that Canada's 'antiquated system of government ... leaves cities largely unrepresented and constantly begging for funds.' He linked Toronto's crisis to a severe fiscal gap in Canadian federation: 'Torontonians pay out $4 billion more in taxes every year than Ottawa and Queen's Park put back in.' Given Toronto's situation in 2002, he expressed 'little wonder there is a growing "charter" movement within the GTA to declare "independence" or seek separate provincial status.' He then declared: 'The *Star* is launching a crusade for a new deal for cities' (Honderich 2002).

Broadly speaking, the term 'New Deal for Cities' refers to a new status for urban municipalities in Canadian federalism. It refers to a variety of reform proposals that would empower municipal governments within the intergovernmental system. It has been used to describe new fiscal, legal, political, and even constitutional relationships between municipalities and upper levels of government.

According to Heather Murray (2004), conceptualizations of 'urban autonomy' range from independent city-state status on the one hand to empowerment through greater tri-level intergovernmental entanglement on the other (2004, 7). Roger Keil and Douglas Young (2003) observed in their study of the politics of municipal autonomy in Toronto:

'We have found no "structured coherence" to this debate but a colorful, yet emerging politics of municipal autonomy' (2003, 95). They found that local leaders perceive 'municipal autonomy' to be the 'solution' to many of the most pressing urban challenges in Toronto: 'For some, [municipal autonomy] meant that greater autonomy might increase business prospects; *for others, it meant that Toronto might be in a better position to deal with the integration and settlement of its large immigrant population*' (ibid., 96; emphasis added).

The New Deal campaign frames the Toronto regime's response to Canada's changing national infrastructure. Local leaders who advocate a New Deal for Cities share the objective of trying to increase the *capacity* of urban governments to address place-specific policy challenges. The inclusion of immigrant settlement policy in this frame has important implications for local capacity building to address multiculturalism in Toronto.

The New Deal for Cities movement in Toronto can be traced to the intersection of several decisions made by upper levels of government in the mid-1990s. The limited literature on this subject suggests that 'the forced amalgamation of Toronto in 1997–1998, coupled with the downloading of social services ... has inspired the current charter debate in Toronto' (ibid., 89). Lorinc (2006) offers a similar account of the origins of the urban autonomy movement: 'In Ontario in the late 1990s, the triple whammy of amalgamation, downloading, and laissez-faire planning so alarmed many prominent urbanists that a group, including Jane Jacobs and philanthropist Alan Broadbent, put together a proposal for a "Greater Toronto Charter"' (2006, 204). The timing of these provincial decisions coincided with federal decisions to reduce provincial transfers for social spending. The contours of what Thomas Courchene calls 'hourglass federalism' were emerging by the mid to late 1990s (Courchene 2005a). Changes in fiscal federalism rooted in federal decisions created new incentives for local political mobilization. One could argue that these decisions had been inspired by global economic imperatives and by the global diffusion of neoliberal ideas.

But in Ontario it was provincial decisions that provided the most immediate incentive for political mobilization in Toronto. The Harris government's decision to forcibly amalgamate the constituent municipalities of Toronto inspired a powerful middle-class citizens' movement against the merger. That movement, Citizens for Local Democracy (C4LD), drew from a middle-class reformist movement that had emerged in the 1970s (Boudreau 2006, 42; Keil and Young 2003, 89).

For C4LD, the issue was not simply local democracy; it was also a tax revolt in response to fiscal downloading (Siemiatycki 1998, 5).

At first, C4LD did not seek the support of immigrants and ethnoracial minorities. Originally, it was a mobilization of 'white, British Toronto' (ibid., 5; Horak 1998). However, in response to this gap, a coalition called New Voices of the New City emerged under the leadership of the Council of Agencies Serving South Asians (CASSA). The coalition developed after CASSA made a submission to the province regarding how amalgamation would affect immigrants. In the summer of 1997, local leaders of New Voices of the New City pulled together a coalition of sixty-three immigrant-serving community organizations. Recent immigrant groups dominated this alliance, most of whose members resided outside central Toronto, in three inner suburbs: North York, Scarborough, and Etobicoke. 'A civic alliance of this scale was unprecedented' among Toronto's immigrant communities (Siemiatycki et al. 2003, 440). The co-chair of the alliance noted that 'most of the 63 [participating] groups had never come together voluntarily on any issue' (Fernando 1997, in ibid., 33).

Amalgamation provided an incentive for new immigrant organizations to cooperate. In doing so they inadvertently built bridges linking Toronto's diverse immigrant communities. The coalition facilitated Toronto's efforts to overcome the complex collective-action problems that had long been a feature of its multiracial ethnic configuration. 'Paradoxically, then, the creation of the megacity of Toronto – denounced for undermining local democracy – stimulated unprecedented civic mobilization among immigrant and visible minority communities' (ibid., 441).

Toronto Councillor Shelley Carroll suggested that the political mobilization surrounding amalgamation precipitated a broader change in Metro Toronto's political culture: 'Amalgamation turned suburbanites into activists ... And the citizens actually amalgamated faster than the politicians, because we started banding together with the other cities to protest this thing, accidentally amalgamating ourselves' (Carroll 2004, interview).

C4LD and New Voices of the New City failed to block Toronto's amalgamation. But after amalgamation, C4LD channelled its energies towards new goals, which included secession from the Province of Ontario (Keil and Young 2001). To this end, it entered into coalitions with municipal bureaucrats and politicians, business leaders, academics, the

Federation of Canadian Municipalities, the Toronto Board of Trade, and the Toronto Environmental Alliance (Keil and Young 2003; Boudreau 2006). Some C4LD members even formed a Committee for the Province of Toronto (Boudreau 2006, 170). Local political leaders expressed support for this radical reform option. Mel Lastman, the first mayor of the amalgamated City of Toronto, made the headlines when, at an international meeting in Florida in the fall of 1999, he declared that Toronto should be a province.

Amalgamation's assault on local democracy was a key mobilizing factor for Toronto's urban autonomy movement, which soon turned its attention to reforms implying a lesser degree of autonomy – for example, the establishment of a city charter (ibid.). C4LD's activities spilled over into what would become a 'charter movement,' and, more broadly, a New Deal for Toronto movement.

Immigrant and visible-minority mobilization in response to amalgamation also contributed to a rescaling process in the City of Toronto: 'In the politics of amalgamation in Toronto, immigrant communities were less concerned with preserving the jurisdictional status quo than attempting to assure that an enlarged city government was responsive to their distinct concerns' (Siemiatycki et al. 2003, 441). For instance, Uzma Shakir, a well-known leader in the immigrant settlement sector in Toronto, describes the issue this way:

> One of the reasons I do believe that municipalities must play a role is because people settle in municipalities, they live in cities, they don't live in jurisdictions, they don't define themselves in constitutional, jurisdictional things, they don't care. Immigrants have come here, they have come to Canada, they have chosen Toronto or Mississauga or Markham and that's where they're living their lives and the nearest level of government to them is the municipal government. Ironically, the municipal government is the least resourced as well. So just the demographic pressures alone have to be taken into consideration and we haven't even begun to talk about issues of diversity and racialization and all of that. But even just the massive growth ... A city must have the resources to be able to deal with it so that it can continue to build its infrastructure – social as well as physical – its economic base. (Shakir 2003, interview)

Shakir's comments clearly reflect the connections that local leaders in Toronto make among immigration, growth, and economic development. Her comments also reflect the shift that Julie-Anne Boudreau (2006) describes in civil society towards territorial rather than sector-specific strategies.

C4LD mobilized in the absence of an opening in the political opportunity structure at the provincial level (Horak 1998, 21). Yet the prevailing closed and hostile climate did not deter local leaders in Toronto from pursuing greater empowerment for the city after amalgamation. In the absence of political opportunities at the provincial level, local leaders turned their attention to building local capacity and local support for greater municipal autonomy as well as to seeking political opportunities at the federal level. Their strategies became 'polyscalar,' showing little regard for the constitutional/jurisdictional status quo as reflected in the constitutional doctrine that municipalities are 'creatures of provinces' – a doctrine that, as Warren Magnusson (2005) argues, portrays municipal autonomy in an unnecessarily limited way and that forbids federal–municipal relationships.

Mobilizing support for a city charter became one of the most important strategies at the provincial scale. It is always difficult to trace the precise origins of an idea, but some local leaders identify the origins of the city charter idea to a conference held in October 1997. Alan Broadbent had organized the event, called Ideas That Matter, to discuss the work of Jane Jacobs. This event created a great deal of interest in cities. Then in spring 1999, Broadbent convened a group of local elites to discuss the place of cities in Canada and commissioned papers on issues of importance to cities. The meeting was held at the Royal Alexandra Theatre in Toronto – a 'donation'[11] made by Broadbent's friend and well-known Toronto businessman David Mirvish – to develop a Toronto and cities agenda (Broadbent 2003, 1). The outcome of the meeting was the publication of a series of articles in a book titled *Toronto: Considering Self-Government* (2000).

After the meeting, Broadbent invited participants who wanted to continue the discussion into his boardroom. According to Broadbent, about twenty-five people attended. Participants included Jane Jacobs, former mayors (Crombie, Sewell, and Hall), journalists (Michael Valpy, Colin Vaughan, and Richard Gwyn), urban scholars (Patricia McCarney, Meric Gertler, Carl Amrhein, and Enid Slack) a former Ontario deputy minister (Don Stevenson), and a former Toronto councillor (Richard Gilbert). This meeting produced the Greater Toronto Area Charter (2001), modelled on city charters adopted by the European Union and the Federation of Canadian Municipalities. Article Two of this charter lists responsibilities that its drafters believe should be transferred to the region; they include immigrant and refugee settlement (Broadbent Group 2005, 40). As Broadbent recalls: 'It was not very difficult to come up with these powers; they are roughly the powers of a province, in the Canadian context.' In his view, the greater challenge was political – or 'what to do with it' (Broadbent 2003).

The drafters then brought the charter to a committee meeting of GTA mayors and regional chairs. That committee endorsed the charter. The challenge now became a political one. We see a hint of radicalism in Toronto's political culture in Broadbent's recollection of ideas for building political support for urban autonomy: 'Still, while we had been able to put some principles and ideas on the table, nothing much had changed. So we began to wonder what we could do next. Jane [Jacobs] and I had thought from the beginning that ultimately these matters would come down to politics. We had thought about various ways to excite the politics around cities. Colin Vaughan, the late CITY-TV journalist, wanted some dramatic civil disobedience: he kept mentioning the Boston Tea Party. Good television, I suppose, and TV did miss the original' (ibid.).

The solution arrived at was to convene the big-city mayors across Canada to develop a political strategy to influence the federal government. This political strategy was inspired by the advice of Privy Council Office (PCO) staff, who told Jacobs in a meeting that 'the federal government might pay a lot more attention to these [the urban agenda] issues if there seemed to be some political imperative behind them' (ibid.). Thus, the charter's drafters invited the then mayors of Vancouver (Philip Owen), Toronto (Mel Lastman), Montreal (Pierre Bourque), Winnipeg (Glen Murray), and Calgary (Al Duerr) to a meeting in Winnipeg in May 2001 (Broadbent 2008, 10). They called this group the 'C5' or 'Charter 5.'

Another important element of their strategy was to include important leaders in civil society. Each city was to be represented by its mayor and by prominent organizations in civil society representing a variety of sectors, including the social sector, the business sector, and labour. According to Broadbent (2003), 'over time, these representatives have tended to include the representatives of the United Way, as the proxy for the social sector, and the Board of Trade or Chamber of Commerce, as the business proxy, and either heads of regional labour councils or other recognized leaders.' The civil-society participants in the coalition called themselves 'C5 Civil' (ibid.). The C5 Mayors and C5 Civil met for the first time in Winnipeg in May 2001 (ibid.) and then again in

Vancouver in January 2002 and in Montreal in June 2002 (Broadbent 2008, 11). However, after those meetings it would be difficult to bring the mayors and civil society together. Mel Lastman cancelled a meeting that was to be held in Toronto in early 2003 (Broadbent 2003, 3). According to Broadbent, the C5 may have been perceived as a threat by some mayors, because 'its agenda risked their supplicant relations with senior governments [and their] allegiance to the Federation of Canadian Municipalities which has been the traditional vehicle for the city voice' (ibid., 3). C5 Civil met twice in Toronto in 2003 before the short-lived coalition faded (Broadbent 2008).

A full account of the cross-provincial political coalitions that local leaders developed to push the urban agenda onto the federal agenda is beyond the scope of this study. Let it be mentioned, though, that shortly after their initial meeting, a meeting of the C5 Civil, the C5 Mayors, and the Federation of Canadian Municipalities' Big Cities Caucus was held. Paul Martin, who would later become prime minister, was invited to speak (Broadbent 2003).

A publication of The Broadbent Group summarizes the origins of what would become the New Deal for Cities movement: '[Jane] Jacobs, together with businessman and philanthropist Alan Broadbent, initiated a process to bring together five of Canada's largest cities to discuss their mutual needs for greater power and autonomy. The "C5" Mayors began meeting, joined by leaders from the business, labour and civil society from each city, to discuss the unique needs of Canada's largest and most economically vibrant urban regions. These events, together with the sustaining efforts of the Federation of Canadian Municipalities, created the momentum for what has become known as "A New Deal for Cities"' (The Broadbent Group 2005).

Broadbent acknowledges the role played by other agents in promoting ideas of urban autonomy nationally. These others include boards of trade (especially in Calgary and Toronto), the Toronto United Way, the Conference Board of Canada (under Anne Golden), the Winnipeg Branch of the Canadian Union of Public Employees (under Paul Moist), the Canada West Foundation under Roger Gibbins, and the TD Bank (Don Drummond and Derek Burleton worked on the issue), as well as the TCSA under the chairmanship of David Pecaut (Broadbent 2008, 12).

Mayor Lastman had cancelled the proposed C5 meeting in 2003. Even so, under his leadership the City of Toronto began pressing the federal and provincial governments for a New Deal for the City of Toronto. In 2000, Toronto passed a resolution calling for a new relationship with the Province of Ontario. In 2001 the city launched its website to promote urban autonomy – the Canada's Cities campaign – and began strengthening and consolidating its relations with civil society in order to achieve a new relationship with upper levels of government. For instance, in 2002 the city facilitated the development of a powerful element within Toronto's urban empowerment coalition – the TCSA. According to David Crombie, a former Toronto mayor and original summit co-chair, the TCSA emerged as a result of a leadership vacuum at upper levels of government and a confluence of interests and goals between Mayor Lastman and prominent local leaders in civil society. Lastman and several local leaders viewed this summit as an opportunity to take on the province and the federal government; but the coalition also appealed to those who simply wanted to tackle important issues facing the City of Toronto (Crombie 2004, interview). In other words, the coalition appealed to local leaders who wanted to pool local resources to build local capacity as well as to those who wanted to address city concerns through new relationships with upper levels of government. The alliance began in Toronto and has since become a regional coalition.

A central ambiguity in the New Deal for Toronto movement concerns the scale at which increased local autonomy should be granted. For instance, this is a central theme in a document prepared by New Deal advocates titled *Towards a New City of Toronto Act* (Broadbent Group 2005). In *The Limits of Boundaries* (2008), Andrew Sancton uses Toronto's urban autonomy movement as a case study of how the difficulties associated with drawing boundaries of political units undermine the case for city-state status for cities.

In the GTA there is a gap in regional governing institutions.[12] This gap in formal institutional mechanisms at the local level[13] has left room for the development of flexible governance arrangements in the region.

Over time, the urban empowerment coalition in Toronto expanded dramatically. Royson James, a well-known *Toronto Star* columnist on urban issues, noted in an article on 15 June 2005, that Toronto Act Now[14] and the University of Toronto[15] had joined the 'ongoing advocacy [efforts] of the Toronto Board of Trade and the TCSA ... One gets the idea that there is unanimity on the issue.' In his view, of the forty-four city councillors on Toronto Council, 'at least 40 believe in "the cause"' (James 2005). The framework of a New Deal for Cities reflected a broad intersectoral agreement on the solution to many of Toronto's current policy challenges – including immigrant settlement challenges.

The rise of the New Deal for Cities movement in Toronto coincided with the availability of new institutional resources. One unintended consequence of Toronto's amalgamation was that it created a powerful new political unit that could more effectively challenge provincial power. After all, 'with a population of 2.5 million people, Toronto is larger than six provinces; with a budget of $7 billion, Toronto also spends more than those six provinces' (Slack 2005, 1). Furthermore, more citizens directly elect the Mayor of Toronto than any other political leader in Canada. This lends an important political resource to Toronto's mayor – democratic legitimacy.

By increasing the fiscal, territorial, and demographic significance of the new City of Toronto, amalgamation also strengthened governance relationships in the city. Given its size and scope, the new city attracts the attention of powerful organizations in Canada to a greater extent than in the past. Toronto councillor Brian Ashton explains:

> With amalgamation, you created a political unit that had the heft and the size and the issues to establish a new platform. A lot of organizations, associations, that would never have stood on this floor to come and lobby (and I say that in a positive way) council members about issues, suddenly realized 'Oh, my God, we've got this big level of government, we've got a huge purchasing power, it's the media capital of the country,' and certain by-laws that they're capable of passing, even the narrowest of regime responsibilities that we can have, have a significant impact on the rest of the country … So suddenly even national associations are coming back here to say that we've got to have some say in local by-laws. This also makes the local population of stakeholders and interest groups realize that let's come together because we have a megaphone that allows us to advocate our issues on the national scene. So there's a huge political dynamic shift taking place. (Ashton 2004, interview)

The new 'platform' for local organizations that the amalgamated city provided local leaders was coupled with a general retrenchment in provincial government spending and a withdrawal from policy areas related to immigrant settlement and multiculturalism policy. Despite the increasing numbers of immigrants arriving in Toronto, the Harris government was unable to secure a federal–provincial agreement on settlement funding to address the fiscal gap between the numbers of immigrants the province receives and the federal settlement dollars flowing into the province. Ashton describes the lack of leadership from the province as well as the tri-level nature of the political dynamics:

> In a strange way, the province shunned a lot of the responsibilities that should have remained if not gravitated to the provincial level. Particularly when it comes to immigration, I think they were more or less saying that that is a federal issue. Oh, wait a minute, nobody asked them, well, if it's a federal issue, and the feds don't do something, then the city's going to be left as a last resort. Of course, along comes amalgamation, the city gets left as a last resort, you get a huge advocacy opportunity, you get a political paradigm shift starting to happen. So, in a fashion, the province under the Harris government started to move in a direction that made it redundant, unnecessary, merely a banker. (Ashton 2004, interview)

The municipal autonomy debate in Toronto is complex, as are its cross-provincial (or national) dimensions. The urban autonomy movement has a 'multitude of origins' (Keil and Young 2003, 89). What is most important to understand here is that downloading and amalgamation combined to change patterns of mobilization in Toronto. The broad elite consensus among local leaders concerning the 'solution' to what John Honderich described as Toronto's profound 'urban malaise' has been striking. These new forms of political mobilization developed in the context of an antagonistic provincial political opportunity structure. Thus, local leaders' strategies became polyscalar as they turned their efforts to building local and regional alliances and lobbying the federal government. They challenged the institutional and constitutional norms of Canadian federalism. Toronto's regime thus became less concerned with the jurisdictional status quo than with ensuring that the city's public-policy challenges were addressed.

Immigrant settlement concerns have been linked to the New Deal consensus, giving momentum to local capacity building in this area in Toronto. In fact, as early as September 2000, the Maytree Foundation convened a conference to discuss the implications of a Greater Toronto Charter for the immigrant settlement sector. This meeting, styled Forum Towards a Greater Toronto Charter: Implications for Immigrant Settlement, was held at the Metro Central YMCA. The Maytree Foundation has since commissioned papers on how greater autonomy for the city might affect the immigrant settlement sector (See Simich 2000; McIsaac 2003). These papers tend to support a greater role for the city in immigrant settlement. The Toronto Region Immigrant Employment

Council (TRIEC) is the best example of how tying municipal responsiveness to immigrants and ethnocultural minorities to this broader New Deal coalition has increased the city's power to address some of the primary concerns of immigrants in Toronto.

Toronto's territorial strategies with respect to immigrant settlement seem to have influenced Vancouver's approach. In May 2005 (after interviews were conducted in Greater Vancouver) former mayor Larry Campbell (2002–5) established a working group on immigration in Vancouver as part of a strategy of the 'Charter 5 (C5)' coalition of cities to negotiate new relationships with upper levels of government in immigration policy (City of Vancouver 2005, 3). At the time the interviews were conducted in Greater Vancouver, local leaders tended to define the issue of municipal responsiveness to immigrants and ethnocultural minorities as responding to 'diversity, not immigration,' because immigration was a federal and provincial jurisdiction. Framing the issue as an 'immigration issue' is a territorial strategy to empower municipalities; it is also part of a larger 'movement' to obtain more resources from upper levels of government. However, unlike in Toronto, in Vancouver such territorial strategies do not seem to percolate up from 'the ground' and do not seem to dominate the agenda, since the provincial and federal governments have been relatively responsive to the sector there.

Immigrant Settlement and the New Deal for Cities Movement: The Emergence of TRIEC

Multiculturalism policies are part of a multifaceted urban agenda within the City of Toronto's urban regime. The TCSA is a key node in the regime and within its municipal autonomy movement. This alliance has established several policy priorities, one of which is becoming a centre of excellence in integrating immigrants.[16] To this end the alliance established the Toronto Region Immigrant Employment Council (TRIEC) to help deal with one of the most important challenges facing newcomers – access to employment. Participation in this local initiative is highly intersectoral. TRIEC's membership includes assessment-service providers, community organizations, employers, foundations, labour unions, occupational regulatory bodies, and post-secondary institutions, in addition to representatives of all levels of government – federal, provincial, and municipal.[17] The Maytree Foundation, Citizenship and Immigration Canada (CIC), Canadian Heritage, and Human

Resources and Skills Development Canada all provide operating funds to TRIEC.

Local policy networks interested in immigrant employment issues have coalesced in this alliance, creating increased capacity in immigrant employment at the city and regional levels. TRIEC has launched a number of its own programs, including a mentorship program whereby immigrants are placed with private-sector and government organizations to gain Canadian experience in their field of expertise. By 2007, TRIEC had placed more than 2,600 mentees in partner organizations; 80 per cent of these people are now employed, with 85 per cent of these employed in their field of choice (TRIEC 2007, 9). TRIEC partners with other organizations to provide resources to both immigrants and employers to facilitate the integration of skilled immigrants into the Toronto workforce. TRIEC is growing significantly. In 2006 it spent just over $1.3 million on its activities. By 2007 this figure was close to $2.3 million (ibid., 21).

Municipal governance arrangements seem to affect the likelihood and nature of cooperative efforts at the regional level. TRIEC was launched in Toronto, which is the only municipality that has been responsive to the concerns and preferences of immigrants and ethnocultural minorities in the GTA sample and whose official motto is 'Diversity Our Strength.' Local leaders from Toronto are overrepresented in TCSA and TRIEC because of Toronto's high level of community capacity – including in immigration and settlement. However, the effort is meant to be regional.

In 2004 the chairs of regional governments in the GTA that have high levels of diversity (Peel and York regions), as well as a representative of one municipality within each region, were asked to participate in TRIEC. Mississauga was the clear choice in Peel, and Markham in York. Tellingly, both regional chairs and Mayor Cousens of Markham accepted the invitation to participate, but Mayor McCallion of Mississauga declined. According to a local leader close to the process, McCallion refused to participate until the federal government reimbursed Peel Region for the costs of immigration incurred by municipalities owing to sponsorship breakdowns (community leader 2004, interview). According to McCallion, she did not participate in TRIEC because its mandate was too limited (McCallion 2004, interview). Mayor Susan Fennel of Brampton accepted the invitation to participate in McCallion's place. TRIEC's 2007 annual report listed Mayor Susan Fennel of Brampton,

Mayor David Miller of Toronto, Bill Fisch, the Regional Chair and CEO of York Region, Emil Kolb, Regional Chair of Peel, and Shirley Hoy, the CAO of the City of Toronto, as the municipal partners (TRIEC 2007, 19).

British Columbia: The Absence of Mobilization for Municipal Autonomy

In B.C. an urban autonomy movement or coalition is absent (Murray 2004, slide 18). As we saw above, some downloading to municipalities has occurred in B.C. but this has not been coupled with a large-scale amalgamation of its oldest city (Vancouver) with the surrounding suburban municipalities. Furthermore, B.C. did not engage in a disentanglement exercise as Ontario had during the Harris years. Municipalities and municipal communities in B.C. did not experience the same level of stress as municipalities in Ontario under the Harris and Eves governments (Sancton 2000; Murray 2004; MacMillan 2006).

Also, municipal regional-governance institutions have been relatively stable in B.C. The GVRD, established in 1965, took steps in 1996 to 'promote multiculturalism in Greater Vancouver' through its Livable Region Strategic Plan (Edgington and Hutton 2002, 10). However, the multiculturalism goals of this regional plan do not seem to have been implemented.

According to Verna Semotuk, a senior planner with the GVRD's Policy and Planning Department, the GVRD does not play a role in multiculturalism and immigration. There are no directions from the GVRD board in these policy fields (Semotuk 2005, interview). Nevertheless, the GVRD board has sometimes raised immigration issues in discussions with upper levels of government (ibid.). Priority areas include increased funding for settlement services, credential recognition, and establishing a mechanism whereby several regional districts could collaborate to disperse immigrants (ibid.). The latter priority is especially interesting. Given the social differences among the twenty-one municipal governments in the GVRD, the policy challenge has been framed as a question of 'dispersal.'

The GVRD does not play an ongoing role in coordinating settlement services or in encouraging multiculturalism policy development at the local level. Vancouver councillor David Cadman, who sat on five GVRD committees during the municipal term before his interview, remarked: 'I can count on one hand the number of times that we've had an immigration theme discussed at the regional level' (Cadman 2004, interview). He offered some reasons for this omission, including the fact that directors[18] of the GVRD tend to feel that immigration is outside their jurisdiction and do not want to assume the costs of programs that are the responsibility of upper levels of government. As he puts it: 'But I've got to say in all honesty, for many directors there's not a lot of enthusiasm in getting into that area, because it's feeling like, "Well, hang on, we're moving into an area that's really not our jurisdiction and the extent to which we fill that vacuum will mean that we'll be paying for that." And really, that should be the senior levels of government that are doing it and aren't' (ibid.).

Furthermore, owing to their limited resources, municipalities in Greater Vancouver prefer to focus on advocacy surrounding policy areas for which they have a clear mandate (ibid.). Thus the federally focused New Deal for municipalities in Greater Vancouver is about 'getting the federal and provincial governments on board around infrastructure issues' (ibid.). It seems that the institutions of the GVRD (now Metro Vancouver) might be impeding regional coordination on immigrant-related issues because they have formally removed this issue from the regional agenda. This might be because of the diversity of municipal governments that are governed by the district, many of which are not immigrant magnets, unlike the municipalities in our sample.[19] Other policy issues for which the Regional District has a clear mandate take precedence.

But the main reason why new governance arrangements in Greater Vancouver have failed is that the provincial–municipal relationship is a strong. Local leaders can turn to the province for support in many of their capacity-building initiatives. For instance, as a Laurier Institution report describes (and as some interviewees mentioned), the province has launched a program called Employment Access for Skilled Immigrants (EASI) that is similar to the TRIEC initiative (Laurier Institution 2005, 10). However, this initiative is provincial and does not focus on the Greater Vancouver region specifically.

The Province of British Columbia supports the goals of the City of Vancouver's proactive 'growth machine.' As Vancouver's mayor in the mid-1980s, Mike Harcourt actively cultivated connections with Asia (Mitchell 2004, 57). Provincial politicians began courting investment from Hong Kong in the early 1980s (ibid., 56). When he became premier in 1991, Harcourt returned to Hong Kong to address fears that a more left-leaning NDP government would be unfriendly to investors (ibid., 57). Furthermore, the province strengthened its multiculturalism policy efforts in the 1990s.

The positive relationship between the City of Vancouver and the province is evident in how the position of mayor of Vancouver has become a stepping stone to the premiership. Gordon Campbell, the current premier, was a mayor of Vancouver in the 1990s. In B.C. the positive relationship between the province and municipalities, coupled with provincial support for the city's growth agenda, has meant that local leaders have not mobilized around the issue of municipal autonomy.

Urban Regime 'Spillover': From Urban to Polyscalar Regimes

The new governance arrangements that have emerged in the GTA and the activities of the C5 and C5 Civil are examples of what urban scholar Julie-Anne Boudreau calls 'polyscalar' social mobilization (Boudreau 2006). Changes in the intergovernmental system created incentives for the mobilization of local networks in Toronto and expanded the scope of participation in Toronto's regime. The political mobilization in response to changes in the national infrastructure in the late 1990s was so powerful that it spilled over into regional governance regimes and into what one might call a 'polyscalar regime.' The Toronto regime's coalition-building activities and efforts to lobby upper levels of government produced positive externalities in the region. These efforts have contributed to capacity building in immigrant integration in the region. These political developments provide the context for understanding the dynamic nature of Toronto's regime as well as for policy change in immigrant settlement and multiculturalism policy at the municipal level in Toronto.

'Polyscalar' political coalitions have some of the characteristics of urban regimes in Stone's conception. According to Stone (1989), informal urban regime relationships are driven by two needs: '(1) *institutional scope* (that is, the need to encompass a wide enough scope of institutions to mobilize the resources required to make and implement governing decisions) and (2) *cooperation* (that is, the need to promote enough cooperation and coordination for the diverse participants to reach decisions and sustain action in support of those decisions)' (1989, 9; emphasis added).

These needs arise owing to a municipal government's limited resources and the fragmentation of power in local communities. The polyscalar dimension of Toronto's urban regime addresses the *fragmentation* of the municipal system in Canada as well as the absence of regional institutions in the GTA. The Federation of Canadian Municipalities (FCM) represents Canadian municipalities collectively, but because its membership is so diverse, it does not provide targeted representation for Canada's largest cities. Thus locally based leaders in urban centres have developed cross-provincial coalitions in response to the inability of formal institutions to address their policy concerns. The FCM's Big City Mayors Caucus partly addresses this gap. Indeed, this caucus has been an important advocate for municipal empowerment. Nevertheless, with twenty-two members, it is unable to represent some of the specific needs of Canada's largest centres (Broadbent 2008, 9). Furthermore, the FCM represents only municipal governments. Local leaders in Toronto and in other large Canadian cities have recognized that capacity building for a New Deal requires regime relationships that include civil-society elites. In order to increase municipal policy capacity by pressuring upper levels of government to pool resources more evenly with them, local leaders have expanded political institutions so that they include civil-society organizations. These coalitions are polyscalar insofar as they address gaps in formal institutions at a variety of scales – municipal, regional, cross-provincial, and multilevel.

Furthermore, as with urban regimes, local leaders are building these coalitions specifically to address the policy challenges facing Canada's urban centres – that is, to bring about what Clarence Stone (1989) calls 'enacted change.' Jane Jacobs, Alan Broadbent, and other local leaders in Toronto deliberately created coalitions that included leaders of civil society with the goal of changing the intergovernmental system. Locally based urban-regime relationships that bridge the public and private sectors became an important resource in the efforts of the C5 mayors to pressure upper levels of government to respond to their concerns. The cross-provincial alliance included the C5 Mayors and the C5 Civil.

Urban regimes are also long-lasting. In this regard, the literature is unclear on how long-lasting an urban coalition must be to constitute a 'regime.' In their review of the urban regime literature, Mossberger and Stoker (2001) describe regime relationships as 'relatively stable arrangements that can span a number of administrations' (813) and as a 'longstanding pattern of cooperation rather than a temporary coalition' (829). They also argue that 'on the issue of stability ... short-term collaboration may be described as an emerging regime or a failed regime, depending on the context' (830). Similarly, Christopher Leo (2003) argues that 'the necessary degree of dominance and stability [for a local coalition] to be termed a regime is an empirical one, and a judgment call as well, since there is and probably can be no precise determination of how much dominance and stability is required in order to justify the "regime" label' (2003, 346). The question of stability (and dominance) is therefore subject to interpretation. That said, in order to constitute a regime, the coalition must include the business community, must contribute to the city's capacity to govern, and must include policy agendas that can be traced to the composition of the urban coalition (Mossberger and Stoker 2001, 829). Regional-governance arrangements such as TRIEC and other TCSA initiatives continue in Toronto. However, the C5 Mayors and C5 Civil were short-lived (Broadbent 2008, 11). As Broadbent noted above, this may be because informal alliances compete with an organization that has already been established to serve this purpose – the Big City Mayors Caucus. Existing institutions and organizations can select certain group 'out.'

Local leaders seeking increased urban autonomy for the City of Toronto seem to be employing several political strategies, many of which are related to urban regime building. In fact, building urban regimes at the local level itself constitutes one of the central strategies of the New Deal for Cities movement. The Toronto regime seems to be employing five related strategies:

1 Encouraging municipalities to act like governments rather than as wards of provinces.
2 Building high-powered coalitions of municipal leaders and elites in civil society.
3 Educating and engaging the public in the New Deal debate.
4 Using diverse strategies in efforts to lobby upper levels of government.
5 Addressing unfunded municipal mandates proactively.

Collectively, these strategies reflect the resources and participants in the Toronto regime. The first was inspired in part by a speech that Jane Jacobs made at the first C5 meeting in which she urged the five mayors to break their 'learned dependency' on upper levels of government and begin acting like true government leaders. This strategy involves insisting on a government-to-government relationship between the City of Toronto and upper levels of government (Abrahams 2004, interview). For instance, the City of Toronto used this strategy when it withdrew from the Association of Ontario Municipalities (AMO) during negotiations with the provincial and federal governments over the distribution of the federal Gas Tax Transfer, a new municipal transfer that was introduced by Ottawa during the Martin government.

The second strategy involves creating urban regime relationships and expanding them to include a variety of scales to form polyscalar regimes. The third is possible because of the Toronto regime's diverse and intersectoral participation. For instance, representatives of the Toronto Board of Trade handed out leaflets in subway stations in Toronto that informed residents of the fiscal gap between the taxes Torontonians pay to upper levels of government and the services they receive. Fourth, public education campaigns are important. All of the coalitions described above – including the TCSA – are engaged in such campaigns. The City of Toronto's Canada's Cities Campaign website is also designed to educate the public ('Canada's Cities: Unleash Our Potential').

The fifth strategy involves municipal governments developing local subcoalitions or 'subregimes' to pool capacity to achieve locally significant public policy goals outside their formal mandates. Julia Koschinsky and Todd Swanstrom (2001) have coined the term 'subregime' to describe coalitions of policy subsystems at the city level in housing and community development. However, they do not use the regime literature in the same way as most urban regime theorists. In particular, they do not view urban regimes as unified across issues, preferring to apply Norton Long's concept (1968) of 'ecology of games' when describing urban policy subsystems. Furthermore, they argue that the relationship between a subregime and the overall regime – whether the former is independent of the latter or subordinate to it – is an empirical question (2001, 116). In Toronto, such alliances or subregimes exist in several areas of priority. TRIEC is an important example of a 'subregime' in immigrant employment. Nevertheless, unlike some 'subregimes' that Koschinsky and Swanstrom describe in housing policy, this initiative is strongly integrated with the overall citywide regime. It is a part of a rescaling process that ties immigration and economic development together into a new kind of immigration-driven 'growth machine' (Molotch and Logan 1987) and an urban autonomy movement.

In *A New City Agenda* (2004), John Sewell[20] recommends this strategy to local leaders. His book is a sort of manifesto for the New Deal for Cities agenda, one that recommends steps that cities should take in a number of policy areas, including immigrant settlement. Specifically, he recommends that municipal leaders launch new and innovative programs and policies and ask for money from upper levels of government after the fact. He uses Toronto's innovation in affordable housing policy in 1945, which resulted in the federal National Housing Act (1949), as an example of this strategy. In Sewell's words:

> Cities must adopt a new strategy. They must define very clearly the programs they know they are *capable of delivering* and that have popular support, and then set to work delivering them. The key is to do enough groundwork at the city level so the *public understands the need for programs and supports the city politicians in their push to get authority and finances for them* ... This book advocates that Toronto – and other cities – begin to develop and implement specific programs because just talking generally about the need for money and power is not a strategy. Cities must ask for particular pieces of legislation that allow them to carry out programs they identify as their mandate, and they must be precise about the monies needed. (2004, 91; emphasis added)

Doing the 'groundwork' to establish the conditions for greater urban autonomy involves developing productive urban coalitions. This is precisely the strategy that has been employed in the immigrant settlement sector with the development of TRIEC. Caroline Andrew (2001) also recommends that local leaders develop governance arrangements as a strategy to become more effective and empowered within the intergovernmental system (2001, 108). As she concludes her article, which makes the case for the importance of cities and municipalities: 'We should not ignore the cities, but this will happen only after they demonstrate to us and to the other levels of government that we cannot ignore them' (ibid., 110).

TRIEC: A Regional and Multilevel Subregime in Immigrant Settlement Policy

Urban scholarship in the late 1990s (Leo 1998; Clarke 1999) uncovered regional regimes in American urban areas (Mossberger and Stoker 2001). For instance, Christopher Leo (1998) uncovered a growth containment regime in Portland, Oregon, that was regional in scope. Similarly, TRIEC constitutes a regional subregime that is highly integrated with the City of Toronto's citywide regime. This subregime has multilevel dimensions, given that civil servants at all three levels of government participate in its working groups and the federal government now funds some of its operating costs. TRIEC also differs from the growth containment regime in Portland insofar as its cooperation is not sustained by formal, public institutions at the regional level. Rather, TRIEC developed from the bottom up. Also, it covers the entire GTA, and municipal governments and civil-society leaders participate on a voluntary basis. TRIEC's boundaries are flexible. As such, the informal governance arrangement avoids the difficult task of defining the dynamic city-region's boundaries in a formal way – a central impediment to the urban autonomy project in 'Toronto' in Andrew Sancton's (2008) view.

TRIEC's purpose is to create policy capacity in the area of immigrant integration into the economy through public–private collaboration. It arose out of a need to bring together a diverse set of actors to deal with barriers to immigrant integration into the labour market. Thus, according to urban regime theory (Stone 1989, 6), it arose out of incentives for local leaders to extend the 'institutional scope' of municipal institutions and for 'cooperation' as a result of the fragmentation of interests that affect or have a stake in immigrant employment. This initiative coordinates the resources that must be brought to bear to address the challenge effectively.

The urban regime concept also contributes to our understanding of how TRIEC's informal governance arrangements are maintained. Stone (1989) explains how regimes are developed and maintained by their ability to provide 'selective incentives' to their partners. According to regime theory, there is a systemic bias towards the preferences of the business community, because of uneven access to selective material incentives (which are generally considered most conducive to the maintenance of collective action). Immigrant employment was chosen as the first concrete policy step in the TCSA's goal of achieving excellence in immigrant settlement because it was one on which a broad cross-sector alliance – including the business community – could agree. However, employment is not a central priority only of the business community. As community leader Amanuel Melles observes, employment cuts across many communities and is an issue on which a multicultural population can agree (Melles 2003, interview). Similarly, Debbie Douglas, the executive director of OCASI, mentioned that a broad consensus exists that labour market integration is central to immigrant integration and that this unites TRIEC's diverse participants. Though she recognizes that TRIEC focuses on a particular type of labour market integration (namely, the integration of highly skilled immigrants), given the community consensus on this issue and its importance to successful immigrant settlement, it is important for OCASI to participate in this initiative (Douglas 2003, interview).

More broadly, local leaders in multiple sectors acknowledge that they share a common and interdependent economic fate. The business leaders who participate in TRIEC recognize that the business community has a stake in immigrant incorporation into Toronto's economy. As Dominic D'Alessandro noted in a letter to Prime Minister Paul Martin: 'The Conference Board of Canada calculates that not recognizing immigrants' learning and credentials costs our economy somewhere between $3-billion and $5-billion annually. As a businessman, a private citizen and an immigrant, I see this as a critical issue both for maximizing the economic potential of Canada and for successful nation-building. This is why I am chairing the Toronto Region Immigrant Employment Council' (D'Alessandro 2004).

As urban regime theory implies, the policy preferences of various sectors cannot be assumed. Rather, policy outputs are negotiated within the regime and elites' preferences are shaped by their interactions with other leaders. In Toronto, as in American cities, the systemic bias towards the business community's preferences is strong, but dismissing local politics as only minor variations on growth politics would be misleading. As Stone (1989) observed with respect to the business community in Atlanta: 'Participation in the governing task and the quest for allies ... had an effect on the business elite, broadening its understanding of what constitutes a favorable economic climate' (1989, 195).

In Toronto, regime participants seem to understand the interdependence of economic-growth objectives and multiculturalism policy goals. Local leaders consider the successful integration of immigrants to be one key to a favourable economic climate. Thus, public-education campaigns have become an important strategy in the regime's quest for allies. The importance of politics to urban regime development is clear when one recognizes how many social issues can be framed as economic-development issues. The business case for immigration has made multiculturalism initiatives a priority within Toronto's regime.

As urban regime theory would predict, TRIEC is supported by relationships of the type emphasized in Stone's earlier work. Business supports the council because it would like to see the region and its business community benefit from highly skilled immigrant labour. Immigrant organizations participate because the programs provide immigrants with mentorship opportunities and, ultimately, jobs. The regional regime is also maintained through networks at the municipal level, with Toronto's network playing an especially strong role in this respect.

Clarence Stone's (2001) more recent work draws on the literature dealing with problem definition and issue framing to explain regime decline in Atlanta. He invites urban scholars 'to consider how issue concerns come to be specified as purposes, and how they are *linked*, *enlarged*, and *refined* for action' (2001, 20; emphasis added). Toronto's regional regime in immigrant employment is linked to an *enlarged* coalition fighting for a New Deal for the city. Immigrant integration into Toronto's economy and the continued attraction and retention of immigrants are *linked* to the city's growth objectives. The goal of immigrant integration has been *refined* to focus on immigrant employment, an issue on which there is broad intersectoral and intercommunity consensus. A focus on immigrant employment has facilitated collective action in Toronto's highly diverse, 'multiracial' context.

The New Deal frame also serves a regime maintenance function. Broadly speaking, it serves as a rallying point for local leaders across the GTA. Toronto's regime illustrates the power of the 'gravitational pull' of regime coalitions that Stone (1989) describes in *Regime Politics:*

> As the circle of cooperating allies grows, its effectiveness becomes cumulatively greater. The governing coalition has a kind of gravitational pull; as its own weight increases, its capacity to attract other civic entities increases. As the network of civic cooperation makes its presence felt, others realize that cooperation pays and noncooperation does not. That's what 'go along to get along' means, and it constitutes an effective form of discipline based on the selective-incentive principle. So long as the network of cooperators is large enough in its own right and has no rival network to serve as an alternative, there are significant gains to be made by going along and opportunity costs to be paid for opposition. (1989, 193)

The level of consensus on the goal of achieving a New Deal for Toronto is high, and local leaders have an incentive to support this political project. The broader purpose of the New Deal coalition has contributed to the weight behind calls for new intergovernmental relationships in immigration policy. It has also created an incentive for the immigrant settlement sector to accept a focus on immigrant employment rather than other issues such as affordable housing or poverty – to 'go along to get along.'

Changing the National and Provincial Infrastructures: The Influence of the Toronto Regime

Policy developments in Ontario's municipal system since the Liberals under Dalton McGuinty were elected and at the federal level under former prime minister Paul Martin (2003–6) suggest that regime development at the local level is not necessarily a 'constitutional cul-de-sac,' as David Crombie said it might be (Crombie 2003, interview). Leaders in Toronto's regime have played a significant role in federal and provincial responsiveness to both Toronto's urban agenda and the urban agenda in Canada more generally.

This policy movement at the federal level began in 2003, when Paul Martin's Liberal government (2003–6) created a Cities Secretariat within the Privy Council Office and appointed John Godfrey as Parliamentary Secretary on the New Deal file. The Martin government's major initiatives in 2004 were a 100 per cent rebate for municipalities on the Goods and Services Tax and the appointment of John Godfrey as Minister of State for Infrastructure and Communities. In its 2005 budget the government committed itself to providing municipalities with $5 billion through the Gas Tax Transfer (GTT) over five years (2005 to 2010). The federal government signed gas tax agreements with all provinces in Canada. Toronto was actually included as a partner in the Canada–Ontario agreement.

The politics and policy outputs surrounding the GTT reflect the influence of Toronto's regime. The federal government announced that this money was to be allocated to municipalities – with the provinces as intermediaries – on a per capita basis for municipal projects that encourage environmental sustainability.[21] This way of distributing the funds places large urban centres at a disadvantage.[22] In recognition of the biases of the proposed allocation formula, the federal government decided to transfer an additional $310 million to municipalities, earmarked for municipal transit. This money would be transferred according to a formula preferred by leaders in Toronto and Canada's other major urban centres – the 'ridership formula,' which acknowledges that residents outside central cities benefit from public transit systems within central cities. For instance, residents of Mississauga, Brampton, and Markham benefit from investments in public transit in Toronto. These policy developments are at least partly a result of advocacy on the part of Toronto's regime and the C5.

In Ontario, the politics surrounding the funding formulas for the GTT reflect changes in the Toronto regime's goals, which themselves reflect a new consensus that the city should be treated as an 'order of government' and not as a municipal 'creature.' The City of Toronto withdrew from the Association of Ontario Municipalities (AMO)[23] and insisted on negotiating the transfer on a government-to-government basis. The result was that Toronto became a partner in the federal government's agreement with Ontario (Canada, Ontario, The Association of Ontario Municipalities, The City of Toronto 2005).

In Ontario, the province was not involved in the allocation of the GTT. The AMO manages the distribution of funds and reporting requirements for all municipalities in Ontario except Toronto. The City of Toronto deals directly with the federal government. This development in fiscal relationships – and the accompanying political relationships – is unprecedented in Canada and illustrates the capacity of fiscal federalism to provide flexibility within Canada's constitutionally entrenched division of powers.

The policy effects of Toronto's regime are also evident in recent developments in Ontario's municipal system. In 2003, political leadership changed at all three levels with the election of a new mayor in Toronto, David Miller, who promised to advocate for a New Deal for his city; the election of a Liberal government under Dalton McGuinty in Ontario; and a change in leadership of the Liberal Party at the federal level, when Paul Martin took over as prime minister from Jean Chrétien. Local leaders viewed these political developments as the simultaneous opening of what John Kingdon (1995) calls 'policy windows' at all three levels of government. Philip Abrahams, the City of Toronto's Manager of Intergovernmental Relations, described this new political environment as an 'alignment of the stars' (Abrahams 2003, interview).

According to Kingdon (1995), ideas make it onto governmental agendas when three policy streams intersect: the policy problem stream, the policy solution stream, and the political stream. In Toronto before 2003, leaders in Toronto's urban regime identified many policy 'problems' that needed to be addressed. Their preferred solution to many of these – one of which was immigrant integration – became a New Deal for Toronto. Thus, with two streams tightly 'coupled,' regime entrepreneurs were seeking political opportunities. Changes in political leadership provided these opportunities.

Several important developments have occurred since 2003. In December 2005 the Ontario government introduced Bill 53, the Stronger City of Toronto for a Stronger Ontario Act (hereafter the Stronger City

Act). This bill passed third reading, was given Royal Assent on 12 June 2006, and came into effect in January 2007. This act establishes a more permissive legal framework for establishing by-laws in certain areas; it also delegates increased authority to the city in land-use planning, roads, housing, governance, integrity and accountability, and enforcement (e.g., of fines for contravening municipal by-laws). In addition, the act gives the city new taxation and fiscal authority. For instance, the city can now raise new excise taxes, including liquor and cigarette taxes, and it can now undertake tax increment financing (TIF)[24] to encourage development.

There is some debate over whether the act goes far enough in terms of empowering the City of Toronto because it does not give the City of Toronto authority to raise revenue through taxes that would result in a significant tax yield. Nevertheless, whereas the act did not give the City of Toronto the authority to raise significant new own-source revenues, it is path-breaking insofar as it explicitly recognizes the city's authority to negotiate and enter into agreements with the federal government. This new authority creates a framework for the city to play a policy-making role in areas of particular concern to it.

The act could also have important fiscal consequences. The City of Toronto could negotiate new fiscal transfers from the federal government. Donald Lidstone, a constitutional lawyer and expert in municipal law in Vancouver, describes the Stronger City Act as a 'constitutional milestone [that] will help cities in the rest of Canada in their quest for palpable recognition as an order of government under our constitutional regime' (Lidstone 2005).

The Stronger City Act could have significant implications for Toronto's role in immigrant settlement policy. On 14 December 2005, during a news conference with Ontario's premier and municipal affairs minister after the bill was introduced in the provincial legislature, Mayor David Miller noted that 'for the first time ever, a municipal government – the City of Toronto – will be allowed to negotiate agreements with other governments ... [For example,] as Canada's leading receptor of immigrants, the city will be able to deal directly with the federal government in preparation for the arrival of newcomers to this country' (Miller 2004). The new act recognized in provincial legislation what had already become political practice in Toronto's powerful urban regime. The City of Toronto is now dealing directly with the federal government on the GTT.

Of particular note is the Canada–Ontario Immigration Agreement, which was finalized on 21 November 2005. That agreement increases federal funding of immigrant settlement services in Ontario from about $800 per immigrant to about $3,400.[25] It is the first Canadian immigration agreement to establish a partnership with municipalities.[26] It recognizes the exceptional nature of Toronto[27] as an important immigrant-receiving centre; it also acknowledges the City of Toronto's capacity in immigrant settlement: 'The City of Toronto in particular has developed experience, expertise and community infrastructure to respond sensitively to the social and economic integration needs and potential of immigrants' (Article 5.3.1). Article 5.3.2 commits Canada and Ontario to signing a Memorandum of Understanding (MOU) with the City of Toronto with respect to new intergovernmental relationships in immigration policy. This MOU will 'be consistent with federal, provincial and Toronto commitments to a New Deal for Cities and Communities which involves a seat at the table for municipalities on national issues most important to them' (Article 5.3.3b). This agreement provides the foundation for a New Deal in immigration and settlement policy for the City of Toronto.

Since the signing of the Canada–Ontario Agreement, the province has launched several programs to facilitate immigrants' access to the labour market. These programs reflect TRIEC's priorities. In June 2006, at TRIEC's hireimmigrants.ca seminar – part of that group's 'A World of Experience' week – Queen's Park announced that it would be introducing the Fair Access to Regulated Professions Act that same month. On 12 December 2006 the Fair Access to Regulated Professions Act, 2006, received Royal Assent. This act addresses barriers in thirty-four regulated professions in Ontario, including medicine, accountancy, law, and engineering. In addition, the province has established an internship program for foreign-trained professionals in ministries and Crown agencies – the first provincial program of its kind. (Participants in TRIEC had already launched such mentorship programs at the municipal level and in the private sector.)

On 23 January 2006 a Conservative minority government under Stephen Harper was elected at the federal level without a single elected seat in the urban cores of Canada's three largest cities – Toronto, Vancouver, and Montreal.[28] Harper introduced a 'new' model of federalism – 'open federalism,' which stresses federal respect for provincial jurisdiction. Some proponents of a federal urban agenda fear that the Harper government will ignore the needs of cities, because municipalities are a provincial responsibility. However, this need not be the case. It is possible for the federal government to develop an *urban* agenda without developing relationships with *municipalities* that bypass provinces. It can do so by adopting an 'urban lens' on national policy making, thereby addressing policy areas the consequences of which are primarily urban (Berdahl 2006).

Immigration policy is a good example of a concurrent federal–provincial jurisdiction the consequences of which are mainly urban. At this writing it seems that the Conservative government will continue to address the needs of Canada's immigrant-magnet cities. The Harper government has honoured the Canada–Ontario Immigration Agreement (Canadian Press 2006). The McGuinty and Harper governments seem to have made it a priority to facilitate the integration of immigrants into the economy by addressing barriers to the recognition of foreign credentials. This is an important priority for Toronto's governing regime as well.

Concluding Thoughts

This chapter has demonstrated the important influence of the intergovernmental context on local governance. What Sellers calls 'national infrastructures' affect the dynamics of urban coalitions or 'regimes.' In Toronto, broad changes in fiscal federalism that originated at the federal level intersected with the Ontario Conservative government's decision to disentangle services, download some of the province's fiscal burden onto municipalities, and amalgamate Metro Toronto. The intersection of these changes in the national infrastructure resulted in the mobilization and consolidation of a powerful and wide-ranging network of regime relationships in Toronto around a new policy frame – a New Deal for Toronto. These networks spilled over to multiple scales, including the GTA, cross-provincially, and across levels of government.

Some of the causal factors of regime dynamics that Karen Mossberger and Gerry Stoker (2001) identified in their review of the regime theory literature – including economic restructuring, political mobilization, and changes in federal grant policies – interact in complex ways. Including the concept of national infrastructure in urban regime analyses is a useful way of understanding the complex interdependence of urban governance and the intergovernmental system. Changes in fiscal federalism, combined with a volatile and strained provincial–municipal relationship between the Province of Ontario and the City of Toronto, created incentives for Toronto's regime to focus its attention on the federal government. In the midst of what local leaders perceived as an 'urban crisis' created by the province's exacerbation of urban problems, Toronto's regime partners launched a national campaign for a New Deal for Cities to deal with a more fundamental city problem – the lack of sufficient political, legislative, and fiscal autonomy. The Toronto regime also increased its efforts to build local capacity in areas of particular concern to the city. Furthermore, in the absence of regional institutions, Toronto's leaders took leadership on behalf of the region and began searching for regional allies. This resulted in flexible new regional-governance institutions that incorporated upper levels of government in local capacity-building initiatives, such as TRIEC.

The dynamics of urban regime building in Toronto have important implications for the comparative study of urban regime maintenance and change. This case illustrates the importance of the national infrastructure to local governance. Given the fragmentation of power in urban systems, local leaders have incentives to build public–private coalitions to pool policy capacity. However, the Toronto case demonstrates that imbalances in the distribution of resources exist both in civil society and in the intergovernmental system. Regime theory teaches us that selective financial incentives are especially important to regime maintenance and development. In civil society, the business community possesses a disproportionate share of these important incentives. In the intergovernmental system, upper levels of government possess a greater ability to tax and spend than municipal governments. Therefore, incentives exist to include resource-rich leaders in the public *and* private sectors in urban regimes.

This chapter also illustrates the role of institutional change in regime dynamics. The amalgamation of Metro Toronto had an important effect on patterns of mobilization in that city. Before amalgamation, urban scholars described the city as a progressive 'middle-class regime.'[29] The secondary literature suggests that amalgamation broadened participation in governance. Toronto is the only city in the sample in which a strong urban empowerment coalition emerged, and the only city that was amalgamated. This suggests that amalgamation may have played an important role in urban regime dynamics in Toronto.

Both C4LD and New Voices for a New City mobilized during amalgamation and contributed to a rescaling process in the federation. New Voices for a New City mobilized an unprecedented number of immigrants and immigrant organizations in the core city and the suburbs under one banner. The business community, too, seemed to play a more important role in Toronto's governance after amalgamation. Also, the new City of Toronto provided a stronger platform for national organizations, given its size and budget and the mayor's electoral legitimacy.

The importance of downloading, disentanglement, and amalgamation to the mobilization of Toronto's regime around a New Deal with upper levels of government is evident when we compare regime goals in the two most proactive municipalities in the sample – Toronto and Vancouver. A central difference between these two municipalities was that in Toronto, the issue was framed as one of immigration and diversity management, whereas in Vancouver, local leaders were responding to diversity rather than to immigration. In Toronto the issue was framed as an immigration issue because of the Toronto regime's determination to develop new relationships among the three levels of government in immigration policy. The City of Toronto framed its response to its changing ethnoracial demographics as a response to 'immigration' to assert its role in this constitutionally defined area of responsibility.

When we compare Toronto to Markham, Brampton, and Mississauga, we also see the independent importance of local-governance arrangements (including amalgamation) to the way in which local leaders respond to changes in national infrastructure. In Toronto, the scope of the governance regime, combined with the catalyst of amalgamation, led local leaders to challenge upper levels of government in the face of changes in the national infrastructure that were detrimental to the city's well-being.

The Toronto case also reveals the interdependence of local pooling arrangements through urban regimes and capacity building in the intergovernmental system as a whole. Toronto's regime was able to effect change in the intergovernmental system because of its powerful local regime coalition. However, at an even more fundamental level, past decisions of upper levels of government influenced the level of resources that the Toronto regime was able to bring to bear on the intergovernmental system as well as who was included in the governance arrangements. For instance, past funding decisions by upper levels of government in the immigrant settlement and other related sectors concentrated resources in the City of Toronto and influenced local-governance arrangements there. Thus, when the intergovernmental context changed, it provided incentives for the mobilization of the strong civil society in Toronto that upper levels of government had contributed to building. Toronto's regime then successfully lobbied for changes to Canada's national infrastructure. Toronto has become a partner in several intergovernmental agreements and is now governed by a City Charter.

Toronto's ability to respond to immigrants has been enhanced by how the issue has been framed. Local leaders linked immigration and multiculturalism challenges to economic-development issues and prosperity in the city. Because immigrant settlement goals are important priorities in Toronto's local-governance regime, these issues were also an important part of local leaders' calls for a New Deal for Cities. Immigrants' needs and preferences were tied to this powerful regime and national movement, resulting in important new local capacity-building efforts in immigration policy – through the leadership of TRIEC – and the negotiation of an agreement on immigration between Ontario and the federal government that includes municipalities as partners. Toronto is singled out in this agreement because of the lobbying efforts of its regime. The federal and provincial governments now recognize the local capacity that the City of Toronto brings to policy discussions on immigration.

The Toronto case also provides evidence that something more fundamental may be at play. Local leaders would like greater levels of support on the part of upper levels of government. Yet to varying degrees, they also want to exercise greater authority at the local level and want to see this authority decentralized in a formal way. Calls for decentralization of decision making encompass a number of areas, but immigration and settlement policy are always on the list. Toronto's urban autonomy movement and polyscalar forms of political mobilization have been shaped by the incentives provided by the intergovernmental context but can also be situated within broader, ongoing rescaling processes (Boudreau 2006).

The intergovernmental context does not *determine* local policy outputs and urban regime dynamics. Rather, it provides a climate of incentives within which local leaders choose how to respond. Local leaders decide which issues will be included in their priorities, which goals they will pursue, who they will seek as allies in their policy endeavours, and how they will frame policy goals; they also choose strategies for attracting regime participants and maintaining the regime. That said, the intergovernmental context provides an important element in local leaders' context of policy choice.

This chapter and chapter 6 have demonstrated that local-governance arrangements are shaped both from below and from above. In their efforts to build productive public–private coalitions in multiculturalism policy, local leaders are constrained by the ethnic configuration of their municipal population and by the distribution of resources in the intergovernmental system.

Yet within these constraints, local politics matters. The importance of local agency is most evident in Toronto. The city was able to overcome the challenges associated with collective action in a multiracial context because of the presence of exceptional leaders in the public and private sectors. Then, when local leaders realized that local pooling was insufficient to meet their policy goals, they broadened their base of regime support by seeking new allies, launched campaigns to educate the public about Toronto's plight, and began lobbying upper levels of government for a New Deal. These patterns of mobilization did not occur due to an opening in the 'political opportunity structure.' Rather, local leaders mobilized to challenge political opportunity structures that were closed and even hostile to local concerns. The Toronto case demonstrates just how much local agency can matter.

The final chapter summarizes the lessons of this book and discusses what we have learned about the possibility of progressive multiculturalism policy at the local level. What explains municipal responsiveness to immigrants and ethnocultural minorities in the GTA and Greater Vancouver? What theoretical lessons can we draw from this study?

8 Municipal Multiculturalism Policies and the Capacity to Manage Social Change

Canadian municipalities are playing a new and largely unacknowledged role in immigrant settlement and integration. As municipal governments are at the forefront of social change, this book suggests that municipalities in immigrant-magnet city-regions can play an important role in fostering what Mario Polèse and Richard Stren (2000) call 'socially sustainable growth' by steering the development of civil society in the direction of social inclusion and interethnic harmony (2000, 16). Municipalities in these highly diverse city-regions can do so through a variety of local multiculturalism initiatives.

Municipalities can influence the direction of change in civil society in the same way as upper levels of government – that is, by funding community organizations that foster interethnic harmony and encourage interethnic equity. But there are also some actions that only municipalities can take. For instance, only they can serve as public and democratically elected bridges among immigrant groups as well as between immigrants and the long-standing population. They can also lead by example in the community by adapting their own corporations to ensure diversity. Furthermore, because of their ability to collaborate with the local community, municipalities have the potential to be innovative in their responses and to tailor multiculturalism policies to the local context.

The variation in levels of municipal responsiveness documented in this study indicates that not all municipalities actively embrace Canada's countrywide model of official multiculturalism. Indeed, some municipalities may be contributing to *dis*harmony by failing to respond to social change and by fostering a sense of exclusion from local governance among immigrants and ethnocultural minorities. Moreover, since municipal governments are relatively weak, they may not be able to direct the evolution of civil society and adapt their corporate structures as effectively as they might if they had more resources.

Canadian municipalities fall into a threefold typology of municipal responsiveness – 'responsive,' 'somewhat responsive,' and 'unresponsive.' Municipalities vary in the extent to which they are *comprehensive* in their responses to immigrants and ethnocultural minorities, as well as in their *policy styles* – they may be *proactive, reactive*, or *inactive* in the multiculturalism policy field.

How does one define responsiveness in an ethnically diverse context? Local leaders of settlement and ethnic organizations perceive the normative framework of Canada's official policy of multiculturalism as 'responsive' to the concerns of immigrants and ethnocultural minorities. They do not consider a laissez-faire public role in ethnocultural relations sufficient, and they expect local public officials to actively address cultural barriers to services and to participation in municipal decision making. In other words, in Canadian cities, immigrants seem to consider the *multicultural citizenship framework* equitable. Municipal multiculturalism policies are resources for immigrants in the integration process.

In the absence of a clear provincial mandate, one might expect variation in municipal approaches to managing diversity; even so, the wide range in responsiveness among the eight cases is somewhat puzzling in light of the constraints on municipal governments' resources imposed by legal frameworks that limit how they can raise revenue. Furthermore, as the lowest level of government in a fragmented system, municipal governments find themselves in a position where they must compete for residents and business investment (Peterson 1981). Neither Ontario nor British Columbia mandates a municipal role in multiculturalism policy, nor do these provinces provide grants to municipalities to accommodate and manage multiculturalism. Thus municipal governments in both provinces are acting without a clear provincial mandate in this area and without additional public resources. They are very much on their own in multiculturalism policy making.

Why do some municipalities nonetheless choose to adapt their services and governance structures to accommodate immigrants and ethnocultural minorities? How do they develop the capacity to respond to such dramatic change in the ethnic composition of their populations? This final chapter begins by summarizing the theoretical lessons of this book. It then proposes a framework for advancing the comparative study of the impact of immigration on the municipal governance of multiculturalism. It concludes by offering lessons for the multiculturalism policy maker.

Towards a Comparative Theory of the Impact of Immigration on City Governance

A multitude of factors influence a municipality's responsiveness to immigrants and ethnocultural minorities. Alan DiGaetano and Elizabeth Strom (2003) suggest an integrated approach to governance that incorporates the three dominant approaches to the study of comparative politics – structure, culture, and rationality. This book, too, proposes an 'integrated approach' – albeit a somewhat differently integrated approach – to the comparative study of multiculturalism policy making at the local level. To understand the complexities of the politics of multiculturalism in Canadian cities, we must combine the insights of urban regime theory, a social diversity perspective, and structural factors, including both institutional perspectives and global rescaling processes. Together, these theoretical ideas incorporate structure, culture, and rationality into urban analysis.

Urban regime theory has been the chief analytical lens through which the complexities of multiculturalism policy making at the local level have been explored. This theory draws one's attention to the important roles played by leadership, relationships among local leaders, and local resources. Though some have questioned the applicability of this theory to Canada, the urban regime concept is amenable to an integrated approach to the study of urban governance. At root, regime theory is a rational choice model of political behaviour in that it views local politics as a problem of collective action. According to regime theorists, the fundamental question in urban politics is how to achieve and sustain cooperation to produce the capacity to develop and implement local agendas. Building on the pioneering work of Marion Orr (1999), we have seen that ethnospecific social capital matters to whether an immigrant group will be included in urban governance arrangements. Moreover, the ethnic configuration of a municipality structures the development of social capital and the intersection of this form of capital with economic capital, besides shaping new political cleavages.

The urban regime concept also incorporates elements of a structural, political economy approach to local power structures through its contention that there is a systematic bias towards including the business community. This bias arises owing to the business community's disproportionate share of material resources in local communities and its ability to offer selective incentives to maintain cooperation. For a full understanding of policy and governance dynamics, we must also place our analyses of the local-level politics of immigration in an intergovernmental context as well as a global one (Léo 1998).

An Urban Regime Framework

In chapters 4 and 5, a framework was developed for looking at the eight cases through an urban regime lens. The factors to which urban regime theory draws one's attention are indeed important to local capacity building in multiculturalism policy. The two most *responsive* municipalities – Toronto and Vancouver – have built lasting regime coalitions that have absorbed multiculturalism policy concerns. *Somewhat responsive* municipalities vary in the extent to which lasting public–private relationships have emerged in response to changes in the ethnocultural demographics of their populations. In the two most responsive of the somewhat responsive municipalities in the sample – Markham and Richmond – socially productive relationships have fostered positive race relations between the largely Chinese immigrant community and the long-standing (largely White) community. Productive coalitions in specific policy arenas have also developed in Surrey and Coquitlam at the departmental level. In Mississauga and Brampton – the *unresponsive* municipalities in the sample – immigrant and ethnocultural minorities do not seem to be represented in the cities' governing arrangements.

A number of factors to which the urban regime concept draws our attention seem to be correlated with municipal responsiveness to immigrants and ethnocultural minorities. First, in responsive municipalities, local political leaders and civil servants acknowledge a municipal role in immigration and settlement even though this role is not mandated provincially. In other words, *political leadership* matters to municipal responsiveness. As Toronto councillor Kyle Rae put it, his city's role in diversity management and in immigration and settlement policy has been 'generated by activist councilors [who] push the envelope because of need' (Rae 2004, interview).

Another factor that influences municipal responsiveness to immigrants and ethnocultural minorities is *how civil society is organized*. A well-organized and well-resourced immigrant settlement sector enhances the possibility for strong leadership in immigrant settlement and other multiculturalism policy goals. For instance, municipal public officials can draw from social networks and from the expertise of the community in policy development in their efforts to engage immigrants and ethnocultural minorities in municipal affairs. Also, a well-organized immigrant community can pressure the municipality to respond and can hold the municipality accountable for its failure to implement policies effectively.[1]

The *organization and orientation of the business community* also matters to municipal responsiveness. In responsive municipalities, the business community acknowledges the importance of immigration to local economic well-being and is willing to pool resources to help the municipal government adapt its services and governance structures. In some cases the immigrant community constitutes either a new business establishment or important developers in the municipality.

Also affecting municipal responsiveness to immigrants and ethnocultural minorities is how multiculturalism policy goals are *framed*. Framing determines who participates in local governance. When immigration is tied to economic-development objectives – as in Toronto and Vancouver – the municipality is more likely to include immigrants and ethnocultural minorities in local governance. In chapter 7 we saw how tying immigrant settlement concerns to the broader New Deal for Cities policy frame led to a greater level of responsiveness to immigrants and ethnocultural minorities in the GTA and also broadened participation in the regime. The importance of framing to municipal responsiveness to immigrants and ethnocultural minorities underscores the significance of both local leadership and political agency. Local media can also play an important role in framing the multiculturalism issue and in mediating local community debates over multiculturalism.

A Social Diversity Interpretation

In chapter 6 we saw that a municipality's ethnoracial configuration contributes to an explanation of the likelihood that a municipality will respond to social change by adapting its services and governance structures to accommodate immigrants and ethnocultural minorities. Bifurcated, biracial municipalities are more likely to be responsive to ethnocultural change than highly heterogeneous, multiracial municipalities. There are several reasons for this. First, the *collective action problem is more easily overcome in bifurcated municipalities*. In addition, a dynamic of community *backlash* followed by a counter-reaction from the immigrant community is more likely to develop in municipalities where there is a large concentration of a single ethnic group. In other words, in biracial municipalities it is more likely that the community will agree that there is a race relations problem. Finally, it is easier for local leaders to respond to a single immigrant group than to a multiplicity of groups.

Nevertheless, one finds that while the ethnic configuration of the municipality matters, so does the *distribution of resources within the municipality and among ethnic groups*. Biracial municipalities in which the ethnocultural minority is Chinese were more responsive than those in which the ethnocultural minority was South Asian. It seems that in municipalities where Chinese immigrants predominate, there is also a powerful Chinese business community that is willing to facilitate municipal responsiveness by donating or pooling its resources.

The ethnic distribution of resources in the municipality also affects how the long-standing community perceives large-scale immigration. In several biracial municipalities with a concentration of Chinese immigrants, we found that the ability of the Chinese business community and of certain wealthy Chinese residents – for instance, in Vancouver's upscale neighbourhood of Kerrisdale – to redefine or challenge the cultural norms of the community or neighbourhood led to resistance to the changes and backlash from some members of the long-standing community.

Also, in the municipalities with large Chinese immigrant populations, an extensive network of Chinese-specific institutions has developed. The Chinese community seems to possess a great deal of intragroup social capital. In contrast, in Surrey – the biracial municipality in the sample with a large South Asian population – the immigrant community seems to be more divided. Furthermore, Surrey's business community does not seem to be pressuring the municipality to adapt its services.

Why do Chinese immigrant communities seem to have higher levels of social capital than South Asian immigrant communities? Clearly, Statistics Canada's category 'South Asian' is too imprecise. Studies will be need to be conducted to explore whether some immigrant communities have greater levels of social capital than others, and why, and to study the intracommunity dynamics of immigrant communities. Immigrant communities that fail to overcome the first-order collective action problem will not be included in urban regimes.

Backlash against immigration is less likely in heterogeneous municipalities – those in which the perception is missing that a single immigrant group is redefining the municipality's cultural norms. Together, all of the above factors suggest that one can expect action in multiracial municipalities only where political will exists alongside strong leadership.

Urban Regimes, National Infrastructures, and Regime 'Spillover'

We saw in chapter 7 that the intergovernmental context also matters to urban regime dynamics. Its effects, however, are mediated by existing urban-governance arrangements. We can incorporate the intergovernmental system into our understanding of regime dynamics by examining how policy decisions by upper levels of government provide *incentives for mobilization* as well as how those decisions *affect the distribution of resources* at the local level. Furthermore, achieving certain locally important policy goals requires the cooperation of upper levels of government. In that sense, upper levels of government are potential regime partners.

In B.C., a general climate of respect for municipal autonomy, a relative lack of downloading, and the province's willingness to address multiculturalism concerns all meant that a locally based autonomy movement did not emerge in the city-region. The province supports the City of Vancouver's economic-development aspirations and its focus on attracting investment from the Pacific Rim. Indeed, in the 1980s two mayors of the City of Vancouver who supported the local Pacific Rim consensus rose to become premier in the 1990s and early 2000s.[2] The need to develop economic ties with Pacific Rim countries is evident to a broad group of public officials at the provincial and municipal levels as well as to local leaders in civil society.

In Toronto the intergovernmental system has structured the choices of local leaders but has not determined the nature of local governance. In the mid-1990s the provincial government's decision to disentangle services and reduce social spending as well as to amalgamate Metro Toronto drove local regime leaders to reach a consensus on the need for greater autonomy for Toronto. A powerful and extensive network of relationships in that city converged around the concept of a New Deal for the city. The Toronto regime's immigrant settlement goals are tied to this new frame. These networks spilled over to multiple scales, including those of the GTA, cities in other provinces, and higher levels of government. In TRIEC we saw evidence of a regional subregime and subsequently a multilevel one. The results of advocacy efforts of leaders in this powerful subregime were apparent when a new Canada–Ontario Immigration Agreement (2005) was negotiated so as to include Toronto as a partner.

The findings in chapter 7 have important implications for urban regime analysis. Urban regime strategies and dynamics in Toronto point to a central deficiency with the current urban-regime concept: its theoretical focus on the political economy of cities at the expense of the intergovernmental context. For urban regime theorists the central puzzle at the local level is how local leaders are able to achieve the capacity to govern. Local governments in both Canada and the United States face enormous resource constraints, given their junior position in the intergovernmental system. Urban regime theorists point out accurately that local politics matter because, in a fragmented urban system, it is local coalition building that creates the capacity to govern at this level. Nevertheless, as the Toronto case suggests, one cannot take for granted that the desire to build capacity at the local level will be limited to municipal–civil society relationships and that local leaders will be satisfied with the resources available through such governance arrangements. The Toronto case suggests that one cannot take for granted that local leaders will accept the existing distribution of legislative authority and fiscal resources.

Canada's urban scholars have made little use of regime analysis to understand local policy making. The dominant focus has been on the intergovernmental context – on the provincial–local relationship in particular. The highly limited nature of municipal autonomy in Canadian municipal systems would lead one to expect that those municipal leaders who want to develop the capacity to address local challenges would turn to upper levels of government for help instead of developing urban regimes. Furthermore, given the constrained position of municipal governments in the Canadian intergovernmental system, one might deduce that elite leaders in civil society – in the business community in particular – have little incentive to develop ongoing relationships with municipal governments. This intergovernmental reality has led well-known Canadian urban scholar Andrew Sancton (1993) to predict that 'the concept of urban political regimes is unlikely to be of much assistance in analyzing Canadian urban politics because massive provincial influence makes business involvement in such regime politics unnecessary' (1993, 20, in Urbaniak 2003, 11, and 2005, 6).

Since the early 1990s the empirical terrain of local politics in Canada has shifted dramatically. Andrew Sancton (1993) may have been correct in his assessment of the general orientation of local business communities in 1993. The Canadian literature now portrays municipal policy making differently; in doing so it is drawing our attention to the importance of governance arrangements to local capacity building and representation (Andrew 2001; Andrew, Graham, and Philips 2002) and to the multilevel dimensions of local policy making (Leo 1997; Leo 2006; Young and Leuprecht 2006). This book suggests that the urban regime concept adds to our understanding of the relationship between urban governance arrangements and the intergovernmental system in Canada by highlighting the systemic bias of local governance coalitions towards the business community at the municipal level. Multilevel arrangements could alter this bias by substituting public resources for private ones at the municipal level, and by altering the evolution of local civil societies more generally by developing capacity in non-business sectors in local communities. But to the extent that urban and multilevel governance arrangements are developing as a consequence of global economic processes of rescaling that many urban scholars have identified, a political economy perspective of governance has become more valuable. As local leaders perceive a greater need to compete and become more assertive, urban-regime building will become an increasingly common strategy of local-capacity building. Also, upper levels of government may become more willing to support local initiatives by developing a variety of multilevel arrangements. Furthermore, the two processes are related.

This study indicates that a focus on the formal place of municipal institutions within the intergovernmental system misses important ways in which local factors and politics are important. Municipal leaders in Canada do indeed participate in the formation of urban regimes. That the Canadian literature has long focused on the intergovernmental context highlights a serious deficiency in urban regime theory – it fails to address the strong possibility that local leaders will turn to upper levels of government to address their resource needs.

The Toronto case highlights the *interdependence* of the two capacity-building strategies. In Toronto, local leaders built a strong urban regime with broad intersectoral participation not only to pool resources across sectors but also to create the political momentum to extract a greater share of resources from upper levels of government. One might even argue that the participation of civil society and high-profile business and financial institutions – including the TD Bank – has been the single most important factor in Toronto's success at achieving a New Deal with upper levels of government, albeit a limited one (Feldman 2006).

In *A New City Agenda* (2004), John Sewell advocates local innovation and capacity building – what this study calls regime building – as a strategy for demonstrating to upper levels of government the powers a city should have: 'As city government becomes clear about the powers it should have to improve the quality of life for citizens, and as it *takes steps to implement its pursuit of increased authority and effectiveness*, momentum will inevitably develop' (2004, 92; emphasis added). Developing multiculturalism policies and other initiatives to facilitate the integration of immigrants in Toronto without the express authority to do so or the necessary financial resources from the province is one of the best examples of this strategy in practice. The Toronto case demonstrates that when upper levels of government fail to take leadership on important policy questions such as immigrant settlement and multiculturalism policy, and when municipal governments push their limits to fill the gap, community agencies and citizens begin to support a rescaling of statehood to the municipal level, because it is accessible and is considered responsive to their concerns. One of the most striking findings in this study is that *settlement leaders in Toronto support decentralization of responsibility for settlement to the municipal level.*

Business participation in the Toronto regime is especially puzzling in light of Toronto's reputation, among governments in the region, for progressive left-wing politics. The simplest explanation is that local business leaders recognize that 'place matters' and that upper levels of government have not done enough to address the place-specific consequences of their policy decisions. In Toronto the immediate impacts of downloading and disentanglement enabled the New Deal idea to resonate with a broad cross-section of the city's leaders. These measures in tandem broadened the business community's ideas concerning what constitutes a favourable economic climate. In the absence of leadership from upper levels of government, municipal officials and leaders in civil society built urban regime coalitions.

Thus, institutional change can affect urban regime dynamics. Amalgamation had a powerful effect on *patterns of mobilization* in Toronto. It seems to have broadened participation in governance; and as Toronto councillor Shelley Carroll put it, it also seems to have 'turned suburbanites into activists' (Carroll 2004, interview). It also led to an *unprecedented level of cooperation among immigrant groups in the city* under the banner of New Voices for the New City. Toronto is the only city in the sample in which a strong urban empowerment coalition emerged; it is also the only city that was amalgamated. This suggests that amalgamation played an especially important role as a catalyst for ongoing 'rescaling' through political mobilization at the local level.

So it is possible that the changes in the intergovernmental system in Ontario in the 1990s led to the development of the first 'regime' in Toronto after amalgamation. More generally, one might connect the development of urban regimes in Canada to processes of rescaling as the municipal level becomes more important as an arena of growth politics. Vancouver's proactive Pacific Rim initiatives began in 1980 when Mike Harcourt was mayor, and they continue to this day. Furthermore, there is now evidence that, in cooperation with leaders in civil society, municipal leaders in Vancouver have just recently begun to carve out a role in 'immigration policy' instead of limiting the municipality's role to responding to 'diversity.' It seems that local growth coalitions in both cities have become more assertive – a change that could be related to broader global economic processes.

The capacity-building strategies of the leaders in one case – the City of Toronto – illustrate most clearly the *importance of local agency* to local multiculturalism policy making. Vancouver's responsiveness to immigrants and ethnocultural minorities is perhaps more expected than Toronto's in light of the dominant theoretical tenets of the urban politics literature described above. In Vancouver the connection between economic-development objectives and immigration is obvious, given the importance of Chinese immigrant investors and developers to its economy. What is more, the province supports the city's efforts to develop the city as a Pacific gateway. Members of the largely Chinese immigrant community are key players in the city's growth coalition. By actively fostering connections between Vancouver and cities in the Pacific Rim, Vancouver's political leaders are behaving as one would expect in the growth machine literature.

Multiculturalism policy development in Vancouver seems to be at least partly a by-product of the city's economic-development objectives. Local political leaders have decided to foster economic and social linkages with the Pacific Rim, and they consider multiculturalism policies a natural complement to this objective. Predictably, the new community dynamics have been managed within the Pacific Rim consensus among local elites. For instance, ethnocultural tensions relating to cultural preferences in housing have created incentives for local leaders to adopt multiculturalism policies, since those tensions challenge the dominant consensus concerning the desirability of growth – the consensus of the local 'growth machine.' But at the same time, ethnic conflict has helped push the issue of race relations onto the municipal agenda.

Toronto's responsiveness is less easily explained. We know that *resources* in Toronto provide a significant part of the explanation. However, we also saw the power of exceptional *local leaders* and of broad cooperation in the face of a complex collective-action problem. In this way, Toronto is *exceptional* among the cases. Toronto is the only city in the sample with a highly heterogeneous, multiracial ethnic configuration that has been responsive to its immigrant and ethnocultural minority population. Immigrant groups in Toronto have managed to overcome the complex first-order collective-action problem that is inherent in Toronto's ethnoracial configuration as well as in the intersection of its ethnoracial configuration and community resources, even though Toronto's ethnic configuration is much more complex than Mississauga's or Brampton's – the two other multiracial municipalities in the sample.

Toronto's exceptionality emerged again in chapter 7. In response to downloading by upper levels of government and a forced municipal merger, local leaders pulled together the city's networks to challenge upper levels of government. These patterns of mobilization developed in the context of a closed and even hostile provincial political-opportunity structure (Horak 1998). The Toronto case demonstrates just how much local agency can matter. Local leaders resisted changes in the intergovernmental system through regime building. Faced with a hostile province, they went against constitutional convention and lobbied the federal government. Thus, the Toronto case illustrates how resources matter to the ability of local leaders to mount successful resistance.

The strength of local leaders, the resources of those leaders, and the extent to which leaders support the decentralization of responsibility to the City of Toronto are all unique to the city. The city's main newspaper, the *Toronto Star,* is a crusader on behalf of the city. Many Toronto city councillors are activist, notwithstanding their institutional and fiscal constraints. The leadership and capacity in the immigrant settlement sector is unparalleled in Canada. Even the business community – as represented by the Toronto Board of Trade – uses activist strategies on behalf of the city to advocate for a New Deal for Toronto. Leaders of some of Canada's most powerful financial institutions, such as the TD Bank, support a New Deal. Extremely powerful leaders, such as philanthropist and entrepreneur Alan Broadbent, are willing to commit their personal resources to the cause of empowering Toronto. Broadbent, a leader of Toronto's New Deal movement, is also personally committed to the goal of increasing the city's capacity to integrate immigrants in an equitable way. Dominic D'Alessandro, President and CEO of Manulife Financial, another financial powerhouse in Canada, is one of the chairs of TRIEC.

Toronto is also exceptional insofar as it is the only city in the sample in which leaders in the immigrant settlement sector expressed support for a stronger municipal role in immigrant settlement and multiculturalism policy development. Like leaders in other sectors, they support a decentralization of the Canadian federation that would give the City of Toronto more autonomy. Meanwhile, power and 'statehood' have been rescaled to a certain extent to informal networks in civil society – to the city's urban regime.

Toronto provides an example of how political agency limits our ability to generalize in the empirical world. As such, it warrants further attention, which might shed light on the circumstances in which local leaders can overcome constraints – both those in the immediate social context and those imposed by the intergovernmental system. The case could tell us something about the potential for progressive politics at the local level. But in the absence of a strong local (i.e., municipal) state, the arrangements developed in civil society are – as urban regime theory suggests – biased towards the business community. Toronto was able to overcome its constraints in part owing to perceived economic imperatives on the part of the business community and the willingness of powerful business leaders to contribute personal resources to sustain cooperation and contribute to capacity building in immigrant settlement. Immigrant integration is a priority because the economic case was made for this policy goal through the process of regime building.

Furthermore, changes in the global economic system may be pushing local leaders to organize in Toronto. There may be an 'invisible hand' at work – that is, the hand of a changing capitalism that is rescaling statehood. From this perspective, changes in the intergovernmental system threatened Toronto's position in a highly competitive global system of cities. Certainly the literature on rescaling would tie the orientation of local leaders and patterns of political mobilization to global changes in capitalism and a corresponding rescaling of statehood. Indeed, there is evidence that in Toronto something fundamental is at work: local leaders are not simply asking upper levels of government to develop place-sensitive, urban policies. Rather, there seems to be a fundamental desire to decentralize power to the local level.

Local leaders in Toronto perceive the city as the fundamental unit of economic, political, and cultural organization. New governance arrangements have developed at multiple scales – and at 'jumped scales,' in the case of the Charter 5. These types of coalitions are consistent with what Neil Brenner (2004) describes as emerging 'state spaces' in Europe. Though the activities of the Charter 5 were not sustained, a lasting form of 'new regionalism' has developed in the GTA. Some of these developments are what one would expect now that statehood is being 'rescaled' and the metropolitan scale is becoming more important. Julie-Anne Boudreau (2006) has tied these 'polyscalar' forms of political mobilization to global rescaling processes.

Yet it is difficult to draw firm conclusions about globalization's impact on the two city-regions in this book. Globalization has not affected patterns of governance at the city level in a uniform way. Provincial–municipal relationships in B.C. are more stable and historically more respectful of municipal autonomy than those in Ontario. To the extent that globalization is changing patterns of mobilization in Canada's city-regions, its effects are being mediated through past and existing institutional arrangements, both formal and informal. Changes in the intergovernmental context are the proximate structural cause of the differences between the two city-regions.

Thus, we are seeing political mobilization at multiple scales. To a certain extent, we are also seeing a state of territorial flux in the institutionalization of political authority, which is consistent with theories of rescaling. But based on the Canadian evidence, it is difficult to generalize about rescaling processes. Local regimes and the intergovernmental context are mediating the effects of global rescaling.

Local leaders acknowledge that cities are now competing with one another on a global scale. One can identify strong 'growth machines' in the two central cities. However, since growth machines are standard in the United States and Canada, we would need to examine both of these cities through time to see how new these processes are. Local leaders in both Vancouver and Toronto seem to govern for growth, as the growth machine literature would predict. What may be new is the assertiveness of local leaders, their orientation to the global scale, and the emerging consensus that cities are the engines of economic growth in the global context. Also new is the extent to which global processes of immigration are central to growth agendas.

In the absence of government intervention, the degrees of freedom for progressive local policy making may be narrow. Local governance arrangements in support of progressive politics may happen only on the margins. Multiculturalism policies complement growth objectives. What are the implications for the direction of both discourse and action in multiculturalism policy making? It could be that multiculturalism policies will address the concerns of immigrants and ethnocultural minorities only to the extent that they are crucial to maintaining the city's economic competitiveness. However, this book also illustrates the potential of local growth agendas to unite highly diverse populations. We need to look for patterns cross-nationally. How, if at all, are immigration and multiculturalism policy responses contributing to rescaling processes? Are cities competing for immigrants? Are they actively trying to court skilled immigrants? What does this mean for progressive policy making? When it comes to urban governance, which immigrants and ethnocultural minorities are 'in' and which are 'out'? Future research should explore the extent to which immigrants are achieving greater levels of inclusion in urban governance arrangements, because attracting and settling them successfully supports neoliberal growth agendas. We must also know more about the extent to which local governance arrangements support an equitable local multicultural citizenship.

A Cross-National Framework of the Urban Governance of Multiculturalism

These questions all point to the value of developing a cross-national, integrated framework for the urban governance of immigration. Researchers could address the following questions. Has the attraction and retention of immigrants through multiculturalism initiatives become a global strategy in cities' efforts to compete? To what extent have innovative governance arrangements emerged in immigrant settlement to complement the economic-development (i.e., growth) objectives of urban regimes? To what extent do immigrants and ethnocultural minorities benefit from the connection between economic-development goals and successful immigrant settlement? What are the motivations for multiculturalism initiatives, and what types of exchanges do local regime participants make in this area? To what extent do multiculturalism policies at the local level in Canada reflect Canada's national policy context and its commitment to official multiculturalism?

The ways in which some local leaders in Canada tie immigration and successful immigrant settlement to growth objectives are not simply a matter of implementing national policy, since this book identifies considerable variation among local multiculturalism initiatives *within* Canada. Multiculturalism objectives make it onto local agendas through political processes. Thus it is better to view official multiculturalism, along with the norms and ideas that underpin it, as a resource for would-be regime entrepreneurs at the local level. A 'national multiculturalism infrastructure' is part of the 'national infrastructure' that shapes local policy making in Canada. As Irene Bloemraad (2006) found in her comparative study of immigrant integration in Canada and the United States, the Canadian federal government's multiculturalism policies and initiatives confer both material and symbolic benefits to immigrants that are important to successful settlement.

In chapter 6 we saw that the development of urban regimes is shaped by changing patterns of social diversity. Here, too, one can see the potential for a cross-national framework of research. If the ethnic configuration of a municipal unit is significant to its policy outputs, regime development, and form of political pluralism in the ways described above, one should find patterns cross-nationally as well. The degree to which Canadian municipalities vary in their adoption of multiculturalism policies or frameworks suggests that the national policy context is not entirely decisive. The application of Hero's perspective to Canada is, of course, based on this premise. His study's findings add yet more weight to the findings discussed here, which have been based on a limited sample of eight cities. The two studies as well as the 'ethnoburb' literature point to the value of a cross-national, integrated social diversity perspective.

There is theoretical work to be done to develop this framework. Most fundamentally, the categories of ethnic configurations developed in this book differ from Hero's in ways that reflect the available data in Canada, as well as this study's focus on high-immigration centres. Furthermore, historical differences in race relations and patterns of immigration in Canada and the United States have the potential to confound comparisons between the two countries. Nevertheless, in their comparative study of policy making in Canada and the United States, Keith Banting, George Hoberg, and Richard Simeon (1997) argue that patterns of social pluralism in the two countries are converging. According to them, 'new' sources of diversity (feminist and multicultural) represent a point of *societal convergence* between Canada and the United States, in contrast to earlier elements of *societal divergence* as represented by 'language, race, and class' (1997, 399). Ultimately, we will want to know how new sources of social pluralism intersect with these older divisions to shape local political dynamics and governance arrangements.

Hero (1998) suggests that one of the main contributions of the social diversity perspective is that it offers a clearer and more precise explanation of change than the dominant theoretical approaches to the study of state politics in the United States (1998, 10). As immigration continues to change the face of the United States and Canada, we may see more convergence in race relations in the two countries and thus also more opportunities for theoretical cross-fertilization.

To what extent is Hero's typology compatible with the typology of ethnic configurations developed in this book? Using Hero's conceptualization, the two categories – multiracial and biracial – would be subsumed under the category 'bifurcated,' in the sense that both types include high proportions of ethnoracial minorities. What evidence is there to suggest that all of the Canadian municipalities discussed in this book should be considered bifurcated? 'Multiracial' municipalities are characterized by a *limited pluralism with some competition* in suburban municipalities and by a highly *dynamic form of pluralism* in Toronto. As such, they share some features of Hero's 'heterogeneous' political subunits, which have moderate levels of ethnoracial minorities and high proportions of 'White ethnics' and which are characterized by a 'competitive pluralism.' Thus, though the biracial and multiracial categories focus on 'visible minorities' to the exclusion of 'White ethnics,'[3] one still finds competition among groups (albeit in a limited form in suburban municipalities) and a dynamic form of 'competition' in Toronto.

To a certain extent this finding calls into question Hero's bifurcated category, which groups together all ethnoracial minorities. In other words, Hero's categories do not adequately capture the possibility of competition among ethnoracial minorities.[4] In addition, from a theoretical perspective it is unclear why a combination of what Hero calls 'White ethnics' and ethnoracial minorities would lead to greater competition than a mixture of ethnoracial minorities and what he considers 'Whites.' Together, however, the *largely limited pluralism in the multiracial suburban municipalities* and the *limited pluralism in biracial municipalities* suggest that his hypothesis of the type of pluralism one would expect in bifurcated municipalities is confirmed to some degree.

According to Hero's conceptualization, a bifurcated context 'leads to hierarchical or limited pluralism' owing to the history of race relations in the United States. The form of pluralism that characterizes bifurcated locales is, in his words, 'historically manifested in various legal and political constraints ... Despite major social and political change during the last generation, this condition continues, albeit in modified form' (1998, 16). This inference seems to have been developed with the historical experience of African Americans (and perhaps also Hispanics) in mind, which limits the applicability of this category to Canada. Furthermore, unless we assume that all immigrant racial minorities will experience the same discrimination and hierarchy as an arguably exceptional racial-minority group – African Americans – it is unclear why one would expect limited pluralism also to exhibit hierarchy.[5] From a theoretical perspective, though it seems logical to expect a more limited form of political pluralism in less diverse locales, it is unclear why one should expect a hierarchical pluralism in many of the American locales that Hero would consider bifurcated.

Hero acknowledges that the historical experience of minority groups differs, but he also argues that 'there is enough similarity within groups and enough differences across groups as delineated to support the designations and arguments made' (ibid., 8). He explains that he made the choice to oversimplify ethnic categories for the 'sake of clarity and parsimony' (ibid., 151).

The extent to which hierarchy exists depends on *the power and resources* of the ethnoracial minorities in the community. In addition, the growing literature on 'ethnoburbs' supports the finding of this study that one must incorporate a political economy perspective to understand changing community dynamics as well as the ways in which immigrants' resources structure community reactions and debates (Li 2007). The concentrated settlement of *highly powerful* immigrants capable of changing the cultural face of a locale seems to matter in many ways; for example, it generates a reaction on the part of long-standing residents (i.e., it intensifies feelings of cultural threat), and it affects the immigrants themselves (i.e., they have more power and resources to mobilize and to influence policy making).

There is arguably a greater cultural distance (including language differences) between many immigrant groups than between long-standing residents in the United States and African Americans. Hero notes that the states that have received a large number of Asian and Hispanic immigrants are the bifurcated states that adopted English-only measures in the 1980s and 1990s (Citrin et al. 1990, in ibid., 109).

In the two Canadian 'ethnoburbs' discussed above – Markham and

Richmond – we saw that the reactionary debates about English in Asian malls, which triggered community debate about multiculturalism, ultimately led to greater *responsiveness* on the part of the municipality. One might argue that this is consistent with Hero's finding that though bifurcated and homogeneous states are more likely to adopt official-English measures, bifurcated states generally tend to produce better policy outcomes for ethnoracial minorities than either homogeneous or heterogeneous states.

It is notable that though formal official-English initiatives were not put forward by the long-standing residents, there was conflict over the lack of English-language signage in Asian malls. In addition, except for Toronto, biracial municipalities have been more responsive than multiracial municipalities to ethnocultural diversity. These findings suggest that *ethnic concentration* and the *economic power* of the immigrant community both matter. Absolute and relative numbers are only one type of resource in a community. Economic and social capital matter as well.

A political economy perspective may allow researchers to connect common global processes – migration of human and financial capital – to specific settlements in countries. For instance, the 'ethnoburb' literature connects new forms of suburban settlement to global processes. Such settlements exist in many countries. Thus, in some respects one might expect a convergence in the types of social diversity that countries are facing in light of common global processes. That said, the politics of multiculturalism are affected by historical experiences with 'race relations' as well as by patterns of immigrant settlement. Documented common cross-national experiences with immigration suggest the value of comparison.

This book raises questions with respect to what one might expect of institutional goals and public policies in contexts with varying types of social diversity. Hero suggests that in heterogeneous environments 'there is a need to arbitrate or broker social heterogeneity and complexity,' and that in bifurcated environments, 'government is expected to interfere little with existing stratified conditions, themselves the product of institutions and social relations historically defined in racial/ethnic terms' (ibid., 20). Except in the City of Toronto, local leaders in Canada's heterogeneous *multiracial* municipalities were unresponsive to immigrants and ethnoracial minorities; they also failed to develop informal governance institutions capable of bridging the public–private divide to broker social change. In Canada it was the *biracial*–bifurcated locales that were more likely to intervene to 'broker social heterogeneity.' Vancouver intervened proactively, but has also had to intervene reactively to broker heterogeneity. Richmond and Markham were pressured to intervene in reaction to race relations crises as well as to pressure from socially and economically powerful Chinese immigrant communities.

It is possible that Hero's expectations regarding institutional and policy purposes hold for bifurcated municipalities in which the dominant minority is African American. If so, the hierarchical pluralism that Hero observes would be structured by a historical legacy of stratified social conditions and past institutions. However, these conditions do not seem to apply to either Canadian municipalities or American 'ethnoburbs.'[6] We need to know more about the orientations of municipal governments towards managing and accommodating diversity and how those orientations are related to patterns of social diversity, to the political economy of cities, and to how these two factors intersect.

In this study the importance of the business community's orientation towards multiculturalism objectives and its composition stands out in part because of the importance of the informal arena – and therefore of civil society – to representations of immigrants in governance arrangements and multiculturalism policy development. A glaring puzzle remains in these cases: Why are so few immigrants elected to local councils? This is especially perplexing in the biracial municipalities, where the immigrant community is concentrated. Why hasn't the municipality's ethnic configuration exerted its causal effect through this arena? In Stone's (1989) seminal study, this was the primary arena through which Atlanta's Black community became part of governance there. The ability to control council was an important resource for the Black community, and the White business community needed its support in order to pursue its urban renewal agenda. This is what created the opportunities for cooperation that resulted in Atlanta's biracial regime (ibid.). Perhaps there will be greater levels of responsiveness to immigrants and ethnocultural minorities in Canada once more progress is made in the electoral arena. The democratic process is a resource that immigrant communities can access without being part of the city's economic elite. Even so, as urban regime theory teaches, electoral resources are only one resource among many.

Just as municipal institutions are embedded in other institutional contexts, so are social contexts embedded in other social contexts defined at other scales – municipal, regional, provincial, national, and international. For instance, immigrant settlement patterns are more concentrated in B.C. than in Ontario. Could this be why B.C. has been more consistently responsive to immigrants and ethnocultural minorities? Conversely, is Canada more receptive to immigration because, as a country, it is not biracial (an ethnic configuration that tends to incite backlash at the local level)?

Political theorist Will Kymlicka (2008) has described Canada as something of an exception among countries in its receptiveness to immigrants and ethnocultural minorities. According to him, two factors associated with an unwillingness to accommodate them are the concentration of a single group and an uncontrolled border. These two conditions exist at a smaller scale in biracial municipalities in the sample. As Paul Peterson (1981) emphasizes in *City Limits*, all cities have uncontrolled borders. For him, this is a key reason why one should expect city governments to focus on economic-development policies and to avoid redistribution. In his formulation, cities compete for business investment as well as for residents who will contribute to the city's economic well-being. However, another implication of cities' open borders is that local leaders cannot control immigrant settlement patterns and, ultimately, the pace of the cultural evolution of local communities. As Kymlicka (2008) observes at the country level, backlash has indeed occurred in biracial municipalities in reaction to concentration. Nevertheless, in Canada, the local responses of biracial communities to change seem to be manageable. Kymlicka's argument that Canada's ethnic configuration has contributed to public support for multiculturalism initiatives, coupled with the findings discussed here, suggests that we need to understand how social contexts at different scales are related as well as how government institutions and policies at all levels affect the local governance of immigration. In other words, research attention ought to be paid to how discourses about immigration and multiculturalism as well as policies at other scales are translated to the local level, where immigration and ethnoracial diversity are experienced on a day-to-day basis.

The case of Surrey suggests that intragroup dynamics are also important and that other forms of diversity – in this case *religious diversity* – should be incorporated into a social diversity interpretation of politics. Internal diversity within socially constructed ethnoracial communities may serve as a barrier to community mobilization as well as to the creation of institutions and services in civil society. If majority constructions of race and difference matter, then Surrey ought to be considered a biracial municipality from the perspective of long-standing residents. In other words, the long-standing community may be reacting to large numbers of an ethnoracial group as if that group is cohesive and culturally homogeneous (and thus a potential threat to the municipality's cultural norms). To develop useable categories of social diversity, we would need to understand more about how immigrant communities construct their identities and about how *majority communities* construct the social diversity of their communities in their imaginations.

Cross-national comparisons of the impact of immigration on cities that take into account the intersection of these factors could lead to the development of a powerful framework for predicting policy outcomes based on demographic change. Hero's social diversity interpretation has the benefit of providing a parsimonious explanation of the relationship between the ethnic configuration of all American states and variation in both aggregated and disaggregated measures of policy outcomes, as well as other dependent variables. His theory also has a high level of *generalizability* within the United States. However, in theory building, one must often sacrifice a certain degree of parsimony for accuracy.

If urban scholars are to develop a cross-national research agenda that compares the responsiveness of cities to immigrants and ethnocultural diversity, they must explore the extent to which common categories of political subunits and associated forms of 'political pluralism' can be developed. In this process, we might ask whether other forms of diversity should be taken into account such as *socio-economic* and *religious* diversity. In addition, we must also examine how *patterns of resource distribution* in civil society affect the local governance of immigration and ethnoracial diversity. The social diversity perspective must also incorporate a political economy perspective that takes seriously the role of the business community in urban governance. It may be necessary to sacrifice a degree of *parsimony* in order to extend the theoretical framework cross-nationally. But in return, urban scholars will be rewarded with greater degrees of *generalizability* and *accuracy*.

The social diversity perspective offers the potential to *predict* the development of new political dynamics on the basis of tracking demographic change that results from migration and immigration. The social diversity interpretation is a 'clear' and 'precise' way of theorizing change in a variety of areas of importance to political scientists (Hero 1998, 10). Its theoretical potential is even greater in high-immigration countries in which ethnic configurations are especially dynamic. Moreover, by taking an integrated approach to local governance, we can observe how governments might intervene in managing ethnocultural relations.

To what extent can the social diversity interpretation shed light on the politics of multiculturalism policy development at the local level in Canada? There is convincing evidence that the ethnic configuration of political societies matters. Building on Hero's work, this book has laid the foundation for the development of a social diversity framework through which to examine how immigrant settlement patterns have contributed to urban regime building and the politics of multiculturalism at the local level. It has developed two new categories of ethnic configurations – biracial and multiracial – and has presented evidence that many aspects of the social contexts inherent in these analytical constructs matter to urban regime building and ultimately to multiculturalism policy development.

This book has also proposed an integrated approach to studying the relationship between social diversity and regime building and demonstrated the value of looking at the findings through a regime lens. This lens was something of a microscope in chapters 4 and 5. In chapter 6 the field of view widened to place urban regimes in their social context. In chapter 7 the lens became a telescope for viewing the intergovernmental and even global spheres; that chapter also provided a fuller account of the systemic sources of power to which urban regime theory draws one's attention. This framework has provided a complete picture of the local governance of multiculturalism.

Both large-scale and smaller case studies of political subunits are valuable in theory building. Case studies have the benefit of describing the nature of political pluralism in a more accurate and convincing way; they also allow us to refine categories and explore the causal mechanisms that establish the correlations in larger-scale studies such as Hero's. The comparative case-study method has the benefit of allowing the researcher to observe how causal mechanisms operate through time (by examining sequences) and through space (by comparing cases). Together, comparative methods and large-scale statistical methods could lead to a powerful explanatory framework for understanding one of the most significant policy challenges of our time – the politics of immigration and multiculturalism in urban places.

Making Multiculturalism Work at the Local Level: Lessons for the Policy Maker

Though the multiculturalism policy challenges in the eight cities are similar, they also have place-specific dimensions. The spatial differences in multiculturalism policy challenges mean that all levels of government – including municipal governments – must tailor their policies to the local context. As Neil Bradford has put it succinctly: 'place matters' (Bradford 2002). Because it matters, municipal governments as the governments closest to citizens should be important policy players in multiculturalism policy development at all levels. They should be responsible for adapting their *own* services and governance structures to diversity; indeed, they could offer input on the needs and preferences of their diverse populations that would be valuable to provincial and federal policy makers who are interested in adapting services and governance structures.

Because of the place-specific nature of multiculturalism policy challenges, decentralizing responsibility for elements of multiculturalism policy making to municipal governments – at least in immigrant-magnet cities – may be the most effective way of achieving the goals of the federal Multiculturalism Act (1988). Funding decisions about community-based organizations and immigrant settlement programs could be made municipally, for instance. Throughout the 1990s the responsibility for immigrant settlement was increased in many provinces. Since municipalities are clearly playing a role in the immigrant settlement experience, intentionally or not, the question becomes: Should (and could) municipalities' role in this area be formalized?

At the same time, the variation documented in this study suggests that we should be cautious about how this role is formalized. Clearly, not all local leaders have the same political will to lead in this important policy area. Furthermore, municipalities vary in their capacity to play an effective role in multiculturalism initiatives. To play a more extensive and effective role in managing multiculturalism, municipalities would need additional resources from upper levels of government. Adding multiculturalism initiatives to the services funded by the property tax could fuel a backlash against immigration. Therefore, provincial or federal grants to municipalities would be needed to address the financial implications of a formalized municipal role.

Upper levels of government can continue to play a role in steering the direction of municipal responsiveness to immigrants and ethnocultural minorities by contributing to the development of civil society. For instance, Canadian Heritage and Citizenship and Immigration Canada have played central roles in funding organizations with which municipalities can pool resources to achieve multiculturalism objectives. They have contributed to rescaling processes through initiatives such as the

Metropolis Project, which strengthens local networks and policy capacity. The federal government has been an important player in developing Canada's national 'multiculturalism infrastructure.'

Nevertheless, the distribution of resources is uneven among Canada's city-regions. The bulk of settlement resources are in urban core cities, yet more and more immigrants are settling in suburbs. This unevenness must be addressed. Increased community capacity building in suburban municipalities might well increase municipal responsiveness to immigrants and ethnocultural minorities. From the federal government's perspective, this represents one way to affect municipal governance without interfering with provincial jurisdiction.

This book's findings also suggest that in the long run, multiracial municipalities may need more resources than biracial ones. In multiracial municipalities, because their populations are more diverse, it is more costly to build ethnospecific social capital by funding community organizations. Also, community engagement and bridging is more complex.

Though it might be easier to help immigrant communities build ethnospecific social capital in biracial cities, additional resources would then have to be spent on other initiatives – specifically, ones that address the possible social isolation of immigrants owing to the development of extensive separate institutions in civil society. And, governments might then have to devote resources to initiatives that manage the reaction of long-standing residents to immigrant concentration. A clear policy implication of this study is that multiculturalism policies must also address the concerns of the non-immigrant population.

This book advocates an integrated academic approach to the study of multiculturalism in cities. So as one might expect, it argues that policy responses must be coordinated across levels of government and sectors. Essentially, we must ask the same question Christopher Leo asks in his 2006 article: How can Canada develop policies that are national in scope yet also allow sufficient flexibility for local variation? Leo contends that local communities and municipalities ought to play a larger role in policy making. However, he favours flexible, informal, multilevel governance arrangements over reforms such as city charters that would formalize municipal authority (of likely only a select group of cities).

Flexibility is certainly a virtue; but such arrangements would be subject to unilateral change by not just one (the province) but two upper levels of government. Furthermore, formalizing a municipal role in multiculturalism policy making and immigrant settlement would not preclude the development of creative multilevel arrangements. Indeed, to contribute to the effectiveness of immigrant settlement and diversity management nationally, municipalities in immigrant magnet city-regions would have to be able to exchange information and collaborate with upper levels of government.

It is worth thinking about how a role in multiculturalism initiatives could be formalized without constraining local innovation. Should a broad sphere of jurisdiction in immigrant settlement and other multiculturalism-related initiatives be written into provincial Municipal Acts? Should these responsibilities be incorporated into legislatively tailored city charters in the immigrant-magnet municipalities? In other words, would legislative asymmetry be better? Should formal mechanisms be put in place to coordinate multiculturalism initiatives and to encourage exchanges of best practices at the metropolitan level in fragmented city-regions?

Imagining a strengthened municipal role both in terms of formal delegation of responsibilities and in terms of empowerment within the intergovernmental system also raises this question: How representative are the decision makers such as municipal mayors and other elected officials? As this book has laid out, local councils do not mirror the ethnoracial demographics of their populations. In the interest of supporting the development of more multicultural local democracies, Myer Siemiatycki's (2006) suggestion that provinces should extend the municipal franchise to newcomers ought to be seriously considered. This could be an essential ingredient in strengthening local democracy and in 'managing diversity' in the community. Unrepresentative councils can be disconnected from immigrant communities and unfamiliar with the challenges they face when settling in cities and accessing services. However, this book also points to the importance of a broader conception of representation, since governance also matters, not merely government's formal institutions. How civil societies develop and who is included in the various agencies, boards, and commissions in cities, how chambers of commerce integrate a diverse business community, and how voluntary organizations themselves are governed will shape whose voices are heard in local decision making. Employment equity at the municipal level is also important to governance. The benefits of more equal representation and participation are a question not only of democratic legitimacy but also of effectiveness in managing community change.

If measures to empower municipalities and to strengthen the representativeness of local councils were taken, then empowering immigrant-magnet municipalities within Canadian federalism could itself become a sort of 'multiculturalism policy.' It would strengthen the representation of ethnoracial minorities in the Canadian system of governance as a whole. Conversely, it is worth asking what the repercussions would be if Canada's largest and most diverse municipalities were to retain little power in a situation in which racial minorities were the majorities and concentrated in relatively few cities in Canada. Could strengthening the general place of municipalities within federalism be important to Canada's multiculturalism model of citizenship?

The challenge constitutes what federalism scholar Richard Simeon (2003) describes as the formula needed for a successful federation. One must simultaneously 'build out' by empowering subunits and 'build in' to the overall intergovernmental system. What is clear is that the issue cannot be overlooked. Whether formalized or not, there is no question that municipalities in Canada's immigrant-magnet city-regions play important roles in immigrant settlement and in managing profound changes in their communities. Managing international migration is one of the central governance challenges of the twenty-first century. As the democratic governments closest to Canada's multicultural communities, municipalities play a central part in negotiating multicultural citizenship in Canada.

Postscript

Canada's immigrant-magnet city-regions are highly dynamic social, political, and economic environments. Since the field research was conducted for this book, I have become aware (on an ad hoc basis) of new municipal initiatives in the eight urban and suburban communities. I would like to mention some of these responses here, with the caveat that it is impossible to evaluate these initiatives systematically without conducting further interviews. Furthermore, this list of recent initiatives is by no means exhaustive. All municipalities in the sample seem to have been experimenting with new ways to accommodate immigrants in their populations.

Some of the new initiatives in the least responsive of the municipalities in the sample are especially interesting and warrant further research and policy attention. As many immigrants settle directly in cities and chose to live in suburbs, it will be important to understand how suburban municipalities in immigrant-magnet city-regions develop the capacity to respond to their diverse populations. Civil societies are less developed and resourced than in central cities. In the interest of developing socially sustainable cities, attention ought to be paid to how Canada's most important immigrant suburbs or 'ethnoburbs' can be supported in their efforts to develop the capacity to manage social change.

The City of Surrey has appointed its first Multicultural Advisory Committee, with Councillor Mary Martin as chair. The committee will be composed of community members. It will develop a strategic plan on ethnocultural inclusiveness for the city as well as create a festival to celebrate Surrey's diversity (City of Surrey 2006). This committee could provide citywide leadership on multiculturalism efforts, which in the past have tended to be centred in the Parks, Recreation, and Culture Department. In addition, the committee could be an important means of integrating multiculturalism-related concerns into a citywide governance arrangement. In essence, it could integrate Surrey's 'subregime' into the city's overall governance arrangement.

In 2006, Coquitlam re-established its Multiculturalism Advisory Committee as a separate committee. Recall that this committee was integrated into the Liveable Cities Advisory Committee between 2003 and 2005. The new committee is composed of two council members as well as representatives of the community. Its mandate is to raise awareness about multiculturalism issues, to serve as a place to dialogue on those issues, to address barriers to citizen involvement, and to help implement the city's multiculturalism policy (City of Coquitlam 2006). Since its establishment, the committee seems to have been active. In addition, interestingly, the city seems to be seeking federal aid in its multiculturalism policy-making efforts. Specifically, it has been awarded a grant from Heritage Canada for $150,000 to meet its mandate, which gives it a larger budget than those of other advisory committees discussed in this book, which receive funding only from municipal councils (Coquitlam 2007a, 3).

To put things in perspective, the Richmond Intercultural Advisory Committee operated without a budget for its first two years (2002–3) and then received a budget of $5,000 in 2004 (Sherlock 2004, interview) – a small fraction of the budget of the equivalent committee in Coquitlam. Though the federal government plays an important role in funding community organizations, this development reflects the potential of a more direct federal role in municipal multiculturalism initiatives. The federal government has also named Coquitlam a 'cultural capital' of Canada for its current diversity initiatives and for its plans to undertake new multiculturalism initiatives (Coquitlam Now 2008). Coquitlam also participates on a Provincial Joint Steering Committee for multiculturalism and immigration (City of Coquitlam 2007b). One of Coquitlam's new strategies for building capacity in immigrant settlement appears to focus on fostering relationships of different sorts with upper levels of government.

Even the 'unresponsive' municipalities discussed in this book seem to be making efforts to adapt their services to increase immigrant and ethnocultural-minority access to municipal services. Brampton has launched a pilot project for multilingual telephone services – a simultaneous-interpretation service for residents and the business community.

As the media release acknowledges, fully one-third of residents who contact the city have a first language other than English or French (City of Brampton 2006). And in 2007 the city approved a more comprehensive Multilingual Program that is establishing a policy for translating written communications as well (City of Brampton 2007). It is worth noting that Brampton's seemingly increased level of responsiveness to immigrants and ethnocultural minorities coincides with its shift to what this book calls a biracial ethnic configuration.

Similarly, Mississauga seems to developing governance arrangements to increase immigrants' access to services and to facilitate their economic integration. Mississauga's website now reports that it has been participating in a number of TRIEC initiatives since 2005 (City of Mississauga, 'City of Mississauga Integrates New Canadians into the Workforce'). Furthermore, that website now lists community organizations that can provide translations of information about Mississauga services in eleven languages (City of Mississauga, 'City Hall Languages'). Though this initiative does not appear to involve the dedication of city resources, it suggests that as a corporation, Mississauga is developing ongoing relationships with leaders in the immigrant settlement and related sectors – an important first step towards developing a corporate strategy to increase city access and equity in terms of services and governance. Furthermore, that these arrangements are being advertised on the city's website indicates that Mississauga is moving towards a more multicultural model of citizenship.

Cultural and linguistic diversity has made it more of a challenge for Mississauga and Brampton to respond effectively and equitably than had their immigrant populations been dominated by a single group. Also, the downloading that happened in Ontario beginning especially in the mid-1990s as well as the lack of an immigration agreement in Ontario meant that the province's settlement sector was underfunded. Furthermore, as noted in chapter 4, though many immigrants now settle directly in the 'outer suburbs' such as Mississauga, Brampton, and Markham in the GTA, the spatial pattern of resource distribution is uneven. These structural forces led to Fair Share movements in Peel and York regions, in which Mississauga and Brampton and Markham respectively are located. As suburban immigrant destinations such as these continue to adapt their services and governance structures to immigration, it will be important to ensure that immigrant-serving community organizations in those locales receive the funding they need in order to influence municipal policy making and participate in local governance. The spatial/scalar distribution of resources matters both across levels of government and within city-regions.

This book has discussed differences in how diversity policies have been framed in Toronto and Vancouver. Local leaders in Toronto framed the municipality's role as not only 'diversity management' but also 'immigration and settlement.' Local leaders in Vancouver stressed that the municipality was responding to diversity (and not developing 'immigration' policies, which were viewed as the responsibility of upper levels of government). This book has noted evidence of a convergence in this respect when former Mayor Larry Campbell established a mayor's working group on immigration and settlement in 2005. Since the interviews were conducted, Vancouver appears to have become more assertive in its role in immigration and settlement. Former Vancouver mayor Sam Sullivan (2005–8), on being elected in 2005, re-established the Mayor's Working Group on Immigration in 2006, renaming it the Mayor's Task Force on Immigration (MTFI). The report of the task force acknowledges the municipality's role in immigrant settlement and describes a need to articulate this role to upper levels of government: 'While Federal and Provincial governments are responsible for immigration policy and the funding and delivery of key programs, all three levels of government are actually involved in providing support and services to newcomers. Cities need to articulate their roles and concerns to senior government, and request financial support in providing locally-based integration programs for newcomers' (City of Vancouver 2007, 2). It also stresses the need for consultation with the community and for exchange of best practices with other cities (ibid, 1). The task force membership was multisectoral and included the business community. In addition to the mayor, the task force included leaders of immigrant settlement agencies, an academic, a representative of the Asia Pacific Foundation of Canada, and a representative of the Vancouver Board of Trade. Baldwin Wong, the city's Multicultural Social Planner, was the staff liaison (ibid, 17). The task force represents another example of the city's 'urban regime' in action.

There is also evidence of multisector alliance building and of increased assertiveness of a broad range of local leaders in immigrant-related matters. For instance, one recommendation of the City of Vancouver's Mayor's Task Force on Immigration is to have the 'mayor convene a Summit meeting with key business leaders, employer and sectoral groups to discuss the feasibility of launching a multi-sectoral Immigrant Employment initiative' (ibid., 1). A multisector alliance also appears to have developed in Peel Region, the region in which both Mississauga and Brampton are located. The Peel Newcomer Strategy Group emerged in 2005 (Peel Newcomer Strategy Group website, 'Our History'). These initiatives are reminiscent of the way in which the Toronto City Summit Alliance and its offspring, the Toronto Region Immigrant Employment Council, developed in Toronto.

There is also some new evidence of the impact of global networks on multiculturalism policy making in some of the Canadian municipalities in this sample. For instance, Toronto and Coquitlam (and possibly other municipalities in the sample) have joined an international Coalition of Municipalities Against Racism and Discrimination that UNESCO has pioneered. This initiative is yet another manifestation of polyscalar forms of networking and mobilization and possibility another sign of the ongoing processes of rescaling of political authority. The effectiveness of such initiatives warrants further study, as does their ultimate significance.

Appendix
List of Interviews

Note: All interviews were conducted by Kristin Good and were in-person unless otherwise indicated.

Abrahams, Phillip. Manager, Intergovernmental Relations, City of Toronto, 18 February 2004.
Ahn, Simon. Library Trustee, City of Coquitlam, 17 June 2004.
Ashton, Brian. Councillor, City of Toronto, 13 October 2004.
Augimeri, Maria. Councillor, City of Toronto, 20 September 2004.
Barnes, Linda. Councillor, City of Richmond, 29 April 2004.
Basi, Ravi. Multicultural Outreach Librarian, Newton Library, Surrey Public Library, 20 May 2004.
Bray, Keith. President and CEO, Markham Board of Trade, 12 January 2005.
Brown, Susan. Senior Policy Adviser, Labour Force Development, Economic Development, Culture and Tourism, City of Toronto, 4 December 2003.
Buss, Greg. Chief Librarian, Richmond Public Library, 17 May 2004.
Cadman, David. Councillor, City of Vancouver, 26 April 2004.
Carroll, Shelley. Councillor, City of Toronto, 21 September 2004.
Casipullai, Amy. Policy and Public Education Coordinator, Ontario Council of Agencies Serving Immigrants (OCASI), 26 November 2003.
Cavan, Laurie. Manager of Community and Leisure Services, Parks, Recreation and Culture, City of Surrey, 20 May 2004.
Chatterjee, Alina. Director of Development/Community Engagement, Scadding Court Community Centre, Toronto, 15 December 2003.
Chan, Sherman. Director, MOSAIC, Vancouver, 23 April 2004.
Chaudhry, Naveed. Executive Director, Peel Multicultural Council (PMC), 10 December 2003.
Cheng, Ansar. Director of Settlement, S.U.C.C.E.S.S., 3 May 2004.

Clapman, Ward. Superintendent, Officer in Charge, RCMP, Richmond City Detachment, 27 May 2004.
Crombie, David. CEO, Canadian Urban Institute, Toronto, 4 February 2004.
Crowe, Terry. Manager of Planning, City of Richmond, 29 April 2004.
Dinwoodie, Murray D. General Manager, Planning and Development, City of Surrey, 22 June 2004.
Douglas, Debbie. Executive Director, OCASI, Toronto, 28 November 2003.
Dunn, Sam. Project Coordinator, Best Practices for Working with Homeless Immigrants and Refugees, Access Alliance Multicultural Community Health Centre, Toronto, 3 November 2003.
Feldman, Mike. Deputy Mayor, City of Toronto, 23 September 2004.
Fennel, Susan. Mayor, City of Brampton, 20 September 2004.
Fisch, Bill. Regional Chair and CEO, Regional Municipality of York, 13 April 2004.
Gibson, Grant D. Councillor, City of Brampton, 9 September 2004.
Gill, Charan. Executive Director, PICS, Surrey, 9 February 2005, telephone interview.
Gill, Warren. VP Development, Simon Fraser University, 17 June 2004.
Green, Jim. Councillor, City of Vancouver, 26 April 2004.
Hall, Suzan. Councillor, City of Toronto, 14 January 2004.
Hall, Ric. Superintendent, Officer in Charge, Coquitlam Detachment, RCMP, June 2004.
Hansen, David. RCMP, Richmond City Detachment, 19 May 2004.
Hardy, Bruce. Executive Director, OPTIONs, Surrey, 9 February 2005, telephone interview.
Harrison, Karen. Director of Library Services, Coquitlam, 9 June 2004.
Hewson, Lauren. Committee Clerk, City of Coquitlam, 10 June 2004.
Houlden, Melanie. Deputy Chief Librarian, Library Administration, Surrey Public Library, 20 May 2004.
Huhtala, Kari. Senior Policy Planner, Policy Planning, Urban Development Division, City of Richmond, 18 February 2005, telephone interview.
Iannicca, Nando. Councillor, City of Mississauga, 19 January 2004.
Innes, Rob. Deputy City Clerk, City of Coquitlam, 10 June 2004.
Jamal, Audrey. Executive Director, Canadian Arab Federation (CAF), Toronto, 2 December 2003.
Jeffrey, Linda. MPP Brampton-Centre, February 2004.
Jones, Jim. Regional Councillor, City of Markham, 27 February 2004.
Jones, Warren. City Manager, City of Coquitlam, 10 June 2004.
Keung, Nicolas. Immigration/Diversity Reporter, *Toronto Star*, January 2005.
Kingsbury, Jon. Mayor, City of Coquitlam, 24 June 2004.

Kohli, Rajpal. Adviser, Equal Employment Opportunity Program, City of Vancouver, 28 April 2004.

Lee, Rose. Policy Coordinator, Diversity Management, Strategic and Corporate Policy/Healthy Cities Office, City of Toronto, 17 November 2003.

Leiba, Sheldon. General Manager, Brampton Board of Trade, 27 May 2005, telephone interview.

Louis, Tim. Councillor, City of Vancouver, 11 June 2004.

Magado, Marlene. Chair, Markham Race Relations Committee, Markham, 17 February 2004. Follow-up interview in October 2004.

Manning, Garnett. Councillor, City of Brampton, 5 October 2004.

McCallion, Hazel. Mayor, City of Mississauga, 13 April 2004.

McCallum, Doug. Mayor, City of Surrey, 20 May 2004.

McIsaac, Elizabeth. Program Manager, Maytree Foundation, Toronto, 26 January 2004.

McKitrick, Annie. Executive Director, Surrey Social Futures; School Trustee, Surrey; Member of RIAC, 27 May 2004.

Melles, Amanuel. Executive Director, Family Neighborhood Services, Toronto, 21 January 2004.

Merryweather, Brian. Manager, Human Resources, Human Resources Division, Finance, Technology, and Human Resources, City of Surrey, 22 June 2004.

Mihevc, Joe. Councillor, City of Toronto, 5 December 2003.

Mital, Umendra. City Manager, City of Surrey, 22 June 2004.

Moscoe, Howard. Councillor, City of Toronto, 23 September 2004.

Moore, Elaine. Regional Councillor, City of Brampton, 7 September 2004.

Nuss, Marie. Executive Director, Brampton Neighbourhood Resource Centre, Brampton, 19 February 2004.

Pantalone, Joe. Councillor and Deputy Mayor, City of Toronto, 13 September 2004.

Rae, Kyle. Councillor, City of Toronto, 23 September 2004.

Richmond, Ted. Coordinator, Children's Agenda Program, Laidlaw Foundation, 7 January 2004.

Sales, Jim. Commissioner of Community Services, City of Markham, 2 March 2004.

Sanghera, Balwant. Chair of RIAC, President of Multicultural Concerns Society of Richmond, 4 May 2004.

Schroeder, Scott. Community Coordinator of Diversity Services, City of Richmond, 29 April 2004.

Seepersaud, Andrea. Executive Director, Inter-Cultural Neighbourhood Social Services (ICNSS), Mississauga, 27 November 2004.

Semotuk, Verna. Senior Planner, GVRD, Policy and Planning Department, 16 February 2005, telephone interview.

Shakir, Uzma. Executive Director, Council of Agencies Serving South Asians (CASSA) and President of OCASI, Toronto, 18 December 2003.

Sherlock, Lesley. Social Planner, City of Richmond, 29 April 2004.

Spaxman, Ray. President, Spaxman Consulting Group, May 2004.

Stobie, Charles. Vice-President, Government Relations, Mississauga Board of Trade, 7 January 2005, telephone interview.

Taranu, Alex. Manager, Urban Design and Public Buildings, City of Brampton, 29 July 2004.

Taylor, Margot. Senior Adviser, Employment, Human Resources Division, City of Surrey, 22 June 2004.

Taylor, Susan. Director, Human Services Planning, Planning and Development Services Department, York Region, 23 January 2004.

Thiessen, Peter (Cpl). NCO i/c Communications Media Relations, RCMP, Richmond City Detachment, 19 May 2004.

Townsend, Ted. Manager, Communications and Corporate Programs, City of Richmond, 10 May 2004.

Usman, Khalid. Councillor, City of Markham, 2 March 2004.

Vander Kooy, Magdelena. District Manager, East Region, Toronto Public Library, City of Toronto, 9 December 2003.

Vescera, Mauro. Program Director, Vancouver Foundation, 11 March 2005, telephone interview.

Villeneuve, Judy. City Councillor, City of Surrey, 22 June 2004.

Welsh, Timothy. Program Director, Affiliation of Multicultural Societies and Service Agencies of B.C. (AMSSA), 13 May 2004.

Wong, Baldwin. Multicultural Social Planner, City of Vancouver, 3 May 2004.

Wong, Denzil. Councillor, City of Toronto, 27 September 2004.

Wong, Milton. Chancellor, Simon Fraser University, Vancouver, 12 January 2005, telephone interview.

Woodman, Lesley Ann. Executive Director, Surrey-Delta Immigrant Services Society, Surrey, 24 June 2004.

Woodsworth, Ellen. City Councillor, City of Vancouver, 26 May 2004.

Woroch, Patricia. Executive Director, Immigrant Services Society of British Columbia, 21 June 2004.

Anonymous Participants

Board Members of an immigrant-serving organization, Mississauga, 24 November 2003.

Four Mississauga civil servants, City of Mississauga, 27 November 2003.

Toronto civil servant, Toronto Public Health, 9 December 2003.
Executive director of an immigrant-serving organization, York Region, 16 December 2003.
Markham civil servant, Human Resources, City of Markham, 2 March 2004.
Municipal civil servant, Regional Municipality of Peel, 9 March 2004.
B.C. civil servant, MCAWS, Government of British Columbia, 1 March 2005, telephone interview.

Notes

1: The Municipal Role in 'Managing' Multiculturalism

1 Some scholars and practitioners distinguish the 'multicultural' model of citizenship from the traditional American 'melting pot' model of integration, in which the onus is on immigrants to adapt to the receiving culture. However, empirical research suggests that the differences between the Canadian and American approaches to accommodating ethnocultural differences are much more nuanced (see Hero 1998, 2007; Hero and Preuhs 2006; Banting et al. 2006).

2 This word is in quotation marks to acknowledge the socially constructed nature of 'race.' The data concerning the 'racial' composition of the Canadian population are based on the Canadian-made concept of 'visible minorities,' which has a racial basis. Federal policies and Statistics Canada have adopted the 1986 Employment Equity Act's definition of 'visible minorities': 'persons, other than Aboriginal people, who are non-Caucasian in race or non-white in colour.' Data on Canada's 'visible minority' population are collected in order to support the implementation of Canada's official multiculturalism.

3 According to Statistics Canada, '[a] census metropolitan area (CMA) … is formed by one or more adjacent municipalities centred on a large urban area (known as the urban core). A CMA must have a total population of at least 100,000, of which 50,000 or more must live in the urban core … To be included in the CMA … other adjacent municipalities must have a high degree of integration with the central urban area, as measured by commuting flows derived from census place of work data' (Statistics Canada n.d.a.).

4 This term is synonymous with the Vancouver CMA and the territory

covered by Metro Vancouver (the former Greater Vancouver Regional District). Recently, the GVRD board (which governs Metro Vancouver) changed its name to Metro Vancouver 'to achieve greater recognition for who [they] are and what [they] do and to give [their] board greater ability to influence at the local, national and international levels' (Metro Vancouver, 'Frequently Asked Questions'). Metro Vancouver is a political decision-making body and corporate entity under which three other corporate entities operate: the Greater Vancouver Regional District, the Greater Vancouver Water District, and the Greater Vancouver Sewerage and Drainage District. Metro Vancouver also owns the Metro Vancouver Housing Corporation (ibid.). The decision-making body is still legally called the GVRD. This book uses both Greater Vancouver and the Vancouver CMA to refer to the geographical city-region. Since it is more straightforward, the former is used unless the book refers to data that Statistics Canada collected, in which case it uses Statistics Canada's concept of CMA. It uses the term GVRD to refer to the corporate entity that continues to exist but has been renamed Metro Vancouver, since it reflects the language of the interviewees and of the secondary literature at the time of the interviews and research.

2: Linking Urban Regime Theory, Social Diversity, and Local Multiculturalism Policies

1 Mossberger and Stoker (2001) make a similar point with respect to shifting academic attention to the broader concept of 'policy networks' rather than regimes. In their view, 'the cost of moving to a higher level of abstraction is to embrace a concept that says even less about how or why coordination may occur' (2001, 821).
2 Elazar (1966, 1970) identifies three political cultures within states – 'moralistic,' 'individualistic,' and 'traditionalistic' – to explain differences in state politics and policies. Hero's theoretical framework complements Elazar's insofar as what he calls 'homogeneous' states tend to be 'moralistic,' heterogeneous states tend to be 'individualistic,' and 'bifurcated' states tend to be 'traditionalistic' (Hero 1998, 9).
3 These data were collected from municipal websites in June 2004 and based on the interviews conducted for this book.
4 Historically, municipal legislation has been highly *restrictive*, permitting municipalities to exercise only those powers for which they can identify express authority in provincial legislation. The two central legal concepts in the current debate about the degree of autonomy delegated to munici-

pal governments by provinces – *natural person powers* and *spheres of jurisdiction* – are explained by Tindal and Tindal in the most recent edition of their *Local Government in Canada* (2009, 180, Box 6.4).

5 As the City Solicitor at the City of Toronto explains, a municipal charter 'codifies the laws applicable to the particular city and contains powers and responsibilities not given to other municipalities in the province' (City Solicitor, City of Toronto 2000, 3). City charters recognize the distinctiveness of urban centres and are a form of asymmetrical provincial-municipal relationship.

6 Donald Lidstone's comment is posted on the City of Toronto's website under 'Comments on the City of Toronto Act.'

7 Like Canadian scholars, within the American context some urban-regime theorists argue that other levels of government influence and participate in urban regimes (Jones and Bachelor 1993; Burns 2002). Like provincial leaders, state political officials participate in local governance arrangements in American cities (Burns 2002).

8 A caveat is in order here: some municipalities with significantly lower immigration rates than the cases discussed here have begun to develop multiculturalism policies in their efforts to attract and retain immigrants.

3: A Comparative Overview of Municipal Multiculturalism Policies

1 Robert Putnam (1993) uses these evaluative criteria in his study of the institutional performance of regional institutions in Italy (1993, 65).

2 The three main settlement programs have been the Immigrant Settlement and Adaptation Program (ISAP), the Host Program, and the Language Instruction for Newcomers Program (LINC).

3 The book focuses on policy *outputs* rather than policy *outcomes* in measuring municipal responsiveness for the same reason as Putnam did: 'social outcomes are influenced by many factors besides governments' (ibid., 66). It is difficult to establish a direct causal relationship between a particular policy and its impact or outcome. In addition, even if one could isolate the causal effect of a policy, there might be a 'time lag' before one is able to observe the effect of the policy. The potential outcomes of many policies that are important to immigrants and ethnocultural minorities are difficult to measure, and the specific goals of policies (and therefore also definitions of outcomes) can change over time. In addition, the potential outcomes of policies are numerous and might have an impact at many levels in society (individual, institutional, community, etc.). The issue of 'who decides' what constitutes a valid measurement of a policy outcome is also an important

question, especially in multiculturalism policy. Detractors of multiculturalism policies sometimes cite the difficulties of measuring their societal impact (according to their personal ideas of what constitutes a valid 'policy outcome') as a justification for not expending resources on such policies. Because of these difficulties, the analysis focuses on policy outputs – on the practices, initiatives, and written policies of municipal governments. In addition, the book evaluates the implementation of multiculturalism policies with the help of interviews conducted with leaders in the immigrant settlement sector and leaders of ethnocultural minority organizations.

4 Where possible, interviews were conducted with civil servants in police services, public-health departments, library-service agencies, parks and recreation-services departments, and planning departments. Also, data on the responsiveness of other departments were collected on a more ad hoc basis. The goal was to ascertain the level of activity of a range of municipal departments and agencies.

5 This categorization builds on Marcia Wallace and Frances Frisken's (2000) typology of possible policy responses to immigration and ethnocultural diversity. As with their typology, the policy categories described here overlap somewhat.

6 Their typology, in turn, builds on Ellen Tate and Louise Quesnel's (1995) categories of 'proactive' and 'reactive' policy styles.

7 See Appendix for a list of people interviewed for this study.

8 While this manuscript was being written, the City of Toronto's website was checked several times. A Diversity Advocate had not yet been appointed. Then on 18 August 2008, the website indicated that Joe Mihevc had taken the position.

9 This is a measure developed by Statistics Canada. The measure constitutes 'a well-defined methodology which identities those who are substantially worse off than the average' (Fellegi 1997). Though anti-poverty groups use it as a measure of poverty, Statistics Canada researchers emphasize that poverty is a relative concept whose definition varies by society and that the agency does not consider it a measure of poverty (ibid.).

10 The RIAC comprises a councillor who serves as a liaison to council, sixteen council-appointed voting members (including six citizens), four representatives of the Richmond Community Services Advisory Council, two youth representatives, and representatives from School District 38, the Royal Canadian Mounted Police, Richmond Health Services, and the Ministry of Children and Family Development.

11 The way in which police services are administered varies across municipalities. The City of Toronto and the City of Vancouver have independent

police forces. Outside Toronto, in the GTA, police services are a regional responsibility. Thus, Peel Region administers the police services of Mississauga and Brampton, and York Region administers Markham's police services. In Greater Vancouver, suburban municipalities can choose between creating an independent police force or contracting police services to the federal RCMP. Richmond, Surrey, and Coquitlam have all pursued the latter option. Under this arrangement, the RCMP detachment is bound by federal rules but is also contractually obligated to the municipality. Under the contract, municipalities fund police services and in this way shape RCMP detachments in significant ways. RCMP detachments must be responsive to the concerns of municipal officials, who otherwise might opt for an independent police force. At the same time, municipalities that contract their police services out to the RCMP are constrained in ways that municipalities with independent police forces are not. For instance, municipalities with their own forces can decide to hire officers from ethnoracial-minority communities; in contrast, the federal government decides who will be assigned to the detachment. Thus, to diversify RCMP staff, RCMP superintendents (i.e., the leaders of RCMP detachments) must show creative leadership and commitment.

12 The federal RCMP has mounted a nationwide effort to recruit visible minorities to its training centre in Regina ('RCMP No Longer Colour Blind' 2003, 1). This initiative was motivated in part by the RCMP's philosophical shift towards 'community policing,' according to which the focus of policing moves from enforcement towards building relationships of trust between communities and the police.

13 It was unclear whether the drive-by shootings to which this interviewee was referring were acts of interracial violence.

14 The Surrey-Delta Immigrant Services Society has been renamed DiverseCity community resources society since the interviews were conducted in the region in 2004.

15 The full name of this organization is OPTIONS: Services to Communities Society. Though it appears to be an acronym, it is not.

16 It is instructive that this community leader referred to the City of Surrey as 'we.' It sounds as if he is an 'insider' in Surrey's governance arrangements, while other organizations such as SDISS are 'outsiders.'

17 In general, the interviews reveal that lone visible-minority representatives on councils are often highly knowledgeable about the concerns of immigrants and visible minorities. Also, several non–visible-minority councillors observed that when visible minorities are elected to council, they often serve as bridges between ethnocultural minorities and the council.

18 MPP Linda Jeffrey also mentioned that the city has been criticized for not hiring members of the Sikh community.

19 To use a minor example of how decision makers' assumptions can make some groups feel included or excluded in city events, the city's choice of a rock 'n' roll band for its New Year's Eve celebration appealed much more to 'White' residents than to Blacks and South Asians. Also, Councillor Manning mentioned what he called a 'cookie cutter' photo of exclusively White people that was used for the cover of Brampton's *Recreation Guide*. Apparently, some citizens were so angry about this photo that they called the city 'screaming' (Manning 2004, interview). In his view, all city commissioners should be required to attend community events to learn about Brampton's diverse communities so that they can bring this knowledge to the city.

20 The Town of Markham also participates in the Toronto Region Immigrant Employment Council (TRIEC), a regional coalition in the Greater Toronto Area (GTA) that was created to address barriers to immigrants' integration into the economy. The Town of Markham also mentors immigrants through TRIEC's 'Career Bridge' program, which places immigrants in participating city bureaucracies and in private-sector jobs to help them gain Canadian experience.

21 Hero (1998) found that American 'Official English' policies were most common in two types of political units – homogeneous and bifurcated (1998, 108). Hérouxville falls into the former category.

4: Determinants of Multiculturalism Policies in the Greater Toronto Area

1 Geographer Michael Doucet (2001) found that Toronto is indeed one of the most diverse cities in the world but that the UN did not in fact make such a declaration.

2 Dillon's rule refers to a decision made by Judge John F. Dillon in 1868 in *Merriam vs. Moody's Executors*: 'A municipal corporation possesses and can exercise the following powers and no others; first, those granted in express words; second, those necessarily implied or necessarily incident to the power expressly granted; third, those absolutely essential to the declared objects and purposes of the corporation – not simply convenient but indispensable; and fourth, any fair doubt as to the existence of a power is resolved by the courts against the corporation' (Tindal and Tindal 2004, 196).

3 'Home rule' provisions, which permit local governments to act in any area that is not explicitly prohibited by state legislation, are common in the western states of the United States (ibid., 197–8).

4 The term 'access and equity' was the dominant way of describing the multiculturalism policies of the City of Toronto and Metropolitan Toronto before amalgamation. Since amalgamation the policy discourse has shifted towards 'managing diversity,' as reflected in the name of the city's primary institutional support for its multiculturalism policies – the Diversity Management and Community Engagement Unit.

5 On 31 October 2008 the government of Ontario announced that it would progressively upload the cost of social services from municipalities to the province. The process will be complete by 2018 (Government of Ontario 2008).

6 Bramptonians elect five city councillors and five regional councillors (who sit on Peel Regional Council) to represent them at the municipal level. Elaine Moore is currently a regional councillor but has also served as a city councillor (for three years) and as a Peel District School Board Trustee (for eleven years). In Mississauga, city councillors sit on both the local city council and Peel Regional Council. The mayors of both Brampton and Mississauga sit on both the local and regional councils.

7 Linda Jeffrey mentioned Brampton's Emergency Preparedness Plan as an important exception to this rule.

8 Lower-tier municipalities in the two-tiered structures in the GTA offer many services of potential importance to the immigrant settlement process – for instance, recreation services and library services.

9 In American 'strong mayor' systems, mayors exercise powers similar to those held by the president at the national level, such as the power to appoint their own administration and to exercise a veto (Banfield and Wilson 1963, 80).

10 Municipal councils vary somewhat in structure across the country; one result is that mayors have different degrees of power. Some municipalities have 'executive committees' chaired by the mayor, which concentrate power to a certain extent. However, in the absence of strong political parties, the decisions of the executive are subject to an 'undisciplined' council. Few Canadian cities have political parties at the local level, and where they do exist, the parties are not as strong as at the provincial and federal levels. Former Montreal mayor Jean Drapeau's (1960–86) Civic Party is an exception in this respect. For an overview of municipal governing structures in Canada, see Tindal and Tindal, *Local Government in Canada* (2009), ch. 8.

11 The Town of Milton posted figures taken from *Novae Res Urbis* – GTA Edition, 18 and 25 July, to show where it stands relative to other municipalities in the GTA. The municipalities with the lowest residential property-tax

rates in the GTA in 2008 were Toronto, Milton, Caledon, and Mississauga (Town of Milton website).

12 As David Crombie put it, Mel Lastman was not a 'tenter' (by which he meant that Lastman did not bring people together in cooperative efforts) (Crombie 2004, interview).

13 Toronto's council is four times the size of Mississauga's. There are eleven council members in Mississauga (including the mayor); in Toronto there are forty-five (including the mayor).

14 Though Toronto does not have a party system, there are informal divisions between 'left' and 'right' as well as numerous other cleavages on council.

15 Grantmakers Concerned with Immigrants and Refugees provides resources to private foundations that are trying to effect change in areas of importance to immigrants and refugees. See GCIR, 'About GCIR.'

16 Dominic D'Alessandro, President and CEO of Manulife Financial, and Diane Bean, Senior Vice-President of Corporate Human Resources at Manulife Financial, co-chair TRIEC.

17 People for Education is a charitable organization of parents who work to preserve a fully funded public-education system in Ontario.

18 Citizens for Local Democracy, popularly known as 'C4LD,' was a coalition that developed to protest the Conservative government's decision to amalgamate Metro Toronto and its constituent municipalities.

19 This group of residents was not formally organized, but acted through social networks and individually.

20 At the city level, the link between diversity and economic development has been influenced by the writings of Richard Florida. Florida (2002) argues that a city's economic fortunes are tied to its ability to attract the ethnically and sexually diverse 'creative class.' For example, he links a city's level of tolerance of gays and of ethnic minorities to its ability to thrive in a global economy. In Toronto's economic-development office, Florida's best-selling book, *The Rise of the Creative Class* (2002), is on staff bookshelves (Brown 2003, interview); it has become something of a 'Bible' for them.

21 Pal's (1995) study focuses on the power dynamics between civil society and the federal government – a level of government with significantly more resources than municipal governments.

5: Determinants of Multiculturalism Policies in Greater Vancouver

1 The Downtown Eastside is a high-needs neighbourhood in Vancouver with high rates of poverty, drug addiction, and crime.

2 Safe injection sites for severe cases of heroin addiction were established in

Vancouver as part of a project called the North American Opiate Medications Initiative (NAOMI). See http://www.city.vancouver.bc.ca/fourpillars/newsletter/Mar04/NAOMI.htm.

3 Canada's 38th general election was held on 28 June 2004. Judy Villeneuve was interviewed on 22 June 2004, six days before that election. At the time of the interview cited here, local leaders were anticipating movement on the federal government's 'New Deal for Cities and Communities' agenda should a Liberal government be elected.

4 In Greater Vancouver, councils are elected through an at-large system. Therefore, city councillors do not necessarily represent all neighbourhoods in a municipality. In fact, the entire council could be composed of councillors who live in the same neighbourhood.

5 Kingsbury mentioned Japanese people. However, according to the 2001 Census, there are 850 residents out of a total of 112,890 in Coquitlam who identify as Japanese (Statistics Canada 2002).

6 The 2001 Census recorded 112,890 residents in Coquitlam. Yet in 2002 only 57,222 people were eligible to vote in Coquitlam's election. Data concerning ineligible residents were not collected; still, one might hypothesize that many were foreign-born residents. Also, of the 57,222 eligible voters, a mere 15,969 cast ballots – a turnout rate of 27.7 percent of residents eligible to vote. Thus only 14.1 percent of the population determined the election results (City of Coquitlam, 'Official Results …').

7 Daniel Chiu received 5,723 votes in the 2002 municipal election. The eight councillors who were successful received between 6,163 and 8,732 votes. Chiu was tenth on the list (ibid.).

8 The two councillors who were elected for the first time in 2002 were Fin Donnelly and Barry Lynch.

9 Since the interviews were conducted, the party system has changed. A new party, Vision Vancouver, has emerged from 'COPE Classic' supporters. The NPA elected six members in the 2005 election along with its candidate for mayor (Sam Sullivan). Vision Vancouver elected four candidates, and COPE elected one member. Then in 2008, led by the Vision Vancouver mayoral candidate (current Mayor Gregor Robertson), a coalition of Vision Vancouver, COPE, and the Vancouver Green Party was victorious. The City of Vancouver's responsiveness to multiculturalism continues as parties change and new electoral coalitions emerge on council.

10 Harcourt would later become an NDP Premier of British Columbia (1991–6). He was named Prime Minister Martin's special adviser on cities in December 2003. In this role he travelled across the country to consult with Canadians on the state of Canada's cities. His research produced a

report titled *From Restless Communities to Resilient Places* (External Advisory Committee 2006).

11 In other words, this idea was diffused from the City of Toronto to the City of Vancouver through the C5 Forum, a coalition of mayors and civil-society leaders from five cities – Vancouver, Calgary, Winnipeg, Toronto, and Montreal – that lobbied the federal government for more resources for cities. This forum is discussed further in chapter 8.

12 *Directory: B.C. Multicultural, Anti-Racism, Immigrant, and Community Service Organizations* – a document published by the B.C. government which lists many of the most prominent immigrant-settlement organizations in Greater Vancouver – lists many more in Vancouver than elsewhere in the city-region.

13 The S.U.C.C.E.S.S. website now indicates that government funding constitutes 71 per cent of its budget (accessed 13 May 2009). The figure cited in the text was on the website during the research phase of this project (2003–5).

14 These funds are now administered by the province under the Agreement for Canada–British Columbia Cooperation on Immigration 2004.

15 The Vancouver Foundation's mission includes but is broader than simply addressing challenges related to ethnocultural diversity. For more information, see Vancouver Foundation website, 'About Us.'

16 This fact was reported in an article in 2004. However, the article does not specify which year *Time* voted Fung one of Canada's most powerful individuals.

17 The 'long-standing community' is primarily White and of European background.

18 Since the interviews were conducted, MOSAIC has been listed as a member of AMSSA. AMSSA's website was accessed on 25 July 2008. When the website was accessed on 17 May 2009, MOSAIC was no longer listed as a member (AMSSA, 'AMSSA Membership').

19 OCASI represents the settlement sector in Ontario. However, it is located in Toronto and appears to adopt strategies that reflect Toronto's more contentious political culture.

20 The University of British Columbia, Simon Fraser University, and the University of Victoria participate in this research alliance in British Columbia. One of the primary goals of the alliance is to foster the development of action-oriented research through partnerships between government and community organizations.

21 In *Crossing the Neoliberal Line* (2004), Katharyne Mitchell describes these controversies in detail, arguing that they reflect the neoliberal transforma-

tions of states that result when global migratory processes disrupt existing forms of liberal ideology (2004, 3).

22 Vancouver is an exception in this respect. In its municipal charter it lists social planning as a municipal responsibility.

23 The authors cite an interview with Kari Huhtala, who was also interviewed for this study.

24 Along with the cases in this study, Burnaby is one of Canada's most significant immigrant-receiving suburban municipalities. In 2006 its population was 200,855 and its foreign-born and visible-minority populations represented 50.6 and 55.4 percent of the population respectively (Statistics Canada 2007b).

25 The term 'Asian' is somewhat misleading. In Greater Vancouver, local people tend to refer to Chinese organizations and cultural amenities as 'Asian' and to 'South Asian' organizations and cultural amenities as 'Indian,' 'East Indian,' or 'Indo-Canadian' (Villeneuve 2004, interview).

26 There is a great deal of academic debate concerning the characteristics of 'global cities,' 'word-class cities,' and 'world cities.' See Caroline Andrew and Patrick Smith (1999) for an overview of this debate among Canadian urban scholars. Few would place Vancouver in the category of 'global city' as defined by the comparative literature. Following Saskia Sassen's *The Global City* (2001), the term generally refers to the top tier in the international hierarchy of cities – that is, the cities (including New York, London, and Tokyo) that control the global economy.

27 The present research did not uncover evidence of a direct relationship between VanCity and the City of Vancouver. VanCity has ongoing relationships with immigrant-settlement agencies, which in turn have lasting relationships with the city (and with Baldwin Wong, the city's multicultural social planner, in particular).

28 Crowe distinguished between 'facilitating' and 'leading.' However, facilitating is also a form of leadership. In fact, urban-regime theory notes the cooperative dimension of leadership. Urban-regime theory implies 'co-leadership.'

29 In reality, this would have been beyond the scope of municipal authority.

30 This concept is new in Richmond. It was developed in Quebec as a normative alternative to the concept of multiculturalism. The terms 'multiculturalism' and 'interculturalism' mean different things to different people, but according to some, the latter term emphasizes cultural integration to a greater extent than the former.

31 According to the 2006 Census, 57 per cent of Surrey's immigrant population does not speak English at home. Punjabi is by far the most common

non-English language in Surrey, with almost 43,000 Surrey residents speaking this language at home. It is followed by Chinese languages (just over 11,000 residents), Korean (just over 5,000), Hindi (just over 5,000), and Tagalog (just under 5,000) (British Columbia 2008).

32 The municipal legislation that now governs Coquitlam is the Community Charter (2003).This legislation lists 'fostering the economic, social and environmental well-being of its [a municipality's] community' as one of four municipal purposes (Community Charter, SBC 2003, c. 26, Part II, Division I, 7(d)). http://www.bclaws.ca/Recon/document/freeside/-%20C%20-/Community%20Charter%20%20SBC%202003%20%20c.%2026/00_Act/03026_02.xml#section7

33 This does not mean that local leaders do not recognize the importance of positive intercultural relations to creating a climate conducive to business prosperity and growth.

6: The Relationship between Urban Regimes, Types of Social Diversity, and Multiculturalism Policies

1 Hero deduces the three forms of political pluralism from the literature on state politics and from his theoretical framework. However, he does not describe the political dynamics within states in any detail to illustrate the nature of these forms of pluralism in practice.

2 The term 'race' is used in the categories, because they were developed on the basis of Statistics Canada's 'visible minority' category, which emphasizes colour and 'race,' defining visible minorities as 'non white in colour and non Caucasian in race.' However, many categories do not constitute what is commonly considered 'racial' categories. For instance, the categories 'Chinese' and 'South Asian' are arguably national and geographical constructions, not 'racial' ones.

3 However, multiracial municipalities share a common feature with Hero's (1998) heterogeneous configuration as well – the presence of a moderate number of 'white ethnics.'

4 These areas are Vancouver's eastside (including Strathcona, Grandview–Woodland, and South Main Street), Kerrisdale–Oakridge–Shaughnessy, Richmond, an area that includes parts of Delta and northwest Surrey, and an area in the Tri-Cities (Coquitlam, Port Coquitlam, Port Moody) (Hiebert 2003, 8–9).

5 David Edgington, Michael Goldberg, and Thomas Hutton (2003) also make this observation with respect to Richmond.

7: Multiculturalism and Multilevel Governance: The Role of Structural Factors in Managing Urban Diversity

1 Karim Karim (2008) refers to Canada's 'legislative and bureaucratic infrastructure' to support pluralism in a recent paper produced for the Global Centre for Pluralism's inaugural round table.

2 As Sellers (2002b) explains: 'An infrastructure may impose not only specific or direct effects on the conditions that elites and activists most immediately involved in governance face but diffuse or indirect effects through influences on preferences of political constituencies, consumers, and firms' (2002b, 623).

3 Mike Harris led the Conservative government from 1995 until April 2002, when Ernie Eves replaced him as leader. Eves was defeated by the Liberals under Dalton McGuinty in 2003. This chapter focuses on the Harris era, because many of the decisions that most affected urban-regime dynamics in Toronto were made during Harris's tenure as premier.

4 The Canada Assistance Plan was a shared-cost program between the federal and provincial governments to provide social services. This program was launched in 1966.

5 The Established Programs Financing grant was a block grant under which the federal government transferred money to the provinces to fund post-secondary education and health. This program was launched in 1977.

6 The CHST reduced funding for social services, post-secondary education, and health from $18.3 billion in 1995–6 to $12.5 billion in 1997–8 (Courchene 2005b, 4).

7 For example, the federal government introduced the Millennium Scholarship Program and the Child Tax Benefit.

8 Examples include the GST exemption, the federal gas-tax transfer (GTT), and homelessness grants under the Supporting Communities Partnership Initiative (SCPI).

9 In his view, in order to adjust to the new, knowledge-based economy paradigm, 'Ottawa transferred aspects of the old-paradigm of nation building (forestry, mining, energy etc.) to provinces, presumably in part to make room on the federal policy plate for new-paradigm policies and programs,' many of which imply new fiscal relationships with Canada's largest city-regions (Courchene 2005a, 13–14).

10 The Conservative party used this political slogan to describe its platform during the 1995 election. The 'revolution' involved a variety of neoliberal-inspired reforms, including budgetary reforms, dramatic cuts to social programs, downloading to cities, and reductions in income taxes.

11 An example of how local civil-society leaders pooled resources to develop the capacity to pressure upper levels of government to give Toronto a new status within Canadian federalism.
12 In this context, 'regional governing institutions' refer to corporate institutions for governing the entire metropolitan area or the Greater Toronto Area, and not to the regional municipalities that constitute the second tier of suburban municipalities in the Greater Toronto Area, such as Peel Region.
13 The province has since taken a leadership role in land-use planning policy.
14 Toronto Act Now is a 'new coalition of social advocacy and environmental groups [that] entered the fray … to ensure the voice of ordinary people is at least heard' (James 2005).
15 Dr Patricia McCarney, one of the local leaders who participated in the drafting of the Greater Toronto Area Charter (2001), is the Director of the Global Cities Program at the University of Toronto's Munk Centre for International Studies.
16 Others include crafting a new fiscal deal for cities, improving Toronto's physical infrastructure, reviving tourism in Toronto, creating a world-class research alliance, investing in people (through early childhood education, public education, and post-secondary education), strengthening the social and community infrastructure, and supporting the arts and cultural industries. See Toronto City Summit Alliance, 'Enough Talk: An Action Plan for the Toronto Region.' http://www.torontoalliance.ca/docs/TCSA_report.pdf
17 See TRIEC, 'About Us,' for a list of participants.
18 Regional Districts in British Columbia are governed by a board of directors 'comprising councillors appointed by and from the councils of incorporated municipalities within its boundaries and representatives elected from the population of the unorganized areas' (Tindal and Tindal 2004, 87). To account for differences in the size of constituent municipalities, a system of weighted voting exists with respect to budgetary decisions and when decisions are being made that will only affect a particular area of the district (ibid.). Regional districts do not levy their own taxes, but bill municipal governments for the services they provide their members. In the case of unincorporated municipalities, they bill the province (ibid., 88).
19 Murray's hypothesis that the GVRD might impede the development of a strong urban-autonomy movement owing to the fact that it brings together a diverse group of competing interests supports this interpretation (Murray 2004, slide 20).
20 John Sewell is a former Toronto mayor, a journalist, and a city activist.
21 Eligible projects included infrastructure projects in water, waste water,

solid-waste management, community energy, capacity building, and transit; and – in communities with populations of less than 500,000 – roads and bridges, presumably because in smaller communities basic infrastructure is less developed.

22 For instance, the population of Mississauga is similar to that of Winnipeg. However, because Mississauga is a suburban municipality that borders Toronto, its needs are very different from Winnipeg's. Winnipeg is the main urban centre in Manitoba and is much older. Its infrastructure is also much older. Furthermore, Mississauga's location in the GTA means that residents of Mississauga benefit from Toronto's infrastructure, including its public-transit system. Thus, the City of Toronto's portion of the GTT will benefit residents of the entire region who work in or visit Toronto. In the Greater Toronto Area, municipalities that border Toronto can 'free ride' on the city's infrastructure investments.

23 The AMO 'is a non-profit organization representing almost all of Ontario's 445 municipal governments and provides a variety of services and products to members and non-members' (AMO 2008). Its equivalent organization at the countrywide level is the Federation of Canadian Municipalities (FCM).

24 'Tax increment financing (TIF) is an economic development tool that municipalities can use to stimulate private investment and development in targeted areas by capturing the increased tax revenue generated by the private development itself and using the tax revenues to pay for public improvements and infrastructure necessary to enable development' (Bond n.d., 1).

25 See the Ministry of Citizenship and Immigration, Government of Ontario Fact Sheet, on the Canada–Ontario Immigration Agreement on the Ministry's website.

26 Annex F of the agreement deals with 'Partnerships with Municipalities.' The Annex recognizes that 'Canada and Ontario share a mutual interest in fostering partnerships and the participation of municipal governments and community and private sector stakeholders in immigration' (1.1). Priorities for tri-level partnerships include the following: information sharing and consultation (4.1), the attraction and retention of immigrants (4.2), and immigrant settlement and integration (4.3). The agreement will be implemented with the assistance of a Municipal Immigration Committee (5.2). According to the terms of the agreement, this new administrative structure is 'consistent with federal, provincial and AMO commitment to a New Deal for Cities and Communities which involves a seat at the table for municipalities on national issues most important to them' (Canada–Ontario Immigration Agreement, 2005).

27 Article 5.3 of Annex F acknowledges that 'in the past five years, up to 50% of all immigrants to Canada have arrived annually in the Toronto census metropolitan area alone and that this creates particular challenges in terms of maintaining services to all residents of this area while ensuring the efficient integration of newcomers into the community' (5.3.1) (ibid.).
28 Former Conservative cabinet minister David Emerson was elected in 2006 in the riding of Vancouver-Kingsway as a Liberal and crossed the floor after the election.
29 See for instance Horak (1998).

8: Municipal Multiculturalism Policies and the Capacity to Manage Social Change

1 Irene Bloemraad (2006) argues that funding for immigrant-settlement organizations is an important part of the explanation for Canada's greater level of success in immigrant-naturalization rates and levels of political incorporation relative to the United States.
2 Mike Harcourt was Mayor of Vancouver between 1980 and 1986 and Premier of British Columbia from 1991 to 1996. The current premier, Gordon Campbell, was Mayor of Vancouver from 1986 to 1993 and has been Premier of B.C. since 2001.
3 'White ethnics' were not included in the categorization, though interestingly, they tend to be more numerous in multiracial municipalities.
4 Hero (1998) nevertheless acknowledges: 'There is, of course, extensive inter- and intra-group complexity, and there also may be interminority political competition' (1998, 11).
5 Hero does not define what he means by hierarchical pluralism. However, it appears to mean that racial minorities have a say but an unequal one in relation to Whites. In his other work, Hero (1992) has developed the concept of 'two-tiered pluralism' to express a similar idea.
6 This is not to say that race-based inequalities do not exist in Canada. Rather, the question needs to be examined empirically. The only ethnoracial 'group' in Canada that approximates the African-American experience is Aboriginal peoples.

References

Abu-Laban, Yasmeen. 2008. 'Pluralism as Process: The Role of Liberal Democratic Institutions.' Unpublished paper presented at the inaugural round table of the Global Centre for Pluralism, Ottawa, 13 May.

– 1997. 'Ethnic Politics in a Globalizing Metropolis: The Case of Vancouver.' In *Politics of the City: A Canadian Perspective,* ed. Timothy L. Thomas. Scarborough: Nelson. 77–96.

Abu-Laban, Yasmeen, and Christina Gabriel. 2002. *Selling Diversity: Immigration, Multiculturalism, Employment Equity, and Globalization*. Peterborough: Broadview.

Access Alliance Multicultural Community Health Centre. 2003 (March). 'Best Practices for Working with Homeless Immigrants and Refugees.' Toronto. http://atwork.settlement.org/sys/atwork_library_detail.asp?doc_id=1003145

Advisory Panel on the Fiscal Imbalance. 2006. 'Reconciling the Irreconcilable: Addressing Canada's Fiscal Imbalance.' Report submitted to the Council of the Federation, 31 March 2006. http://www.councilofthefederation.ca/pdfs/Report_Fiscalim_Mar3106.pdf

Ail, Josh, Dawn Dobson-Borsoi, and Rob Eley. n.d. 'Closing the Cultural Gap: The City of Surrey.' Report commissioned by the City of Surrey. Burnaby: B.C. Institute of Technology.

AMO (Association of Municipalities of Ontario). 'A Guide to AMO Member Benefits and Services.' http://www.amo.on.ca/AM/Template.cfm?Section=About_AMO&Template=/CM/ContentDisplay.cfm&ContentID=152512

AMSSA. 'AMSSA Membership.' http://www.amssa.org/members/index.htm

Andrew, Caroline. 2001. 'The Shame of (Ignoring) the Cities.' *Journal of Canadian Studies* 35, no. 4: 100–10.

Andrew, Caroline, Katherine A. Graham, and Susan D. Philips, eds. 2002. *Urban Affairs: Back on the Policy Agenda*. Montreal and Kingston: McGill–Queen's University Press.

Andrew, Caroline, and Patrick Smith. 1999. 'World-Class Cities: Can or Should Canada Play?' In *World-Class Cities: Can Canada Play?* ed. Caroline Andrew, Pat Armstrong, and André Lapierre. Ottawa: University of Ottawa Press. 5–25.

Bailey, Robert W. 1999. *Gay Politics, Urban Politics: Identity and Economics in the Urban Setting*. New York: Columbia University Press.

Banfield, Edward C., and James Q. Wilson. 1963. *City Politics*. Cambridge, MA: Harvard University Press.

Banting, Keith, George Hoberg, and Richard Simeon, eds. 1997. *Degrees of Freedom: Canada and the United States in a Changing World*. Montreal and Kingston: McGill–Queen's University Press.

Banting, Keith, Richard Johnston, Will Kymlicka, and Stuart Soroka. 2006. 'Do Multiculturalism Policies Erode the Welfare State? An Empirical Analysis.' In *Multiculturalism and the Welfare State: Recognition and Redistribution in Contemporary Democracies*, ed. Keith Banting and Will Kymlicka. New York: Oxford University Press. 49–91.

Basran, Gurcharn S., and B. Singh Bolaria. 2003. *The Sikhs in Canada: Migration, Race, Class, and Gender*. New Dehli: Oxford University Press.

Belgrave, Roger. 1995. 'Bell Supporters Come Out in Force.' *Markham Economist and Sun*, 30 August.

Bell, Carole. 1995. Letter to the Editor. *Markham Economist and Sun*, 16 August.

Berdahl, Loleen. 2006. 'How the Harper Government Can Court Canada's Big Cities.' *Calgary Herald*, 26 February.

Bloemraad, Irene. 2006. *Becoming a Citizen: Incorporating Immigrants and Refugees in the United States and Canada*. Berkeley: University of California Press.

Bond, Kenneth W. n.d. 'Tax Increment Financing – Can You? Should You?' Squire Sanders Legal Counsel Worldwide. http://www.nysedc.org/memcenter/TIF%20Paper.pdf

Boudreau, Julie-Anne. 2006. 'Intergovernmental Relations and Polyscalar Social Mobilization: The Cases of Montreal and Toronto.' In *Canada: The State of the Federation 2004*, ed. Robert Young and Christian Leuprecht. Montreal and Kingston: McGill-Queen's University Press. 161–80.

Bradford, Neil. 2002. 'Why Cities Matter: Policy Research Perspectives for Canada.' Canadian Policy Research Networks (CPRN) Discussion Paper no. F23, June. http://www.cprn.org/doc.cfm?doc=168&l=en

Brampton Board of Trade. http://www.bramptonbot.com

– 'Mandate of Multiculturalism Committee.' http://www.bramptonbot.com/committees.htm

Brenner, Neil. 2004. *New State Spaces: Urban Governance and the Rescaling of Statehood*. Oxford: Oxford University Press.
Bridson-Boyczuk, Karen. 2004. 'Business Leaders Singing City's Praises: Business Leaders Still Bullish about City.' *Mississauga News*, 27 November.
Broadbent, Alan. 2008. *Urban Nation: Why We Need to Give Power Back to the Cities to Make Canada Strong*. Toronto: HarperCollins.
– 2003. 'New Structures/New Connections.' Speech to Conference on Municipal–Federal–Provincial Relations. Queen's University, Institute of Intergovernmental Relations, 9–10 May, Kingston. http://www.queensu.ca/iigr/conf/Arch/03/03-2/Broadbent.pdf
– 2001. 'The Philanthropic Contract: Mutual Benefit for the Public Good.' Ottawa: Caledon Institute of Social Policy, June. http://www.scribd.com/doc/8177190/Alan-Broadbent-The-Philanthropic-Contract-2001
Broadbent Group. 2005. *Towards a New City of Toronto Act*. Toronto: Zephyr.
Browning, Rufus P., Dale R. Marshall, and David H. Tabb. 1984. *Protest Is Not Enough: The Struggle of Blacks and Hispanics for Equality in the United States*. Berkeley: University of California Press.
Burns, Peter. 2002. 'The Intergovernmental Regime and Public Policy in Hartford, Connecticut.' *Journal of Urban Affairs* 24, no. 1: 55–73.
Cairns, Alan C. 1985. 'The Embedded State: State-Society Relations in Canada.' In *State and Society: Canada in Comparative Perspective*, ed. Keith Banting. Toronto: University of Toronto Press, in cooperation with the Royal Commission on the Economic Union and Development Prospects for Canada and the Canadian Government Publications Centre. Ottawa: Supply Services Canada. 53–86.
'Canada's Cities: Unleash Our Potential.' http://www.canadascities.ca/index.htm
Canada-Ontario Immigration Agreement. 2005. 'Annex F,' Citizenship and Immigration, http://www.cic.gc.ca/EnGLIsh/department/laws-policy/agreements/ontario/ont-2005-annex-f.asp
Canada, Ontario, Association of Municipalities of Ontario, and City of Toronto. 2005. Agreement for the Transfer of Federal Gas Tax Revenues under the New Deal for Cities and Communities. 17 July. http://www.infc.gc.ca/altformats/pdf/gtf-fte-on-eng.pdf
Canadian Press. 2006. 'Queen's Park Welcomes Immigration Funding from Feds,' *Toronto Star*, 15 December. http://www.thestar.com/article/153151
Carabram. http://www.carabram.org
Carens, Joseph H. 2000. *Culture, Citizenship, and Community: A Contextual Exploration of Justice as Evenhandedness*. Oxford: Oxford University Press.
Carassauga. http://www.carassauga.mantis.biz

CERIS. 'Governance Board and Directors.' http://ceris.metropolis.net/frameset_e.html

Chin, Joe. 2008. '"Forget about Being Debt-Free": Hazel.' *Mississauga News*, 12 June. http://www.mississauga.com/article/15194

Chong, Dennis. 1991. *Collective Action and the Civil Rights Movement*. Chicago: University of Chicago Press.

CIC (Citizenship and Immigration Canada). 2002. *Facts and Figures 2002: Immigration Overview*. http://www.cic.gc.ca/english/pub/facts2002/immigration/immigration_4.html

Citrin, Jack, Beth Reingold, and Donald P. Green. 1990. 'The "Official English" Movement and the Symbolic Politics of Language in the United States.' *Western Political Quarterly* 43, no. 3: 535–60.

City of Brampton. 2007. 'City of Brampton Approves Multilingual Program.' Media release 07–006, 17 January 2007. http://www.brampton.ca/media-releases/07-006.pdf

– 2006. 'Brampton Launches Pilot Project for Multilingual Telephone Services.' Media release 06-017, 27 January 2006. http://www.city.brampton.on.ca/media-releases/06-017.tml

City of Coquitlam. 2007a. 'Multiculturalism Advisory Committee.' Minutes – Regular Committee Meeting, 6 November 2007. http://www.coquitlam.ca/NR/rdonlyres/99B3D5CD-D1AD-474C-A2AF-CD77724D0013/76366/CITYDOCS616297v1RC_COTW_feb252008_59911.PDF

– 2007b. 'To: Acting City Manager. From: Manager Corporate Planning. Subject: Participation on Provincial Joint Steering Committee on Multiculturalism. For: Committee of the Whole.' 4 October. File 01-0620-01/000/2007-1.

– 2006. 'Multiculturalism Advisory Committee – Terms of Reference.' Policy and Procedure Manual, ch. 5. November. File 01-0540-00/01-002/1, Doc. 457607 vol. 1.

– 2003. 'Liveable Communities Advisory Committee.' Policy and Procedure Manual, 3 November, rev. 19 January 2004, File 01-0540-20/536/2003-1, Doc 213113.

– 2002. 'Official Results of the 2002 Local Government Election.' http://www.coquitlam.ca/City+Hall/City+Government/Election/_Official+Results+of+the+2002+Local+Government+Election.htm

City of Mississauga. 'City Hall – Languages.' http://www.mississauga.ca/portal/cityhall/languages

– 'City of Mississauga Integrates New Canadians Into the Workforce.' http://www.mississauga.ca/portal/cityhall/pressreleases?pdf

City of Richmond. 2002. 'Proposed Richmond Intercultural Advisory Committee.' Report to General Purposes Committee. File 4055-01. 7 February.

– 1996. 'Advisory Committee on Intercultural Relations – Minutes.' 18 April 1996.
– 1994a. 'Memorandum re: Report on Multiculturalism.' File 0100-E6. 7 January.
– 1994b. 'Report to Committee of the Whole re: Ethnic Relations.' File 0100-E6. 15 August.
City of Surrey. 2006. News release: 'Mayor Dianne Watts appoints Coun. Mary Martin as Surrey's first Multicultural Advisory Committee Chair.' 5 December. http://www.surrey.ca/Whats+New/News+Releases/Old+--+2006/December/Mayor+Dianne+Watts+appoints+Coun.+Marvin+Hunt+Chair+of+Intergovernmental+Advisory+Committee.htm
– 1997. 'Task Force Report on Intercultural Inclusivity: Reaching Out in Surrey.' Parks and Recreation Department. January.
City of Toronto. 'Diversity Advocate.' http://www.toronto.ca/diversity/advocate.htm
– 2006. Backgrounder: '2006 City Operating Budget of $7.6 Approved by Council.' 30 March. http://www.toronto.ca/budget2006/pdf/2006operatingbackgrounder_march30_final.pdf
– 2004. 'Review of the Implementation of Recommendations of the Final Report of the Task Force on Community Access and Equity.' Clause embodied in Report no. 3 of the Audit Committee, as adopted by the Council of the City of Toronto at its meeting on 18–20 May 2004.
– 2001. 'Immigration and Settlement Policy Framework.' Clause embodied in Report no. 4 of the Community Services Committee, as adopted by the Council of the City of Toronto at its meeting held on 30 May–1 June 2001.
City of Vancouver. 'Multiculturalism and Diversity: Role of Social Planning in Multiculturalism and Diversity,' http://vancouver.ca/commsvcs/socialplanning/initiatives/multicult/civicpolicy.htm
– n.d. 'Building Inclusive Community: Diversity Policy and Practice in the City of Vancouver.'
– 2007. 'Report of the Mayor's Task Force on Immigration,' November. http://vancouver.ca/commsvcs/socialplanning/initiatives/multicult/PDF/0711_MTFI_report.pdf
– 2005. 'Administrative Report.' Mayor's Working Group on Immigration in consultation with Director of Social Planning. http://vancouver.ca/ctyclerk/cclerk/20051004/documents/rr1.pdf
– 1999. 'Administrative Report'. RTS No. 943, CC File No. 1304.
– 1997. 'Equal Employment Opportunity Program: Progress Report 1986–1996.' May.

City Solicitor, City of Toronto. 2000. 'Powers of Canadian Cities – The Legal Framework.' June (updated October 2001). http://www.canadascities.ca

Clarke, Susan E. 1999. 'Regional and Transnational Regimes: Multi-Level Governance Processes in North America.' Paper presented at the annual meeting of the American Political Science Association, September, Atlanta.

Clarke, Susan E., and Gary L. Gaile. 1998. *The Work of Cities*. Minneapolis: University of Minnesota Press.

Clutterbuck, Peter, and Rob Howarth. 2002. 'Toronto's Quiet Crisis: The Case for Social and Community Infrastructure Investment.' Research Paper no. 198, University of Toronto, November. http://www.urbancentre.utoronto.ca/pdfs/curp/Clutterbuck_198_Toronto.pdf

Coquitlam Now. 2008. 'It's Official – Coquitlam Is a Cultural Capital of Canada,' http://www2.canada.com/coquitlamnow/news/story.html?id=dd6644c0-8034-410d-9696-c5124ef195a1&k=63027

Courchene, Tom. 2007. 'Global Futures for Canada's Global Cities.' Montreal: Institute for Research on Public Policy. *Policy Matters* 8, no. 2 (June). http://www.irpp.org/wp/archive/wp2005-03.pdf

– 2005a. 'City State and the State of Cities: Political-Economy and Fiscal Federalism Dimensions.' Montreal: Institute for Research on Public Policy. http://www.irpp.org/wp/archive/wp2005-03.pdf

– 2005b. 'Vertical and Horizontal Fiscal Imbalances: An Ontario Perspective.' Background notes for a presentation to the Standing Committee on Finance, House of Commons, 4 May.

Dahl, Robert A. 1961. *Who Governs? Democracy and Power in an American City*. New Haven: Yale University Press.

D'Alessandro, Dominic. 2004. 'Capitalize on Immigrants' Promise.' *Globe and Mail*, 15 September.

Dewing, Michael, and Marc Leman. 2006. 'Canadian Multiculturalism.' Library of Parliament, p. 23. http://www.parl.gc.ca/information/library/PRBpubs/936-e.htm

DiGaetano, Alan, and Elizabeth Strom. 2003. 'Comparative Urban Governance: An Integrated Approach.' *Urban Affairs Review* 38, no. 3: 356–93.

Diverse City. 'The Greater Toronto Leadership Project.' http://www.diversecitytoronto.ca

Doucet, Michael. 2001. 'The Anatomy of an Urban Legend: Toronto's Multicultural Reputation.' CERIS working paper no. 16. April. http://ceris.metropolis.net/Virtual%20Library/WKPP%20List/WKPP2001/CWP16_Doucet_final.pdf

– 1999. 'Toronto in Transition: Demographic Change in the Late Twentieth Century.' CERIS working paper no. 6. May. http://ceris.metropolis.net/research-policy/wkpp_list.htm

Edgington, David W., Michael A. Goldberg, and Thomas Hutton. 2003. 'The Hong Kong Chinese in Vancouver.' Research on Immigration and Integration in the Metropolis (RIIM), Vancouver Centre of Excellence, working paper series no. 03–12. http://mbc.metropolis.net/research/working/2003.html

Edgington, David W., Bronwyn Hanna, Thomas Hutton, and Susan Thompson. 2001. 'Urban Governance, Multiculturalism, and Citizenship in Sydney and Vancouver.' Research on Immigration and Integration in the Metropolis (RIIM), Vancouver Centre of Excellence, working paper series no. 01–05. http://mbc.metropolis.net/research/working/2001.html

Edgington, David W., and Thomas A. Hutton. 2002. 'Multiculturalism and Local Government in Greater Vancouver.' Research on Immigration and Integration in the Metropolis (RIIM), Vancouver Centre of Excellence, working paper series no. 02–06. http://mbc.metropolis.net/research/working/2002.html

Elkin, Stephen. 1987. *City and Regime in the American Republic*. Chicago: University of Chicago Press.

Fainstein, Norman I., and Susan S. Fainstein. 1986. *Restructuring the City*. New York: Longman.

FCCM (Federation of Chinese Canadians in Markham). http://www.fccm.ca/index2.html

Feldman, Lionel. 2006. 'Urban Policy Making: Substance, Process Strategy in the Quest for a New Deal for Toronto.' Speech delivered at the Local and Urban Politics section (CPSA) dinner at annual meeting of the Canadian Political Science Association, 3 June, Toronto.

Fellegi, Ivan P. 1997. 'On Poverty and Low Income.' Statistics Canada.

Ferman, Barbara. 2002. 'Broadening the Franchise: Charter Reform and Neighborhood Representation in Los Angeles.' Prepared for the John Randolph Haynes and Dora Haynes Foundation Conference on Reform, L.A. Style: The Theory and Practice of Urban Governance at Century's Turn. University of Southern California School of Policy, Planning, and Development, Los Angeles, 19–20 September 2002.

– 1996. *Challenging the Growth Machine: Neighborhood Politics in Chicago and Pittsburgh*. Lawrence: University Press of Kansas.

Florida, Richard. 2002. *The Rise of the Creative Class*. New York: Basic.

Francis, Diane. 2001. 'Cities Fight for Fair Refugee Policy.' *National Post*, 15 May.

Frisken, Frances, and Marcia Wallace. 2003. 'Governing the Multicultural City-Region.' *Canadian Public Administration* 46, no. 2: 153–77.

– 2000. 'The Response of the Municipal Public Service Sector to the Challenge of Immigrant Settlement.' Rev. May 2002. http://www.settlement.org/downloads/Municipal_Sector.pdf

Funders' Network on Racism and Poverty. http://www.rapnet.ca

Garber, Judith A., and David L. Imbroscio. 1996. 'The Myth of the North American City Reconsidered: Local Constitutional Regimes in Canada and the United States.' *Urban Affairs Review* 31, no. 5: 595–624.

GCIM (Global Commission on International Migration). 2005. 'Migration in an Interconnected World: New Directions for Action.' 88. http://www.gcim.org/attachements/gcim-complete-report-2005.pdf

GCIR (Grantmakers Concerned with Immigrants and Refugees). 'About GCIR.' http://www.gcir.org/about

Gherson, Giles. 2005. 'Push Pedal to Metal on New Deal Campaign.' *Toronto Star*, 5 February, A2.

Gillespie, Kerry. 2004. 'A Star Recruit to Fight for Cities: Honderich Hired as Mayor's Envoy – He'll Champion Urban "New Deal."' *Toronto Star*, 27 May, B1.

Goldberg, Michael A, and John Mercer. 1986. *The Myth of the North American City: Continentalism Challenged*. Vancouver: UBC Press.

Good, Kristin. 2007. 'Urban Regime Building as a Strategy of Intergovernmental Reform: The Case of Toronto's Role in Immigrant Settlement.' Paper presented at the Canadian Political Science Association Annual Conference, Saskatoon, May 30–June 1.

– 2006. 'Multicultural Democracy in the City: Explaining Municipal Responsiveness to Immigrants and Ethnocultural Minorities.' PhD diss., Department of Political Science, University of Toronto.

– 2005. 'Patterns of Politics in Canada's Immigrant-Receiving Cities and Suburbs: How Settlement Patterns Shape the Municipal Role in Multiculturalism Policy.' *Policy Studies* 26, nos. 3 and 4: 261–89.

– 2004. 'Explaining Municipal Responsiveness to Immigration: An Urban Regime Analysis of Toronto and Mississauga.' Centre for Urban and Community Studies, Research Paper no. 199, December 2004.

Government of British Columbia. February 2008. '2006 Census Fact Sheet: Surrey.' Welcome BC: Multiculturalism and Immigration Branch, February. http://www.llbc.leg.bc.ca/public/PubDocs/bcdocs/440579/surrey2006.pdf

Government of Canada. 2003. *Multiculturalism: Respect, Equality, Diversity: Program Guidelines*. July. Ottawa: Canadian Heritage.

Government of Canada. External Advisory Committee on Cities and Communities. 2006. Final Report: *From Restless Communities to Resilient Places: Building a Stronger Future for All Canadians*. http://www.civicgovernance.ca/files/uploads/munities_to_resilient_places_Mike_Harcourt_0.pdf

Government of Ontario. 2008. 'Fact Sheet: Uploading Ontario Works.' http://

www.news.ontario.ca/mah/en/2008/10/fact-sheet-uploading-ontario-works.html

Graham, Katherine A., Susan D. Philips, and Allan M. Maslove. 1998. *Urban Governance in Canada: Representation, Resources, and Restructuring*. Toronto: Harcourt Canada.

GVRD (Greater Vancouver Regional District), Policy and Planning Department. 2003. '2001 Census Bulletin no. 6 – Immigration.' February.

Hansen, Darah. 2003. 'RCMP No Longer Colour Blind.' *Richmond News*, 22 November, pp. 1, 3.

Harding, Alan. 1995. 'Elite Theory and Growth Machines.' In *Theories of Urban Politics*, ed. David Judge, Gerry Stoker, and Harold Wolman. Thousand Oaks: Sage. 35–53.

Hiebert, Daniel. 2003. 'Are Immigrants Welcome? Introducing the Vancouver Community Studies Survey.' RIIM working paper no. 03-06. March. http://mbc.metropolis.net/Virtual%20Library/2003/wp03-06.pdf

Hero, Rodney E. 2007. *Racial Diversity and Social Capital: Equality and Community in America*. New York: Cambridge University Press.

– 2003. 'Multiple Theoretical Traditions in American Politics and Racial Policy Inequality.' *Political Research Quarterly* 56, no. 4 (December): 401–8.

– 1998. *Faces of Inequality*. New York: Oxford University Press.

– 1992. *Latinos and the U.S. Political System: Two-Tiered Pluralism*. Philadelphia: Temple University Press.

Hero, Rodney E., and Robert R. Preuhs. 2006. 'Multiculturalism and Welfare Policies in the U.S.A.: A State-Level Comparative Analysis.' In *Multiculturalism and the Welfare State: Recognition and Redistribution in Contemporary Democracies*, ed. Keith Banting and Will Kymlicka. New York: Oxford University Press. 121–51.

Honderich, John. 2002. 'Wanted: A New Deal for Canada's Cities.' *Toronto Star*, 12 January, A1.

Horak, Martin. 1998. 'The Power of Local Identity: C4LD and the Anti-Amalgamation Mobilization in Toronto.' Centre for Urban and Community Studies, Research Paper no. 195.

Horan, Cynthia. 2002. 'Racializing Regime Politics: Innovations in Regime Theory.' *Journal of Urban Affairs* 24, no. 1: 19–33.

Huhtala, Kari. 2004. 'Richmond – A City of Cultural Fusion and Change.' Information leaflet, 17 February.

Hunter, Floyd. 1953. *Community Power Structure: A Study of Decision Makers*. Chapel Hill: University of North Carolina Press.

Hurst, Lynda. 2005. 'How the Doctrine of "Frenchness" Failed.' *Toronto Star*, 10 November, A6.

Hutton, Thomas. 1998. 'The Transformation of Canada's Pacific Metropolis: A Study of Vancouver.' Montreal: Institute for Research on Public Policy.

Imbroscio, David L. 1997. *Reconstructing City Politics: Alternative Economic Development Regimes*. Thousand Oaks: Sage.

Inglis, Christine. 1996. 'Multiculturalism: New Policy Responses to Diversity.' MOST Policy Papers no. 4. Paris: UNESCO, http://www.unesco.org/most/pp4.htm#practice

IOM (International Organization for Migration). 'Regional and County Figures.' http://www.iom.int/jahia/Jahia/pid/255

Ip, David. 2005. 'Contesting Chinatown: Place-Making and the Emergence of "Ethnoburbia" in Brisbane, Australia.' *GeoJournal* 64, no. 1: 63–74.

James, Royson. 2005. 'New Deal Gathers Steam: Miller Pleads with Councillors to Mobilize Constituents – City, Province to Hold Public Meetings on Issues Next Week.' *Toronto Star*, 15 June, B1.

Jones, Brian D., and Lynn W. Bachelor. 1993. *The Sustaining Hand: Community Leadership and Corporate Power*, 2nd rev. ed. Lawrence: University Press of Kansas.

Judge, David. 1995. 'Pluralism.' In *Theories of Urban Politics*, ed. David Judge, Gerry Stoker, and Harold Wolman. Thousand Oaks: Sage. 13–34.

Karim, Karim H. 2008. 'Recognizing Difference: How Has Multiculturalism Actually Worked?' Unpublished paper presented at the inaugural round table of the Global Centre for Pluralism, Ottawa, 13 May.

Karyo Communications. 2000. 'Intercultural Marketing Plan.' Surrey Parks, Recreation, and Culture, City of Surrey, 1 November.

Keil, Roger, and Douglas Young. 2003. 'A Charter for the People? A Research Note on the Debate about Municipal Autonomy in Toronto.' *Urban Affairs Review* 39, no. 1: 87–102.

Kingdon, John. 1995. *Agendas, Alternatives, and Public Policies*. New York: Longman.

Kipfer, Stefan, and Roger Keil. 2002. 'Toronto Inc? Planning the Competitive City in the New Toronto.' *Antipode* (March): 227–64.

Kolb, Emil. 2004. Letter to the Honourable Judy Sgro, Office of the Chair, Regional Municipality of Peel, 19 August.

Koschinsky, Julia, and Todd Swanstrom. 2001. 'Confronting Policy Fragmentation: A Political Approach to the Role of Housing Nonprofits.' *Policy Studies Review* 18, no. 4: 111–27.

Krivel, Peter. 1995. 'Councillor Sparks "Racism" Protest.' *Toronto Star*, 21 August, A6.

Kymlicka, Will. 2008. 'Canadian Pluralism in Comparative Perspective.' Unpublished paper presented at the inaugural round table of the Global Centre for Pluralism, Ottawa, 13 May.

– 2003. 'Being Canadian.' *Government and Opposition* 38, no. 3: 357–85.
– 1998. *Finding Our Way: Rethinking Ethnocultural Relations in Canada*. Toronto: Oxford University Press.
– 1995. *Multicultural Citizenship*. Toronto: Oxford University Press.
Lauria, Mickey, ed. 1997. *Reconstructing Urban Regime Theory: Regulating Urban Politics in a Global Economy*. Thousand Oaks: Sage.
Laurier Institution. 2005. 'Diversity 2010: Leveraging the 2010 Olympic and Paralympic Games to Brand and Make Greater Vancouver a Truly Multicultural Community.' Reference WD 2351, 31 March 2005. http://www.thelaurier.ca/pdf/diversity_2010.pdf
Leo, Christopher. 2006. 'Deep Federalism: Respecting Community Difference in National Policy.' *Canadian Journal of Political Science* 39, no. 3: 481–506.
– 2003. 'Are There Urban Regimes in Canada? Comment on Timothy Cobban's "The Political Economy of Urban Development: Downtown Revitalization in London, Ontario, 1993–2002."' *Canadian Journal of Urban Research* 12, no. 2: 344–8.
– 1998. 'Regional Growth Management Regime: The Case of Portland, Oregon.' *Journal of Urban Affairs* 20: 363–94.
– 1997. 'City Politics in an Era of Globalization.' In *Reconstructing Urban Regime Theory: Regulating Urban Politics in a Global Economy*, ed. Mickey Lauria. Thousand Oaks: Sage.
Levi, Margaret. 1996. 'Social Capital and Unsocial Capital: A Review Essay of Robert Putnam's *Making Democracy Work*.' *Politics and Society* 24 (March): 45–55.
Ley, David. 2007. 'Multiculturalism: A Canadian Defence.' Research on Immigration and Integration in the Metropolis, Vancouver Centre of Excellence, working paper no. 07–04. http://mbc.metropolis.net/Virtual%20Library/2007/WP07-04.pdf
Ley, David, Daniel Hiebert, and Geraldine Pratt. 1992. 'Time to Grow Up? From Urban Village to World City, 1966–91.' In *Vancouver and its Region*, ed. Graeme Wynn and Timothy Oke. Vancouver: UBC Press, 234–66.
Ley, David, Peter Murphy, Kris Olds, and Bill Randolph. 2001. 'Immigration and Housing in Gateway Cities: The Cases of Sydney and Vancouver.' RIIM, Vancouver Centre of Excellence, working paper no. 01–03. http://mbc.metropolis.net/Virtual%20Library/2001/wp0103.pdf
Li, Wei. 2007. 'Introduction: Asian Immigration and Community in the Pacific Rim.' In *From Urban Enclave to Ethnic Suburb: New Asian Communities in Pacific Rim Countries*, ed. Wei Li. Honolulu: University of Hawaii Press. 1–22.
– 1999. 'Building Ethnoburbia: The Emergence and Manifestation of the Chinese Ethnoburb in Los Angeles' San Gabriel Valley.' *Journal of Asian American Studies* 2, no. 1: 1–28.

– 1997. 'Ethnoburb versus Chinatown: Two Types of Urban Ethnic Communities in Los Angeles.' *Cybergeo: European Journal of Geography* (online). http://www.cybergeo.eu/index1018.html

Lichbach, Mark Irving, and Alan S. Zuckerman, eds. 1997. *Comparative Politics: Rationality, Culture, and Structure*. Cambridge: Cambridge University Press.

Lidstone, Donald. 2005. 'Comments on the City of Toronto Act.' http://www.toronto.ca/mayor_miller/torontoact_comments

Lijphart, Arend. 1975. 'The Comparable-Cases Strategy in Comparative Research.' *Comparative Political Studies* 8, no. 2: 158–77.

Lim, April, Lucia Lo, Myer Siemiatycki, and Michael Doucet. 2005. 'Newcomer Services in the Greater Toronto Area: An Exploration of the Range and Funding Sources of Settlement Services.' CERIS working paper no. 35, January. http://ceris.metropolis.net/Virtual%20Library/WKPP%20List/WKPP2005/CWP35_Lim-Lo-Siemiatycki_Final.pdf

Link, Michael W., and Robert W. Oldendick. 1996. 'Social Construction and White Attitudes toward Equal Opportunity and Multiculturalism.' *Journal of Politics* 55, no. 1: 191–206.

Lipsky, Michael M. 1976. 'Toward a Theory of Street-Level Bureaucracy.' In *Theoretical Perspectives on Urban Politics*, ed. Willis D. Hawley and Michael Lipsky. Englewood Cliffs: Prentice-Hall. 196–213.

Lo, Lucia, Lu Wang, Shuguang Wang, and Yinhuan Yuan. 2007. 'Immigrant Settlement Services in the Toronto CMA: A GIS-Assisted Analysis of Supply and Demand.' CERIS working paper no. 59, July. http://ceris.metropolis.net/Virtual%20Library/WKPP%20List/WKPP2007/CWP59.pdf

Logan, John R., and Harvey L. Molotch. 1987. *Urban Fortunes: The Political Economy of Place*. Berkeley: University of California Press.

Long, Norton E. 1968. 'The Local Community as an Ecology of Games.' In *The Search for Community Power*, ed. Willis D. Wawley and Frederick M. Wirt. Englewood Cliffs: Prentice-Hall.

Lorinc, John. 2006. *The New City: How the Crisis in Canada's Urban Centres Is Reshaping the Nation*. Toronto: Penguin.

MacMillan, Melville. 2006. 'Municipal Relations with the Federal and Provincial Governments: A Fiscal Perspective.' In *Canada: The State of the Federation 2004*, ed. Robert Young and Christian Leuprecht. Montreal and Kingston: McGill–Queen's University Press.

Magnusson, Warren. 2005. 'Are Municipalities Creatures of the Provinces?' *Journal of Canadian* Studies 39, no. 2: 5–29.

– 1985. 'The Local State in Canada: Theoretical Perspectives.' *Canadian Public Administration* 28, no. 4: 575–99.

Markham Board of Trade. 1995. Media release. 29 August.

Mascoll, Philip, and Jim Rankin. 2005. 'Racial Profiling Exists.' *Toronto Star,* 31 March, A1, A20.

Mayor's Advisory Committee. 1996. 'Working Together Towards Better Understanding and Harmony in the Town of Markham.' June.

Maytree Foundation. 2004. 'Annual Report.' Toronto. http://maytree.com/PDF_Files/AnnualReport2004.pdf

McIsaac, Elizabeth. 2003. 'Nation Building through Cities: A New Deal for Immigrant Settlement in Canada.' Ottawa: Caledon Institute of Social Policy. http://epe.lac-bac.gc.ca/100/200/300/caledon_institute/nation_building/553820436.pdf

McRoberts, Kenneth. 1997. *Misconceiving Canada: The Struggle for National Unity*. Toronto: Oxford University Press.

Metropolis. http://canada.metropolis.net/index_e.html

Metro Vancouver. 'Frequently Asked Questions': Facts and Figures. http://www.metrovancouver.org/about/Pages/faqs.aspx

Miller, David. 2004. 'Mayor's Remarks at the News Conference with the Premier and the Minister of Municipal Affairs and Housing at the Introduction of the New City of Toronto Act.' http://www.toronto.ca/mayor_miller/speeches/cta_remarks.htm

Ministry of Citizenship and Immigration, Government of Ontario. 'Canada and Ontario sign historic immigration agreement.' http://www.news.ontario.ca/archive/en/2005/11/21/c768397cd-27996-tpl.html

Mississauga News. http://www.mississauga.com

Mitchell, Katharyne. 2004. *Crossing the Neoliberal Line: Pacific Rim Migration and the Metropolis*. Philadelphia: Temple University Press.

Molotch, Harvey. 1976. 'The City as a Growth Machine.' *American Journal of Sociology* 82, no. 2: 309–30.

Mossberger, Karen, and Gerry Stoker. 2001. 'The Evolution of Urban Regime Theory: The Challenge of Conceptualization.' *Urban Affairs Review* 36, no. 6: 810–35.

Murray, Heather. 2006. 'Rethinking Intergovernmental Relations in Canada? An Analysis of City–Provincial Relations in Winnipeg and Vancouver.' Paper prepared for the annual conference of the Canadian Political Science Association, York University, Toronto, 1 June. http://www.cpsa-acsp.ca/papers-2006/Murray-Heather.pdf

– 2004. 'The Urban Autonomy Debate in Canada: The Rhetoric, the Rationale, and the Reality.' Presentation prepared for the Local Government Institute and School of Public Administration Seminar Series, University of Victoria, 22 October. http://publicadmin.uvic.ca/cpss/lgi/pdfs/murray_oct22_presentation.pdf

Nevitte, Neil. 1996. *The Decline of Deference: Canadian Value Change in Cross-National Perspective*. Mississauga: Broadview.

Olds, Kris. 2001. *Globalization and Urban Change: Capital, Culture, and Pacific Rim Megaprojects*. Oxford: Oxford University Press.

Olson, Mancur. 1965. *The Logic of Collective Action*. Cambridge, MA: Harvard University Press.

Ornstein, Michael. 2006. 'Ethno-Racial Groups in Toronto: A Demographic and Socio-Economic Profile 1971–2001' (Ornstein Report update). Institute for Social Research, York University, January. http://www.isr.yorku.ca/download/Ornstein--Ethno-Racial_Groups_in_Toronto_1971-2001.pdf

– 2000. 'Ethno-Racial Inequality in the City of Toronto: An Analysis of the 1996 Census' (Ornstein Report). Prepared for the Access and Equity Unit Strategic and Corporate Policy Division, Chief Administrator's Office, City of Toronto, May. http://www.toronto.ca/diversity/pdf/ornstein_fullreport.pdf Ornstein. 2000.

Orr, Marion. 1999. *Black Social Capital: The Politics of School Reform in Baltimore, 1986–1998*. Lawrence: University Press of Kansas.

Orr, Marion, and Valerie C. Johnson. 2008. *Power in the City: Clarence Stone and the Politics of Inequality*. Lawrence: University Press of Kansas.

Orr, Marion, and Gerry Stoker. 1994. 'Urban Regimes and Leadership in Detroit.' *Urban Affairs Quarterly* 30, no. 1: 48–73.

Owens, Michael. 2001. 'Pulpits and Policy: The Politics of Black Church-Based Community Development in New York City, 1980–2000.' PhD diss., State University of New York, Albany.

Pal, Leslie. 1995. *Interests of State: The Politics of Language, Multiculturalism, and Feminism in Canada*. Montreal and Kingston: McGill–Queen's University Press.

Peel Newcomer Strategy Group. 'Our History.' http://peelnewcomer.org/index.php?option=com_content&task=view&id=16&Itemid=167

Peterson, Paul E. 1981. *City Limits*. Chicago: University of Chicago Press.

Pierre, Jon. 2005. 'Comparative Urban Governance: Uncovering Complex Causalities.' *Urban Affairs Review* 40, no. 4: 446–62.

Pierson, Paul. 1993. 'When Effect Becomes Cause: Policy Feedback and Political Change.' *World Politics* 45 (July): 595–28.

Poirier, Christian. 2006. 'Ethnocultural Diversity, Democracy, and Intergovernmental Relations in Canadian Cities.' In *Canada: The State of the Federation 2004*, ed. Robert Young and Christian Leuprecht. Montreal and Kingston: McGill–Queen's University Press.

Polèse, Mario, and Richard Stren. 2000. 'Understanding the New Sociocultural Dynamics of Cities: Comparative Urban Policy in a Global Context.' In *The*

Social Sustainability of Cities: Diversity and the Management of Change, ed. Mario Polèse and Richard Stren. Toronto: University of Toronto Press. 3–38.

Przeworski, Adam, and Henry Teune. 1970. *The Logic of Comparative Social Inquiry*. New York: John Wiley.

Putnam, Robert. 2007. 'E Pluribus Unum: Diversity and Community in the 21st Century.' 2006 Johan Skytte Prize Lecture. *Scandinavian Political Studies* 30, no. 2: 137–74.

– 1993. *Making Democracy Work: Civic Traditions in Modern Italy*. Princeton: Princeton University Press.

Pynn, Larry. 1997. 'Richmond's Asian Malls Want to Attract All Races.' *Vancouver Sun*, 6 December, B1.

Queen, Lisa. 1995. 'York's Cultural Segregation Increases Threat of Conflict.' *The Era-Banner*, 25 June.

'RCMP No Longer Colour Blind' [editorial]. 2003. *Richmond News*, 22 November.

RIAC (Richmond Intercultural Advisory Committee). 2004. '2004–2010 Richmond Intercultural Strategic Plan and Work Program.' Prepared with the support of Larry Axelrod, PhD, Project Consultant, The Neutral Zone Coaching and Consulting Services, and the City of Richmond. January. http://www.richmond.ca/__shared/assets/2004_-_2010_Richmond_Intercultural_Strategic_Plan_and_Work_Program9791.pdf

Rowe, Mary W. 2000. *Toronto: Considering Self-Government*. Toronto: Ginger.

Sancton, Andrew. 2008. The *Limits of Boundaries: Why City-Regions Cannot Be Self-Governing*. Montreal and Kingston: McGill-Queen's University Press.

– 2006. 'Why Municipal Amalgamations? Halifax, Toronto, Montreal.' In *Canada: The State of the Federation 2004*, ed. Robert Young and Christian Leuprecht. Montreal and Kingston: McGill–Queen's University Press. 119–37.

– 2002. 'Municipalities, Cities, and Globalization: Implications for Canadian Federalism.' In *Canadian Federalism: Performance, Effectiveness, and Legitimacy*, ed. Herman Bakvis and Grace Skogstad. Toronto: Oxford University Press. 261–77.

– 2000. 'The Municipal Role in the Governance of Canadian Cities.' In *Canadian Cities in Transition*, 2nd ed., ed. Trudi Bunting and Pierre Filion. Toronto: Oxford University Press. 425–42.

– 1994. 'Mayors as Political Leaders.' In *Leaders and Leadership in Canada*, ed. Maureen Mancuson, Richard Price, and Ronald Wayenberg. Toronto: Oxford University Press. 174–89.

Sandercock, Leonie. 1998. *Towards Cosmopolis: Planning for Multicultural Cities*. Chichester: Wiley.

Sartori, Giovanni. 1970. 'Concept Misformation in Comparative Politics.' *American Political Science Review* 64, no. 2: 1033–53.

Sassen, Saskia. 2001. *The Global City: New York, London, and Tokyo*. 2nd ed. Princeton: Princeton University Press.
– 1998. 'Whose City Is It? Globalization and the Formation of New Claims.' In *Globalization and Its Discontents: Essays of the New Mobility of People and Money*, ed. Saskia Sassen. New York: New Press. xix–xxxvi.
Sellers, Jeffrey M. 2005. 'Re-placing the Nation: An Agenda for Comparative Urban Politics.' *Urban Affairs Review* 40, no. 4: 419–45.
– 2002a. *Governing from Below: Urban Regions and the Global Economy*. Cambridge: Cambridge University Press.
– 2002b. 'The Nation-State and Urban Governance: Toward a Multilevel Analysis. *Urban Affairs Review* 37, no. 5: 611–41.
Sewell, John. 2004. *A New City Agenda*. Toronto: Zephyr.
Siegel, David. 2005. 'Municipal Reform in Ontario: Revolutionary Evolution.' In *Municipal Reform in Canada: Reconfiguration, Re-empowerment, and Rebalancing*, ed. Joseph Garcea and Edward LeSage. Toronto: Oxford University Press. 127–48.
Siemiatycki, Myer. 2006. 'The Municipal Franchise and Social Inclusion in Toronto: Policy and Practice.' Inclusive Cities Canada.
– 1998. 'Immigration and Urban Politics in Toronto.' Paper presented to the Third International Metropolis Conference, Israel, 29 November–3 December. http://www.international.metropolis.net/events/israel/papers/Siemiatycki.html
Siemiatycki, Myer. 2006. 'The Municipal Franchise and Social Inclusion in Toronto: Policy and Practice: Executive Summary.' Inclusive Cities Canada and the Community Social Planning Council of Toronto (CSPC), October. http://www.cdhalton.ca/pdf/icc/ICC_Municipal_Franchise_and_Social_Inclusion_in_Toronto.pdf
Siemiatycki, Myer, Tim Rees, Roxana Ng, and Khan Rahi. 2003. 'Integrating Community Diversity in Toronto: On Whose Terms?' In *The World Is a City*, ed. Paul Anisef and Michael Lamphier. Toronto: University of Toronto Press. 371–456.
Simeon, Richard. 2003. 'Federalism and Decentralization in Canada.' Paper presented at the 2nd International Conference on Decentralization. http://www.forumfed.org
Simich, Laura. 2000. *Towards a Greater Toronto Charter: Implications for Immigrant Settlement*. For the Maytree Foundation, Toronto: Zephyr.
Slack, Enid. 2005. 'Easing the Fiscal Restraints: New Revenue Tools in the City of Toronto Act.' Munk Centre for International Studies, University of Toronto, 21 February. http://ideas.repec.org/p/ttp/itpwps/0507.html
Smith, Patrick. 1992. 'The Making of a Global City: Fifty Years of Constituent

Diplomacy – the Case of Vancouver.' *Canadian Journal of Urban Research* 1, no. 1: 90–112.

Smith, Patrick, and Kennedy Stewart. 2006. 'Local Whole-of-Government Policymaking in Vancouver: Beavers, Cats, and the Mushy Middle Thesis.' In *Canada: The State of the Federation 2004*, ed. Robert Young and Christian Leuprecht. Montreal and Kingston: McGill–Queen's University Press. 251–72.

Statistics Canada. 2008a. *Canada's Ethnocultural Mosaic, 2006 Census.* Cat. no. 97-562-X. Ottawa: Minister of Industry, April. http://www12.statcan.ca/english/census06/analysis/ethnicorigin/pdf/97-562-XIE2006001.pdf

– 2008b. '2006 Census: Ethnic Origins, Visible Minorities, Place of Work and Mode of Transportation.' *The Daily,* 2 April, cat no. 11-001-XIE, Ottawa: Minister of Industry. http://www.statcan.gc.ca/daily-quotidien/080402/dq080402-eng.pdf

– 2007a. '2006 Census: Immigration, Citizenship, Language, Mobility, and Migration.' *The Daily,* 4 December, cat. no. 11-001-XIE, Ottawa: Minister of Industry. http://www.statcan.gc.ca/daily-quotidien/071204/dq071204-eng.pdf

– 2007b. '2006 Community Profiles.' 2006 Census, 13 March, cat. no. 92-591-XWE. Ottawa: Minister of Industry. http://www12.statcan.ca/english/census06/data/profiles/community/Index.cfm?Lang=E

– 2007c. Immigration in Canada: A Portrait of the Foreign-born Population, 2006 Census, cat, no. 97-557-XIE, December. Ottawa: Minister of Industry. http://www12.statcan.gc.ca/english/census06/analysis/immcit/pdf/97-557-XIE2006001.pdf

– 2005. 'Study: Canada's Visible Minority Population in 2017.' *The Daily*, 22 March cat. no. 11-001-XIE, Ottawa: Minister of Industry. 6–7. http://www.statcan.gc.ca/daily-quotidien/050322/dq050322b-eng.htm

– 2004. 'Study: Immigrants in Canada's Urban Centres.' *The Daily*, 18 August, cat. no. 11-001-XIE. Ottawa: Minister of Industry. http://www.statcan.gc.ca/daily-quotidien/040818/dq040818-eng.pdf

– 2003. *2001 Census: Analysis Series. Canada's Ethnocultural Portrait: The Changing Mosaic.* Cat. no. 96F0030XIE2001008, 23 January. Ottawa: Minister of Industry. http://www12.statcan.gc.ca/english/census01/products/analytic/companion/etoimm/pdf/96F0030XIE2001008.pdf

– 2002. 'Community profiles: 2001 Census.' 27 June. Cat. no. 93F0053XIE. http://www12.statcan.ca/english/Profil01/CP01/Index.cfm?Lang=E

– n.d.a. 'Census metropolitan area (CMA) and census agglomeration (CA),' in Census Dictionary, website: http://www12.statcan.ca/english/census06/reference/dictionary/geo009.cfm

– n.d.b. 'Selected Religions by Immigrant Status and Period of Immigration, 2001 Counts, for Census Subdivisions (Municipalities) with 5,000-plus Population – 20% Sample.' http://www12.statcan.ca/english/census01/products/highlight/religion/Page.cfm?Lang=E&Geo=CSD&Code=5915004&View=3b&Table=1&StartRec=1&Sort=2&B1=5915004&B2=Counts

Stoker, Gerry. 1998. 'Governance as Theory: Five Propositions.' *International Social Science Journal* 50, no. 155: 17–28.

Stoker, Gerry, and Karen Mossberger. 1994. 'Urban Regime Theory in Comparative Perspective.' *Environment and Planning C, Government and Policy* 12: 195–212.

Stone, Clarence. 2008. 'Urban Regimes and the Capacity to Govern: A Political Economy Approach.' In *Power in the City: Clarence Stone and the Politics of Inequality*, ed. Marion Orr and Valerie C. Johnson. Lawrence: University Press of Kansas. 76–107.

– 2005. 'Looking Back to Look Forward: Reflections on Urban Regime Analysis.' *Urban Affairs Review* 40, no. 3: 309–41.

– 2001. 'The Atlanta Experience Re-examined: The Link between Agenda and Regime Change.' *International Journal of Urban and Regional Research* 25, no. 1: 20–34.

– 1989. *Regime Politics: Governing Atlanta 1946–1988*. Lawrence: University Press of Kansas.

Stone, Clarence, Jeffrey R. Henig, Bryan D. Jones, and Carol Pierannunzi. 2001. *Building Civic Capacity: The Politics of Reforming Urban Schools*. Lawrence: University Press of Kansas.

Stone, Clarence N., Robert K. Whelan, and William J. Murin. 1986. *Urban Policy Politics in a Bureaucratic Age*. Englewood Cliffs: Prentice-Hall.

S.U.C.C.E.S.S. 'Donations.' http://www.successbc.ca/eng/component/option,com_wrapper/Itemid,127

Tate, Ellen, and Louise Quesnel. 1995. 'Accessibility of Municipal Services for Ethnocultural Populations in Toronto and Montreal.' *Canadian Public Administration* 38, no. 3: 325–51.

TCSA (Toronto City Summit Alliance). 2003. 'Enough Talk: An Action Plan for the Toronto Region.' April. http://www.torontoalliance.ca/docs/TCSA_report.pdf

– 'Emerging Leaders Network.' http://www.torontoalliance.ca/tcsa_initiatives/eln

Tindal, C. Richard, and Susan Nobes Tindal. 2009. *Local Government in Canada*, 7th ed. Toronto: Nelson.

Tindal, C. Richard, and Susan Nobes Tindal. 2004. *Local Government in Canada*, 6th ed. Toronto: Nelson.

Town of Markham. 2005. 'Markham Race Relations Committee.' http://www.markham.ca/Markham/Departments/Council/StdCmte/MRRC.htm
– 2002. 'Report to Finance and Administration re: Multilingual Interpretive Services.' 4 November.
Town of Milton. 2008. 'Milton GTA Tax Comparison.' http://www.milton.ca/ecodev/TAX_COMPARISON.pdf
TRIEC (Toronto Region Immigrant Employment Council). 'About Us – Council Members.' http://www.triec.ca/about/TRIEC/council
– 2007. 'Strength in Collaboration.' Annual Review. http://www.triec.ca/files/47/original/TRIEC_2007_AnnualReview.pdf
Uitermark, Justus, Ugo Russi, and Henk Van Houtum. 2005. 'Reinventing Multiculturalism: Urban Citizenship and the Negotiation of Ethnic Diversity in Amsterdam.' *International Journal of Urban and Regional Research* 29, no. 3: 622–40.
Urbaniak, Tom. 2005. 'Beyond Regime Theory: Mayoral Leadership, Suburban Development, and the Politics of Mississauga, Ontario.' PhD diss., University of Western Ontario.
– 2003. 'Regime Theory and the Politics of Mississauga, 1960–1976.' Paper presented at the annual Canadian Political Science Association Meeting, Halifax.
Vancouver Foundation. 'About Us.' http://www.vancouverfoundation.bc.ca/about/index.htm
Verma, Sonia. 2002. 'Newspaper a Fearless Crusader: Star Has a Storied History of Taking Up a Good Cause.' *Toronto Star,* 2 November, A1.
Walker, Jack L. 1983. 'The Origins and Maintenance of Interest Groups in America.' *American Political Science Review* 77 (June): 390–406.
Wallace, Marcia, and Frances Frisken. 2000. 'City-Suburban Differences in Government Responses to Immigration in the Greater Toronto Area.' Centre for Urban and Community Studies, University of Toronto, research paper no. 197.
White, Stephen. 1992a. 'Brampton Aims to Eliminate Racism.' *Toronto Star*, 17 September, BR4.
– 1992b. 'Multicultural Agencies to Cut Duplication.' *Toronto Star*, 17 September, BR4.
Young, Mary Lynn. 2004. 'Ethnic Media Blooms in Wake of Tsunamis.' workopolis.com, 30 December.
Young, Robert, and Christian Leuprecht, eds. 2006. *Canada: The State of the Federation 2004.* Montreal and Kingston: McGill–Queen's University Press.

Index

Studies in Comparative Political Economy and Public Policy

1 *The Search for Political Space: Globalization, Social Movements, and the Urban Political Experience* / Warren Magnusson

2 *Oil, the State, and Federalism: The Rise and Demise of Petro-Canada as a Statist Impulse* / John Erik Fossum

3 *Defying Conventional Wisdom: Free Trade and the Rise of Popular Sector Politics in Canada* / Jeffrey M. Ayres

4 *Community, State, and Market on the North Atlantic Rim: Challenges to Modernity in the Fisheries* / Richard Apostle, Gene Barrett, Peter Holm, Svein Jentoft, Leigh Mazany, Bonnie McCay, Knut H. Mikalsen

5 *More with Less: Work Reorganization in the Canadian Mining Industry* / Bob Russell

6 *Visions for Privacy: Policy Approaches for the Digital Age* / Edited by Colin J. Bennett and Rebecca Grant

7 *New Democracies: Economic and Social Reform in Brazil, Chile, and Mexico* / Michel Duquette

8 *Poverty, Social Assistance, and the Employability of Mothers: Restructuring Welfare States* / Maureen Baker and David Tippin

9 *The Left's Dirty Job: The Politics of Industrial Restructuring in France and Spain* / W. Rand Smith

10 *Risky Business: Canada's Changing Science-Based Policy and Regulatory Regime* / Edited by G. Bruce Doern and Ted Reed

11 *Temporary Work: The Gendered Rise of a Precarious Employment Relationship* / Leah Vosko

12 *Who Cares? Women's Work, Childcare, and Welfare State Redesign* / Jane Jenson and Mariette Sineau with Franca Bimbi, Anne-Marie Daune-Richard, Vincent Della Sala, Rianne Mahon, Bérengère Marques-Pereira, Olivier Paye, and George Ross

13 *Canadian Forest Policy: Adapting to Change* / Edited by Michael Howlett

14 *Knowledge and Economic Conduct: The Social Foundations of the Modern Economy* / Nico Stehr

15 *Contingent Work, Disrupted Lives: Labour and Community in the New Rural Economy* / Anthony Winson and Belinda Leach

16 *The Economic Implications of Social Cohesion* / Edited by Lars Osberg

17 *Gendered States: Women, Unemployment Insurance, and the Political Economy of the Welfare State in Canada, 1945–1997* / Ann Porter

18 *Educational Regimes and Anglo-American Democracy* / Ronald Manzer

19 *Money in Their Own Name: The Feminist Voice in Poverty Debate in Canada, 1970–1995* / Wendy McKeen

20 *Collective Action and Radicalism in Brazil: Women, Urban Housing, and Rural Movements* / Michel Duquette, Maurilio de Lima Galdino, Charmain Levy, Bérengère Marques-Pereira, and Florence Raes

21 *Continentalizing Canada: The Politics and Legacy of the Macdonald Royal Commission* / Gregory J. Inwood

22 *Globalization Unplugged: Sovereignty and the Canadian State in the Twenty-First Century* / Peter Urmetzer

23 *New Institutionalism: Theory and Analysis* / Edited by André Lecours

24 *Mothers of the Nation: Women, Family, and Nationalism in Twentieth-Century Europe* / Patrizia Albanese

25 *Partisanship, Globalization, and Canadian Labour Market Policy: Four Provinces in Comparative Perspective* / Rodney Haddow and Thomas Klassen

26 *Rules, Rules, Rules, Rules: Multi-Level Regulatory Governance* / Edited by G. Bruce Doern and Robert Johnson

27 *The Illusive Trade-off: Intellectual Property Rights, Innovation Systems, and Egypt's Pharmaceutical Industry* / Basma Abdelgafar

28 *Fair Trade Coffee: The Prospects and Pitfalls of Market-Driven Social Justice* / Gavin Fridell

29 *Deliberative Democracy for the Future: The Case of Nuclear Waste Management in Canada* / Genevieve Fuji Johnson

30 *Internationalization and Canadian Agriculture: Policy and Governing Paradigms* / Grace Skogstad

31 *Military Workfare: The Soldier and Social Citizenship in Canada* / Deborah Cowen

32 *Public Policy for Women: The State, Income Security, and Labour* / Edited by Marjorie Griffin Cohen and Jane Pulkingham

33 *Smiling Down the Line: Info-Service Work in the Global Economy* / Bob Russell

34 *Municipalities and Multiculturalism: The Politics of Immigration in Toronto and Vancouver* / Kristin R. Good